Metals and Oxidative Damage
in Neurological Disorders

Metals and Oxidative Damage in Neurological Disorders

Edited by

James R. Connor
Pennsylvania State University
Hershey, Pennsylvania

Plenum Press • New York and London

The Library of Congress Cataloging-in-Publication Data

On file

Cover micrographs: Astrocytes loaded with two dyes: Fura 2, which fluoresces green upon coupling with calcium, and a yellow dye that is an indicator of mitochondrial membrane potential. *Front cover:* Normal astrocytes showing dye localized to normally polarized mitochondria. *Back cover:* Astrocytes treated with an oxidizing agent causing loss of mitochondrial membrane potential, thus almost no yellow fluorescence is visible; the calcium indicator, however, is still prominent. The photomicrographs were prepared by Sara Robb-Gaspers and James Connor with assistance from Greg Young.

ISBN 0-306-45534-X

© 1997 Plenum Press, New York
A Division of Plenum Publishing Corporation
233 Spring Street, New York, N.Y. 10013

http://www.plenum.com

10 9 8 7 6 5 4 3 2 1

Printed in the United States of America

The pleasure of an accomplishment is magnified when it can be shared with people you love. I am blessed that the joys I have known have been magnified three-fold. As such, I lovingly and gratefully dedicate this book to my wife Judy, my daughter Jennifer, and my son Jonathan.

Contributors

Michael Aschner   Department of Physiology and Pharmacology, Bowman Gray School of Medicine, Winston–Salem, North Carolina 27157

George Bartzokis   Department of Psychiatry, UCLA, Los Angeles, California 90024; and the Research Service and the Psychiatry Service, West Los Angeles Veterans Affairs Medical Center, Los Angeles, California 90073

M. Flint Beal   Neurology Service, Massachusetts General Hospital, Boston, Massachusetts 02114

Mark W. Becher   Department of Pathology and the Neuropathology Laboratory, The Johns Hopkins University School of Medicine and School of Hygiene and Public Health, Baltimore, Maryland 21205

David R. Borchelt   Department of Pathology and the Neuropathology Laboratory, The Johns Hopkins University School of Medicine and School of Hygiene and Public Health, Baltimore, Maryland 21205

Lucie I. Bruijn   Department of Biological Chemistry, The Johns Hopkins University School of Medicine and School of Hygiene and Public Health, Baltimore, Maryland 21205; *current address:* Departments of Neurology and Neuroscience, the Ludwig Institute, University of California at San Diego, La Jolla, California 92093

Jean Lud Cadet   Molecular Neuropsychiatry Section, NIH/NIDA, Intramural Research Program, Baltimore, Maryland 21224

Elsbeth G. Chikhale   Neurochemistry and Brain Transport Section, Laboratory of Neurosciences, National Institute on Aging, National Institutes of Health, Bethesda, Maryland 20892

Don W. Cleveland   Department of Biological Chemistry, The Johns Hopkins University School of Medicine and School of Hygiene and Public Health, Baltimore,

Maryland 21205; *current address:* Departments of Neurology and Neuroscience, the Ludwig Institute, University of California at San Diego, La Jolla, California 92093

James R. Connor   George M. Leader Family Laboratory for Alzheimer's Disease Research, Department of Neuroscience and Anatomy, Pennsylvania State University College of Medicine, M. S. Hershey Medical Center, Hershey, Pennsylvania 17033

Neal G. Copeland   Mammalian Genetics Laboratory, ABL-Basic Research Program, NCI-Frederick Cancer Center Research and Development, Frederick, Maryland 21702

Valeria C. Culotta   Departments of Biochemistry and Environmental Health Sciences, The Johns Hopkins University School of Medicine and School of Hygiene and Public Health, Baltimore, Maryland 21205

Deborah A. Dawson   Stroke Branch, NINDS, National Institutes of Health, Bethesda, Maryland 20892

Michael Gassen   Department of Pharmacology, Bruce Rappaport Family Research Institute, Faculty of Medicine, Technion, Haifa 31096, Israel

Yelena Glinka   Department of Pharmacology, Bruce Rappaport Family Research Institute, Faculty of Medicine, Technion, Haifa 31096, Israel

Paul F. Good   Department of Pathology and Fishberg Research Center for Neurobiology, Mount Sinai School of Medicine, New York, New York 10029

Edward D. Hall   CNS Diseases Research, Pharmacia and Upjohn, Inc., Kalamazoo, Michigan 49001

Nancy A. Jenkins   Mammalian Genetics Laboratory, ABL-Basic Research Program, NCI-Frederick Cancer Center Research and Development, Frederick, Maryland 21702

J. G. Joshi   Department of Biochemistry, University of Tennessee, Knoxville, Tennessee 37996

Michael K. Lee   Department of Pathology and the Neuropathology Laboratory, The Johns Hopkins University School of Medicine and School of Hygiene and Public Health, Baltimore, Maryland 21205

C. Warren Olanow   Department of Neurology, Mount Sinai School of Medicine, New York, New York 10029

Charles Palmer   Division of Newborn Medicine, Department of Pediatrics, Pennsylvania State University College of Medicine, M. S. Hershey Medical Center, Hershey, Pennsylvania 17033

Carlos A. Pardo   Department of Pathology and the Neuropathology Laboratory, The Johns Hopkins University School of Medicine and School of Hygiene and Public Health, Baltimore, Maryland 21205

Daniel P. Perl   Department of Pathology and Fishberg Research Center for Neurobiology, Mount Sinai School of Medicine, New York, New York 10029

Ananda S. Prasad   Division of Hematology–Oncology, Department of Internal Medicine, Wayne State University School of Medicine, and Harper Hospital, Detroit, Michigan 48201

Donald L. Price   Departments of Pathology, Neurology, and Neuroscience, and the Neuropathology Laboratory, The Johns Hopkins University School of Medicine and School of Hygiene and Public Health, Baltimore, Maryland 21205

Joseph R. Prohaska   Department of Biochemistry and Molecular Biology, School of Medicine, University of Minnesota–Duluth, Duluth, Minnesota 55812

Olivier Rabin   Neurochemistry and Brain Transport Section, Laboratory of Neurosciences, National Institute on Aging, National Institutes of Health, Bethesda, Maryland 20892

George V. Rebec   Program in Neural Science, Department of Psychology, Indiana University, Bloomington, Indiana 47405

Sara J. Robb-Gaspers   George M. Leader Family Laboratory for Alzheimer's Disease Research, Department of Neuroscience and Anatomy, Pennsylvania State University College of Medicine, M. S. Hershey Medical Center, Hershey, Pennsylvania 17033

Leslie A. Shinobu   Neurology Service, Massachusetts General Hospital, Boston, Massachusetts 02114

Sangram S. Sisodia   Departments of Pathology and Neuroscience and the Neuropathology Laboratory, The Johns Hopkins University School of Medicine and School of Hygiene and Public Health, Baltimore, Maryland 21205

Quentin R. Smith   Neurochemistry and Brain Transport Section, Laboratory of Neurosciences, National Institute on Aging, National Institutes of Health, Bethesda, Maryland 20892; *current address:* Department of Pharmaceutical Sciences, Texas Tech University HSC, Amarillo, Texas 79106

G. T. Vatassery   Research Service and the Geriatric Research Education and Clinical Center (GRECC), Veterans Affairs Medical Center, Minneapolis, Minnesota 55417; and Department of Psychiatry, University of Minnesota, Minneapolis, Minnesota 55455

Philip C. Wong   Department of Pathology and the Neuropathology Laboratory, The Johns Hopkins University School of Medicine and School of Hygiene and Public Health, Baltimore, Maryland 21205

Zhou-Shang Xu   Department of Biological Chemistry, The Johns Hopkins University School of Medicine and School of Hygiene and Public Health, Baltimore, Maryland 21205; *current address:* The Worcester Foundation for Experimental Biology, Shrewsbury, Massachusetts 01545

Moussa B. H. Youdim   Department of Pharmacology, Bruce Rappaport Family Research Institute, Faculty of Medicine, Technion, Haifa 31096, Israel

Preface

The purpose of this book is to bring together scientists and clinicians interested in oxidative injury in the nervous system but whose approaches to investigation and treatment design vary widely. Indeed the goal of this book is to show that the investigative approaches and potential therapeutic interventions perhaps do not vary as widely as some may think. I think that the readers of this book will not read it from front to back, but will pick chapters of interest. Thus, the chapters are organized to contain information that is essential to understanding basic aspects of oxidative injury, and thus have some redundancy. However, within the context of each chapter the reader should hopefully find impetus and direction to go on to another chapter.

The book is divided into three sections. The first section contains reviews of metals and their role in generating oxidative injury. Iron is considered in three of these chapters because of its relative abundance in the brain and its potency in inducing free radicals. The second section focuses on mechanisms by which the brain attempts to protect itself from oxidative injury. Some of these mechanisms have the potential to be protective in some situations and potentially damaging in others. The third section contains the clinical diseases in which oxidative injury is known to contribute to the pathogenic process. This section ends with a chapter on antioxidant therapeutic strategies in neurological disorders. Finally, a common pathway is proposed for oxidative injury in the brain. We have argued that whatever the mechanism by which the oxidative stress is triggered, eventually a cascade will be initiated that will result in cell death. This cascade would be consistent in each of the disease processes and would be independent of the metal(s) that induced the process.

We hope this book serves as a reference for those entering the field and brings together individuals already working in the field from seemingly disparate approaches. We hope it serves as a catalyst to clinicians to consider oxidative stress as a component of many neurological diseases and consider metal regulation as an important component in their patients' diets.

James R. Connor

Hershey, Pennsylvania

Contents

Chapter 1
Iron and Neurotransmitter Function in the Brain
Yelena Glinka, Michael Gassen, and Moussa B. H. Youdim

1.1.	Introduction	1
1.2.	Biosynthesis and Metabolism of Neurotransmitters	3
	1.2.1. Catecholamine Neurotransmitters	3
	1.2.2. GABA	3
	1.2.3. Serotonin (5-HT)	4
1.3.	The Regulation of Axonal Growth	4
1.4.	Axonal Transport and Iron	5
1.5.	Translocation of Neurotransmitters: Storage, Release, Reuptake	6
	1.5.1. Uptake of Neurotransmitters into Vesicles	6
	1.5.2. Neurotransmitter Release	7
	1.5.3. Neurotransmitter Reuptake	8
1.6.	Receptors, Metabolic Consequences of the Reception, Backward Signaling	8
	1.6.1. Regulation of Receptor Expression	9
	1.6.2. Influence of Iron on Second Messenger Systems	10
	1.6.3. Iron Interaction with Protein Phosphorylation	11
	1.6.4. Allosteric Control of Receptor Sensitivity and Desensitization of Receptors	12
1.7.	Influence of Neural Signaling on the Iron Regulation in the Cell	13
1.8.	Intercellular Interactions in the CNS	14
1.9.	Conclusion	15
1.10.	References	15

Chapter 2

**Evidence for Iron Mismanagement in the Brain
in Neurological Disorders**

James R. Connor

2.1. Introduction 23
2.2. Regional and Cellular Distribution of Iron in the Brain 24
    2.2.1. Aging and Alzheimer's Disease.......................... 25
    2.2.2. Parkinson's Disease 26
2.3. Acquisition of Iron... 27
2.4. Iron Mobilization... 29
2.5. Intracellular Iron Management 30
2.6. Iron Regulatory Proteins (IRPs) 32
2.7. Iron and Multiple Sclerosis.................................. 33
2.8. Summary 33
2.9. References 34

Chapter 3

Magnetic Resonance Imaging of Brain Iron

George Bartzokis

3.1. Introduction 41
3.2. MRI Evaluation of Tissue Iron 43
    3.2.1. T_2 Is a Nonspecific Measure of Iron 43
    3.2.2. Specific MRI Measures of Ferritin 44
    3.2.3. Initial Clinical Research Applications of the FDRI Method ... 45
    3.2.4. Future Directions of *in Vivo* Evaluation of Tissue Iron with
        MR 51
3.3. Conclusions 51
3.4. References 52

Chapter 4

Neurochemical Roles of Copper as Antioxidant or Prooxidant

Joseph R. Prohaska

4.1. Introduction 57
4.2. Copper Distribution 58
4.3. Neurochemical Functions of Copper.......................... 60
    4.3.1. Cuproenzymes... 60

4.3.2. Copper Complexes 63
4.3.3. Prooxidant Properties 64
4.3.4. Antioxidant Roles................................... 65
4.4. Copper Deficiency.. 65
    4.4.1. Dietary Copper Deficiency........................... 65
    4.4.2. Genetic Copper Deficiency 66
4.5. Copper Toxicity.. 67
    4.5.1. CNS Toxicity 67
    4.5.2. Genetic Diseases................................... 68
4.6. Cu,Zn-Superoxide Dismutase 70
4.7. Summary 70
4.8. References 71

Chapter 5
Manganese Neurotoxicity and Oxidative Damage
Michael Aschner

5.1. Introduction 77
5.2. Manganese Transport in the CNS........................... 78
5.3. Manganese Neurotoxicity: What Possible Mechanisms?........... 80
    5.3.1. Manganese Neurotoxicity—A Dual Spectrum 80
    5.3.2. Mechanisms Associated with Manganese-Induced
           Neurotoxicity 81
    5.3.3. Manganese as an Antioxidant and Brain Protectant.......... 87
5.4. Summary and Future Directions 88
5.5. References 89

Chapter 6
The Role of Zinc in Brain and Nerve Functions
Ananda S. Prasad

6.1. Introduction 95
6.2. Zinc and Cell Division 96
6.3. Transport of Zinc.. 96
6.4. Biochemical Role of Zinc................................. 97
6.5. Zinc and Free Radicals 99
6.6. Zinc and Brain Function in Animals and Humans................ 100
6.7. Zinc and Epilepsy 101
6.8. Zinc and Nerve Functions................................. 102

6.9. Summary 107
6.10. References 108

Chapter 7
Delivery of Metals to Brain and the Role of the Blood–Brain Barrier
Quentin R. Smith, Olivier Rabin, and Elsbeth G. Chikhale

7.1. Introduction 113
7.2. Circulation in Serum and Blood 114
7.3. Blood–Brain Barrier 115
7.4. Transport Methods and Kinetic Analysis 116
7.5. Rates of Metal Transport into Brain and CSF 119
7.6. Transport Mechanisms... 121
    7.6.1. Iron 121
    7.6.2. Manganese 124
    7.6.3. Zinc... 124
    7.6.4. Calcium.. 125
    7.6.5. Toxic Metals (Aluminum, Lead, Mercury) 126
7.7. Changes with Diet, Disease, and Development 126
7.8. Conclusions 127
7.9. References 127

Chapter 8
Ferritin: Intracellular Regulator of Metal Availability
J. G. Joshi

8.1. Introduction 131
8.2. Formation of Holoferritin and Loading of Iron into Apoferritin....... 133
8.3. Reductive Release of Iron.................................... 133
8.4. Ferritin and Binding of Nonferrous Metal Ions 134
    8.4.1. Ferritin and Beryllium 135
    8.4.2. Ferritin and Zinc.................................... 137
    8.4.3. Ferritin and Cadmium 138
    8.4.4. Ferritin and Manganese 138
    8.4.5. Ferritin and Arsenic 138
    8.4.6. Ferritin and Aluminum................................ 139
    8.4.7. Aluminum and Free Radicals........................... 140
8.5. Aluminum, Iron, Free Radicals, and the Formation of Amyloid
    Plaques Characteristic for AD................................ 141
8.6. The Role for Novel Ferritin H-Chain mRNAs.................... 142
8.7. References 143

Chapter 9
Ascorbate: An Antioxidant Neuroprotectant and Extracellular
Neuromodulator
George V. Rebec

9.1. Introduction 149
9.2. Uptake into Brain Tissue.................................. 150
9.3. Release into Extracellular Fluid 151
    9.3.1. Modulation by Dopamine............................. 152
    9.3.2. The Nigro-Thalamo-Cortical System.................... 153
    9.3.3. Heteroexchange at the Glutamate Transporter 154
    9.3.4. Other Considerations 155
9.4. Neuroprotectant Effects.................................... 155
    9.4.1. Degeneration of Dopaminergic Neurons.................. 155
    9.4.2. Glutamate-Induced Neurotoxic Effects 159
    9.4.3. Neurotoxic Mechanisms in Schizophrenia and Tardive
         Dyskinesia.. 159
9.5. Clinical and Functional Implications 161
    9.5.1. Dopaminergic Modulation 161
    9.5.2. Glutamatergic Modulation 164
9.6. Conclusions 164
9.7. References 165

Chapter 10
Vitamin E: Neurochemical Aspects and Relevance
to Nervous System Disorders
G. T. Vatassery

10.1. Introduction 175
10.2. Familial Ataxia with Vitamin E Deficiency 176
10.3. Peripheral Nerve Damage 177
10.4. Abetalipoproteinemia...................................... 177
10.5. Aging 178
10.6. Parkinson's Disease 178
10.7. Alzheimer's Disease 179
10.8. Alteration of Neurotoxicity by Vitamin E 180
10.9. Tardive Dyskinesia and Vitamin E 180
10.10. Vitamin E in Other Neurological Conditions..................... 181
10.11. Vitamin E in Cell Proliferation and Growth in the Nervous System ... 182
10.12. Neurobiology of Vitamin E: Experimental Data from Animals 182
10.13. Conclusions 184
10.14. References 185

Chapter 11
Nitric Oxide and Oxidative Damage in the CNS
Deborah A. Dawson

11.1. Nitric Oxide: Synthesis and Inhibition   189
    11.1.1. NO Synthases: Constitutive and Inducible Enzymes   189
    11.1.2. NO Synthases: Localization within the CNS   190
    11.1.3. NO Synthases: Inhibition of NO Production.   190
11.2. Mechanisms of NO Toxicity   191
    11.2.1. Oxidative Damage: Direct and by Combination with
        Superoxide   191
    11.2.2. Oxidative Damage: Alterations in Iron Homeostasis   192
    11.2.3. Inhibition of Respiration and DNA Synthesis.   193
11.3. Role of NO in CNS Disorders   194
    11.3.1. NO and Glutamate Toxicity   194
    11.3.2. Stroke   194
    11.3.3. Neurodegenerative Disorders   196
11.4. Summary and Conclusions   199
11.5. References   199

Chapter 12
Iron and Oxidative Stress in Neonatal Hypoxic–Ischemic Brain Injury:
Directions for Therapeutic Intervention
Charles Palmer

12.1. Introduction to Oxidant Stress and Secondary Brain Injury   205
12.2. Reactive Oxygen Species (ROS): An Overview   206
12.3. Vascular Injury   209
    12.3.1. The Primary Insult   209
    12.3.2. Reperfusion: Sources of ROS   209
    12.3.3. Effects of ROS on the Microvasculature.   210
    12.3.4. Neutrophils   210
    12.3.5. Antioxidants and Microvascular Injury   211
12.4. Free-Radical-Mediated Injury to Brain Parenchyma   212
    12.4.1. Actions on Cellular Function   213
    12.4.2. Impaired Energy Metabolism and ROS.   214
    12.4.3. Elevated Intracellular Calcium and ROS.   214
    12.4.4. Excitotoxic Injury and ROS.   215
    12.4.5. Mitochondrial Dysfunction and ROS   215
    12.4.6. Hydroxyl Radicals: From Iron or Nitric Oxide?   216
    12.4.7. Oxidant Stress in Young Neurons and Oligodendrocytes.   216

12.4.8. Microglia as Source of ROS 216
12.4.9. Apoptosis and ROS 217
12.5. Iron... 217
12.5.1. Susceptibility of Immature Brain to Iron-Mediated Injury..... 217
12.5.2. Sources of Iron in Cerebral Ischemia 218
12.5.3. Iron Accumulation in the Post-Ischemic Brain............. 219
12.6. Nitric Oxide and the Immature Brain........................... 220
12.6.1. NO: Injury Mechanisms............................... 220
12.6.2. NO and Regulation of Iron Metabolism 221
12.6.3. NO Synthase Inhibition and Neuroprotection.............. 222
12.7. Appendix: Rescue Therapies................................... 223
12.8. References 225

Chapter 13
**The Role of Oxidative Processes and Metal Ions in Aging and
Alzheimer's Disease**
Leslie A. Shinobu and M. Flint Beal

13.1. Age-Related Changes in Cellular Energy Metabolism............... 237
13.1.1. Introduction.. 237
13.1.2. The Mitochondria and Oxidative Phosphorylation 238
13.1.3. Age-Related Changes in Mitochondrial DNA Structure....... 240
13.1.4. Age-Related Changes in Mitochondrial Function 242
13.2. Oxidative Damage and Alzheimer's Disease 244
13.2.1. Introduction.. 244
13.2.2. Risk Factors for Alzheimer's Disease 244
13.2.3. Evidence for Impaired Energy Metabolism in Alzheimer's
        Disease 245
13.2.4. Evidence of Mitochondrial Dysfunction in Alzheimer's
        Disease 245
13.2.5. Evidence for Oxidative Stress in Alzehimer's Disease........ 247
13.2.6. Peripheral Markers of Oxidative Damage in Alzheimer's
        Disease 250
13.2.7. Specific Roles for Reactive Oxygen Species in the
        Pathogenesis of Alzheimer's Disease 251
13.3. The Role of Metal Ions in Aging and Alzheimer's Disease 255
13.3.1. Aluminum.. 256
13.3.2. Copper... 257
13.3.3. Iron 257
13.3.4. Zinc 259

13.4. Summary 260
13.5. References 261

Chapter 14
Oxidative Stress with Emphasis on the Role of LAMMA in Parkinson's Disease
Paul F. Good, Daniel P. Perl, and C. Warren Olanow

14.1. Introduction 277
14.2. Oxidative Stress.. 278
14.3. Evidence of Oxidative Stress and Oxidative Damage in Parkinson's
      Disease.. 280
14.4. Microprobe Studies of Trace Elements in Parkinson's Disease........ 283
      14.4.1. The Laser Microprobe 283
      14.4.2. Laser Microprobe Studies in Parkinson's Disease 284
14.5. Parkinsonism Associated with Accumulation of Trace Metals 287
14.6. Therapeutic Implications... 289
14.7. References 290

Chapter 15
Perspectives on the Mechanisms of Familial Amyotrophic Lateral Sclerosis Caused by Mutations in Superoxide Dismutase 1
David R. Borchelt, Philip C. Wong, Mark W. Becher, Lucie I. Bruijn, Don W. Cleveland, Neal G. Copeland, Valeria C. Culotta, Nancy A. Jenkins, Michael K. Lee, Carlos A. Pardo, Donald L. Price, Sangram S. Sisodia, and Zhou-Shang Xu

15.1. Introduction 295
15.2. Clinical and Neuropathological Phenotypes of FALS 296
15.3. Etiological Factors/Mechanisms in ALS and FALS 298
15.4. Functions of Cu/Zn SOD1 300
15.5. Properties of FALS Mutant SOD1 302
15.6. Transgenic Models of SOD1-Linked FALS 303
15.7. Nature of the Toxic Property.. 305
15.8. Factors Governing the Severity of Disease 307
15.9. Conclusions 307
15.10. References 308

Chapter 16
Tardive Dyskinesia and Oxidative Stress
Jean Lud Cadet

16.1. Introduction 315
16.2. Description 315
16.3. Neuropathological Findings Associated with Neuroleptic Use 316
16.4. Models of Tardive Dyskinesia................................. 317
16.5. Free-Radical Hypothesis of Tardive Dyskinesia................... 318
16.6. Clinical Implications of the Free-Radical Hypothesis 320
16.7. Conclusion 320
16.8. References 321

Chapter 17
Antioxidant Therapeutic Strategies in CNS Disorders
Edward D. Hall

17.1. Introduction 325
17.2. Chemistry of Lipid Peroxidation.............................. 326
17.3. Potential Mechanisms of Lipid Peroxidation Inhibition............. 327
  17.3.1. Inhibition of Oxygen Radical Formation................... 327
  17.3.2. Enzymatic Radical Scavenging 328
  17.3.3. Chemical Radical Scavenging 328
  17.3.4. Peroxynitrite Scavenging 329
  17.3.5. Peroxyl Radical Scavenging............................ 329
  17.3.6. Iron Chelation....................................... 330
  17.3.7. Membrane Stabilization 330
17.4. Tirilazad: An Example of a Multi-Mechanistic Antioxidant
      Neuroprotective 330
  17.4.1. Tirilazad Lipid Peroxidation-Inhibiting Mechanisms 332
  17.4.2. Protection of Endothelial Function 333
  17.4.3. Protection of Calcium Homeostatic Mechanisms 333
  17.4.4. Tirilazad Clinical Neuroprotection Trials 334
17.5. 2-Methylaminochromans 334
17.6. Pyrrolopyrimidines... 335
17.7. References 336

Chapter 18
Oxidative Stress-Induced Cell Damage in the CNS:
A Proposal for a Final Common Pathway
Sara J. Robb-Gaspers and James R. Connor

18.1. Introduction 341
18.2. Chemistry of Free Radicals.................................. 341

18.3. Unique Vulnerability of the Brain.   342
18.4. Mitochondria as Mediators of Oxidative Damage in Astrocytes.   344
18.5. Summary   347
18.6. References   349

Appendix
Accepted Biomarkers of Oxidative Damage in Tissues
Leslie A. Shinobu and M. Flint Beal   353

Index   357

Chapter 1

Iron and Neurotransmitter Function in the Brain

Yelena Glinka, Michael Gassen, and Moussa B. H. Youdim

1.1. INTRODUCTION

Iron involvement in basic cell functions as well as in many human diseases has been described in excellent reviews (see, for example, Lauffer, 1992). Neural cells, like any other cell type, require iron for DNA synthesis, mitochondrial respiration, and other vitally important reactions. They are, on the other hand, susceptible to iron toxicity caused by iron overload. But the most remarkable characteristic of neural cells is observed only on the metacellular level: Neurons are organized in a network, which is used for collecting and analyzing of incoming information and producing an adequate response. This response can include regulation of non-neuronal systems.

The complicated network of neural cells also includes interactions between neurons and various types of glial cells. Neurotransmitters and ionic currents are the central means of rapid inter- and intracellular communication for passing information between neurons. Additionally, all members of the net, neurons and glia cells, interact by a variety of other factors (growth factors, cytokines) for mutual support, regulation, and protection. The complete process of transmission includes biosynthesis and transport of neurotransmitters, loading of vesicles with neurotransmitters for storage and controlled release, and, finally, binding of the messengers to receptors on the post-synaptic cell causing a sequence of metabolic events there. Within this process are multiple options for processing of the transmitted information (see Figure 1.1).

Yelena Glinka, Michael Gassen, and Moussa B. H. Youdim   Department of Pharmacology, Bruce Rappaport Family Research Institute, Faculty of Medicine, Technion, Haifa 31096, Israel.
Metals and Oxidative Damage in Neurological Disorders, edited by Connor. Plenum Press, New York, 1997.

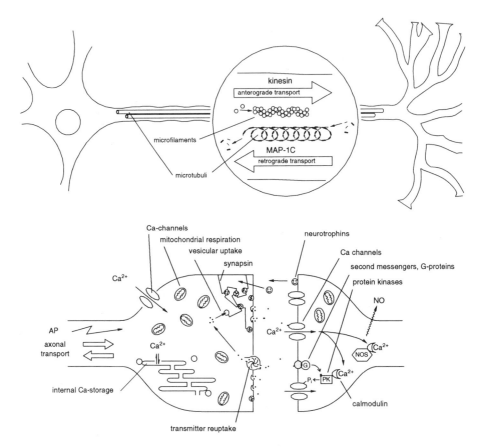

FIGURE 1.1. Sites of iron influence on neural transmission: (a) along the axon; (b) pre- and postsynaptically.

After binding, neurotransmitters have to be removed quickly by reuptake into the presynaptic neuron and/or by metabolism involving glia cells. Backward signals from the post-synaptic cell help to establish and maintain the synaptic connection. High specificity of the signal reception and response is provided by a unique neurite architecture, including vesicles traveling along axons in both directions, synapses with receptors, neurotransmitters, and rapid reuptake systems into the presynaptic and glial cells. The selection of proper direction of axon outgrowth is crucial to establish the correct logical architecture in the nervous system. All these processes and their possible interference with iron will be reviewed below with more or less detail depending on the current knowledge of each subject. We will also discuss possible mechanisms for how neurotransmitter-stimulated processes can mediate iron uptake and storage.

1.2. BIOSYNTHESIS AND METABOLISM OF NEUROTRANSMITTERS

1.2.1. Catecholamine Neurotransmitters

The common biosynthetic precursor of the catecholamine neurotransmitters is the amino acid L-tyrosine. 4-Hydroxylation by a specific tyrosine hydroxylase requires Fe^{2+}, O_2, and tetrahydrobiopterine, and is the rate-determining step of the reaction sequence. Decarboxylation of the product L-DOPA is afforded by DOPA decarboxylase, a pyridoxal phosphate containing enzyme, and leads to dopamine (DA) formation. Further β-hydroxylation of DA yields norepinephrine (NE). Tyrosine hydroxylase is, on the one hand, feedback inhibited by low concentrations of DA, but it can be activated by cAMP-dependent phosphorylation only in the presence of bound DA (Ribeiro et al., 1992). Stoichiometric addition of Fe^{2+} and DA significantly reduced phosphorylation of the enzyme and thus its non-regulated activation (Almas et al., 1992). Iron function in regulation of protein phosphorylation will be discussed more generally later in section 1.6.

Activity of tyrosine hydroxylase is reduced by 50% in the brains of patients with Parkinson's disease. The enzyme can be stimulated 13-fold by addition of iron (1 mM), but the responsiveness decreases slightly, to a factor of 11, for the enzyme from Parkinsonian brains (Rausch et al., 1988).

Nutrient iron deficiency increases the turnover of catecholamine neurotransmitters in rats. An increase of the main metabolites, dihydroxyphenylacetic acid (DOPAC) and homovanillic acid (HVA) (Beard, 1987; Beard et al., 1988; 1990; 1994), is possibly also related to a defect in catecholamine uptake (Beard et al., 1994). Altered monoamine metabolism is likely to influence the redox balance of the neuron, since NE, DA, as well as serotonin (5-HT), are more active antioxidants than their metabolites DOPAC, HVA, and 5-hydroxyindolylacetic acid (5-HIAA) in preventing damage by free radicals generated with hydrogen peroxide $(H_2O_2)/Fe^{2+}$ (Liu and Mori, 1993). Other data suggests however, that catecholamines enhance xanthine-oxidase-generated oxidative stress by releasing Fe^{3+} from ferritin (Allen et al., 1994).

1.2.2. GABA

The GABAergic neuronal system is of great pharmacological importance since it is the target for benzodiazepines and barbiturates, two widely used classes of drugs. GABA is the main inhibitory neurotransmitter in the CNS and occurs in high concentrations (in micromole per gram quantities) in the basal ganglia, especially in the substantia nigra and the striatum, as well as in the hypothalamus. GABA cannot cross the blood–brain barrier and is synthesized exclusively in the neurons by the "GABA-shunt," a side path of the Krebs cycle (see Figure 1.2).

The GABAergic pathways are affected directly by iron and also by reactive oxygen species (ROS) formed by iron-catalyzed reactions. Both chronic iron deficiency and acute iron overload disturb the balance of GABAergic processes. In iron-deficient rats, the activity of glutamate dehydrogenase and GABA transaminase (GABA-T) are decreased. Isocitrate dehydrogenase, an enzyme of the Krebs cycle, is also affected. The activities of

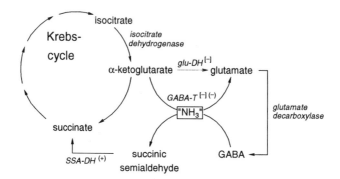

FIGURE 1.2. Influence of iron and cobalt on the activity of GABA-shunt enzymes; GABA-T, GABA trans-aminase; SSA-DH, succinic semialdehyde dehydrogenase; glu-DH, glutamate dehydrogenase; [−], decrease in iron deficiency; (+)(−), in/decrease with acute Fe or Co overload (Taneja *et al.*, 1986; Pathak *et al.*, 1984).

glutamate decarboxylase and succinic semialdehyde dehydrogenase however, are unaffected (Taneja *et al.*, 1986). Acute focal iron and cobalt overload, as used to induce epilepsy in rats, leads to an increase of succinic semialdehyde dehydrogenase (SSA-DH), whereas GABA-T is decreased in the cobalt focus only (Pathak *et al.*, 1984). Injection of iron strongly impairs the release of GABA, which has been attributed to iron induced LPO. The GABA uptake proved to be relatively insensitive, leading to an overall lack of inhibitory neurotransmission (Zhang *et al.*, 1989). The authors relate the pathogenesis of iron-induced epilepsy to this GABA imbalance. This decrease of GABA can last until 6 hrs after the injection (Shiota *et al.*, 1989). The inhibitory effect of iron on GABA release was also demonstrated in cortical cell culture (Swaiman and Machen, 1985). Somewhat contradictory to the results described above is the finding that LPO slows down the uptake of GABA into synaptosomes (Rafalowska *et al.*, 1989). Interestingly, only the presynaptic processes like biosynthesis, storage, uptake, and release, as well as metabolism are affected by peroxide; there is no change in postsynaptical GABA response (Pellmar, 1987).

1.2.3.   Serotonin (5-HT)

5-HT is generated in two enzymatic steps starting from the essential amino acid tryptophane. The rate-determining step, just like in the biosynthesis of catecholamines, is the aromatic hydroxylation catalyzed by the heme enzyme tryptophane hydroxylase. This enzyme is dependent on iron and consequently inhibited by iron chelators (Kuhn *et al.*, 1980; Waldmeier *et al.*, 1993). Furthermore, changes in the cellular redox status, as induced by molecular oxygen, sulfhydryl, or disulfide reagents, also inactivate the enzyme, which has been attributed to modification of thiol side chains and iron sites (Kuhn *et al.*, 1980).

1.3.   THE REGULATION OF AXONAL GROWTH

Axonal outgrowth and targeting are dependent on both the properties of the presynaptic neuron itself and on a set of surface receptors of the surrounding cells. The choice of the proper direction is of vital importance for the presynaptic neuron. Usually the first axon growing from a ganglion (the so-called pioneering axon) is followed by the others, which can use the first one as a guidance cue, but the selected direction may be modified by non-neuronal cells of the environment (Garriga *et al.*, 1993). The moving trajectory of the growth cone of the pioneering neuron is guided by specific glycoproteins: cadherins, integrines, fibronectin, laminin (Gomez and Letourneau, 1994), polysialic acids (Tang *et al.*, 1994), F-actin (Lin and Forscher, 1993), and UNC-5, a receptor of immunoglobulin super-family (Hamelin *et al.*, 1993).

The synthesis and surface exposition of the environmental cues are regulated by mechanisms that will not be discussed here. Oxidative modification of the cell surfaces may result in a loss of guidance cues. As fibronectin binds iron (Sinosich *et al.*, 1983) and is also subject to iron-induced oxidative dimerization (Vartio and Kuusela, 1991), it must be considered whether this can bring about a destination error of axonal outgrowth. The Fe^{2+}-induced covalent modification of actin by cytosolic catecholamines and 5-HT (Velez Pardo *et al.*, 1995) can have the same consequences, as actin plays an important role in shaping the axon in response to environmental cues.

The mechanism of axonal sprouting is common to all eukaryons that form outgrowing tips. Elongation of the tip is associated with cytosolic Ca^{2+} and proton gradients. The magnitude of the gradient was found to correlate with the rate of growth, and inhibition of Ca^{2+}-fluxes by lanthanum abolished the pH gradient and inhibited tip elongation (Gibbon and Kropt, 1994). As Fe^{2+}, Co^{2+}, and Ni^{2+} block Ca channels (Winegar and Lansman, 1991), they can most likely also influence axon development.

1.4.   AXONAL TRANSPORT AND IRON

Active transport along the axon is not only an important supply route to bring cellular components and neurotransmitter precursors to the nerve terminals, but also provides a major tool of chemical communication for the neuronal cell. Anterograde vesicular flow is used for neurotransmitter signaling, and retrograde transport is employed by responses of postsynaptic cells via neurotrophins (Chao, 1994), cytokines, and probably by other back-ward-directed chemical signals like nitric oxide or arachidonic acid (Davis and Murphey, 1994). The viability of the presynaptic neuron depends on its successful dialogue with the postsynaptic cell: If it is interrupted, e.g., by experimental axotomy, the presynaptic neuron does not survive. The involvement of impaired axonal transport in the pathogenesis of Alzheimer's disease (AD) has recently been discussed (Richard *et al.*, 1989; Caporaso *et al.*, 1994). The myelin sheath seems to protect the axon from outside damage, but myelin itself is very susceptible to iron-induced oxidation (Beard *et al.*, 1993). This effect may be involved in pathogenic processes like multiple sclerosis or demyelinating neuropathies.

The inner axonal space can be susceptible to the deleterious effects of elevated cytosolic iron concentration. It contains several structural elements whose function may interfere with iron. Thus reorganization and disruption of the actin cytoskeleton caused by its phosphorylation by protein kinase C (PKC) depends on its activation by a metal ion, probably zinc (Hedberg *et al.*, 1991). The covalent modification of actin by cytosolic catecholamines as well as 5-HT, which is strongly enhanced by Fe^{2+} (Velez Pardo *et al.*,

1995), may probably lead to the same result. Axonal transport proceeds along microtubules built up from tubulin subunits. The integrity of the axon is supported by a stiff microtubular structure built up of α- and β-subunits. Tubulin aggregation requires Mg-GTP (Monasterio and Timasheff, 1985). Most GTPases are activated by Mg^{2+} or Mn^{2+} (Higashijima *et al.*, 1987; Manne and Kung, 1985). Thus, other divalent cations, and possibly Fe^{2+}, might be competitive inhibitors of GTPases in respect to the cofactor metal, and iron overload may interfere with the building of the axonal inner structure. Tubulin can be specifically modified by DA binding catalyzed by prostaglandin H synthase (Mattammal *et al.*, 1995). Availability of cytosolic DA for this presumably deleterious reaction may be provided by impairment of vesicular storage of this catecholamine or of its synthesis and degradation. Both processes appear to be related to iron deregulation.

Several other proteins are involved in axonal transport: Kinesin is responsible for anterograde flow, and microtubule-associated proteins (MAP1 and MAP2) for retrograde flow. MAP2 and its kinase are able to bind iron (Sanchez *et al.*, 1995), but with unknown effect on their functions. Both kinesin and MAPs split ATP, and hence everything mentioned above about ATPases inhibition by iron (see section 1.5) may be effective here. The structure of these proteins is probably closer to myosin than to the other ATPases. It is interesting that myosin is also inhibited by several divalent cations and undergoes thiol crosslinking in the presence of iron and H_2O_2 (Nimura *et al.*, 1987).

Although there is no direct data on the role of iron in axonal transport in the literature, the above observations allow the expectation of interference between iron and axonal transport. To support this conclusion it is worth mentioning that binding to the D_1 receptor stimulates the expression of MAP2, and the receptor itself is regulated by iron (Ben-Shachar *et al.*, 1993) (see section 1.2).

1.5. TRANSLOCATION OF NEUROTRANSMITTERS: STORAGE, RELEASE, REUPTAKE

1.5.1. Uptake of Neurotransmitters into Vesicles

Synthesized in the cytosol, neurotransmitters (DA, NE, GABA, 5-HT) are taken up into membrane vesicles for storage and anterograde transport by energy-dependent processes. Hence, energy depletion may be expected to impair uptake and transport. Although the mitochondrial respiratory chain, the main brain source of ATP, is constructed of iron-dependent enzymes, moderate iron deficiency does not bring about a mitochondrial failure in the brain (Evans and Mackler, 1985; Beard *et al.*, 1993), and thus is unlikely to affect vesicular uptake of neurotransmitters. According to the accepted model, a specific carrier protein takes catecholamines into vesicles; this active transport is driven by a pH gradient. Only the deprotonated amines can pass the membrane, the more acidic conditions inside the vesicles prevent the escape of transmitters. Since catecholamines are iron chelators (Liu and Mori, 1993), elevation of the cytosolic iron concentration may decrease their availability for binding to the carrier protein and consequently increase the cytosolic concentration of the transmitter. Iron-stimulated oxidative stress may also bring about damage to the vesicular membrane and leakage of the neurotransmitter to the cytosol (Yu *et al.*, 1993). Although there is nothing known about iron inhibiting the uptake of neurotransmitter into vesicles, maintenance of the proton gradient might be impaired by iron inhibition of ATPase, which maintains this gradient.

Non-compartmentalized catecholamine neurotransmitter, for example DA or NE, may be autoxidized or metabolized by monoamine oxidase in the cytosol, increasing the cytosolic pool of free radicals and hydrogen peroxide (Cohen and Werner, 1994), or may interact with cytosolic enzymes. When reaching the mitochondrial matrix they will inhibit mitochondrial respiration (Ben-Shachar et al., 1994). The products of iron-catalyzed catecholamine oxidation can bind covalently to proteins, as established for catechol-o-methyltransferase, actin, and G proteins. This modification interferes with binding of nucleotides by the proteins and is expected to impair their function (Velez Pardo et al., 1995). Increased DA concentration enables formation of DA–DNA adducts, catalyzed by prostaglandin H synthase, and this can impair binding of promotors or inhibitors of transcription (Mattammal et al., 1995).

1.5.2. Neurotransmitter Release

Release of neurotransmitter from vesicles is induced by an action potential reaching the nerve terminal. The membrane depolarization causes Ca^{2+} entry into the cytosol, which starts a series of events resulting in the exocytosis of the transmitter to the synaptic cleft. Release is preceded by mobilization of the loaded vesicles, which are bound by cytoskeleton elements such as synapsin and spectrin. Phosphorylation of these structural proteins by a Ca^{2+}/calmodulin (CaM)-dependent protein kinase enables the vesicles to diffuse freely to the synaptic cleft. Hydrolysis of the phosphorylated sites by a Mn^{2+}-dependent phosphoprotein phosphatase can prevent the release of vesicles. Iron-induced inhibition of the phosphatase (Tipper et al., 1986) may thus maintain a higher level of phosphorylation of synapsin and avoid inhibition of this process. The average effect of iron will depend on the balance between the protein kinase and phosphatase activities. Spectrin crosslinking and consequently affected function appears to be metal dependent (Girotti et al., 1986). Since spectrin participates in vesicular transport, release of neurotransmitters, and adhesion of the axonal growth cone, its modification might cause deregulation of these processes. Whether this can take place naturally is still unknown, but some functional relation of spectrin to pathogenic processes in the brain is illuminated by the finding of accumulated products of spectrin degradation in degenerating neurons of AD brains (Masliah et al., 1991). Membrane fusion is a Mg-ATP-consuming process (Hurtley, 1993) and involves GTP-binding proteins (Higashijima et al., 1987), and as such it may also be regarded as a target for iron effect.

Divalent cations other than Ca^{2+} can probably release neurotransmitters. This was demonstrated for divalent lead, which released NE by a mechanism analogous to that of Ca^{2+} (Tomsig and Suszkiw, 1993).

Unlike Ca^{2+}-induced transmitter release, K^+-evoked glutamate, GABA, DA, NE, and acetylcholine release, independent of external Ca^{2+}, appears to be unrelated to a vesicular system. Transmitter molecules are thought to be released directly from the cytosol by reverse function of their reuptake transporters, which is promoted by a raised internal sodium concentration. This mechanism functions with delay, continuously and slowly as compared to vesicular release induced by external Ca^{2+}. However, this process is thought to be also Ca^{2+}-dependent, but instead of glutamate-induced influx from the extracellular space, the Ca^{2+} is possibly mobilized from internal stores, such as the mitochondria or the endoplasmatic reticulum (Adam-Visi, 1992). Hence this neurotransmitter release does not appear to support neuronal communication, but rather to be a false signal, although there is a possibility that it can play some trophic role with an impaired axonal function. This mechanism is especially important in GABAergic synapses, where up to 30% of K^+-stimulated transmitter release from cortical slices was independent of external Ca^{2+}, and it also plays some role in catecholaminergic terminals. The glutamate release by glial cells was also found to be independent of external Ca^{2+} (Adam-Visi, 1992) (see also section 1.6).

Membrane damage of vesicles and mitochondria by iron-induced lipid peroxidation may contribute to increase the unspecific outflow of neurotransmitters from the nerve terminal by both increasing the pool of free cytosolic transmitters and raising the level of cytosolic Ca^{2+}. Finally K^+-evoked release of neurotransmitters is independent of external Ca^{2+} and may be activated by raising the extracellular K^+ concentration, which is regulated by astrocytes. Their function is sensitive to iron toxicity (see section 2.6).

1.5.3. Neurotransmitter Reuptake

Reuptake of neurotransmitters into presynaptic cells is driven by a specific transporter and is coupled by Na^+-dependent ATP cleavage. A variety of transport ATPases: Na^+/K^+-, Ca^{2+}/Mg^{2+}-, Ca^{2+}-, H^+- and CaM-dependent, as well as F1-ATPase-dependent from the respiratory chain are inhibited by iron (Anghileri *et al.*, 1993; Korge and Campbell, 1993; Lippe *et al.*, 1991; Rohn *et al.*, 1993; Li *et al.*, 1995; Leclerc *et al.*, 1991; Braughler *et al.*, 1985). In most cases the inhibition is brought about by an iron-induced oxidation of the enzyme's SH groups. But at least for Na^+/K^+-ATPase, which is inhibited by several divalent cations, such as Be^{2+}, Zn^{2+}, Cu^{2+}, Cd^{2+}, and Fe^{2+}, it has been proposed that the inhibitory mechanism does not involve oxidation (Price and Joshi, 1984). However, in spite of all this data the effect of iron on the reuptake mechanisms has so far never been directly examined.

Elevated cytosolic iron appears to diminish the efficiency of vesicular loading, storage, and reuptake, and its effect is rather negative. Mobilization of the vesicles and neurotransmitter release might be modified by elevated iron. The neurotransmitter fraction released in a non-specific, Ca^{2+}-independent manner, which may provide false information in the communicative network, is expected to increase when iron is raised to a certain toxic level. This conclusion is supported by the observation that synaptosomal DA uptake and release are modified by iron-induced and NO-mediated peroxidation (Ramassamy *et al.*, 1994). For other neurotransmitters this conclusion is more speculative since direct experimental data is lacking, and may be regarded as a question waiting for an answer.

1.6. RECEPTORS, METABOLIC CONSEQUENCES OF THE RECEPTION, BACKWARD SIGNALING

The effectiveness of neurotransmission is influenced by several factors: the quantity of receptor protein, metabolic events triggered by the binding of the transmitter at the receptor, allosteric regulation of the receptor function, and covalent modification (e.g., phosphorylation) of the receptor. The impaired synaptic plasticity for instance, that is characteristic for the aged brain can be understood in terms of insufficient receptor expression and receptor coupling to G proteins (Pedigo, 1994). Another intracellular response to transmitter binding is modulation of the activity of cytosolic enzymes by Ca^{2+} entrance. It activates either phospholipases or nitric oxide production (Ferrendelli et al., 1974), which stimulates guanylate cyclase (Daniel et al., 1993) and an ADP-ribosyltransferase (Schuman et al., 1994). Several critical points can be found in this sequence of metabolic events that might allow modulation by iron or other cations. In this chapter, we will look at whether metal ions can disturb the neurotransmitter control of the metabolic events in the postsynaptic cell.

1.6.1. Regulation of Receptor Expression

The sensitivity of the postsynaptic membrane to neurotransmitters is mainly determined by the density of receptors, and thus by their expression and degradation. The expression of receptor proteins was shown to be regulated by neural activity via a variety of first, second, and third messengers, as well as post-transcriptional mechanisms (Laufer and Changeux, 1989).

On this level, a close interdependence of different neurotransmitter systems can be observed. An extensively studied example is the expression of the GABA-A receptor, which is induced by its own agonist, 4,5,6,7-tetrahydroisoxazolo[5,4-c]pyridin-3-ol (Kardos et al., 1994), as well as by an activation of NMDA receptor (Memo et al., 1991). This reaction is modulated by corticosteroids (Orchinik et al., 1994). Furthermore, stimulation of the GABA-A receptor induces a receptor-mediated reduction of the GABA-A receptor α-subunit (Montpied et al., 1991). The regulatory mechanisms can be understood in terms of the following model: Ca^{2+} influx through NMDA receptors and through voltage-gated channels evokes synthesis of c-fos and other immediate early genes, which encode transcription factors and control secondary programs of gene expression (Bading et al., 1993). The products of these genes are thought to act as messengers in the coupling of the transsynaptic signal with gene expression in postsynaptic cells. The Ca^{2+} signal is thought to be propagated by a Ca-responsive element or/and a cAMP-responsive element located in the c-fos promoter. The set of stimulated immediate-early genes depends on the neurotransmitter system, the class of its receptor (Carter, 1992), and the region of the brain (Sirinathsinghji et al., 1994). The regulation of receptor expression also includes protein phosphorylation by CaM-dependent protein kinases, serine-threonine and tyrosine kinases (Bading et al., 1993), which are involved in many metabolic cascades.

The classic example of post transcriptional regulation of neurotransmitter receptor expression is the induction of a 35000 KD mRNA binding protein that destabilizes the mRNA of the NE β-receptor. This process is stimulated by β-agonists. Similarly, iron controls the expression of ferritin and the transferrin receptor (TfR) by activation of a regulatory mRNA binding protein (for a review see, e.g., Crichton and Ward, 1992).

Changes of receptor expression by an analogous mechanism may help to explain the decrease of D_2 binding capacity (B_{max}) and the increase of GABA binding in iron-deficient rats (Youdim et al., 1989; Ben-Shachar et al., 1993). It is certainly worthwhile to search for a cytosolic protein that is able to bind to the receptor DNA or mRNA, influences expression of the receptor protein, and is regulated by iron. The D_1 receptor is affected by iron in a different way, here, iron deficiency markedly increases the K_D but leaves B_{max} unchanged.

There is no effect on the NE β- and γ-receptors or on muscarinic cholinergic and 5-HT receptors (Youdim *et al.,* 1989).

1.6.2.   Influence of Iron on Second Messenger Systems

Transmitter binding results in a cascade of metabolic events, triggered either by a direct opening of ion channels (Ca^{2+} for glutamate receptors, Cl^- for GABA receptors) or second-messenger-mediated systems. The release of second messengers like cAMP or inositol-2,4,5-triphosphate involves: (1) GTP stimulation of G-proteins; (2) dissociation of their regulatory α-subunit; and (3) its binding to the specific enzymes—adenylate cyclase, phospholipase C, guanylate cyclase, or protein kinase C, and their activation. Hydrolysis of the GTP molecule in the α-subunit induces dissociation of the complex and terminates enzyme activation. When this hydrolysis is inhibited, the regulatory subunit remains bound to the enzyme, which remains in the activated form (Sitaramayya *et al.,* 1991). Competitive inhibition of Mg^{2+}- or Mn^{2+}-activated GTPase by other cations (Higashijima *et al.,* 1987; Manne and Kung, 1985) may bring about such an effect, which would lead to uncoupling of receptor–transmitter interactions and the second messenger systems. This speculative statement is supported by an observation of an aluminum stimulatory role in G-protein-coupled transmembrane signaling (Haug *et al.,* 1994).

Another example of metal dependent second messenger stimulation is the activation of phospholipase C by Cd^{2+}, Ni^{2+}, Co^{2+}, Fe^{2+}, or Mn^{2+} via G proteins and adenylate cyclase (Smith *et al.,* 1994; Marti *et al.,* 1994). A wide variety of di- and trivalent cations like aluminum, manganese, lead, vanadium, or rare earth elements can interfere with the turnover of cAMP as they bind to CaM (Buccigross and Nelson, 1986; 1988; You *et al.,* 1990; Vig *et al.,* 1991) and alter its capacity to activate phosphodiesterase (Vig *et al.,* 1991). As well as Ca^{2+}, other metal ions, like Al^{3+}, Mn^{2+}, Hg^{2+}, and Cd^{2+}, were shown to interact directly with phosphodiesterase and compete for binding to CaM with Ca^{2+} (Richardt *et al.,* 1985). Other CaM dependent enzymes, such as CaM-protein phosphatase (Goldstein, 1993), protein kinase, phosphodiesterase (Chao, 1994), and NO synthase (NOS), can also be affected. Hence, metal cations are involved in setting the balance between synthetic and degradative processes of second messengers: activation of adenylate cyclase and phosphodiesterase, protein kinase, and protein phosphatase.

Dopaminergic transmission is most likely sensitive to various transition metals, as it is mediated via G-protein dependent second messenger mechanisms. D_1 receptors activate adenylate cyclase, and D_2 receptors inhibit this system (Hayes *et al.,* 1992). They are coupled to G_s and G_i proteins, respectively (Kebabian, 1994). Activation of D_1-like receptors (but not D_2) induced expression of *fos*-like and *fra*-like transcription factors in striatal neurons (Liu *et al.,* 1995). Activation of the D_2-receptor in GABAergic neurons in the basal ganglia inhibited the expression of glutamate dehydrogenase mRNA, as it follows from D_1-induced inhibition of adenylate cyclase via G_i, whereas glutamate exerted a stimulatory effect (Qin *et al.,* 1994).

Metal regulation of gene transcription might be mediated by G proteins. Metal-induced inhibition of GTP turnover might cause potentiation of the metabolic response for the transmission and slowing down of G-protein recycling. This suggestion can explain the differential effect of iron deficiency on adenylate cyclase activity described by Ben-Shachar *et al.* (1993). Although adenylate cyclase is a metal-activated enzyme and may be expected to be activated in the absence of competing cation (Fe), D_1-coupled enzyme activity was significantly decreased and D_2-coupled activity was increased in iron-deficient brains. In the case of the D_2 receptor it was correlated to a decrease of agonist binding and, hence, with less active G_i, but the binding capacity of D_1 receptors was not effected by iron deficiency. This may imply low activatory efficiency of the G_s protein, probably because of too rapid GTP hydrolysis. Similarly, D_2-coupled inhibition of cyclase activity could be less effective for the same reason. This is supported by the finding of Moser and Schuster (1990) that D_1-coupled adenylate cyclase activity is inhibited by iron via G proteins.

There is also evidence for an influence of reactive oxygen species on second-messenger-mediated processes. Hypoxia and reoxygenation of kidney slices leads to increased activity of phospholipase C (Cotterill *et al.*, 1993), the enzyme that starts the phosphoinositide cascade and releases 1,4,5-inositol triphosphate (IP_3) to eventually trigger Ca^{2+} release from the endoplasmatic reticulum, its main intracellular store. *In vitro*, this simulation can be induced by H_2O_2 (Meij *et al.*, 1994) or by the toxic lipid peroxidation product *trans*-4-hydroxynon-2-enal (Rossi *et al.*, 1991). This processes seem to be important in experimental alcoholic liver damage, where levels of arachidonic acid were decreased and the activity of phospholipase C was increased (Nanji *et al.*, 1993), as well as in inflammation (Winrow *et al.*, 1993). Although all these findings have not been confirmed for brain tissue, it seems likely that the same mechanisms can be observed in the CNS and the release of IP_3 is enhanced under conditions of oxidative stress. The binding of IP_3 to its effector sites appears to be regulated in just the opposite way. In gerbil hypocampus, IP_3 binding was seriously diminished after ischemia. This effect could be prevented by phenobarbital and the lipid peroxidation inhibitor KB-5666 (Hara *et al.*, 1991).

The β-receptor adenyl cyclase system of the sarcolemma is irreversibly inhibited in myocardial ischemia. The same effect could be induced *in vitro* by ascorbate/Fe^{2+} and prevented by t-butyl-4-hydroxyanisole (BHA), a powerful antioxidant (Schimke *et al.*, 1992). Iron is a potent enhancer of oxidative stress, as Fe^{2+} is able to convert superoxide ($O_2^{\cdot-}$) and H_2O_2 into the more reactive and more toxic hydroxyl radical ($HO^{\cdot}$). Thus it can play an important role in all the radical-mediated processes described above.

1.6.3. Iron Interaction with Protein Phosphorylation

G protein coupled triggering of adenylate and guanylate cyclases as well as of phospholipase starts several metabolic cascades mediated by protein phosphorylation/ dephosphorylation by protein kinases, which are metal-activated enzymes (Stohs and Bagchi, 1995). Protein kinase C (PKC) is stimulated by Mg^{2+}, Mn^{2+}, and Zn^{2+} (Hedberg *et al.*, 1994), but it is also sensitive to other divalent cations in low concentrations such as iron, cobalt, chromium (Stohs and Bagchi, 1995), and even lead (Goldstein, 1993). Lead activates PKC at picomolar concentrations and furthermore stimulates protein dephosphorylation by CaM-dependent protein phosphatase in the nanomolar range. Thus, the overall balance of phosphorylation is sensitive to lead (Goldstein, 1993). Phosphoprotein phosphatases 1 and 2 as well as CaM-dependent phosphoprotein phosphatase are activated by Mn^{2+} and Ni^{2+} (Seki *et al.*, 1995), which can also shift the balance between phosphorylation and dephosphorylation in a similar way as described for lead.

In high concentrations, divalent cations compete with magnesium for binding with the substrate ATP (Adams and Taylor, 1993), and hence inhibit the kinase reaction. Zinc is necessary for activating translocation of PKC from the cytosol to the membrane, and for aggregation of phospholipids to anchor PKC at the membrane (Maurer *et al.*, 1992), but these functions can be performed by gadolinium (III). Tyrosine kinase is normally activated by Mg^{2+} and Mn^{2+} (Ernould *et al.*, 1993), Hg^{2+} causes irreversible ligand-independent tyrosine phosphorylation and c-*fos* transcription (Tomsig and Suszkiw, 1993).

Furthermore, PKC is activated by iron-catalyzed oxidation of its sulfhydryl groups (Taher *et al.*, 1993) and its expression is upregulated by ferric transferrin, but not by low-molecular-weight iron complexes (Alcantara *et al.*, 1994). Other protein kinases, like cGMP-dependent protein kinase, are also activated by oxidation of their SH groups with S–S bridge formation induced by heavy metals (Landgraf *et al.*, 1991).

With all the data presented here, it becomes clear that the overall balance between protein phosphorylation and dephosphorylation is highly sensitive to metal cations. The degree of phosphorylation will depend on the individual responses of kinases and phosphatases to the different ions, which can therefore affect the activity of enzymes and the sensitivity of the postsynaptic cell to transmitter binding.

1.6.4. Allosteric Control of Receptor Sensitivity and Desensitization of Receptors

Besides the quantity of receptor molecules, the sensitivity of the receptor is regulated by allosteric effectors. Thus, monovalent cations modulate coupling of G proteins to their receptors (Tian and Deth, 1993). The NMDA receptor is activated by glycine (Zorumski and Thio, 1992), and the GABA-A receptor is activated by benzodiazepines and inhibited by zinc (Gyenes *et al.*, 1994). Iron was not an object of these studies, although it can possibly replace Zn^{2+} to either antagonize or mimic its effect.

Receptors can be desensitized in several ways. The most common mechanism involves phosphorylation of the receptor protein by cytosolic protein kinases (Montastruc *et al.*, 1993; Wang *et al.*, 1993). Phosphorylation effects the rate of receptor desensitization as well as the assembly of receptor subunits and its aggregation at the synapse (Swope *et al.*, 1992). It can be regulated via second messengers (see section 1.6.2). Iron's ability to activate phosphorylation and inhibit phosphatase reaction might support desensitization of the receptor (see section 1.6.3).

The AMPA receptor is rapidly desensitized by Ca^{2+}. This appears to be a key mechanism for the short-term regulation of responses mediated by these receptors (Zorumski and Thio, 1992). The molecular mechanism of this type of desensitization is not yet elucidated. Hence possible interference with other metal ions cannot be discussed.

Another variant of desensitization was postulated for the NMDA receptor. Ca^{2+}, which enters the cell via the NMDA glutamate receptors, stimulates the formation of NO. The receptor has an oxidative modulatory site, where cysteine SH residues are accessible to nitrosylation and disulfide crosslinking (Lei *et al.*, 1992). NO^+ donors like sodium nitroprusside can downregulate the NMDA-receptor and limit Ca^{2+}-influx to the cell (Lipton *et al.*, 1993), thus exhibiting neuroprotective properties. The balance between neuroprotection and neurotoxicity depends on the cellular redox environment: If sodium nitroprusside is reduced, e.g. by ascorbate, the ferrous species will release NO instead of NO^+ and consequently be toxic (Lipton and Moser, 1994; Lipton *et al.*, 1994a). With the participation of iron it is possible to postulate the inverse mechanism: A complex between NO and ferric iron can also release NO^+ if one electron is moved from the ligand to the central iron. In general, NO complexes with ferric heme are known to be reactive, they can function as nitrosyl donors to carbon, oxygen, nitrogen, and sulfur moieties (Wade and Castro, 1990). The inhibition of Ca^{2+} influx can be an important factor for the self-regulation of NO formation. SH groups of the receptor protein essential for its activity can be blocked by heavy metals (Hg^{2+}, Zn^{2+}, Cu^{2+}, Mn^{2+}, etc.) with a subsequent inhibition of the binding of the messenger, as demonstrated for porcine brain neurotensin receptor (Carraway et al., 1993). Finally, direct competition of cations for the pore of the ligand-gated channel may be proposed, at least for divalent cations such as Cd^{2+}, which binds to Ca^{2+} channels (Rothman et al., 1991). It has been also demonstrated that NO forms a complex with the heme group in NOS itself (Wang et al., 1994; Griscavage et al., 1994). The inhibition can be reversed by tetrahydrobiopterine, suggesting that NO predominantly coordinates to the ferric enzyme (Griscavage et al., 1994).

1.7. INFLUENCE OF NEURAL SIGNALING ON THE IRON REGULATION IN THE CELL

Neurotransmitters can modulate cellular iron uptake and storage by triggering phosphorylation of an iron responsive element binding protein (IRE-BP). This protein regulates the translation of ferritin and TfR genes by binding to the specific mRNAs in the absence of free iron, and dissociation if the intracellular iron concentration is increased. Phosphorylation of the IRE-BP by PKC increases its binding to mRNA threefold (Eisenstein et al., 1993), thus stimulating TfR and suppressing ferritin synthesis.

The control of cellular iron balance can be severely disturbed by NO. Generally, NO forms complexes with transition metals, especially iron, and thus interferes with many processes where these are involved. Thus it inhibits iron-sulfur proteins such as aconitase or mitochondrial NADH-coenzyme Z reductase and induces a massive loss of iron from the attacked cells. As iron-sulfur enzymes alone could not account for the high loss of iron, it soon became obvious that NO also releases iron bound to ferritin (Payne et al., 1990). Recent results show that NO does not only interfere with the storage of iron by ferritin but also with the expression of this protein, as well as of the TfR (Pantopoulos and Hentze, 1995). The authors demonstrated, for macrophages, that NO induces the binding of an iron-regulatory protein to the iron-responsive elements on the mRNAs for ferritin and the transferrin receptor, even if iron is present. Thus NO stimulates expression of the TfR, even in iron-replete cells, and overrides the cellular controls of iron homeostasis. NO gives an impressive example of neurotransmitter-mediated regulation of iron distribution in the brain. It can explain why glutamatergic activity is linked to accumulation of iron in the basal ganglia (Shoham et al., 1992) and GABAergic neurotransmission to a decrease of the iron concentration in the striatum, globus pallidus, and ventral pallidum (Hill, 1985).

The involvement of various neurotransmitters in the regulation of brain iron levels reflects the close regulatory interdependence of these species.

1.8. INTERCELLULAR INTERACTIONS IN THE CNS

Although collection and processing of information is exclusively the function of the various types of neurons, they require different kinds of support from non-neuronal cells. Astrocytes provide mechanical rigidity to brain tissue, which otherwise would be fluffy if constructed of neurons only. Moreover, they are responsible for metabolic regulations: They absorb excess potassium ions from the extracellular space when neurons fire, and prevent diffusion of neurotransmitters from synaptic cleft. Their ability to meet these requirements is provided by the development of neurotransmitter receptors and is regulated by both internal and external mechanisms (Shao and McCarthy, 1993). Being connected to blood vessels they probably channel the extra potassium to the blood stream, absorb nutrients from blood, and participate in the formation of blood–brain barrier. Oligodendrocytes sheath axons, increasing the rate of the depolarization wave along axons and protecting them mechanically. In the peripheral nervous system they link several axons together to form bundles and regulate the rate of axonal extension. Oligodendrocytes express G proteins, coupled to 5-HT, somatostatin, and muscarinic acetylcholine receptors. They regulate rectifying K^+ inward channels (Karschin *et al.*, 1994). Microglial cells (brain macrophages) are responsible for phagocytosis in the brain (including degenerated neurons and macroglia), and cytotoxic and microbicidal activities. Their action is partially based on ROS, including nitrosyl radicals, and partially on glutamate and interleukin(IL)-1 production (Piani *et al.*, 1994). Glutamate can initiate a cascade of deleterious events in neurons, mediated by calcium, and induce NO synthesis (Dawson *et al.*, 1991). The last compound can release iron from ferritin to disrupt intracellular iron homeostasis and contribute to ROS formation (Lipton *et al.*, 1994a,b; Lipton and Moser, 1994). Nitrosyl radicals and ROS can damage all types of neural cells. The expansion of macrophages is controlled by astrocytes; microglia-derived cytokines induce astroglial scarring and astrocytes produce transforming growth factor (TGF)-β and IL-10, which inhibit the production of other cytokines, release of ROS and nitric oxide and expansion of B and T cells in brain (Piani *et al.*, 1994). Glutamate released by macrophages is absorbed by astrocytes enhanced by cytokine induction. Thus negative-feedback circuits may restrict deleterious action of macrophages (Piani *et al.*, 1994).

Interaction of neural cells, including neuron–neuron interactions, is thought to be mediated by cytokines (IL-1, IL-2, IL-6). They maintain communication between immune, endocrine, and nervous systems. IL-1 is synthesized by both neurons and glial cells; its receptors were found in hippocampal neurons (De Sousa and Cunnigham, 1994). Communicative and protective functions are supposed for the brain-derived neurotrophic factor (BDNF), neurotrophin (NT)-3, and ciliary neurotrophic factor (CNTF). They are expressed by glial cells and target tissues and act both in autocrine and paracrine ways. They attenuate or even prevent neuron injury, support their survival and differentiation, and regulate receptor and neurotransmitter development (Ludlam and Kessler, 1993). Thus, BDNF upregulates DA production in the substantia nigra (Lindsay *et al.*, 1994); fibroblast growth factor (FGF), nerve growth factor (NGF), and insulin-like growth factor (IGF) protect neurons against iron-induced oxidative damage (Zhang *et al.*, 1993); astroglia-derived FGF-2 protects tyrosine-hydroxylase immunoreactive cells against the neurotoxin *N*-methylphenyltetrahydropyridine (MPTP) (Otto and Unsicker, 1993).

The above regulatory loops may be sensitive to iron in several ways:

1.) Astrocytes are susceptible to iron toxicity (Tiffany-Castiglioni *et al.*, 1987), probably because they do not contain ferritin, and therefore cannot bind free iron (Beard *et al.*, 1993). Thus, significant deviations in function of the regulatory loop, including neuron–microglia–astroglia interactions may be expected with an increase of iron content. In such a case microglia activation might be non-balanced with astrocyte response, and neural cells may occur target for the aggression of microglia, with a subsequent loss of neurotransmitter communication.

2.) Cytokines regulate iron metabolism and storage by expression of ferritin, transferrin, and its receptor (Tsuji *et al.*, 1991).

3.) Cytokines and growth factors function through interaction with cytokine receptor superfamily. Binding of these ligands to their receptors is coupled to tyrosine phosphorylation by cytoplasmic Janus kinase (JAK). Activated JAKs phosphorylate cytoplasmic proteins belonging to a family of transcription factors like signal transducers and activators of transcription (Ihle *et al.*, 1992).

Ferric or ferrous iron activates tyrosine kinases (Alcain *et al.*, 1994; Salmeron *et al.*, 1995) and inhibits dephosphorylation by the Mn^{2+} dependent subtype phosphoprotein phosphatase (Tipper *et al.*, 1986), thus providing long-term potentiation of cytokine or growth factor signal. On the other hand, with the phosphatase calcineurin that uses Fe as a cofactor (King and Huang, 1984; Vincent and Averill, 1990), an increase of Fe concentration should have the opposite effect.

1.9. CONCLUSION

In this review, we demonstrated where iron can be involved in neuronal communication. There is a lot of direct and indirect experimental evidence available that shows that iron can affect many steps of this process, starting from biosynthesis of neurotransmitters and including vesicular uptake and storage of the transmitter, axonal transport of the loaded vesicles and their mobilization, transmitter release and reuptake, and reception of the neurotransmitter by postsynaptic cells with a subsequent cascade of metabolic reactions. It is involved in intercellular communication mediated by cytokines and growth factors. Some of the statements above are based on indirect data and have to be confirmed (or rejected) experimentally. In contrast, others are established quite well. Taken together this review is dedicated to iron-regulatory issues, rather than iron's deleterious function in brain and its important role in neural communication.

1.10. REFERENCES

Adam-Visi, V., 1992, External Ca^{2+}-independent release of neurotransmitters. *J. Neurochem.* **58**:395–405.
Adams, J. A., and Taylor, S. S., 1993, Divalent metal ions influence catalysis and active-site accessibility in the cAMP-dependent protein kinase, *Protein Sci.* **2**:2177–2186.

Alcain, F. J., Low, H., and Crane, F. L., 1994, Iron at the cell surface controls DNA synthesis in CCl 39 cells, *Biochem. Biophys. Res. Commun.* **203**:16–21.

Alcantara, O., Obeid, L., Hannun, Y., Ponka, P., and Boldt, D. H., 1994, Regulation of protein kinase C (PKC) expression by iron: Effect on different iron compounds on PKC-β and PKC-alpha gene expression and role of the 5'-flanking region of the PKC-β gene in the response to ferric transferrin, *Blood* **84**:3510–3517.

Allen, D. R., Wallis, G. L., and McCay, P. B., 1994, Catechol adrenergic agents enhance hydroxyl radical generation in xanthine oxidase systems containing ferritin: Implications for ischemia/reperfusion, *Arch. Biochem. Biophys.* **315**:235–243.

Almas, B., Le Bourdelles, B., Flatmark, T., Mallet, J., and Haavik, J., 1992, Regulation of recombinant human tyrosine hydroxylase isozymes by catecholamine binding and phosphorylation. Structure/activity studies and mechanistic implications, *Eur. J. Biochem.* **209**:249–255.

Anghileri, L. J., Maincent, P., and Cordova-Martinez, A., 1993, On the mechanism of soft tissue calcification induced by complexed iron, *Exp. Toxicol. Pathol.* **45**:365–368.

Bading, H., Ginty, D. D., and Greenberg, M. E., 1993, Regulation of gene expression in hippocampal neurons by distinct calcium signaling pathways, *Science* **260**:181–186.

Beard, J., 1987, Feed efficiency and norepinephrine turnover in iron deficiency, *Proc. Soc. Exp. Biol. Med.* **184**:337–344.

Beard, J., Tobin, B., and Smith, S. M., 1988, Norepinephrine turnover in iron deficiency at three environmental temperatures, *Am. J. Physiol.* **255**:R90-6.

Beard, J. L., Tobin, B. W., and Smith, S. M., 1990, Effects of iron repletion and correction of anemia on norepinephrine turnover and thyroid metabolism in iron deficiency, *Proc. Soc. Exp. Biol. Med.* **193**:306–312.

Beard, J. L., Connor, J. D., and Jones, B. C., 1993, Brain iron: location and function, *Prog. Food Nutr. Sci.* **17**:183–221.

Beard, J. L., Chen, Q., Connor, J., and Jones, B. C., 1994, Altered monamine metabolism in caudate-putamen of iron-deficient rats, *Pharmacol. Biochem. Behav.* **48**:621–624.

Ben-Shachar, D., Tovi, A., and Youdim, M. B. H., 1993, Iron regulation of dopaminergic transmission: relevance to movement disorders, in: *Iron in Central Nervous System Disorders* (P. Riederer and M. B. H. Youdim, eds.), Springer-Verlag, Wien, pp. 55–66.

Ben-Shachar, D., Zuk, R., and Glinka, Y., 1994, Dopamine neurotoxicity: inhibition of mitochondrial respiration, *J. Neurochem.* **64**:718–723.

Braughler, J. M., Duncan, L. A., and Chase, R. L., 1985, Interaction of lipid peroxidation and calcium in the pathogenesis of neuronal injury, *Cent. Nerv. Syst. Trauma* **2**:269–283.

Buccigross, J. M., and Nelson, D. J., 1986, EPR studies show that all lanthanides do not have the same order of binding to calmodulin, *Biochem. Biophys. Res. Commun.* **138**:1243–1249.

Buccigross, J. M., and Nelson, D. J., 1988, Interactions of spin-labeled calmodulin with trifluoperazine and phosphodiesterase in the presence of Ca(II), Cd(II), La(III), Tb(III), and Lu(III), *J. Inorg. Biochem.* **33**:139–147.

Caporaso, G. L., Takei, K., Gandy, S. E., Matteoli, M., Mundigl, O., Greengard, P., and De Camilli, P., 1994, Morphologic and biochemical analysis of the intracellular trafficking of the Alzheimer β/A4 amyloid precursor protein, *J. Neurosci,* **14**:3122–3138.

Carraway, R. E., Mitra, S. P., and Honeyman, T. W., 1993, Effects of GTP analogs and metal ions on the binding of neurotensin to porcine brain membranes, *Peptides* **14**:37–45.

Carter, D. A., 1992, Neurotransmitter-stimulated immediate-early gene responses are organized through differential post-synaptic receptor mechanisms, *Brain Res. Mol. Brain Res.* **16**:111–118.

Chao, M. V., 1994, Mechanism of action of neurotrophin receptors, *Neuropsychopharmacology* **10**:109S.

Cohen, G., and Werner, P., 1994, Free radicals, oxidative stress, and neurodegeneration, in: *Neurodegenerative Diseases* (D. B. Calne, ed.), W. B. Saunders, Philadelphia, pp. 139–162.

Cotterill, L. A., Gower, J. D., Clark, P. K., Fuller, B. J., Thorniley, M. S., Goddard, J. G., and Green, C. J., 1993, Reoxygenation following hypoxia stimulates lipid peroxidation and phosphatidylinositol breakdown, *Biochem. Pharmacol.* **45**:1947–1951.

Crichton, R. R., and Ward, R. J., 1992, Structure and molecular biology of iron binding proteins and the regulation of "free" iron pools, in: *Iron and Human Disease* (R. B. Lauffer, ed.), CRC Press, Boca Raton, Florida, pp. 23–75.

Daniel, H., Hemart, N., Jaillard, D., and Crepel, F., 1993, Long-term depression requires nitric oxide and guanosine 3′:5′ cyclic monophosphate production in rat cerebellar Purkinje cells, *Eur. J. Neurosci.* **5:**1079–1082.

Davis, G. W., and Murphey, R. K., 1994, Long-term regulation of short-term transmitter release properties: retrograde signaling and synaptic development, *Trends Neurosci.* **17:**9–13.

Dawson, V. L., Dawson, T. M., London, E. D., Bredt, D. S., and Snyder, S. H., 1991, Nitric oxide mediates glutamate neurotoxicity in primary cortical cultures, *Proc. Natl. Acad. Sci. USA* **88:**6368–6371.

De Sousa, E. B., and Cunnigham, E. T., Jr., 1994, Mapping of interleukin-1 receptors in the central nervous system: autoradiographic localization and *in situ* hybridization studies, *Neuropsychopharmacology* **10:**136S.

Eisenstein, R. S., Tuazon, P. T., Schalinske, K. L., Anderson, S. A., and Traugh, J. A., 1993, Iron-responsive element-binding protein. Phosphorylation by protein kinase C, *J. Biol. Chem.* **268:**27363–27370.

Ernould, A. P., Ferry, G., Barret, J. M., Genton, A., and Boutin, J. A., 1993, Purification and characterization of the major tyrosine protein kinase from the human promyelocytic cell line, HL60, *Eur. J. Biochem.* **214:**503–514.

Evans, T. C., and Mackler, B., 1985, Effect of iron deficiency on energy conservation in rat liver and skeletal muscle submitochondrial particles, *Biochem. Med.* **34:**93–99.

Ferrendelli, J. A., Chang, M. M., and Kinscherf, D. A., 1974, Elevation of cyclic GMP levels in central nervous system by excitatory and inhibitory amino acids, *J. Neurochem.* **22:**535–540.

Garriga, G., Desai, C., and Horvitz, H. R., 1993, Cell interactions control the direction of outgrowth, branching and fasciculation of the HSN axons of Caenorhabditis elegans, *Development* **117:**1071–1087.

Gibbon, B. C., and Kropt, D. L., 1994, Cytosolic pH gradients associated with tip growth, *Science* **263:**1419–1421.

Girotti, A. W., Thomas, J. P., and Jordan, J. E., 1986, Xanthine oxidase-catalyzed crosslinking of cell membrane proteins, *Arch. Biochem. Biophys.* **251:**639–653.

Goldstein, G. W., 1993, Evidence that lead acts as a calcium substitute in second messenger metabolism, *Neurotoxicology* **14:**97–101.

Gomez, T. M., and Letourneau, P. C., 1994, Filopodia initiate choices made by sensory neuron growth cones at laminin/fibronectin borders in vitro, *J. Neurosci.* **14:**5959–5972.

Griscavage, J. M., Fukuto, J. M., Komori, Y., and Ignarro, L. J., 1994, Nitric oxide inhibits neuronal nitric oxide synthase by interacting with the heme prosthetic group. Role of tetrahydrobiopterin in modulating the inhibitory action of nitric oxide, *J. Biol. Chem.* **269:**21644–21649.

Gyenes, M., Wang, Q., Gibbs, T. T., and Farb, D. H., 1994, Phosphorylation factors control neurotransmitter and neuromodulator actions at the gamma-aminobutyric acid type A receptor, *Mol. Pharmacol.* **46:**542–549.

Hamelin, M., Zhou, Y., Su, I. M., Scott, I. M., and Culotti, J. G., 1993, Expression of the UNC-5 guidance receptor in the touch neurons of *C. elegans* steers their axons dorsally, *Nature* **364:**327–330.

Hara, H., Kato, H., Araki, T., Onodera, H., and Kogure, K., 1991, Involvement of lipid peroxidation and inhibitory mechanisms on ischemic neuronal damage in gerbil hippocampus: Quantitative autoradiographic studies on second messenger and neurotransmitter systems, *Neuroscience* **42:**159–169.

Haug, A., Shi, B., and Vitorello, V., 1994, Aluminum interaction with phosphoinositide-associated signal transduction, *Arch. Toxicol.* **68:**1–7.

Hayes, G., Biden, T. J., Selbie, L. A., and Shine, J., 1992, Structural subtypes of the dopamine D_2 receptor are functionally distinct: expression of the cloned D_{2A} and D_{2B} subtypes in a heterologous cell line, *Mol. Endocrinol.* **6:**920–926.

Hedberg, K. K., Birrell, G. B., and Griffith, O. H., 1991, Phorbol ester-induced actin cytoskeletal reorganization requires a heavy metal ion, *Cell Regul.* **2:**1067–1079.

Hedberg, K. K., Birrell, G. B., Mobley, P. L., and Griffith, O. H., 1994, Transition metal chelator TPEN counteracts phorbol ester-induced actin cytoskeletal disruption in C6 rat glioma cells without inhibiting activation or translocation of protein kinase C, *J. Cell Physiol.* **158:**337–346.

Higashijima, T., Ferguson, K. M., Sternweis, P. C., Smigel, M. D., and Gilman, A. G., 1987, Effects of Mg^{2+} and the beta gamma-subunit complex on the interactions of guanine nucleotides with G proteins, *J. Biol. Chem.* **262:**762–766.

Hill, J. M., 1985, Iron concentration reduced in ventral pallidum, globus pallidus, and substantia nigra by GABA-transaminase inhibitor, gamma-vinyl GABA, *Brain Res.* **342:**18–25.

Hurtley, S. M., 1993, Membrane proteins involved in targeted membrane fusion, *Trends Biol. Sci.* **18:**453–455.

Ihle, J. N., Witthuhm, B. R., Ouelle, F. W., Yamamoto, K., Thierfelder, W. E., Kreider, B., and Silvennoinen, O., 1992, Signalling by the cytokine receptor superfamily: JAKs and STATs, *Trends Biol. Sci.* **19:**222–227.

Kardos, J., Elster, L., Damgaard, I., Krogsgaard-Larsen, P., and Schousboe, A., 1994, Role of GABA$_B$ receptors in intracellular Ca^{2+} homeostasis and possible interaction between GABA$_A$ and GABA$_B$ receptors in regulation of transmitter release in cerebellar granule neurons, *J. Neurosci. Res.* **39**:646–655.

Karschin, A., Wischmeyer, E., Davidson, N., and Lester, H. A., 1994, Fast inhibition of inwardly rectifying K$^+$ channels by multiple neurotransmitter receptors in oligodendroglia, *Eur. J. Neurosci.* **6**:1756–1764.

Kebabian, J. W., 1994, Neurotransmitter receptors in neurodegeneration, in: *Neurodegenerative Diseases* (D. B. Calne, ed.), W. B. Saunders, Philadelphia, pp. 119–128.

King, M. M., and Huang, C. Y., 1984, The calmodulin-dependent activation and deactivation of the phosphoprotein phosphatase, calcineurin, and the effect of nucleotides, pyrophosphate, and divalent metal ions. Identification of calcineurin as a Zn and Fe metalloenzyme, *J. Biol. Chem.* **259**:8847–8856.

Korge, P., and Campbell, K. B., 1993, The effect of changes in iron redox state on the activity of enzymes sensitive to modification of SH groups, *Arch. Biochem. Biophys.* **304**:420–428.

Kuhn, D. M., Ruskin, B., and Lovenberg, W., 1980, Tryptophan hydroxylase. The role of oxygen, iron, and sulfhydryl groups as determinants of stability and catalytic activity, *J. Biol. Chem.* **255**:4137–4143.

Landgraf, W., Regulla, S., Meyer, H. E., and Hofmann, F., 1991, Oxidation of cysteines activates cGMP-dependent protein kinase, *J. Biol. Chem.* **266**:16305–16311.

Laufer, R., and Changeux, J. P., 1989, Activity-dependent regulation of gene expression in muscle and neuronal cells, *Mol. Neurobiol.* **3**:1–53.

Lauffer, R. B., 1992, Iron, aging, and human disease: historical background and new hypotheses, in: *Iron and Human Disease* (R. B. Lauffer, ed.), CRC Press, Boca Raton, FL, pp. 1–20.

Leclerc, L., Marden, M., and Poyart, C., 1991, Inhibition of the erythrocyte (Ca^{2+} + Mg^{2+})-ATPase by nonheme iron, *Biochim. Biophys. Acta.* **1062**:35–38.

Lei, S. Z., Pan, Z. H., Aggarwal, S. K., Chen, H. S., Hartman, J., Sucher, N. J., and Lipton, S. A., 1992, Effect of nitric oxide production on the redox modulatory site of the NMDA receptor-channel complex, *Neuron* **8**:1087–1099.

Li, C. Y., Watkins, J. A., Hamazaki, S., Altazan, J. D., and Glass, J., 1995, Iron binding, a new function for the reticulocyte endosome H$^+$-ATPase, *Biochemistry* **34**:5130–5136.

Lin, C. H., and Forscher, P., 1993, Cytoskeleton remodeling during growth cone-target interactions, *J. Cell Biol.* **121**:1369–1383.

Lindsay, R. M., Wiegand, S. J., Altar, C. A., and DiStephano, P. S., 1994, Neurotrophic factors: from molecule to man, *Trends Neurosci.* **17**:182–190.

Lippe, G., Comelli, M., Mazzilis, D., Sala, F. D., and Mavelli, I., 1991, The inactivation of mitochondrial F1 ATPase by H2O2 is mediated by iron ions not tightly bound in the protein, *Biochem. Biophys. Res. Commun.* **181**:764–770.

Lipton, S. A., and Moser, A., 1994, Actions of redox-related congeners of nitric oxide at the NMDA receptor, *Neuropharmacology* **33**:1229–1233.

Lipton, S. A., Choi, Y. B., Pan, Z. H., Lei, S. Z., Chen, H. S., Sucher, N. J., Loscalzo, J., Singel, D. J., and Moser, A., 1993, A redox-based mechanism for the neuroprotective and neurodestructive effects of nitric oxide and related nitroso-compounds, *Nature* **364**:626–632.

Lipton, S. A., Singel, D. J., and Moser, A., 1994a, Nitric oxide in the central nervous system, *Prog. Brain Res.* **103**:359–364.

Lipton, S. A., Yeh, M., and Dreyer, E. B., 1994b, Update on current models of HIV-related neuronal injury: Platelet-activating factor, arachidonic acid and nitric oxide, *Adv. Neuroimmunol.* **4**:181–188.

Liu, F. C., Takahashi, H., McKay, R. D., and Graybiel, A. M., 1995, Dopaminergic regulation of transcription factor expression in organotypic cultures of developing striatum, *J. Neurosci.* **15**:2367–2384.

Liu, J., and Mori, A., 1993, Monoamine metabolism provides an antioxidant defense in the brain against oxidant- and free radical-induced damage, *Arch. Biochem. Biophys.* **302**:118–127.

Ludlam, W. H., and Kessler, J. A., 1993, Leukemia inhibitory factor and ciliary neurotrophic factor regulate expression of muscarinic receptors in cultured sympathetic neurons, *Dev. Biol.* **155**:497–506.

Manne, V., and Kung, H. F., 1985, Effect of divalent metal ions and glycerol on the GTPase activity of H-ras proteins, *Biochem. Biophys. Res. Commun.* **128**:1440–1446.

Marti, H. H., Jung, H. H., Pfeilschifter, J., and Bauer, C., 1994, Hypoxia and cobalt stimulate lactate dehydrogenase (LDH) activity in vascular smooth muscle cells, *Pflugers Arch.* **429**:216–222.

Masliah, E., Hansen, L., Mallory, M., Albright, T., and Terry, R. D., 1991, Abnormal brain spectrin immunoreactivity in sprouting neurons in Alzheimer disease, *Neurosci. Lett.* **129:**1–5.

Mattammal, M. B., Strong, R., Lakshmi, V. M., Chung, H. D., and Stephenson, A. H., 1995, Prostaglandin H synthetase-mediated metabolism of dopamine: Implication for Parkinson's disease, *J. Neurochem.* **64:**1645–1654.

Maurer, M. C., Grisham, C. M., and Sando, J. J., 1992, Activation and inhibition of protein kinase C isozymes alpha and beta by Gd^{3+}, *Arch. Biochem. Biophys.* **298:**561–568.

Meij, J. T., Suzuki, S., Panagia, V., and Dhalla, N. S., 1994, Oxidative stress modifies the activity of cardiac sarcolemmal phospholipase C, *Biochim. Biophys. Acta.* **1199:**6–12.

Memo, M., Bovolin, P., Costa, E., and Grayson, D. R., 1991, Regulation of gamma-aminobutyric acid A receptor subunit expression by activation of N-methyl-D-aspartate-selective glutamate receptors, *Mol. Pharmacol.* **39:**599–603.

Monasterio, O., and Timasheff, S. N., 1985, Role of the dianionic form of the GTP gamma-phosphate in the polymerization process of tubulin, *Arch. Biol. Med. Exp. (Santiago)* **18:**325–329.

Montastruc, J. L., Galitzky, J., Berlan, M., and Montastruc, P., 1993, Mechanism of receptor regulation during repeated administration of drugs, *Therapie* **48:**421–426.

Montpied, P., Ginns, E. I., Martin, B. M., Roca, D., Farb, D. H., and Paul, S. M., 1991, gamma-Aminobutyric acid (GABA) induces a receptor-mediated reduction in $GABA_A$ receptor alpha subunit messenger RNAs in embryonic chick neurons in culture, *J. Biol. Chem.* **266:**6011–6014.

Moser, A., and Schuster, O., 1990, Iron inhibits D-1 dopamine receptor coupled adenylate cyclase via G-proteins in the caudate nucleus of the rat, *Biochem. Biophys. Res. Commun.* **171:**1372–1377.

Nanji, A. A., Zhao, S., Lamb, R. G., Sadrzadeh, S. M., Dannenberg, A. J., and Waxman, D. J., 1993, Changes in microsomal phospholipases and arachidonic acid in experimental alcoholic liver injury: Relationship to cytochrome P-450 2E1 induction and conjugated diene formation, *Alcohol Clin. Exp. Res.* **17:**589–603.

Nimura, E., Miura, K., Shinobu, L. A., and Imura, N., 1987, Enhancement of Ca^{2+}-sensitive myosin ATPase activity by cadmium, *Ecotoxicol. Environ. Safety* **14:**184–189.

Orchinik, M., Weiland, N. G., and McEwen, B. S., 1994, Adrenalectomy selectively regulates GABA A receptor subunit expression in the hippocampus, *Mol. Cell Neurosci.* **5:**451–458.

Otto, D., and Unsicker, K., 1993, FGF-2-Mediated protection of cultured mesencephalic dopaminergic neurons against MPTP and MPP^+: Specificity and impact of culture conditions, non-dopaminergic neurons, and astroglial cells, *J. Neurosci. Res.* **34:**382–383.

Pantopoulos, K., and Hentze, M. W., 1995, Nitric oxide signaling to iron regulatory protein: Direct control of ferritin mRNA translation and transferrin receptor mRNA stability in transfected fibroblasts, *Proc. Natl. Acad. Sci. USA* **92:**1267–1271.

Pathak, D. N., Roy, D., and Singh, R., 1984, Changes in the activity of gamma-amino butyric acid transaminase and succinic semialdehyde dehydrogenase in the cobalt and iron experimental epileptogenic foci in the rat brain, *Biochem. Int.* **9:**59–68.

Payne, M. J., Woods, L. F., Gibbs, P., and Cammack, R., 1990, Electron paramagnetic resonance spectroscopic investigation of the inhibition of the phosphoroclastic system of Clostridium sporogenes by nitrite, *J. Gen. Microbiol.* **136:**2067–2076.

Pedigo, N. W., 1994, Neurotransmitter receptor plasticity in aging, *Life Sci.* **55:**1985–1991.

Pellmar, T. C., 1987, Peroxide alters neuronal excitability in the CA1 region of guinea-pig hippocampus in vitro, *Neuroscience* **23:**447–456.

Piani, D., Constam, D. B., Frei, K., and Fontana, A., 1994, Macrophages in brain: friends or enemies? *News in Physiological Sciences* **9:**80–84.

Price, D. J., and Joshi, J. G., 1984, Ferritin: protection of enzymatic activity against the inhibition by divalent metal ions in vitro, *Toxicology* **31:**151–163.

Qin, Z. H., Zhang, S. P., and Weiss, B., 1994, Dopaminergic and glutamatergic blocking drugs differentially regulate glutamic acid decarboxylase mRNA in mouse brain, *Brain Res. Mol. Brain Res.* **21:**293–302.

Rafalowska, U., Liu, G. J., and Floyd, R. A., 1989, Peroxidation induced changes in synaptosomal transport of dopamine and gamma-aminobutyric acid, *Free Radic. Biol. Med.* **6:**485–492.

Ramassamy, C., Girbe, F., Pincemail, J., Christen, Y., and Costentin, J., 1994, Modifications of the synaptosomal dopamine uptake and release by two systems generating free radicals: Ascorbic acid/Fe^{2+} and L-arginine/NADPH, *Ann. N. Y. Acad. Sci.* **738**:141–152.

Rausch, W. D., Hirata, Y., Nagatsu, T., Riederer, P., and Jellinger, K., 1988, Tyrosine hydroxylase activity in caudate nucleus from Parkinson's disease: Effects of iron and phosphorylating agents, *J. Neurochem.* **50**:202–208.

Ribeiro, P., Wang, Y., Citron, B. A., and Kaufman, S., 1992, Regulation of recombinant rat tyrosine hydroxylase by dopamine, *Proc. Natl. Acad. Sci. USA* **89**:9593–9597.

Richard, S., Brion, J. P., Couck, A. M., and Flament-Durand, J., 1989, Accumulation of smooth endoplasmic reticulum in Alzheimer's disease: New morphological evidence of axoplasmic flow disturbances, *J. Submicrosc. Cytol. Pathol.* **21**:461–467.

Richardt, G., Federolf, G., and Habermann, E., 1985, The interaction of aluminum and other metal ions with calcium-calmodulin-dependent phosphodiesterase, *Arch. Toxicol.* **57**:257–259.

Rohn, T. T., Hinds, T. R., and Vincenzi, F. F., 1993, Ion transport ATPases as targets for free radical damage. Protection by an aminosteroid of the Ca^{2+} pump ATPase and Na$^+$/K$^+$ pump ATPase of human red blood cell membranes, *Biochem. Pharmacol.* **46**:525–534.

Rossi, M. A., Curzio, M., DiMauro, C., Fidale, F., Garramone, A., Esterbauer, H., Torrielli, M., and Dianzani, M. U., 1991, Experimental studies on the mechanism of action of 4-hydroxy-2,3-*trans*-nonenal, a lipid peroxidation product displaying chemotactic activity toward rat neutrophils, *Cell Biochem. Funct.* **9**:163–170.

Rothman, R. B., Reid, A., Mahboubi, A., Kim, C. H., De Costa, B. R., Jacobson, A. E., and Rice, K. C., 1991, Labeling by [3H]1,3-di(2-tolyl)guanidine of two high affinity binding sites in guinea pig brain: Evidence for allosteric regulation by calcium channel antagonists and pseudoallosteric modulation by sigma ligands, *Mol. Pharmacol.* **39**:222–232.

Salmeron, A., Borroto, A., Fresno, M., Crumpton, M. J., Ley, S. C., and Alarcon, B., 1995, Transferrin receptor induces tyrosine phosphorylation in T cells and is physically associated with the TCR ξ-chain, *J. Immunol.* **154**:1675–1683.

Sanchez, C., Diaz-Nido, J., and Avila, J., 1995, Variations in in vivo phosphorylation at the proline-rich domain of the microtubule-associated protein 2 (MAP2) during rat brain development, *Biochem. J.* **306**:481–487.

Schimke, I., Haberland, A., Will-Shahab, L., Kuttner, I., and Papies, B., 1992, In vitro effect of reactive O$_2$ species on the β-receptor-adenylyl cyclase system, *Mol. Cell. Biochem.* **110**:41–46.

Schuman, E. M., Meffert, M. K., Schulman, H., and Madison, D. V., 1994, An ADP-ribosyltransferase as a potential target for nitric oxide action in hippocampal long-term potentiation, *Proc. Natl. Acad. Sci. USA* **91**:11958–11962.

Seki, K., Chen, H. C., and Huang, K. P., 1995, Dephosphorylation of protein kinase C substrates, neurogranin, neuromodulin, and MARCKS by calcineurin and protein phosphatases 1 and 2A, *Arch. Biochem. Biophys.* **316**:673–679.

Shao, Y., and McCarthy, K. D., 1993, Regulation of astroglial responsiveness to neuroligands in primary culture, *Neuroscience* **55**:991–1001.

Shiota, A., Hiramatsu, M., and Mori, A., 1989, Amino acid neurotransmitters in iron-induced epileptic foci of rats, *Res. Commun. Chem. Pathol. Pharmacol.* **66**:123–133.

Shoham, S., Wertman, E., and Ebstein, R. P., 1992, Iron accumulation in the rat basal ganglia after excitatory amino acid injections—dissociation from neuronal loss, *Exp. Neurol.* **118**:227–241.

Sinosich, M. J., Davey, M. W., Teisner, B., and Grudzinskas, J. G., 1983, Comparative studies of pregnancy associated plasma protein-A and alpha 2-macroglobulin using metal chelate chromatography, *Biochem. Int.* **7**:33–42.

Sirinathsinghji, D. J., Schuligoi, R., Heavens, R. P., Dixon, A., Iversen, S. D., and Hill, R. G., 1994, Temporal changes in the messenger RNS levels of cellular immediate early genes and neurotransmitter/receptor genes in the rat neostriatum and substantia nigra after acute treatment with eticlopride, a dopamine D$_2$ receptor antagonist, *Neuroscience* **62**:407–423.

Sitaramayya, A., Marala, R. B., Hakki, S., and Sharma, R. K., 1991, aInteractions of nucleotide analogues with rod outer segment guanylate cyclase, *Biochemistry* **30**:6742–6747.

Smith, J. B., Smith, L., Pijuan, V., Zhuang, Y., and Chen, Y. C., 1994, Transmembrane signals and protooncogene induction evoked by carcinogenic metals and prevented by zinc, *Environ. Health Perspect.* **102**(Suppl 3):181–189.

Stohs, S. J., and Bagchi, D., 1995, Oxidative mechanisms in the toxicity of metal ions, *Free Radic. Biol. Med.* **18:**321–336.

Swaiman, K. F., and Machen, V. L., 1985, The effect of iron on mammalian cortical neurons in culture, *Neurochem. Res.* **10:**1261–1268.

Swope, S. L., Moss, S. J., Blackstone, C. D., and Huganir, R. L., 1992, Phosphorylation of ligand-gated ion channels: A possible mode of synaptic plasticity, *FASEB J.* **6:**2514–2523.

Taher, M. M., Garcia, J. G., and Natarajan, V., 1993, Hydroperoxide-induced diacylglycerol formation and protein kinase C activation in vascular endothelial cells, *Arch. Biochem. Biophys.* **303:**260–266.

Taneja, V., Mishra, K., and Agarwal, K. N., 1986, Effect of early iron deficiency in rat on the gamma-aminobutyric acid shunt in brain, *J. Neurochem.* **46:**1670–1674.

Tang, J., Rutishauser, U., and Landmesser, L., 1994, Polysialic acid regulates growth cone behavior during sorting of motor axons in the plexus region, *Neuron.* **13:**405–414.

Tian, W. N., and Deth, R. C., 1993, Precoupling of $G_i/G_{(o)}$-linked receptors and its allosteric regulation by monovalent cations, *Life Sci.* **52:**1899–1907.

Tiffany-Castiglioni, E., Zmudzki, J., Wu, J. N., and Bratton, G. R., 1987, Effects of lead treatment on intracellular iron and copper concentrations in cultured astroglia, *Metab. Brain Dis.* **2:**61–79.

Tipper, J., Wollny, E., Fullilove, S., Kramer, G., and Hardesty, B., 1986, Interaction of the 56,000-dalton phospho-protein from reticulocytes with regulin and inhibitor 2, *J. Biol. Chem.* **261:**7144–7150.

Tomsig, J. L., and Suszkiw, J. B., 1993, Intracellular mechanism of Pb^{2+}-induced norepinephrine release from bovine chromaffin cells, *Am. J. Physiol.* **265:**C1630-6.

Tsuji, Y., Miller, L. L., Miller, S. C., Torti, S. V., and Torti, F. M., 1991, Tumor necrosis factor-alpha and interleukin 1-alpha regulate transferrin receptor in human diploid fibroblasts. Relationship to the induction of ferritin heavy chain, *J. Biol. Chem.* **266:**7257–7261.

Vartio, T., and Kuusela, P., 1991, Disulfide-bonded dimerization of fibronectin in vitro, *Eur. J. Biochem.* **202:**597–604.

Velez Pardo, C., Jimenez del Rio, M., Pinxteren, J., De Potter, W., Ebinger, G., and Vauquelin, G., 1995, Fe^{2+}-mediated binding of serotonin and dopamine to skeletal muscle actin: resemblance to serotonin binding proteins, *Eur. J. Pharmacol.* **288:**209–218.

Vig, P. J., Ravi, K., and Nath, R., 1991, Interaction of metals with brain calmodulin purified from normal and cadmium exposed rats, *Drug Chem. Toxicol.* **14:**207–218.

Vincent, J. B., and Averill, B. A., 1990, Sequence homology between purple acid phosphatases and phosphoprotein phosphatases. Are phosphoprotein phosphatases metalloproteins containing oxide-bridged dinuclear metal centers? *FEBS Lett.* **263:**265–268.

Wade, R. S., and Castro, C. E., 1990, Redox reactivity of iron(III) porphyrins and heme proteins with nitric oxide. Nitrosyl transfer to carbon, oxygen, nitrogen, and sulfur, *Chem. Res. Toxicol.* **3:**289–291.

Waldmeier, P. C., Buchle, A. M., and Steulet, A. F., 1993, Inhibition of catechol-O-methyltransferase (COMT) as well as tyrosine and tryptophan hydroxylase by the orally active iron chelator, 1,2-dimethyl-3-hy-droxypyridin-4-one (L1, CP20), in rat brain in vivo, *Biochem. Pharmacol.* **45:**2417–2424.

Wang, J., Rousseau, D. L., Abu-Soud, H. M., and Stuehr, D. J., 1994, Heme coordination of NO in NO synthase, *Proc. Natl. Acad. Sci. USA* **91:**10512–10516.

Wang, L. Y., Taverna, F. A., Huang, X. P., MacDonald, J. F., and Hampson, D. R., 1993, Phosphorylation and modulation of a kainate receptor (GluR6) by cAMP-dependent protein kinase, *Science* **259:**1173–1175.

Winegar, B. D., and Lansman, J. B., 1991, Block of current through single calcium channels by Fe, Co, and Ni. Location of the transition metal binding site in the pore, *J. Gen. Physiol.* **97:**351–367.

Winrow, V. R., Winyard, P. G., Morris, C. J., and Blake, D. R., 1993, Free radicals in inflammation: Second messengers and mediators of tissue destruction, *Br. Med. Bull.* **49:**506–522.

You, G. F., Buccigross, J. M., and Nelson, D. J., 1990, Comparison of Ca(II), Cd(II), and Mg(II) titration of tyrosine-99 spin-labeled bovine calmodulin, *J. Inorg. Biochem.* **38:**117–125.

Youdim, M. B. H., Ben-Shachar, D., and Yehuda, S., 1989, Putative biological mechanisms of the effect of iron deficiency on brain biochemistry and behavior, *Am. J. Clin. Nutr.* **50:**607–617.

Yu, M. J., McCowan, J. R., Phebus, L. A., Towner, R. D., Ho, P. P., Keith, P. T., Luttman, C. A., Saunders, R. D., Ruterbories, K. J., and Lindstrom, T. D., 1993, Benzylamine antioxidants: Relationship between structure, peroxyl radical scavenging, lipid peroxidation inhibition, and cytoprotection, *J. Med. Chem.* **36:**1262–1271.

Zhang, Y., Tatsuno, T., Carney, J. M., and Mattson, M. P., 1993, Basic FGF, NGF, and IGFs protect hippocampal and cortical neurons against iron-induced degeneration, *J. Cereb. Blood Flow Metab.* **13:**378–388.

Zhang, Z. H., Zuo, Q. H., and Wu, X. R., 1989, Effects of lipid peroxidation on GABA uptake and release in iron-induced seizures, *Chin. Med. J. (Engl.)* **102:**24–27.

Zorumski, C. F., and Thio, L. L., 1992, Properties of vertebrate glutamate receptors: Calcium mobilization and desensitization, *Prog. Neurobiol.* **39:**295–336.

Chapter 2

Evidence for Iron Mismanagement in the Brain in Neurological Disorders

James R. Connor

2.1. INTRODUCTION

Studies on manipulation of iron in the diet, both during development and in adults, have clearly established that iron is required for normal neurological function. A deficiency in iron availability to the brain may directly affect the general metabolic activity of the brain by decreasing the function of cytochrome oxidase, glucose 6-phosphate dehydrogenase, NADH dehydrogenase, succinic dehydrogenase, and aldehyde dehydrogenase—all of which require iron as an essential cofactor and are relatively elevated in brain (Cammer, 1984). Also, the brain is rich in myelin, and iron is required for the biosynthesis of cholesterol and lipids, which are key components of myelin and whose biosynthesis occurs in the brain at higher levels than in other organs (Larkin and Rao, 1990; Pleasure *et al.,* 1984). Finally, specific neurotransmitter systems such as GABA, dopamine, and norepinephrine require iron for synthesis (see Chapter 1). The link between dopaminergic dysfunction and iron is particularly strong; indeed the substantia nigra (SN), the site of dopaminergic cell bodies in the brain, is an iron-rich area in the brain (Benkovic and Connor, 1993).

A number of significant reviews concerning brain iron and neuropathologies (Koeppen, 1995), brain iron and nutrition (Beard *et al.,* 1993), iron status and cognition (Pollit and Ponpon, 1993), brain iron metabolism and development (Connor, 1994), and iron metabolism in general (Baynes and Bothwell, 1991) have documented the link between this essential element and brain function. There is general consensus that low-

James R. Connor   George M. Leader Family Laboratory for Alzheimer's Disease Research, Department of Neuroscience and Anatomy, Pennsylvania State University College of Medicine, M. S. Hershey Medical Center, Hershey, Pennsylvania 17033.

Metals and Oxidative Damage in Neurological Disorders, edited by Connor. Plenum Press, New York, 1997.

body-iron status influences attentional processes in young people (Pollit and Ponpon, 1993), and more recently evidence has been presented that adults with low iron status have decreased cognitive performance. There is compelling evidence for dysregulation of iron in brain neuropathologies of Alzheimer's disease (AD), Parkinson's disease (PD), multiple sclerosis (MS), and Tardive dyskinesia, Pick's disease, Huntington's disease, Hallervordeen-Spatz, and aceruolplasmia. Iron accumulates in the basal ganglia, an area of the brain controlling movement, in these diseases (Chapter 3). This latter brain region, under normal circumstances, has iron concentrations that are equal to those in liver (Hallgren and Sourander, 1958), so additional iron accumulation with disease would be expected to be associated with symptoms or tissue damage normally associated with iron overload. The reason for iron accumulation in basal ganglia is not known but presumably reflects a loss of iron mobility in brain with disease or increased specific uptake or transport of iron into the basal ganglia. The mode(s) of iron mobilization within and from the brain have received little attention. The turnover of iron in the brain is reportedly slow (Dallman and Spirito, 1977), but this idea has recently been challenged (Dickinson *et al.,* 1996). Given the propensity of the basal ganglia to accumulate iron in numerous neurological diseases, including a defect in ceruoplasmin (probably not unrelated is the ability of ceruloplasmin to function as a ferroxidase), the absence of knowledge regarding iron mobilization within and from the brain represents a significant void in our understanding of iron management in brain. A discussion of iron acquisition by the brain is presented later.

Within the brain as in other organs, there is a system for the acquisition of iron from the plasma pool (transferrin receptor), a mechanism for the mobilization of iron (transferrin), a mechanism for cell specific iron storage (ferritin), and an intracellular protein responsible for monitoring intracellular iron (iron regulatory protein). However, the brain presents unique challenges for the acquisition of iron, as it resides behind a "barrier" so that iron must be transported across the endothelial cells lining the blood vessels. Also, within the brain there is both regional localization of function and specificity, and diversity of function among the different cell types, all of which instill dramatic differences in iron requirements.

It is necessary to maintain iron homeostasis in the brain (and other organs) because iron is considered the most potent agent for induction of oxidative stress. The potential of iron to cause tissue damage lies in its ability to donate an electron in the Fenton reaction and produce the ·OH radical. This ·OH radical is the most reactive of the "free radicals" and subsequent oxidative damage from its generation is likely. Iron is considered the most potent inducer of oxidative damage because it is the most abundant transition metal in biological systems. The dual role of iron as an essential element and potent toxicant has led to the development of the iron maintenance system for stringent regulation of iron bioavailability.

2.2. REGIONAL AND CELLULAR DISTRIBUTION OF IRON IN THE BRAIN

The amount and distribution of iron in the brain is age and region dependent (Hallgren and Sourander, 1958; Connor *et al.,* 1992b; Loeffler *et al.,* 1995). There is regional heterogeneity in the disposition of iron in the brain that is remarkably similar across many species, with the basal ganglia, substantia nigra, and deep cerebellar nuclei particularly rich in iron (Hill and Switzer, 1984; Benkovic and Connor, 1993; Koeppen, 1995). Magnetic resonance imaging (MRI) on living brains of children and adolescents revealed iron-rich areas similar to that shown by the histochemical studies (Aoki *et al.*, 1989, Chapter 3). At the cellular level, the predominant cell containing iron in the mouse, rat, monkey, and human brain are oligodendrocytes. This cell staining is also remarkably constant throughout most brain regions (Benkovic and Connor, 1993; Hill and Switzer, 1984; Connor and Menzies, 1995; Dwork *et al.*, 1988; Dickinson and Connor, 1995; Levine, 1991; Morris *et al.*, 1992b).

The concentration (not amount) of iron is highest at birth, decreases during the first two weeks of life, and then begins a steady increase throughout life (Roskams and Connor, 1994; Hallgren and Sourander, 1958). There is no current understanding of the mechanism by which an iron-rich area in the brain will accumulate more iron than the neighboring brain regions. The substantia nigra, an area of the brain that is distinguished by robust iron accumulation, does not become iron rich until early adolescence in humans or rats (Connor *et al.*, 1995; Aoki *et al.*, 1989). The relationship between iron deficiency, dopaminergic dysfunction, motor impairment, and attentional disorders is clear, and the relatively late postnatal stage for accumulation of iron by the substantia nigra is perhaps the reason the dopaminergic system and attending related behaviors are vulnerable to iron deficiency.

2.2.1. Aging and Alzheimer's Disease

With normal aging, iron reportedly accumulates in brain in both humans and rats (Benkovic and Connor, 1993; Connor *et al.*, 1990). Iron levels in normal human aging brain increase in most areas until about the third decade of life and then reach a plateau (Hallgren and Sourander, 1958). This accumulation of iron in the brain with age in rats and humans does not appear region specific, as all areas studied revealed more iron with age. At the cellular level, some increased iron staining is seen in neurons, but the predominant cells to contain stainable iron are still oligodendrocytes.

In Alzheimer's disease, the cellular distribution of iron is abnormal. Neuritic plaques contain high levels of stainable iron, as do the cells surrounding the plaques (Goodman, 1953; Connor *et al.*, 1992a; Connor, 1993) (Figure 2.1). This relationship between iron and neuritic plaques has been examined by a few laboratories with some striking results. Iron promotes deposition of the Aβ4 peptide fragment (Mantyh *et al.*, 1993), which is supposedly the toxic component of the amyloid protein. Iron levels in the media can affect the expression of the amyloid precursor protein (APP; Bodovitz *et al.*, 1995). A putative iron responsive element (IRE) has been reported on the amyloid precursor protein mRNA (Tanzi and Hyman, 1991) near the site for sequencing the Aβ4 peptide fragment, suggesting a direct role for iron in amyloid production. The protein that binds to the IRE, the iron regulatory protein, has recently been demonstrated in human brain and is discussed below.

In AD, bulk analyses of iron in brain fail to discern differences in iron concentrations between normal and AD brain tissue (Ehmann *et al.*, 1986). However, analyses of discrete areas such as amygdala, hippocampus, nucleus basalis (Thompson *et al.*,

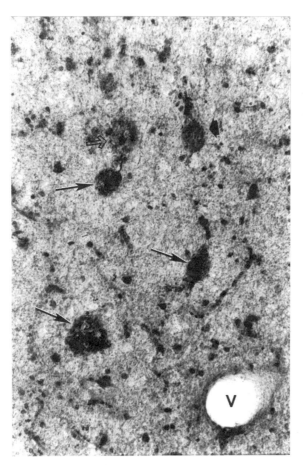

FIGURE 2.1. A light micrograph from the cerebral cortex of a patient who had Alzheimer's disease. The brain was stained with the Perl's histochemical stain for iron and the resulting reaction intensified with diaminobenzidine. Normal cellular iron staining is seen as small black dots throughout the micrograph. In the lower right corner is a blood vessel (V). The large iron stained areas indicated by various arrows are neuritic plaques. Three types of plaques are visible in this micrograph. Mature plaques show densely stained amyloid cores and robust iron staining in the area surrounding the core (black arrows). A second type of plaque contains a densely stained core, but the iron staining surrounding the plaque is relatively light (clear arrow). The third type of plaque stains more diffusely for iron than the others. There is no dense core at the center (arrowhead). This latter plaque is considered an immature form.

1988), globus pallidus, frontal cortex, and motor cortex reveal that iron levels are elevated in AD brains compared to normal. In some regions such as occipital cortex and substantia nigra, iron levels decrease relative to normal in AD, and in caudate-putamen the iron levels are similar between aged and AD populations (Connor *et al.,* 1992b; Loeffler *et al.,* 1995). These data clearly suggest alterations in the iron management system in the brain in AD, particularly in the ability to mobilize iron within and from specific brain regions. Consequences of increased iron accumulation and decreased mobility would include metabolic deficiencies and increased susceptibility to oxidative stress.

2.2.2. Parkinson's Disease

There is considerable evidence that brain iron mismanagement and oxidative damage can be linked to the disease process in PD. There is controversy regarding whether the amount of iron in the substantia nigra is elevated in PD relative to controls (Earle, 1968; Sofic *et al.*, 1988; Dexter *et al.*, 1990; Uitti *et al.*, 1989; Loeffler, 1995). Perhaps more important than the amount of iron is its intracellular distribution in the substantia nigra. In PD, iron accumulates in neuromelanin granules in the cells of the substantia nigra (Good *et al.*, 1992; Youdim *et al.*, 1989) and this accumulation has been suggested to promote oxidant stress (Youdim *et al.*, 1989). The transferrin receptor is present on neuromelanin-containing cells in the human substantia nigra indicating that these cells have a high, constant iron requirement. These cells also immunostain for H-rich ferritin (Connor and Menzies, 1995), further suggesting their capacity for iron storage is minimal and turnover may be high.

2.3. ACQUISITION OF IRON

The process of brain iron acquisition appears highly regulated and does not reflect overall blood–brain barrier permeability (Morris *et al.*, 1992c; Crowe and Morgan, 1992). The brain obtains iron via transferrin receptors expressed on endothelial cells of the brain microvasculature (Fishman *et al.*, 1987; Kalaria *et al.*, 1992; Jefferies *et al.*, 1984). This constant expression of transferrin receptors on brain microvasculature reveals that the brain has a continuous iron requirement. There is no information on the relationship between Tf receptor expression on the microvasculature and the amount of iron within a given brain region. Most studies on iron mobilization and iron uptake in brain have failed to consider the cerebrospinal fluid as an active distributer of iron, despite the accumulation of systemically injected iron in the choroid plexus (Dwork, 1995), the ability of choroid plexus to synthesize and secrete transferrin (Tsutsumi *et al.*, 1989), and the high levels of Tf that are found in CSF (Elovaara *et al.*, 1985; New *et al.*, 1983; Dziegielewska *et al.*, 1980). Recent reports of the presence of receptors for lactoferrin (Faucheux *et al.*, 1995) and ferritin (Hulet *et al.*, 1996) in the brain indicate that the brain can acquire iron through means other than that mediated by transferrin. Brain iron uptake in a hypotransferrinemic strain of mouse is normal (Dickinson *et al.*, 1996) or perhaps even elevated (Simpson *et al.*, 1991), further indicating that methods for iron acquisition by the brain are not limited to transferrin-mediated delivery.

During development, the rate of iron uptake into the brain has been studied in rats and is highest during the second postnatal week (Roskams and Connor, 1994). Under normal dietary conditions in adult rats, uptake of iron into the brain is reportedly homogeneous followed by a redistribution of iron to the basal ganglia (Dwork *et al.*, 1990; Dwork, 1994), which, as mentioned, is an iron-rich area. The rate of uptake of iron into the brain is affected by the iron status of the body. Studies on animals in which dietary iron levels are manipulated demonstrate an increased rate of brain iron uptake when the iron burden is low and a decreased rate when it is high (Taylor *et al.*, 1991).

Transferrin Receptors

The density of Tf receptors on the brain microvasculature is six to ten times higher in adult human brains compared to the brain parenchyma (Kalaria *et al.*, 1992). Autoradiographic studies reveal a heterogeneous distribution of Tf receptors in the adult brain; areas associated with motor function have a relatively higher density than non-motor areas (Mash *et al.,* 1990; Hill *et al.,* 1985). Tf receptors in human brain reportedly codistribute with cytochrome oxidase, indicating a direct positive relationship between iron acquisition by cells and mitochondrial energy activity (Morris *et al.,* 1992a). Immunohistochemical studies on the brain (Giometto *et al.,* 1990; Connor and Menzies, 1995) reveal neurons as the predominant cell type to stain for antibodies to Tf receptors, although immunopositive astrocytes and oligodendrocytes can be seen. During development, Tf receptor expression has been seen on oligodendrocytes (Lin and Connor, 1989) and if the oligodendrocyte population is functionally compromised, the expression of Tf receptors in brain is decreased by 50% (Roskams and Connor, 1992). Recently, Tf receptors have been demonstrated on microglia in brain during the first 10 days of postnatal development (Kaur and Ling, 1992).

In injury and disease there are changes in the expression of transferrin receptors in the brain. Following axotomy, iron uptake and Tf receptor expressions are increased on the cell bodies of the axotomized neurons (Graeber *et al.,* 1989). Tf receptor expression and iron uptake are also elevated in peripheral nerves both at the site of injury and distal to the injury (Raivich *et al.,* 1991). Also, Tf receptor expression is elevated on motor neurons of the mouse mutant for progressive motor neuronopathy (Moos, 1995a). These observations indicate a critical role for iron in the reparative process of neurons. Iron, via Tf, appears to also undergo retrograde axonal transport in motor neurons, suggesting iron (and Tf) are involved in the synergistic relationship between neurons and muscle (Moos *et al.,* 1995b).

Because of the potential role for Tf receptor expression and iron uptake in reparative processes in neurons, it is important to note that the density of Tf receptors decreases in the hippocampus of Alzheimer's diseased individuals (Kalaria *et al.,* 1992). The decrease in Tf receptor density may be related (in a negative feedback relationship) to the accumulation of iron that occurs in this region in AD (Thompson *et al.,* 1988). At the cellular level, Tf receptor expression in human hippocampus normally occurs on pyramidal cells (Connor, 1994). In AD, Tf receptor expression is robust on cells surrounding the neuritic plaques (Connor, 1994; Connor and Menzies, 1995), further supporting a role for iron disruption at the cellular level in AD and suggesting iron may influence cellular amyloid production. The increased Tf receptor expression on cells associated with plaques, coupled with the decrease in total Tf receptor density in AD, indicates the receptors related to normal function (i.e., neuronal iron acquisition) must be decreased. This observation is particularly significant to the novel therapeutic concept of using trophic factors coupled to transferrin or antibodies to Tf receptors as a means of penetrating the blood–brain barrier and gaining access to the brain (Friden *et al.,* 1993; Bickel *et al.,* 1993). Clearly, it is essential to know where the Tf receptor population is present if the deliverance of trophic factors via Tf-Tf receptors is going to promote neuronal growth and development.

In Parkinsonian patients, the density of Tf receptor expression in the substantia nigra is unchanged with the disease (Faucheux *et al.,* 1993) despite the loss of neurons and the accumulation of iron. These observations are difficult to explain and require further study. Neuromelanin-containing cells in the substantia nigra express Tf receptors as do the oligodendrocytes in this region (Connor and Menzies, 1995). The content of iron in the substantia nigra has been shown to correspond to the severity of the symptoms associated with PD but the amount of iron accumulation is controversial, as mentioned earlier. Thus, knowledge as to the mechanism by which iron accumulates in this brain region is essential.

The dopaminergic neurons in the SN project to the caudate and striatum. It has been suggested, but not demonstrated, that iron may be retrogradely transported from the caudate to the SN, and this transport system may be disrupted in PD. Studies on Tf receptor expression in the caudate in PD are controversial, as both an increase (Faucheux et al., 1995) and a decrease (Mash et al., 1991) in the receptor density have been reported. There is also no direct evidence that Tf receptors are associated with dopamine terminals in the caudate; a likely, but not exclusive prerequisite for iron to be taken up by the dopaminergic neurons and retrogradely transported to the cell bodies in the SN. Analysis of MPTP treated animals, a model for neurodegeneration in the SN, shows loss of Tf receptors and loss of binding of a dopamine agonist, suggesting a relationship between dopaminergic terminals and Tf receptor expression (Mash et al., 1991; Faucheux et al., 1995). The MPTP model is an acute event rather than a chronic disease like PD. Thus, the significance of Tf receptor loss in this model is difficult to relate to the PD condition. A lactoferrin receptor has been demonstrated in this brain region and this receptor is increased in the mesencephalon of patients with PD (Faucheux et al., 1995), suggesting iron could accumulate via a lactoferrin pathway. At present, it remains that possibly the critical event in the onset of PD, excess iron accumulation in the SN, is not understood.

2.4. IRON MOBILIZATION

Transferrin has traditionally been considered the key to iron mobilization. Transferrin is made in the liver and secreted for whole-body-iron mobilization. However, transferrin is also made in the brain by oligodendrocytes (Bloch et al., 1985) and choroid plexus (Aldred et al., 1987) and can be secreted from both (Espinosa de los Monteros, 1990; Tsutsumi et al., 1989). Transferrin levels are relatively high in cerebrospinal fluid (Elovaara et al., 1985), especially in perinatal brains (New et al., 1983). Transferrin levels in brain are higher at birth than at any other age (Roskams and Connor, 1994; Connor, 1994). The Tf present in the brain at birth is likely derived from the plasma pool because the blood–brain barrier (BBB) is not complete and brain Tf mRNA expression is low (Levine et al., 1984; Bartlett et al., 1991). Near the end of the first postnatal week there is a precipitous decline of brain Tf levels (Roskams and Connor, 1994), a rapid increase in whole brain Tf mRNA levels (Levine et al., 1984; Bartlett et al., 1991), and the appearance of Tf positive cells in the brain (Connor and Fine, 1987). The brain is the only organ in which a postnatal increase in Tf mRNA has been demonstrated, and the increase in Tf mRNA production by the brain coincides with closure of the BBB and increase in iron uptake (Crowe and Morgan, 1992). The cells responsible for the majority of Tf mRNA production in brain are oligodendrocytes (Bloch et al., 1985; Bartlett et al., 1991; Connor and Menzies, 1996). When the oligodendrocyte population is compromised as in the myelin-deficient *jimpy* mouse, Tf mRNA levels are < 5% of normal (Bartlett *et al.,* 1991). If oligodendrocyte development is normal, but myelin production is abnormal, Tf mRNA expression is unaffected (Connor *et al.,* 1993). Thus, Tf mRNA production and oligodendrocyte maturation are tightly coupled.

The function of the Tf secreted in the brain has not been determined. It was originally considered that iron from plasma was deposited within the endothelial cells in the brain microvasculature and then iron would enter the brain bound to Tf produced by the brain (Fishman *et al.,* 1987). However, recent evidence demonstrates that endogenous brain Tf is not necessary for iron uptake into the brain (Dickinson and Connor, 1995, 1996; Gocht *et al.,* 1993). Tf produced in brain may be involved in the redistribution of iron within the brain. The current thinking is that iron is taken into the brain homogenously and then redistributed to iron-rich areas (Dwork, 1995), but there is very little research in this area. The lack of understanding of iron transport into the brain is a significant detriment to the investigation into the possibility of exploiting the extant delivery system for iron (via Tf) to cross the BBB for delivery of therapeutic trophic factors (Friden *et al.,* 1993; Bickel *et al.,* 1993).

The mobilization of iron in brain is a key component of understanding the contribution that iron is making to pathology in neurological disease. A consistent finding in AD and PD, two of the diseases in which iron accumulation in brain is found and Tf has been examined, is a loss of Tf and a decrease in the ratio of Tf to iron (Loeffler *et al.,* 1995). Thus the potential for iron mobility in brain is clearly diminished in these diseases and is likely related to the excessive iron accumulation. Mechanisms to increase Tf production in brain and consequently iron mobility have not received much attention. Tf protein in brain can be manipulated. Rats exposed to a low iron diet shortly after weaning have a doubling of brain Tf protein content within 14d of dietary treatment, presumably reflecting a need for increased iron mobility associated with lowered iron levels in brain (Chen *et al.,* 1995). It is curious to consider if a diet low in iron could increase brain iron mobility in the adult/aged population.

2.5. INTRACELLULAR IRON MANAGEMENT

Once iron has entered the cell, the protein most responsible for its sequestration is ferritin. The function of ferritin is two-fold: to remove excess cellular iron and, as such, to protect the cell from iron-induced oxidative stress (Balla *et al.,* 1992), and at the same time to keep the iron available should it be needed in subsequent oxidative reactions (see Chapter 8). The 450kD mature ferritin protein consists of varying ratios of two subunits. The two subunits, a heavy (H) chain and a light (L) chain are functionally distinct and their genes reside on different chromosomes (Levi *et al.,* 1992). The heavy chain of ferritin has ferroxidase activity and rapidly converts the soluble ferrous form of iron to the insoluble, stored ferric form. The L chain does not have ferroxidase activity and functions to promote mineralization of iron at the ferritin core. Thus in organs in which iron storage is high (e.g., liver), the L chain of ferritin is the predominant subunit, whereas in organs in which iron turnover is high (e.g., heart) the H subunit of ferritin predominates. In disease states such as hemochromatosis, iron accumulates in the heart and the L subunit of ferritin increases relative to the H subunit as more iron is sequestered (Powell *et al.,* 1974).

In the brain, with its regional and cellular diversity in function and iron requirements, the analysis of ferritin subunits is a challenging and elucidating adventure. At the cellular level, H-ferritin is expressed before L during development and is seen in a subpopulation of oligodendrocytes and select neuronal nuclei (Blissman *et al.,* 1996). The role of H-ferritin in oligodendrocyte development is under study, but it likely serves to provide iron in a biologically retrievable manner inside the cells as relatively high amounts of iron are required for myelin production (Larkin and Rao, 1990). H-ferritin may also be protective in the oligodendrocytes as it increases in these cells in response to hypoxia (Qi and Dawson, 1994) and cytokine stimulation (Sanyal and Szuchet, 1995) consistent with the concept of providing intracellular protection.

The role of H-ferritin in nuclei is also currently under study. The observation that H-ferritin binds to DNA in a saturable manner (Manges and Connor, 1996) and is known to bind the sequence CAGTGC (Broyles *et al.,* 1995) is a completely novel finding and suggests a potential new role for ferritin (presumably limited to the H subunit) in transcriptional regulation or as a protector of DNA from iron-induced DNA cleavage.

In brains of adult monkeys and human beings, a specific cellular distribution of ferritin subunits also exists. The H-ferritin subunit is predominant in pyramidal neurons and the L subunit is predominant in microglial cells. Oligodendrocytes contain both H and L subunits. Astrocytes do not immunostain for either ferritin subunit except in the striatum (an iron-rich area) in which they immunostain robustly for the L-ferritin subunit (Connor *et al.,* 1994; Connor and Menzies, 1995). In AD brain tissue, ferritin immunostaining is robust in the area of neuritic plaques (Grundke-Iqbal *et al.,* 1990; Kaneko *et al.,* 1989; Connor *et al.,* 1992a) even in what appear to be newly forming plaques (Robinson *et al.,* 1995). The association of ferritin with neuritic plaques further supports the argument that iron is involved in the process of plaque formation.

Quantitatively, the H-ferritin subunit predominates in the human brain, although the ratio of H:L ferritin subunits is dependent upon brain region, age, and disease state (Connor *et al.,* 1995b). In those regions (e.g., globus pallidus), where iron concentrations are the highest and, perhaps more significantly, the ratio of Tf/iron is lowest (indicating a relatively low iron mobility) the H:L ratio approaches unity. With normal aging, iron accumulates in the brain and ferritin (both subunits) increases. The increase with age in H ferritin subunits is remarkably consistent (doubles) among different age regions regardless of the amount of iron accumulation, except in the globus pallidus. In this latter brain region, iron accumulation decreases with normal aging whereas H-ferritin increases eight to ten times. The age-related increase in L ferritin subunits is more variable depending on brain region. The specific cellular populations contributing to the changes in ferritin concentration with age have not been examined.

The normal pattern of iron accumulation and concomitant increase in ferritin with age is not seen when Alzheimer's or Parkinson's disease is superimposed on the aging process. Indeed a pattern is emerging in which ferritin concentration and iron accumulation may be related to the severity of the histopathology. In occipital cortex, a region normally spared histopathological changes in AD, ferritin increases without a measurable increase in iron. In the motor and parietal cortices, areas of moderate histopathological change in AD, both iron and ferritin increase with age. Those brain regions, such as the temporal and frontal cortices, in which histopathology is severe are those in which iron accumulates without a concomitant increase in ferritin (Connor et al., 1992b). In Parkinson's disease, where iron has been shown to accumulate in the substantia nigra, ferritin does not increase (Connor et al., 1995b; Dexter et al., 1990; but see Riederer et al., 1989). The consequence of increased cellular iron accumulation without adequate ferritin to sequester it could lead to increased iron availability for induction of free radicals and subsequent oxidative damage. The existence of oxidative damage and a role for oxidative damage in both AD and PD are well documented (Chapters 13 and 14). Thus, we propose that iron accumulation is a, but not the, critical factor in potential for oxidative damage, and suggest that the critical factor is the response of ferritin to the iron accumulation.

2.6. IRON REGULATORY PROTEINS (IRPs)

IRPs are cytoplasmic proteins responsible for the concordant regulation of ferritin and the transferrin receptor. The IRP "senses" intracellular iron levels through a currently unknown mechanism and binds to an iron responsive element on the transferrin receptor mRNA ($3'$ end) stabilizing the message, or to the $5'$ end of ferritin mRNA and blocking translation. Thus, when iron levels in brain are low, the IRP provides a mechanism for increasing Tf receptor protein translation and at the same time blocks the translation of ferritin (Klausner et al., 1993). There are two IRPs, which are known as IRP1 and IRP2 (Guo et al., 1995). The two IRPs are found on two different chromosomes and respond differently to iron availability (Guo et al., 1994). It is not known why two IRPs exist. IRP2 may be a developmentally regulated protein as its mRNA is higher in fetal tissue than the mRNA for IRP1 (Samaniego et al., 1994). IRP1 may be the more stable of the two proteins; IRP1 is normally present in adult mouse liver, and is elevated following liver inflammation. However, when the liver begins to regenerate, IRP2 is increased (Cairo and Pietrangelo, 1994).

Analysis of the IRP in the brain is in its infancy. Both IRP1 and IRP2 are present in mouse brain (Henderson et al., 1993). The regional distribution of IRP1 mRNA has been reported in adult mouse brain and has some overlap with NO synthetase (Jaffrey et al., 1994). IRP1 is also found in astrocytes in rabbits (Koeppen, 1995). We have recently established that IRP1 is the predominant form of IRP in the human brain (Hu and Connor, 1996). There is evidence that the IRP may undergo confirmational change or that the IRP/IRE complex is altered in AD brain tissue (personal observation). The result of this alteration in the IRP or IRP/IRE complex is the formation of an extremely stable bond. The consequence of such a stable bond would be increased stabilization of the Tf receptor mRNA and a blockage of the synthesis of ferritin. The stabilization of the Tf receptor mRNA would generate more Tf receptors and could increase the influx of iron into the cell. At the same time, ferritin synthesis would not increase to sequester the additional iron that is accumulating. This is the scenario that has been reported in

AD (and in PD), suggesting that altered IRP binding activity could be the mechanism responsible for abnormal iron accumulation. This observation is currently under study.

2.7. IRON AND MULTIPLE SCLEROSIS

The predominance of stainable iron is found in oligodendrocytes in the brain, and the only known function of oligodendrocytes is to produce myelin. Thus, it is logical to consider a relationship between disruptions in iron metabolism and dysmyelination. Iron deficiency during development is associated with hypomyelination (Larkin and Rao, 1990). Cerebroside and phosphatidylethanolamine are reduced in the brain following consumption of a diet deficient in iron (Oloyede *et al.,* 1992). In multiple sclerosis, the prototype of demyelinating diseases, cerebroside content of the brain is also decreased, as are other lipids (Wilson and Tocher, 1991). MR analysis of patients with multiple sclerosis reveals iron deposition in the putamen and thalamus, which can be correlated with the degree of white matter abnormality (Drayer *et al.,* 1987).

The relatively high levels of iron in oligodendrocytes may be related to the heightened vulnerability of these cells to oxidative damage (Groit *et al.,* 1990; Kim and Kim, 1991; Oka *et al.,* 1993). Ferritin is enriched in oligodendrocytes presumably to sequester the intracellular iron and limit the availability of iron for oxidative injury. However, substances such as nitric oxide (NO) can remove iron from ferritin (Reif and Simmons, 1990) and NO has been shown to kill oligodendrocytes (Mitrovic *et al.,* 1994, 1995). The effect of NO is postulated to be mediated through free iron (Reif and Simmons, 1990; see also Chapters 11 and 12).

Evidence that oxidative stress could be associated with demyelination come from observations that macrophages from patients with multiple sclerosis have increased production of superoxide radicals and hydrogen peroxide (Fisher *et al.,* 1988). Lipid peroxidation has been demonstrated in MS (Hunter *et al.,* 1985). The release of free radicals from macrophages is part of the microbicidal and tumoricidal activity of these cells, but may exacerbate myelin breakdown in MS lesion sites (Fisher *et al.,* 1988; Johnson *et al.,* 1989). In an animal model for studying demyelination (experimental allergic encephalopathy) there is a decrease in symptomology and pathology following antioxidant or iron chelation therapy (Bowern *et al.,* 1984; Hartung *et al.,* 1988). There has not been clinical interest in examining antioxidants as potential therapeutics in MS despite suggestive evidence that such a pursuit may be fruitful (Levine, 1992).

2.8. SUMMARY

We propose that there is a loss of iron homeostasis in the brain with age that derives from continued age-associated iron accumulation coupled with a relative decrease in Tf production by the brain. The consequence of decreased Tf production relative to the amount of intracellular iron will be decreased iron mobility. Brain iron accumulation relative to Tf production appears greater in disease states. Decreased iron mobility can be managed by increased production of ferritin, which is the response seen in normal aging. However, in disease states, iron accumulates without inducing a sufficient increase in ferritin. The mechanism for uncoupling ferritin synthesis and cellular iron accumulation will likely be found at the level of the intracellular iron regulatory protein, which has only recently begun to be studied in brain. The consequence of iron accumulation without sufficient levels of proteins for mobilization and/or sequestration of iron would be increased production of free radicals and oxidative damage.

ACKNOWLEDGMENTS.   The research discussed in this chapter was funded by the National Institutes of Health, the American Health Assistance Foundation, the Alzheimer's Association, the National Multiple Sclerosis Society, and the Jane B. Barsumian Trust Fund.

2.9. REFERENCES

Aldred, A., Dickson, P., Marley, P., and Schreiber, G., 1987, Distribution of transferrin synthesis in brain and other tissues, *J. Biol. Chem.* **262**:5293–5297.

Aoki, S., Okada, Y., Nishimura, K., Barkovich, A. J., Kjos, B. O., Brasch, R. C., and Norman, D., 1989, Normal deposition of brain iron in childhood and adolescence: MR imaging at 1.5 T, *Radiology* **172**:381–385.

Balla, G., Jacob, H. S., Balla, J., Rosenberg, M., Nath, K., Apple, F., Eaton, J. W., and Vercellotti, J. R., 1992, Ferritin: A cytoprotective antioxidant strategem of endothelium, *J. Biol. Chem.* **267**:18148–18153.

Bartlett, W. P., Li, X.-S., and Connor, J. R., 1991, Expression of transferrin mRNA in the CNS of normal and jimpy CNS, *J. Neurochem.* **57**:318–322.

Baynes, R. D., and Bothwell, T. H., 1991, Iron deficiency, *Ann. Rev. Nutr.* **11**:133–48.

Beard, J. L., Connor, J. R., and Jones, B. C., 1993, Iron in the brain, *Nutr. Rev.* **51**:157–170.

Benkovic, S., and Connor, J. R., 1993, Ferritin, transferrin and iron in normal and aged rat brains, *J. Comp. Neurol.* **338**:97–113.

Bickel, U., Yoshikawa, T., Landaw, E. M., Faull, K. F., and Pardridge, W. M., 1993, Pharmacologic effect in vivo in brain by vector-mediated peptide drug delivery, *Proc. Natl. Acad. Sci. USA* **90**:2618–2622.

Blissman, G., Menzies, S., Beard, J., Palmer, C., and Connor, J., 1996, The expression of ferritin subunits and iron in oligodendrocytes in neonatal porcine brains, *Dev. Neurosci.* (in press).

Bloch, B., Popovici, T., Levin, M. J., Tuil, D., and Kahn, A., 1985, Transferrin gene expression visualized in oligodendrocytes of the rat brain by using in situ hybridization and immunohistochemistry, *Proc. Natl. Acad. Sci. USA* **81**:6706–6710.

Bodovitz, S., Falduto, M. T., Frail, D. E., and Klein, W. L., 1995, Iron levels modulate α-secretase cleavage of amyloid precursor protein, *J. Neurochem.* **64**:307–315.

Bowern, N., Ramshaw, I. A., Clark, I. A., and Doherty, P. C., 1984, Inhibition of autoimmune neuropathological process by treatment with an iron-chelating agent, *J. Exp. Med.* **160**:1532–1543.

Broyles, R. H., Blair, F. C., Kyker, K. D., Kurien, B. T., Steward, D. R., Hala'sz, H., Berg, P. E., and Schechter, A. N., 1995, A ferritin-like protein binds to a highly conserved CAGTGC sequence in the β-globin promoter, in: *Sickle Cell Disease and Thalassemias: New Trends in Therapy,* Volume 234 (Y. Beuzard, B. Lubin, and J. Ross, eds.), Colloque INSERM/John Libbey Eurotext Ldt., pp. 43–51.

Cairo, G., and Pietrangelo, A., 1994, Transferrin receptor gene expression during rat liver regeneration, *J. Biol. Chem.* **269**:6405–6409.

Cammer, W., 1984, Oligodendrocyte associated enzymes, in: *Oligodendroglia* (W. T. Norton, ed.), Plenum Press, New York, pp. 199–232.

Chen, Q., Connor, J. R., and Beard, J. L., 1995, Brain iron, transferrin and ferritin concentrations are altered in developing iron-deficient rats, *J. Nutr.* **125**:1529–1535.

Connor, J. R., 1993, Cellular and regional maintenance of iron homeostasis in the brain: normal and diseased states, in: *Iron in Central Nervous System Disorders* (P. Riederer and M. B. H. Youdim, eds.), Springer-Verlag, Wien, pp. 1–18.

Connor, J. R., 1994, Iron acquisition and expression of iron regulatory proteins in the developing brain: Manipulation by ethanol exposure, iron deprivation and cellular dysfunction, *Dev. Neurosci.* **16**:233–247.

Connor, J. R., Boeshore, K. L., Benkovic, S. A., and Menzies, S. L., 1994, Isoforms of ferritin have a specific cellular distribution in the brain, *J. Neurosci. Res.* **37**:461–465.

Connor, J. R., and Fine, R. E., 1987, Development of transferrin-positive oligodendrocytes in the rat central nervous system, *J. Neurosci. Res.* **17**:51–59.

Connor, J. R., and Menzies, S. L., 1995, Cellular management of iron in the brain, *J. Neurol. Sci.* **134**:33–44.

Connor, J. R., and Menzies, S. L., 1996, Relationship of iron to oligodendrocytes and myelination, *Glia* **17**:83–93.

Connor, J. R., Phillips, T. M., Lakshman, M. R., Barron, K. D., Fine, R. E., and Csiza, C. K., 1987, Regional variation in the levels of transferrin in the CNS of normal and myelin-deficient rats, *J. Neurochem.* **49**:1523–1529.

Connor, J. R., Menzies, S. L., St. Martin, S., and Mufson, E. J., 1990, The cellular distribution of transferrin, ferritin and iron in the human brain, *J. Neurosci. Res.* **27**:595–611.

Connor, J. R., Menzies, S. L., St. Martin, S., Fine, R. E., and Mufson, E. J., 1992a, Altered cellular distribution of transferrin, ferritin and iron in Alzheimer's diseased brains, *J. Neurosci. Res.* **31**:75–85.

Connor, J. R., Snyder, B. S., Beard, J. L., Fine, R. E., and Mufson, E. J., 1992b, The regional distribution of iron and iron regulatory proteins in the brain in aging and Alzheimer's disease, *J. Neurosci. Res.* **31**:327–335.

Connor, J. R., Roskams, A. J. I., Menzies, S. L., and Williams, M. E., 1993, Transferrin in the central nervous system of the *shiverer* mouse myelin mutant, *J. Neurosci. Res.* **36**:501–507.

Connor, J. R., Pavlick, G., Karli, D., Menzies, S. L., and Palmer, C., 1995a, A histochemical study of iron-positive cells in the developing rat brain, *J. Comp. Neurol.* **355**:111–123.

Connor, J. R., Snyder, B. S., Arosio, P., Loeffler, D. A., and LeWitt, P., 1995b, A quantitative analysis of isoferritins in select regions of aged, Parkinsonian and Alzheimer's diseased brains, *J. Neurochem.* **65**:717–724.

Crowe, A., and Morgan, E. H., 1992, Iron and transferrin uptake by brain and cerebrospinal fluid in the rat, *Brain Res.* **592**:8–16.

Dallman, P., and Spirito, R. A., 1977, Brain iron in the rat: Extremely slow turnover in normal rat may explain the long-lasting effects of early iron-deficiency, *J. Nutr.* **107**:1075–1081.

Dexter, D. T., Carayon, A., Vidailhet, M., Ruberg, M., Agid, F., Agid, Y., Lees, A. J., Wells, F. R., Jenner, P., and Marsden, C. D., 1990, Decreased ferritin levels in brain in Parkinson's disease, *J. Neurochem.* **55**:16–20.

Dickinson, T. K., and Connor, J. R., 1994, Histological analysis of selected brain regions of the hypotransferrinemic mouse, *Brain Res.* **635**:169–178.

Dickinson, T. K., and Connor, J. R., 1995, Cellular distribution of iron, transferrin and ferritin in the hypotransferrinemic (Hp) mouse brain, *J. Comp. Neurol.* **355**:67–80.

Dickinson, T. K., Devenyi, A. G., and Connor, J. R., 1996, Distribution of injected iron 59 and manganese 54 in hypotransferrinemic mice, *J. Lab. Clin. Med.* **128**:270–278.

Drayer, B., Burger, P., Hurwita, B., Dawson, D., and Cain, J., 1987, Reduced signal intensity on MR images of thalamus and putamen in multiple sclerosis: Increased iron content? *AJNR* **8**:413–419.

Dwork, A. J., 1994, Effects of diet and development upon the uptake and distribution of cerebral iron, *J. Neurol. Sci.* **134**(supplement):45–51.

Dwork, A. J., Schon, E. A., and Herbert, J., 1988, Nonidentical distribution of transferrin and iron in human brain, *Neuroscience* **27**:333–335.

Dwork, A. J., Lawler, G., Zybert, P. A., Durkin, M., Osman, M., Wilson, N., and Barkai, A. I., 1990, An autoradiographic study of the uptake and distribution of iron by the brain of the *you* rat, *Brain Res.* **518**:31–39.

Dziegielewska, K. M., Evans, C. A. N., Malinowskq, D. H., Mollgård, K., Reynolds, M. L., and Saunders,

N. R., 1980, Blood-cerebrospinal fluid transfer of plasma proteins during fetal development in the sheep, *J. Physiol.* **300:**457–465.

Earle, K. M., 1968, Studies in Parkinson's disease including x-ray fluorescent spectroscopy of formalin fixed brain tissue, *J. Neuropathol. Exp. Neurol.* **27:**1–14.

Ehmann, W. D., Markesbery, W. R., Alauddin, M., Hossain, T., and Brubaker, E. H., 1986, Brain trace elements in Alzheimer's disease, *Neurotoxicology* **7:**197.

Elovaara, I., Icen, A., Palo, J., and Erkinjuntti, T., 1985, CSF in Alzheimer's disease, *J. Neurol. Sci.* **70:**73–80.

Espinosa de los Monteros, A., Kumar, S., and Scully, S., 1990, Transferrin gene expression and secretion by rat brain cells in vitro, *J. Neurosci. Res.* **18:**299–304.

Faucheux, B. A., Hirsch, E. C., Villares, J., Selimi, F., Mouatt-Prigent, A., Javoy-Agid, F., and Agid, Y., 1993, Distribution of 125-I ferrotransferrin binding sites in the mesencephalon of control subjects and patients with Parkinson's disease, *J. Neurochem.* **60:**2338–234.

Faucheux, B. A., Nillesse, N., Damier, P., Spik, G., Mouatt-Prigent, A., Pierce, A., Leveugle, B., Kubis, N., Hauw, J-J., and Agid, Y., 1995, Expression of lactoferrin receptors is increased in the mesencephalon of patients with Parkinson's disease, *Proc. Natl. Acad. Sci. USA* **92:**9603–9607.

Fisher, M., Levine, P. H., Weiner, B. H., Vaudreuil, C. H., Natale, A., Johnson, M. H., and Hoogasian, J. J., 1988, Monocyte and polymorphonuclear leukocyte toxic oxygen metabolite production in multiple sclerosis, *Inflammation* **12:**123–131.

Fishman, J. B., Rubin, J. B., Handrahan, J. V., Connor, J. R., and Fine, R. E., 1987, Receptor mediated uptake of transferrin across the blood brain barrier, *J. Neurosci. Res.* **18:**299–305.

Friden, P. M., Walus, L. R., Watson, P., Doctrow, S. R., Kozarich, J. W., Backman, C., Bergman, H., Hoffer, B., Bloom, F., and Granholm, A.-C., 1993, Blood-brain-barrier penetration and in vivo activity of an NGF conjugate, *Science* **259:**373–377.

Giometto, B., Bozza, F., Argentiero, V., Gallo, P., Pagni, S., Piccinno, M. G., and Tavolato, B., 1990, Transferrin receptor in rat central nervous system: An immunohistochemical study, *J. Neurol. Sci.* **98:**81.

Gocht, A., Keith, A. B., Candy, J. M., and Morris, C. M., 1993, Iron uptake in the brain of the myelin deficient rat, *Neurosci. Lett.* **154:**187–190.

Good, P. F., Olanow, C. W., and Perl, D. P., 1992, Neuromelanin-containing neurons of the substantia nigra accumulate iron and aluminum in Parkinson's disease: A LAMMA study, *Brain Res.* **593:**343–346.

Goodman, L., 1953, Alzheimer's disease: a clinicopathologic analysis of twenty-three cases with a theory of pathogenesis, *J. Nerv. Ment. Dis.* **118:**97–130.

Graeber, M. B., Raivich, G., and Kreutzberg, G. W., 1989, Increase in transferrin receptors and iron uptake in regenerating motor neurons, *J. Neurosci. Res.* **23:**342.

Griot, C., Vandevelde, M., Richard, A., Peterhasn, E., and Stocker, R., 1990, Selective degeneration of oligodendrocytes mediated by reactive oxygen species, *Free Rad. Res. Comms.* **11:**181–193.

Grundke-Iqbal, I., Fleming, J., Tung, Y.-C., Lassman, H., Iqbal, K., and Joshi, J. G., 1990, Ferritin is a component of the neuritic plaque in Alzheimer's dementia, *Acta Neuropathol.* **81:**105–110.

Guo, B., Brown, F. M., Phillips, J. D., Yu, Y., and Leibold, E. A., 1995, Characterization and expression of iron regulatory protein 2 (IRP2), *J. Biol. Chem.* **270:**16259–16535.

Guo, B., Yu, Y., and Leibold, E. A., 1994, Iron regulates cytoplasmic levels of a novel iron-responsive element-binding protein without aconitase activity, *J. Biol. Chem.* **269:**24252–24260.

Hallgren, B., and Sourander, P., 1958, The effect of age on the nonhaemin iron in the human brain, *J. Neurochem.* **3:**41–51.

Hartung, H. P., Schafer, B., Heininger, K., and Toyka, K. V., 1988, Suppression of experimental autoimmune neuritis by the oxygen radical scavengers superoxide dismutase and catalase, *Ann. Neurol.* **23:**453–460.

Henderson, B. R., Seiser, C., and Kuhn, L. C., 1993, Characterization of a second RNA-binding protein in rodents with specificity for iron-responsive elements, *J. Biol. Chem.* **268:**27327–27334.

Hill, J. M., and Swizter, R. C., 1984, The regional distribution and cellular localization of iron in the rat brain, *Neuroscience* **11:**595–603.

Hill, J. M., Ruff, M. R., Weber, R. J., and Pert, C. B., 1985, Transferrin receptors in rat brain: neuropeptide-like pattern and relationship to iron distribution, *Proc. Natl. Acad. Sci. USA* **82:**4553.

Hu, J., and Connor, J. R., 1996, Characterization of the iron regulatory protein in the human brain, *J. Neurochem.* **67**:838–844.

Hulet, S. W., Arosio, P., Debinski, W., Powers, S., and Connor, J. R., 1996, Demonstration and characterization of a ferritin receptor in the brain, *FASEB J.* **10**(3):A251.

Hunter, M. I. S., Nlemadim, B. C., and Davidson, D. L. W., 1985, Lipid peroxidation production and antioxidant proteins in plasma and cerebrospinal fluid from multiple sclerosis patients, *Neurochem. Res.* **10**:1645–1652.

Jaffrey, S. R., Cohen, N. A., Rouault, T. A., Klausner, R. D., and Snyder, S. H., 1994, The iron responsive element binding protein: a target for synaptic actions of nitric oxide, *Proc. Natl. Acad. Sci. USA* **91**:12994–12998.

Jefferies, W. A., Brandon, M. R., Hunt, S. V., Williams, A. F., Gatter, K. C., and Mason, D. Y., 1984, Transferrin receptor on endothelium of brain capillaries, *Nature* **312**:162–163.

Johnson, D., Toms, R., and Weiner, H., 1989, Studies of myelin breakdown in vitro, in: *Myelination and Demyelination* (S. U. Kim, ed.), Plenum Press, New York, pp. 219–236.

Kalaria, R. N., Sromek, S. M., Grahovac, I., and Harik, S. I., 1992, Transferrin receptors of rat and human brain and cerebral microvessels and their status in Alzheimer's disease, *Brain Res.* **585**:87–93.

Kaneko, Y., Kitamoto, T., Tateosjo, J., and Yamaguchi, K., 1989, Ferritin immunohistochemistry as a marker for microglia, *Acta Neuropathol.* **79**:129–136.

Kaur, C., and Ling, E. A., 1995, Transient expression of transferrin receptors and localization of iron in amoeboid microglia in postnatal rats, *J. Anat.* **186**:165–173.

Kim, Y. S., and Kim, S. U., 1991, Oligodendroglial cell death induced by oxygen radicals and its protection by catalase, *J. Neurosci. Res.* **29**:100–106.

Klausner, R. D., Rouault, T. A., and Harford, J. B., 1993, Regulating the fate of mRNA: The control of cellular iron metabolism, *Cell* **72**:19–28.

Koeppen, A., 1995, The history of iron in the brain, *J. Neurol. Sci.* **134**(supplement):1–9.

Koeppen, A. H., and Borke, R. C., 1991, Experimental superficial siderosis of the central nervous system. I. Morphological observations, *J. Neuropathol. Exp. Neurol.* **50**:579–594.

Larkin, E. C., and Rao, G. A., 1990, Importance of fetal and neonatal iron: Adequacy for normal development of central nervous system, in: *Brain, Behaviour and Iron in the Infant Diet* (J. Dobbing, ed.), Springer-Verlag, London, pp. 43–63.

Levi, S., Yewdall, S. J., Harrison, P. M., Santambrogio, P., Cozzi, A., Rovida, E., Albertini, A., and Arosio, P., 1992, Evidence that H-L-chains have cooperative role in the iron uptake mechanism of human ferritin, *Biochem. J.* **288**:591–596.

Levine, M., Tuil, D., and Uzan, G., 1984, Expression of the transferrin gene during development of non-hepatic tissue, *Biochem. Biophys. Res. Comm.* **122**:212.

Levine, S. M., 1991, Oligodendrocytes and myelin sheaths in normal, *quaking* and *shiverer* brains are enriched in iron, *J. Neurosci. Res.* **29**:413–419.

Levine, S. M., 1992, The role of reactive oxygen species in the pathogenesis of multiple sclerosis, *Med. Hypoth.* **39**:271–274.

Lin, H. H., and Connor, J. R., 1989, The development of the transferrin-transferrin receptor system in relation to astrocytes, MBP, and galactocerebroside in normal and myelin deficient rat optic nerves, *Dev. Brain Res.* **49**:281–293.

Loeffler, D. A., Connor, J. R., Juneau, P. L., Snyder, B. S., Kanaley, L., DeMaggio, A. J., Nguyen, H., Brickman, M., and LeWitt, P. A., 1995, Transferrin and iron in normal, Alzheimer's disease and Parkinson's disease brain regions, *J. Neurochem.* **65**:710–716.

Manges, K., and Connor, J. R., 1996, Heavy chain ferritin binds DNA, *Soc. Neurosci. Abstr.* **22**(1):529.

Mantyh, P. W., Ghilardi, J. R., Rogers, S., DeMaster, E., Allen, C. J., Stimson, E. R., and Maggio, J. E., 1993, Aluminum, iron and zinc ions promote aggregation of physiological concentrations of beta-amyloid peptide, *J. Neurochem.* **61**:1171–1174.

Mash, D. C., Pablo, J., Flynn, D. D., Efange, S. M. N., and Weiner, W. J., 1990, Characterization and distribution of transferrin receptors in the rat brain, *J. Neurochem.* **55**:1972–1978.

Mash, D. C., Pablo, J., Buck, B. E., Sanchez-Ramos, J., and Weiner, W. J., 1991, Distribution and number of transferrin receptors in Parkinson's disease and in MPTP-treated mice, *Exp. Neurol.* **114**:73–81.

Mitrovic, B., Ignarro, L. J., Montestruque, S., Smoll, A., and Merrill, J. E., 1994, Nitric oxide as a potential pathological mechanism in demyelination: Its differential effects on primary glial cells *in vitro, Neuroscience* **61**:575–585.

Mitrovic, B., Ignarro, L. J., Vinters, H. V., Akers, M. A., Schmid, I., Uittenbogaart, C., and Merrill, J. E., 1995, Nitric oxide induces necrotic but not apoptotic death in oligodendrocytes, *Neuroscience* **65**:531–539.

Moos, T., 1995a, Increased accumulation of transferrin by motor neurons of the mouse mutant progressive motor neuronopathy (pmn/pmn), *J. Neurocytol.* **24**:389–398.

Moos, T., 1995b, Age-dependent uptake and retrograde axonal transport of endogenous albumin and transferrin in rat motor neurons, *Brain Res.* **672**:14–23.

Morris, C. M., Candy, J. M., Bloxham, C. A., and Edwardson, J. A., 1992a, Distribution of transferrin receptors in relation to cytochrome oxidase activity in the human spinal cord, lower brain stem and cerebellum, *J. Neurol. Sci.* **111**:158–172.

Morris, C. M., Candy, J. M., Oakley, A. E., Bloxham, C. A., and Edwardson, J. A., 1992b, Histochemical distribution of non-haem iron in the human brain, *Acta Anatom.* **144**:235–257.

Morris, C. M., Keith, A. B., Edwardson, J. A., and Pullen, R. G. L., 1992c, Uptake and distribution of iron and transferrin in the adult rat brain, *J. Neurochem.* **59**:300–306.

New, H., Dziegielewska, K. M., and Saunders, N. R., 1983, Transferrin in fetal rat brain and cerebrospinal fluid, *Int. J. Devl. Neurosci.* **1**:369–373.

Oka, A., Beliveau, M. J., Rosenberg, P. A., and Volpe, J. J., 1993, Vulnerability of oligodenroglia to glutamate: Pharmacology, mechanisms and prevention, *J. Neurosci.* **13**:1441–1453.

Oloyede, O. B., Folayan, A. T., and Odutauga, A. A., 1992, Effects of low-iron status and deficiency of essential fatty acids on some biochemical constituents of rat brain, *Biochem. Internat.* **27**:913–922.

Pleasure, D., Kim, S. U., and Silberberg, D. H., 1984, In vitro studies of oligodendroglial lipid metabolism, in: *Oligodendroglia* (W. H. Norton, ed.), Plenum Press, New York, pp. 175–197.

Pollitt, E., and Ponpon, I., 1993, Reversal of developmental delays in iron-deficient anaemic infants treated with iron, *Lancet* **341**:1–4.

Powell, L. W., Alpert, E., Isselbacher, K. J., and Drysdale, J. W., 1974, Abnormality in tissue isoferritin distribution in idiopathic haemochromatosis, *Nature* **250**:333–335.

Qi, Y., and Dawson, G., 1994, Hypoxia specifically and reversibly induces the synthesis of ferritin in oligodendrocytes and human oligodendrogliomas, *J. Neurochem.* **63**:1485–1490.

Raivich, G., Graeber, M. B., Gehrmann, J., and Kreutzberg, G. W., 1991, Transferrin receptor expression and iron uptake in the injured and regenerating rat sciatic nerve, *Eur. J. Neurosci.* **3**:919–927.

Reiderer, P., Sofic, E., Rausch, W.-D., Schmidt, B., Reynolds, G. P., Jellinger, K., and Joudim, M. B. H., 1989, Transition metals, ferritin, glutathione, and ascorbic acid in Parkinsonian brains, *J. Neurochem.* **52**:515–520.

Reif, D. W., and Simmons, R. D., 1990, Nitric oxide mediates iron release from ferritin, *Arch. Biochem. Biophys.* **283**:537–541.

Robinson, S. R., Noone, D. F., Kril, J., and Halliday, G. M., 1995, Most amyloid plaques contain ferritin-rich cells, *Alzheimer's Res.* **1**:191–196.

Roskams, A. J. I., and Connor, J. R., 1992, The transferrin receptor in the myelin deficient (md) rat, *J. Neurosci. Res.* **31**:421–427.

Roskams, A. J., and Connor, J. R., 1994, Iron, transferrin and ferritin in the rat brain during development and aging, *J. Neurochem.* **63**:709–716.

Samaniego, F., Chin, J., Iwai, K., Rouault, T. A., and Klausner, R. D., 1994, Molecular characterization of a second iron-responsive element binding protein, iron regulatory protein 2. Structure, function, and post-translational regulation, *J. Biol. Chem.* **269**:30904–30910.

Sanyal, B., and Szuchet, S., 1995, Tumor necrosis factor α induces the transcription of the H chain ferritin gene in cultured oligodendrocytes, *Soc. Neurosci. Abstr.* **21**:(part 1):3.

Simpson, R. J., Raja, K. B., Halliwell, B., Evans, P. J., Aruoma, O. I., and Konjin, A. M., 1991, Iron speciation in hypotransferrinemic mouse serum, *Biochem. Soc. Trans.* **19**:317S.

Sofic, E., Riederer, P., Heinsen, H., Beckman, H., Reynolds, G. P., Hebenstreit, G., and Youdim, M. B. H.,

1988, Increased iron (III) and total iron content in post mortem substantia nigra of Parkinsonian brain, *J. Neural Transm.* **74:**199–205.

Tanzi, R. E., and Hyman, B. T., 1991, Alzheimer's mutation (letter), *Nature* **350:**564.

Taylor, E. M., Crowe, A., and Morgan, E. H., 1991, Transferrin and iron uptake by the brain: Effects of altered iron status, *J. Neurochem.* **57:**1584–1592.

Thompson, C. M., Marksbery, W. r., Ehmann, W. D., Mao, Y.-X., and Vance, D. E., 1988, Regional brain trace-element studies in Alzheimer's disease, *Neurotoxicology* **9:**1–9.

Tsutsumi, M., Skinner, M. K., and Sanders-Bush, E., 1989, Transferrin gene expression and synthesis by cultured choroid plexus epithelial cells: Regulation by serotonin and cyclic adenosine 3′,5′-monophosphate, *J. Biol. Chem.* **264:**9626–9631.

Uitti, R. J., Rajput, A. H., Rozdilsky, B., Bickis, M., Wollin, T., and Yuen, W. K., 1989, Regional metal concentrations in Parkinson's disease and control brains, *Can. J. Neurol. Sci.* **16:**310–314.

Wilson, R., and Tocher, D. R., 1991, Lipid and fatty acid composition is altered in plaque tissue from multiple sclerosis brain compared with normal brain white matter, *Lipids* **26:**9–15.

Youdim, M. B. H., Ben-Schachar, D., and Reiderer, P., 1989, Is Parkinson's disease a progressive siderosis of substantia nigra resulting in iron and melanin-induced neurodegeneration? *Acta Neurol. Scand.* **126:**47–54.

Chapter 3

Magnetic Resonance Imaging of Brain Iron

George Bartzokis

3.1. INTRODUCTION

The frequency of Alzheimer's Disease (AD) increases dramatically with age. The prevalence of AD among people over 65 years of age is 5–10%. The prevalence is as high as 47% among those over age 85 (Evans *et al.,* 1989). Similarly, small age-related increases in neurofibrillary tangles (NFT) have been observed in brains of nondemented individuals (Price *et al.,* 1991; Morris *et al.,* 1991; Arriagada *et al.,* 1992). NFT density has been specifically associated with both dementia severity (Berg *et al.,* 1993) and cortical atrophy (Huesgen *et al.,* 1993). Based on the age-related increases observed both in AD and NFT, some have speculated that there is a common pathologic process leading to NFT formation in both aging and AD (Arriagada *et al.,* 1992), while others suggest that AD may represent a specific brain vulnerability to age-related oxidation in which iron may play a role (Stadtman, 1990; Smith *et al.,* 1991).

The age-related increases in AD and NFT are paralleled by age-related increases in brain tissue iron in some cortical and basal ganglia structures, and these increases in iron have been shown to continue well into old age (Hallgren and Sourander, 1958; Klintworth, 1973; Bartzokis *et al.,* 1994b). Quantitative studies have demonstrated increased levels of non-heme iron (Hallgren and Sourander, 1960; Ehmann *et al.,* 1986; Dedman *et al.,* 1992; Good *et al.,* 1992) or ferritin (Dedman *et al.,* 1992; Good *et al.,* 1992; Bartzokis *et al.,* 1994a) in the brain of AD patients beyond that observed in normal aging. NFT have been found to contain significantly increased levels of iron and aluminum compared to other parts of the same neuron, and neurons with NFT

George Bartzokis   Department of Psychiatry, UCLA, Los Angeles, California 90024; and the Research Service and the Psychiatry Service, West Los Angeles Veterans Affairs Medical Center, Los Angeles, California 90073.
Metals and Oxidative Damage in Neurological Disorders, edited by Connor. Plenum Press, New York, 1997.

contain markedly increased amounts of iron (4–18 times) compared to neurons from normal brains. Even NFT-free neurons from AD brains may have a 2.5–3 fold increase in iron compared to neurons from normal brains (Good *et al.*, 1992).

A disruption in brain iron homeostasis is also suggested by an increase in iron levels in and near the plaques found in AD brains (Goodman, 1953; Grundke-Iqbal *et al.*, 1990; Connor *et al.*, 1992; see also Chapter 2). This observation suggests that AD-specific increases in brain iron above and beyond normal aging may exist because, unlike NFT, plaques are rarely seen in normal aging (Price *et al.*, 1991; Morris *et al.*, 1991; Berg *et al.*, 1993).

The accumulation of brain iron in aging and AD is probably not simply an epiphenomenon of neurodegeneration where increased concentrations of iron are due to structure shrinkage resulting from cell loss (Bartzokis *et al.*, 1994a, 1994b). Elevated iron levels may therefore have significant pathophysiologic consequences that manifest as different neurodegenerative diseases depending on which brain region is most affected. For example, Huntington's disease (HD) may be characterized by increased iron in the caudate nucleus and Parkinsonian syndromes may have various combinations of iron increases in the substantia nigra and putamen (Dexter *et al.*, 1992; Chen *et al.*, 1993). In fact, iron and free radical neurotoxic processes have been implicated in both Parkinson's disease (PD) (Earle, 1968; Dexter *et al.*, 1989, 1992; Jenner *et al.*, 1992; Chen *et al.*, 1993), and HD (Klintworth, 1973; Dexter *et al.*, 1992; Chen *et al.*, 1993), in addition to AD (Hallgren and Sourander, 1960; Ehmann *et al.*, 1986; Dedman *et al.*, 1992; Good *et al.*, 1992; Bartzokis *et al.*, 1994a).

A non-neurologic example of how ferritin (the principal iron storage protein) may be an important risk factor for tissue injury has recently been published in relation to plasma ferritin and ischemic heart disease. Plasma ferritin, at levels previously regarded as "normal," was shown to be a robust risk factor for ischemic heart disease (Salonen *et al.*, 1992). The authors postulated that the role of iron in ischemic heart disease was to catalyze free radical reactions, resulting in hydroxyl radical formation, lipid peroxidation, and eventual arterial plaque formation.

The relationship between brain iron, aging, and neurodegenerative diseases like AD has remained unclear, in part because of the lack of noninvasive methods to measure brain iron with specificity *in vivo*. This is an important limitation since postmortem studies indicate that, probably due to the blood brain barrier, peripheral markers are not useful in predicting brain tissue iron levels (Hallgren and Sourander, 1958). In addition, postmortem studies and tissue biopsies are constrained by multiple limitations (Griffiths and Crossman, 1993). Thus, a noninvasive method that can evaluate tissue iron with specificity *in vivo* is essential.

In vivo tissue iron quantification could improve our understanding of the pathophysiology of brain aging, AD, and other neurodegenerative diseases, leading to more precise diagnostic assessments. Aid in diagnosis is the most immediate, clinically relevant application of *in vivo* methods of quantifying tissue iron levels and is supported by pilot studies (see below). In addition, given the known neurotoxic consequences of increased tissue iron levels, premorbid iron levels could represent an important risk factor for neurodegenerative disorders.

3.2. MRI EVALUATION OF TISSUE IRON

3.2.1. T_2 Is a Nonspecific Measure of Iron

Magnetic resonance imaging (MRI) can indirectly measure tissue iron through its effect on transverse relaxation times (T_2) and transverse relaxation rates (R_2). T_2 is a time constant (expressed in milliseconds) that defines the exponential decay curve of transverse magnetization of tissues (transverse magnetization = proton density $\times$ $e^{t/T2}$) where t is under experimental control. The T_2 of any tissue is determined by its structural and biochemical properties and therefore, T_2 measurements can provide biochemical information. T_2 and R_2 (a rate expressed in sec^{-1}) are related by the simple formula ($R_2 = 1/T_2 \times 1000$) and will be used almost interchangeably in this chapter.

The possibility of using T_2 measures to investigate tissue iron levels has been largely unrealized (Brooks et al., 1989; Chen et al., 1989, 1993; Drayer, 1989; Kucharczyk et al., 1993). A major problem has been the limited specificity of T_2 shortening as a marker of tissue iron. The lack of specificity is inherent in the measure of T_2 since many tissue characteristics can affect this measure (Malisch et al., 1991; Schenker et al., 1993; Bartzokis et al., 1994a, 1994b). For example, in addition to iron, normal differences in tissue structure and biochemistry can also shorten T_2. In brain, this is most clearly demonstrated by the reduction in the T_2 of white matter (probably caused by the increased viscosity/density of the tightly packed myelinated axons). Thus, white matter has a T_2 similar to caudate and putamen, yet has only about half the iron of these structures (Hallgren and Sourander, 1958).

In addition to the effects that nonpathologic tissue characteristics can have on T_2, many pathologic processes can also profoundly alter this measure. One of the most profound effects pathologic processes can have on T_2 is to increase MR-visible tissue water content (Kamman et al., 1988). Since T_2 of water is over 1000 ms, while T_2 of brain parenchyma is under 100 ms, it is clear that relatively small increases in water content will lengthen the measured T_2 of the tissue and thus mask any T_2 shortening caused by increased tissue iron levels. This strong effect of water on T_2 underlies the clinical reliance on imaging sequences that are sensitive to changes in T_2 and produce "T_2-weighted" images for detecting brain pathology.

It can thus be expected that in normal aging and neurodegenerative diseases like AD, as tissue is lost, the increase in MR-visible water will increase T_2.[†] Age-related increases in brain T_2 have been demonstrated (Jernigan et al., 1991; Bondareff et al., 1988; Besson et al., 1990; Bartzokis et al., 1994b) and have been shown to mask the decrease in T_2 caused by increased iron content (Bartzokis et al., 1994a, 1994b). A recent MR study of HD may be an example of the inconsistencies created by the lack of specificity of T_2 as a measure of tissue iron. HD is known to cause a marked degree of tissue loss in the basal ganglia (Jernigan et al., 1991). This marked tissue destruction, combined with the lack of specificity of T_2 measures, may have contributed to the

---

[†]For brevity, in the remainder of the text "water" will refer to "MR-visible water."

absence of the expected decrease in T_2 suggested by biochemical analysis of the tissue, which showed a three- to sixfold increase in ferritin concentration in the postmortem samples of HD basal ganglia (Chen *et al.*, 1993).

Qualitative and more recent quantitative high-field MRI studies have demonstrated age-related increases in human brain iron through the effects of iron on T_2 in younger individuals (birth to 30 years of age) (Aoki *et al.*, 1989; Olanow *et al.*, 1989; Milton *et al.*, 1991; Pujol *et al.*, 1992; Schenker *et al.*, 1993; Thomas *et al.*, 1993). In these young subjects, the magnitude of the change in iron levels is greatest, climbing rapidly from the near-zero levels at birth (Hallgren and Sourander, 1958). In addition, the confounding effects of neurodegenerative processes observed in older individuals are probably much reduced or absent in these young cohorts. Close examination of the data from these studies reveals that no decreases in T_2 were detected past the age of 30. A single qualitative study reported T_2 changes suggestive of increased putamen iron past age 30, but no increases in the caudate iron were detected (Olanow *et al.*, 1989). Most of these investigators were aware of and discussed the issue of the lack of specificity in such MRI measures, which resulted in relatively poor correlation between MRI measures and published iron levels of brain structures (Milton *et al.*, 1991; Pujol *et al.*, 1992; Schenker *et al.*, 1993; Rutledge *et al.*, 1987).

3.2.2. Specific MRI Measures of Ferritin

Upwards of 90% of non-heme iron in the brain is found in the iron storage protein ferritin (Morris *et al.*, 1992; Floyd and Carney, 1993). Ferritin has been shown to exert a strong magnetic effect that results in marked T_2 shortening *in vitro* and *in vivo* (Gillis and Koenig, 1987; Vymazal *et al.*, 1992; Bartzokis *et al.*, 1993a) and may account for the original observation of shortened T_2 in the basal ganglia (Drayer, 1989).

Early work on the effect of field strength on relaxation times suggested that unlike T_1, which is clearly field-strength dependent, T_2 was independent of field strength (Johnson *et al.*, 1985; Bottomley *et al.*, 1987). This early work was contradicted by subsequent reports of field-strength dependence of tissue T_2 parameters both *in vivo* (Bernardino *et al.*, 1989; Bizzi *et al.*, 1990; Bartzokis *et al.*, 1993a) and *in vitro* (Dockery *et al.*, 1989; Bartzokis *et al.*, 1993a). These contradictory results are not surprising in light of the fact that, to date, the only physiologic compound observed to cause field-dependent T_2 shortening is ferritin, that ferritin levels can vary greatly between tissues, and that ferritin levels can be profoundly age dependent (Hallgren and Sourander, 1958; Bartzokis *et al.*, 1993a).

There are many possible explanations for the field-dependent T_2 shortening caused by ferritin. One explanation is that the field inhomogeneity created by the heterogeneous distribution of paramagnetic ferric iron atoms in the ferritin core increases the observed R_2 to a greater extent in higher- than in lower-field-strength instruments. This same theory has been proposed to explain the field-dependent R_2 increases observed in stationary red blood cells (Brooks and Di Chiro, 1987; Gomori *et al.*, 1987). Other explanations involve special properties that may be unique to iron in a crystalline form. The small microcrystalline ferric oxide structures in the ferritin complex (Theil, 1990) may exhibit a variety of magnetic behaviors, such as ferromagnetism, antiferromagnetism (Blaise *et al.,* 1965; Boas and Window, 1966), and super-paramagnetism (Neel, 1961). Regardless of the mechanism producing the observed field dependence, iron in the form found in ferritin, and in quantities found in normal human brains, contributes markedly and specifically to the field dependence measured with clinical MRI instruments (Bartzokis *et al.,* 1993).

A recently developed method termed the field dependent R_2 increase (FDRI) quantifies tissue ferritin with specificity by capitalizing on this unique effect of ferritin (Bartzokis *et al.,* 1993a). The method involves measuring T_2 in two MR instruments with different magnetic field strengths, determining the transverse relaxation rate [R_2, ($R_2 = 1/T_2$)] in each, and calculating the difference in R_2 (high field R_2 − low field R_2), or FDRI. This method assesses tissue ferritin with specificity by eliminating interfering effects on R_2, such as increased water content or increased cell packing, which cause field-independent changes in R_2. For example, an increase in water content will decrease R_2 in a field-independent fashion, i.e. water reduces R_2 in both high- and low-field-strength instruments. Thus, when the difference in R_2 of the two instruments is examined, these field-independent effects cancel out (Bartzokis *et al.,* 1994a, 1994b). *In vitro* studies demonstrated that iron in the form found in ferritin contributes markedly and specifically to FDRI (Bartzokis *et al.,* 1993a; Vymazal *et al.,* 1996). In addition, the field dependence of the relaxation rate increase caused by ferritin was shown to be linear (Bartzokis *et al.,* 1993a; Vymazal *et al.,* 1996) and the slope per unit iron was shown to be almost identical regardless of the iron loading of ferritin molecules (Vymazal *et al.,* 1996). These observations suggest that FDRI is a specific quantitative measure of the iron contained in tissue ferritin stores.

Additional methods of evaluating different aspects of tissue iron content have been published. For example, direct measures of microscopic field inhomogeneities created by ferritin and other paramagnetic substances can be evaluated using T_2^* measurements, especially at higher field strengths (Ordidge *et al.,* 1994). In addition, where cell-size areas of magnetic susceptibility changes exist (such as macrophages with high iron content), water molecules diffusing near such cells would display T_2 shortening that would be dependent on the time the water molecules have to diffuse through the inhomogeneity (e.g., the interecho time). Thus, using a single field-strength and different echo times may allow isolation of this portion of the iron contribution to the T_2 shortening (Ye *et al.,* 1994).

Used in concert, these *in vivo* techniques could potentially segregate the effects of different states of tissue iron and provide multiple levels of information about differences in iron metabolism between healthy and diseased tissues. Thus, measures of the amount of iron contained in tissue ferritin could be supplemented by estimates of total iron (free plus in ferritin), and, by subtracting ferritin iron from total iron, the amount of free iron.

3.2.3. Initial Clinical Research Applications of the FDRI Method

The results obtained by exploiting the specificity of the FDRI method closely approximate published postmortem data (Hallgren and Sourander, 1958). In addition to the initial study describing the method of quantifying FDRI (Bartzokis *et al.,* 1993a),

four other studies have evaluated the usefulness of the FDRI method in investigating
the role of iron in age-related disorders.

First, age-related increases in brain iron were evaluated by examining 20 normal
males with a mean age of 55.5 years (range of 20–80 years) (Bartzokis *et al.*, 1994b).
The FDRI measured in these subjects was very highly and significantly correlated ($r =
0.96$) with published brain iron levels (Hallgren and Sourander, 1958) for the four
regions examined (caudate, putamen, globus pallidus, and frontal white matter), which
replicated earlier results (Bartzokis *et al.*, 1993a). In addition, as expected from postmortem data (Hallgren and Sourander, 1958), robust age-related increases in FDRI
were observed in the caudate ($r = 0.76$, $p = 0.0001$) and putamen ($r = 0.77$, $p =
0.0001$). Scatterplots of the caudate and putamen data (Figures 3.1 A and B) indicated
that this increase continued into old age. When the correlations were recomputed using
only subjects over 30 years of age, the correlations remained highly significant in the

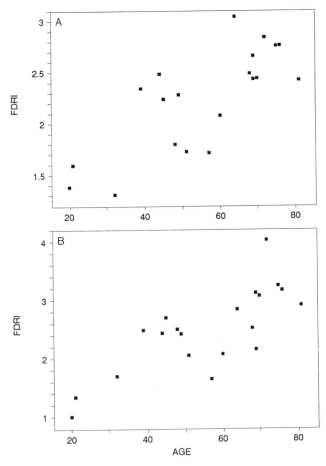

FIGURE 3.1. Scatterplots of age versus (a) caudate and (b) putamen FDRI measures in a sample of 20
normal males.

caudate (r = 0.66, p < .01, df = 16) and putamen (r = .61, p < .01, df = 16). This increase was structure specific since, in this age range, the correlations of age and FDRI in the globus pallidus (r = 0.34) and white matter (r = 0.29) were significantly lower and did not reach statistical significance (p > .05).

The data also showed that age-related increases in tissue iron stores can be quantified *in vivo* despite age-related processes that result in increased brain tissue water content, significantly decreasing R_2 (Jernigan *et al.*, 1991) and canceling out the increase in R_2 caused by iron (Bartzokis *et al.*, 1994a, 1994b). For example, in the FDRI study, significant age-related reductions in R_2 were observed in white matter in both high- (r = −0.5, p = .03) and low-field (r = −0.65, p = .002) instruments. The subtraction of high- from low-field R_2 values in the FDRI method removed this confounding effect, resulting in a similar low age versus white matter iron correlation (Figure 3.2) observed in postmortem data (Hallgren and Sourander, 1958; Klintworth, 1973).

Table 3.1 displays results of multiple regression analyses that evaluated the correlation between R_2 and age when data from the high- and low-field-strength magnets were

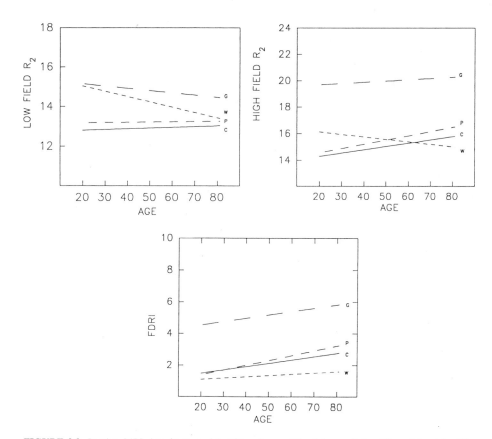

FIGURE 3.2. In vivo MRI data from caudate (C), putamen (P), globus pallidus (G), and frontal white matter (W) from a sample of 20 normal males. Data is displayed as regression lines of age versus R_2, at both high (1.5 Tesla) and low (0.5 Tesla) field strengths, and field-dependent R_2 increase (FDRI).

Table 3.1
Significance of Low- and High-Field R_2 Data Contribution
to the FDRI Age Correlations[a]

Field strength		Caudate	Putamen	Globus pallidus	White matter
0.5 T	B	−26.22	−21.68	−15.23	−15.52
	t	−2.61	−2.15	−2.17	−2.33
	p	.018	.046	.045	.032
1.5 T	B	27.31	19.43	5.85	−0.62
	t	4.72	4.92	1.69	<1
	p	.0002	.0001	.11	.93
Intercept		−19.3	37.2	163.1	283.8

[a]R_2 values for each brain region obtained using the high-(1.5 T) and low-(0.5 T) field-strength MRI instruments.

combined, and estimated the optimal weights to maximize that correlation. The regression coefficients are displayed in this table. In caudate and putamen regions, these weights are roughly equal, opposite in sign, and statistically significant, suggesting that a simple difference score (i.e., the FDRI) is close to the optimal weighting. In the globus pallidus and white matter regions, the multivariate analysis indicated that only low-field R_2 was significantly associated with age. Adding information from the high-field-strength magnet did not improve the correlation, suggesting that the low-field magnet provides a measure of age-related processes that are unrelated to iron stores, but rather processes, such as a diffuse increase in tissue water content, that decrease R_2 with age (Table 3.1 and Figure 3.2).

Second, age-related increases in brain iron were evaluated by examining a smaller ($N = 13$) sample of normal subjects with a younger age distribution (half the sample under 30 years of age with a mean age of 38.5 years and a range of 20–77 years) using new MRI instruments (Bartzokis et al., 1997). This study was conducted to more closely examine the age-related changes in globus pallidus FDRI. Postmortem studies show that the iron levels in the globus pallidus increase steeply with age and plateau after 30 years of age (Hallgren and Sourander, 1958). The MRI data from this study again demonstrated that, in the four regions examined, the FDRI was very highly and significantly correlated ($r = 0.98$) with published brain iron levels (Hallgren and Sourander, 1958). Again, the data showed a robust age-related increase in the FDRI of the caudate ($r = 0.72, p = .009$) and putamen ($r = 0.83, p = .0004$). Finally, the data confirmed that globus pallidus FDRI also increases with age ($r = 0.65, p = .015$) and, as expected, the increase seems to reach a plateau after the age of 30 (Figure 3.3). In an attempt to replicate and evaluate the curvilinearity observed in postmortem data (Hallgren and Sourander, 1958), an exponential function with an intercept of zero (brain iron at birth is very low) was fitted to the globus pallidus data. For the FDRI data, the optimal exponential functions with an intercept at zero yielded a robust ($r = .725$) and highly significant ($p = .005$) correlation between age and FDRI that, as expected from the postmortem data, is an improvement over the linear function ($r = .65, p = 0.015$).

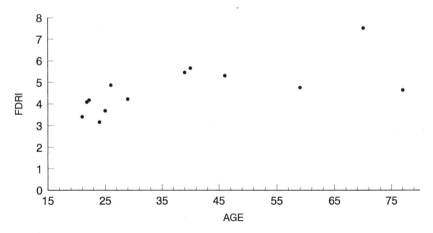

FIGURE 3.3. Scatterplots of age versus globus pallidus FDRI measures in a sample of 13 normal males with a median age of 30.

Third, a pilot study using the FDRI method to compare brain iron stores in five male patients with AD and eight age and sex-matched normal controls revealed increased brain iron in AD patients (Bartzokis *et al.*, 1994a). Subjects with AD met the criteria for probable AD established by the National Institute of Neurological and Communicative Disorders and Stroke, and the Alzheimer's Disease and Related Disorders Association (McKhann *et al.*, 1984). No patient had a history of head trauma with loss of consciousness, a psychoactive substance use disorder, a systemic illness or other neurologic illness that could account for the cognitive impairment, or evidence of significant Parkinsonism or other movement disorder. The clinical neuroimaging studies were essentially normal for age, demonstrating cerebral atrophy, or mild periventricular hyperintensity on the T_2-weighted images. The patients with AD ranged in age between 68 and 78 years (mean = 72.6, sd = 3.97), had mild to severe AD with Mini-Mental State Examination (Folstein *et al.*, 1975) scores ranging between 10–25 (mean = 17.8, sd = 5.45) and a length of illness ranging between 3 and 8 years (mean = 5.2, sd = 1.79).

Because group sizes were different and power to detect differences in variances was low due to the small sample sizes, both parametric (t) and nonparametric (Savage) tests were used in the statistical analysis. FDRI values were higher among AD patients in all three subcortical gray matter regions and amygdala. The difference reached statistical significance in the caudate (t = 2.3, df = 11, p = .04; Savage test χ^2 = 4.76, df = 1, p = .03), globus pallidus (t = 2.4, df = 11, p = .04; Savage test χ = 4.24, df = 1, p = .04), and almost did so in the putamen (t = 1.9, df = 11, p = .08; Savage test χ = 3.24, df = 1, p = .07). In addition, the difference between the AD and normal group reached statistical significance in the amygdala (t = 2.7, df = 11, p = .02; Savage test χ = 4.53, df = 1, p = .023). The FDRI means for frontal white matter were similar in the two groups (t < .5, ns). The increases in FDRI were significantly larger in the caudate, globus pallidus, and amygdala than in the frontal white matter.

Several possible confounding variables, including age, height, and education were examined. Within-group correlations of these measures with FDRI were generally low and nonsignificant, and statistical control of those variables did not affect the results.

We also evaluated whether the increased iron was simply a marker for disease-related cell loss that results in higher iron concentrations (Hallgren and Sourander, 1960), or whether the iron contributes to the pathogenesis of disease processes and aging, as others have suggested (Goodman, 1953; Stadtman, 1990; Connor et al., 1992). In order to assess this, the actual volume of the basal ganglia structures from which FDRI data was obtained was evaluated. A linear relationship was observed for the five AD patients between FDRI and volume of the putamen ($r = 0.8$, df $= 4$, $p = .07$). When volume was statistically controlled in that region, group differences in putamen FDRI no longer approached significance. However, in the globus pallidus and caudate, statistical control of structure volume did not affect the results. Therefore, for the caudate and globus pallidus, the observed changes in these studies cannot be attributed solely to brain atrophy causing an increase in iron concentration.

The fourth study examined 12 additional patients with idiopathic PD in order to further evaluate the specificity and possible diagnostic utility of the FDRI results. All PD patients met accepted criteria (Adams and Victor, 1993) and particular attention was given to exclude patients with "Parkinson's plus" syndrome or with a family history of movement disorders. All patients scored in the normal range (27 or above) on the Mini-Mental State Examination.

Multiple discriminant analysis was used to evaluate the power of FDRI in the caudate nucleus and globus pallidus in order to discriminate among the three diagnostic groups (5 subjects with AD, 12 with PD, and 20 normal controls), with age serving as a covariate. Overall, the statistical model was highly significant ($F = 3.38$, df $= 6,64$, $p = .006$). It successfully discriminated subjects with Alzheimer's disease from those with Parkinson's ($F = 4.61$, df $= 3,32$, $p = .009$) and from normals ($F = 6.99$, df $= 3,332$, $p = .001$); the latter groups did not differ ($F < 1$, ns). Four of the five AD subjects were correctly classified, with one misclassification as a normal. The model did not discriminate PD subjects from normals, but it incorrectly classified only 2 of those 32 individuals as Alzheimer's cases (both were normals). Thus, in this pilot sample, FDRI in two basal ganglia regions achieved sensitivity of .80 and specificity of .94 for the classification "not Alzheimer's."

AD is not a basal ganglia disorder and, unlike the amygdala, the caudate and globus pallidus are not heavily affected by NFT and plaques, but the increased FDRI of these structures appears robust enough to be a useful diagnostic aid. Thus, basal ganglia involvement may occur at subthreshold levels, while related structures, such as the amygdala and the nucleus basalis of Meynert, may be more susceptible, and thus produce clinically relevant signs. Nevertheless, evaluation of the basal ganglia may provide insights into the pathogenetic process of AD as some studies report a more malignant course of AD in subjects with extrapyramidal symptoms (Richards et al., 1993; Soinien et al., 1992; Miller et al., 1991).

3.2.4. Future Directions of *in Vivo* Evaluation of Tissue Iron with MR

Recent improvements in methodology make it possible to optimize gray/white matter contrast and thus permit reliable evaluation of more complex and subtle structures that are highly relevant to AD, such as the parietal cortex, hippocampus, and ventral globus pallidus (Bartzokis *et al.*, 1993b).

Like the amygdala, the hippocampus and parietal cortex are known to be affected by the AD process (Samuel *et al.*, 1991), and increased ferritin levels in the parietal cortex of AD patients have been observed (Dedman *et al.*, 1992). The nucleus basalis of Meynert is also affected in AD, and the number of NFT in this region has been observed to correlate with disease severity and duration (Samuel *et al.*, 1991). This nucleus is not distinct enough to be independently visualized at the macroscopic level but it is very closely associated anatomically with the ventral extension (inferior to the anterior commissure) of the globus pallidus (Doucette *et al.*, 1986; Arendash *et al.*, 1987; Carpenter, 1991).

Evaluation of the ventral globus pallidus may be important because it has among the highest iron concentrations in the brain and, as described above, the portion of the globus pallidus at the level of and just above the anterior commissure was observed to discriminate AD from PD and matched normal controls. Demonstrating increased iron levels in the ventral globus pallidus would strengthen the suggestion that increased globus pallidus iron levels may be a marker of increased iron in other important structures like the nucleus basalis of Meynert. Thus, neurotoxic iron levels may be implicated in the marked and consistent cholinergic deficits observed in AD since this nucleus is the largest source of cholinergic fibers throughout the brain (Arendash *et al.*, 1987; Carpenter, 1991).

3.3. CONCLUSIONS

Using currently available MRI technology, *in vivo* evaluation of the role of iron in neurodegenerative disorders is possible. Recent technical improvements promise methods of obtaining measures on relative iron content of different cellular compartments (e.g. ferritin versus non-ferritin iron). In addition, more accurate measures of even smaller structures of interest, such as the ventral globus pallidus, will be possible.

Although increased brain iron is unlikely to be the etiologic cause of AD and other neurodegenerative diseases, given the known neurotoxic effects of increased iron levels (Halliwell and Gutteridge, 1985, 1988; Floyd and Carney, 1993), reducing iron levels could prove to be an important therapeutic intervention. A clarifying analogy is found in Wilson's disease, where it is therapeutically useful to treat an important factor of the pathogenesis (copper accumulation) while being unable to address the etiology (defective ceruloplasmin).

Serial MRI evaluations could be used to monitor new treatment strategies such as iron chelation (Andersen *et al.*, 1991). Chelation is successfully being used in patients with transfusion hemosiderosis and has been tentatively shown to be effective in the treatment of some patients with AD (McLachlan *et al.,* 1991). Thus, *in vivo* iron quantification could have important and immediate implications for the diagnosis and treatment of patients with AD.

If methods that can quantify tissue iron with specificity *in vivo* are found to be useful in the diagnostic process, larger prospective studies can be contemplated to evaluate the prognostic value of brain tissue iron quantification. Such studies can be carried out by examining asymptomatic high risk populations of subjects with strong family histories of AD or AD variants and/or with the expression of apolipoprotein E4 alleles (Harrington *et al.,* 1994; St. Clair *et al.,* 1994). If tissue ferritin levels prove to be an important risk factor, the FDRI measure could provide new opportunities for rational treatment and prevention strategies for these diseases, including iron chelation (McLachlan *et al.,* 1991) and diet manipulation (Hendrie *et al.,* 1995).

ACKNOWLEDGMENTS.   This work was supported by the Research Service of the Department of Veterans Affairs. The author acknowledges and thanks Jim Mintz, Ph.D., for statistical analysis, and Sun Sook Hwang and Heraclio Avila for assistance in preparing the figures.

3.4. REFERENCES

Adams, R. D., and Victor, M., *Principles of Neurology,* McGraw-Hill, New York, 1993.

Andersen, P. B., Birgegard, G., Nyman, R., and Hemmingsson, A., 1991, Magnetic resonance imaging in idiopathic hemochromatosis, *Eur. J. Haematology* **47:**174–178.

Andorn, A. C., Britton, R. S., and Bacon, R. R., 1990, Evidence that lipid peroxidation and total iron are increased in Alzheimer's brain, *Neurobiol. Aging* **11:**316.

Aoki, S., Okada, Y., Nishimura, K., Barkovich, A. J., Kjos, B. O., Brasch, R. C., and Norman, D., 1989, Normal deposition of brain iron in childhood and adolescence: MR imaging at 1.5 T, *Radiology* **172**(2):381–385.

Arendash, G. W., Millard, W. J., Dunn, A. J., and Meyer, E. M., 1987, Long-term neuropathological and neurochemical effects of nucleus basalis lesions in the rat, *Science* **238:**952–956.

Arriagada, P. V., Marzloff, K., and Hyman, B. T., 1992, Distribution of Alzheimer-type pathologic changes in nondemented elderly individuals matches the pattern in Alzheimer's disease, *Neurol.* **42:**1681–1688.

Bartzokis, G., Aravagiri, M., Oldendorf, W. H., Mintz, J., and Marder, S. R., 1993a, Field dependent transverse relaxation rate increase may be a specific measure of tissue iron stores, *Magn. Reson. Med.* **29:**459–464.

Bartzokis, G., Mintz, J., Marx, P., Osborn, D., Gutkind, D., Chiang, F., Phelan, C. K., and Marder, S. R., 1993b, Reliability of in vivo volume measures of hippocampus and other brain structures using MRI, *Magn. Reson. Imag.* **11:**993–1006.

Bartzokis, G., Sultzer, D., Mintz, J., Marx, P., Phelan, C. K., and Marder, S. R., 1994a, MRI suggests increased brain iron in Alzheimer's disease, *Biol. Psych.* **35:**480–487.

Bartzokis, G., Mintz, J., Sultzer, D., Marx, P., Herzberg, J. S., Phelan, C. K., and Marder, S. R., 1994b, In vivo MR evaluation of age-related increases in brain iron, *Am. J. Neuroradiol.* **15:**1129–1138.

Bartzokis, G., Beckson, M., Hance, D. B., Marx, P., Foster, J. A., and Marder, S. R., 1997, MR evaluation of age-related increase of brain iron in young adult and older normal males, *Magn. Reson. Imag.* **15:**29–35.

Berg, L., McKeel, D. W., Miller, J. P., Baty, J., and Morris, J. C., 1993, Neuropathological indexes of Alzheimer's disease in demented and nondemented persons aged 80 years and older, *Arch. Neurol.* **50:**349–358.

Bernardino, M. E., Chaloupka, J. C., Malko, J. A., Chezmar, J. L., and Nelson, R. C., 1989, Are hepatic and muscle T_2 values different at 0.5 and 1.5 Tesla?, *Magn. Reson. Imag.* **7**:363–367.

Besson, J. A. O., Crawford, J. R., Parker, D. M., Ebmeier, K. P., Best, P. V., Gemmell, H. G., Sharp, P. F., and Smith, F. W., 1990, Multimodal imaging in Alzheimer's disease, *Br. J. Psych.* **157**:216–220.

Bizzi, A., Brooks, R. A., Brunetti, A., Hill, J. M., Alger, J. R., Miletich, R. S., Francavilla, T. L., and Di Chiro, G., 1990, Role of iron and ferritin in MR imaging of the brain, *Radiology* **177**:59–65.

Blaise, A., Chappert, J., and Girardet, J. L., 1965, Observation par mesures magnetique et effet Mossbauer d'un antiferromagnetisme de grains fins dans la ferritine, *Comptes Rendus Hebdomadaires Seanc. Acad. Sci., Paris,* **261D**:2310–2313.

Boas, J. F., and Window, B., 1966, Mossbauer effect in ferritin, *Aust. J. Phys.* **19**:573–576.

Bondareff, W., Raval, J., Colletti, P. M., and Hauser, D. L., 1988, Quantitative magnetic resonance imaging and the severity of dementia in Alzheimer's disease, *Am. J. Psych.* **145**:853–856.

Bottomley, P. A., Hardy, C. J., Argersinger, R. E., and Allen-Moore, G., 1987, A review of 1H nuclear magnetic resonance relaxation in pathology: Are T_1 and T_2 diagnostic?, *Med. Phys.* **14**:1–37.

Brooks, D. J., Luthert, P., Gadian, D., and Marsden, C. D., 1989, Does signal-attenuation on high-field T_2-weighted MRI of the brain reflect regional cerebral water proton T_2 values and iron levels, *J. Neurol. Neurosurg. Psych.* **52**:108–111.

Brooks, R. A., and Di Chiro, G., 1987, Magnetic resonance imaging of stationary blood: A review, *Med. Phys.* **14**:903–913.

Brun, A., and Gustafron, L., 1978, Limbic lobe involvement in presenile dementia, *Arch. Psych. Nervenkr.* **226**:79–93.

Carpenter, M. B., 1991, *Core Text of Neuroanatomy,* William & Wilkins, Baltimore, pp. 383–384.

Chen, J. C., Hardy, P. A., Clauberg, M., Joshi, J. G., Parravano, J., Deck, J. H. N., Henkelman, R. M., Becker, L. E., and Kucharczyk, W., 1989, T_2 values of the human brain: Comparison with quantitative assays of iron and ferritin, *Radiology* **173**:521–526.

Chen, J. C., Hardy, P. A., Kucharczyk, W., Clauberg, M., Joshi, J., Vourlas, A., Dahr, M., and Henkelman, R., 1993, MR of human postmortem brain tissue: Correlative study between T_2 and assays of iron and ferritin in Parkinson and Huntington disease, *Am. J. Neuroradiol.* **14**:275–281.

Coffey, C. E., Figiel, G. S., Djang, W. T., and Weiner, R. D., 1990, Subcortical hyperintensity on magnetic resonance imaging: A comparison of normal and depressed elderly subjects, *Am. J. Psych.* **147**:187–189.

Connor, J. R., Menzies, S. L., St. Martin, S. M., and Mufson, E. J., 1992, A histochemical study of iron, transferrin, and ferritin in Alzheimer's disease brains, *J. Neurosci. Res.* **31**:75–83.

Dedman, D. J., Treffry, A., Candy, J. M., Taylor, G. A. A., Morris, C. M., Bloxham, C. A., Perry, R. H., Edwardson, J. A., and Harrison, P. M., 1992, Iron and aluminum in relation to brain ferritin in normal individuals and Alzheimer's disease and chronic renal-dialysis patients, *Biochem. J.* **287**:509–514.

Dexter, D. T., Wells, F. R., Lees, A. J., Agid, F., Agid, Y., Jenner, P., and Marsden, C. D., 1989, Increased nigral iron content and alterations in other metal ions occurring in the brain in Parkinson's disease, *J. Neurochem.* **52**:1830–1836.

Dexter, D. T., Jenner, P., Schapira, A. H., and Marsden, C. D., 1992, Alterations in levels of iron, ferritin, and other trace metals in neurodegenerative diseases affecting the basal ganglia, *Ann. Neurol.* **32**:S94–S100.

Dockery, S. E., Suddarth, S. A., and Johnson, G. A., 1989, Relaxation measurements at 300 Hz using MR microscopy, *Magn. Reson. Med.* **11**:182–192.

Doucette, R., Fishman, M., Hachinski, V. C., and Mersky, H., 1986, Cell loss from the nucleus basalis of Meynert in Alzheimer's disease, *Can. J. Neurol. Sci.* **13**:435–440.

Drayer, B. R., 1989, Basal ganglia: Significance of signal hypointensity on T_2-weighted images, *Radiology* **173**:311–312.

Duvernoy, H. M., 1988, *The Human Hippocampus: an Atlas of Applied Anatomy.* Springer-Verlag, New York, pp. 135–139.

Earle, K. M., 1968, Studies on Parkinson's disease including X-ray fluorescent spectroscopy of formalin fixed brain tissue, *J. Neuropathol. Exp. Neurol.* **27**:1–14.

Ehmann, W. D., Markesbery, W. R., Alauddin, M., Hossain, T. I. M., and Brubaker, E. H., 1986, Brain trace elements in Alzheimer's disease, *Neurotoxicology* **7**:197–206.

Evans, D. A., Funkenstein, H. H., Albert, M. S., Scherr, P. A., Cook, N. R., Chown, M. J., Hebert, L. E., Hennekens, C. H., and Taylor, J. O., 1989, Prevalence of Alzheimer's disease in a community population of older persons, *JAMA* **18:**2551–2556.

Floyd, R. A., and Carney, J. M., 1993, The role of iron in oxidative processes and aging, *Toxicol. Indust. Health* **9:**197–214.

Folstein, M. F., Folstein, S. E., and McHugh, P. R., 1975, "Mini-mental state": A practical method for grading the cognitive state of patients for the clinician, *J. Psych. Res.* **12:**189–198.

Gillis, P., and Konig, S. H., 1987, Transverse relaxation of solvent protons induced by magnetized spheres: Application to ferritin, erythrocytes, and magnetite, *Magn. Reson. Med.* **5:**323–345.

Gomori, J. M., Grossman, R. I., Yu-Ip, C., and Asakura, T., 1987, NMR relaxation times of blood: dependence on field strength, oxidation state, and cell integrity, *J. Comput. Assist. Tomogr.* **11:**684–690.

Good, P. F., Perl, D. P., Bierer, L. M., and Schmeidler, J., 1992, Selective accumulation of aluminum and iron in the neurofibrillary tangles of Alzheimer's disease: A laser microprobe (LAMMA) study, *Ann. Neurol.* **31:**286–292.

Goodman, L., 1953, Alzheimer's disease: A clinico-pathologic analysis of twenty-three cases with a theory on pathogenesis, *J. Nerv. Ment. Dis.* **117:**97–130.

Griffiths, P. D., and Crossman, A. R., 1993, Distribution of iron in the basal ganglia and neocortex in postmortem tissue in Parkinson's disease and Alzheimer's disease, *Dementia* **4:**61–65.

Grundke-Iqbal, I., Fleming, J., Tung, Y. C., Lassmann, H., Iqbal, K., and Joshi, J. G., 1990, Ferritin is a component of the neuritic (senile) plaque in Alzheimer dementia, *Acta Neuropathologica* **81:**105–110.

Hallgren, B., and Sourander, P., 1958, The effect of age on the non-haemin iron in the human brain, *J. Neurochem.* **3:**41–51.

Hallgren, B., and Sourander, P., 1960, The non-haemin iron in the cerebral cortex in Alzheimer's disease, *J. Neurochem.* **5:**307–310.

Halliwell, B., and Gutteridge, J. M., C., 1985, The importance of free radicals and catalytic metal ions in human diseases, *Molec. Aspects Med.* **8:**89–193.

Halliwell, B., and Gutteridge, J. M., C., 1988, Iron as biological pro-oxidant, *ISI Atlas Sci. Biochem.* **1:**48–52.

Harrington, C. R., Louwagie, J., Rossau, R., Vanmechelen, E., Perry, R. H., Perry, E. K., Xuereb, J. H., Roth, M., and Wischik, C. M., 1994, Influence of apolipoprotein E genotype on senile dementia of the Alzheimer and Lewy body types, *Am. J. Path.* **145:**1472–1484.

Hendrie, H. C., Osuntokun, B. O., Hall, K. S., Ogunniyi, A. O., Hui, S. L., Unverzagt, F. W., Gureje, O., Rodenberg, C. A., Olusegun, B., Musick, B. S., Adeyinka, A., Farlow, M. R., Oluwole, S. O., Class, C. A., Komolafe, O., Brashear, A., and Burdine, V., 1995, Prevalence of Alzheimer's disease and dementia in two communities: Nigerian Africans and African Americans, *Am. J. Psych.* **152:**1485–1492.

Huesgen, C. T., Burger, P. C., Crain, B. J., and Johnson, G. A., 1993, In vitro MR microscopy of the hippocampus in Alzheimer's disease, *Neurology* **43:**145–152.

Jernigan, T. L., Salamon, D. P., Butters, N., and Hesselink, J. R., 1991, Cerebral structure on MRI: Specific changes in Alzheimer's and Huntington's disease, *Biol. Psych.* **29:**55–81.

Jenner, P., Schapira, A. H. V., and Marsden, C. D., 1992, New insights into the causes of Parkinson's disease, *Neurology* **42:**2241–2250.

Johnson, G. A., Hefkens, R. J., and Brown, M. A., 1985, Tissue relaxation time: in vivo field dependence, *Radiology* **156:**805–810.

Kamman, R. L., Go, K. G., Brouwer, W., and Berendsen, H. J. C., 1988, Nuclear magnetic resonance relaxation in experimental brain edema: Effects of water concentration, protein concentration, and temperature, *Magn. Reson. Med.* **6:**265–274.

Klintworth, G. K., 1973, Huntington's chorea: Morphologic contributions of a century, in: *Huntington's Chorea, 1872–1972* (A. Barbeau, T. N. Chase, and G. W. Paulson, eds.), Raven Press, New York, pp. 353–368.

Kucharczyk, W., Henkelman, R. M., and Chen, J., 1993, Brain iron and T2 signal, *Am. J. Neuroradiol.* **14:**1795–1796.

Malisch, T. W., Hedlund, L. W., Suddarth, S. A., and Johnson, G. A., 1991, MR microscopy at 7.0 T: Effects of brain iron, *J. Magn. Reson. Imag.* **1:**301–305.

McKhann, G., Drachman, D., Folstein, M., Katzman, R., Price, D., and Stadlan, E. M., 1984, Clinical diagnosis of Alzheimer's disease: Report of the NINCDS–ADRDA work group under the auspices of Department of Health and Human Services task force on Alzheimer's disease, *Neurology* **34:**939–944.

McLachlan, D. R. C., Dalton, A. J., Kruck, T. P. A., Bell, M. Y., Smith, W. L., Kalow, W., and Andrews, D. F., 1991, Intramuscular desferrioxamine in patients with Alzheimer's disease, *Lancet* **337:**1304–1308.

Miller, T. P., Tinklenberg, J. R., Brooks, J. O., and Yesavage, J. A., 1991, Cognitive decline in patients with Alzheimer disease: Differences in patients with and without extrapyramidal signs, *Alzheimer Disease & Associated Disorders* **5:**251–256.

Milton, W. J., Atlas, S. W., Lexa, F. J., and Mozley, P. D., 1991, Deep gray matter hypointensity patterns with aging in healthy adults: MR imaging at 1.5 T, *Radiology* **181:**715–719.

Mirra, S. S., Heyman, A., McKeel, D., Sumi, S. M., Crain, B. J., Brownlee, L. M., Vogel, F. S., Huges, J. P., van Belle, G., and Berg, L., 1991, The consortium to establish a registry for Alzheimer's disease (CERAD). Part II. Standardization of the neuropathologic assessment of Alzheimer's disease, *Neurology* **41:**479–486.

Morris, C. M., Candy, J. M., Keith, A. B., Oakley, A. E., Taylor, G. A., Pullen, R. G. L., Bloxham, C. A., Gocht, A., and Edwardson, J. A., 1992, Brain iron homeostasis, *J. Inorg. Biochem.* **47:**257–265.

Morris, J. C., McKeel, D. W., Storandt, M., Rubin, E. H., Price, J. L., Grant, E. A., Ball, M. J., and Berg, L., 1991, Very mild Alzheimer's disease: Informant-based clinical, psychometric, and pathologic distribution from normal aging, *Neurology* **41:**469–478.

Neel, L., 1961, Superparamagnetisme des grain tres fins antiferomagnetiques. *Comptes Rendus Hebdomadaires Seanc. Acad. Sci., Paris,* **252B:**4075–4080.

Olanow, C. W., Holgate, R. C., Murtaugh, R., and Martinez, C., 1989, MR imaging in Parkinson's disease and aging, in: *Parkinsonism and Aging* (D. B. Calne, G. Comi, D. Crippa, R. Horowski, and M. Trabucchi, eds.), Raven Press, New York, pp. 155–164.

Ordidge, R. J., Gorell, J. M., Deniau, J. C., Knight, R. A., and Helpern, J. A., 1994, Assessment of relative brain iron concentration using T_2-weighted and T_2*-weighted MRI at 3 Tesla, *Magn. Reson. Med.* **32:**335–341.

Price, J. L., Davis, P. B., and White, D. L., 1991, The distribution of tangles, plaques and related immunohistochemical markers in healthy aging and Alzheimer's disease, *Neurobiol. Aging* **12:**295–312.

Pujol, J., Junque, C., Vendreli, P., Grau, J. M., Marti-Vilalta, J. L., Olive, C., and Gili, J., 1992, Biological significance of iron-related magnetic resonance imaging changes in the brain, *Arch. Neurol.* **49:**711–717.

Richards, M., Folstein, M., Albert, M., Miller, L., Bylsma, F., Lafleche, G., Marder, K., Bell, K., Sano, M., Devanand, D., Loreck, D., Wootten, J., and Stern, Y., 1993, Multicenter study of predictors of disease course in Alzheimer disease (the "Predictors Study"). II Neurological, psychiatric, and demographic influences on baseline measures of disease severity, *Alzheimer Disease & Associated Disorders* **7:**22–32.

Rutledge, J. N., Hilal, S. K., Silver, A. J., Defendini, R., and Fahn, S., 1987, Study of movement disorders and brain iron by MR, *Am. J. Neuroradiol.* **8:**397–411.

Sadeh, M., and Sandbank, U., 1980, Neuraxonal dystrophy and hemosiderin in the central nervous system, *Ann. Neurol.* **7:**286–287.

Salonen, J. T., Nyyssonen, K., Korpela, H., Tuomilehto, J., Seppanen, R., and Salonen, R., 1992, High iron levels are associated with excess risk of myocardial infarction in eastern Finnish men, *Circulation* **85:**803–811.

Samuel, W. A., Henderson, V. W., and Miller, C. A., 1991, Severity of dementia in Alzheimer disease and neurofibrillary tangles in multiple brain regions, *Alzheimer Dis. Assoc. Disord.* **5:**1–11.

Schenker, C., Meier, D., Wichmann, W., Boiesiger, P., and Valavanis, A., 1993, Age distribution and iron dependency of the T_2 relaxation time in the globus pallidus and putamen, *Neuroradiology* **35:**119–124.

Smith, C. D., Carney, J. M., Starke-Reed, P. E., Oliver, C. N., Stadtman, E. R., Floyd, R. A., and Markesberry, W. R., 1991, Excess brain protein oxidation and enzyme dysfunction in normal aging and Alzheimer disease, *Proc. Natl. Acad. Sci. USA* **88:**10540–10543.

Soininen, H., Laulumaa, V., Helkala, E. L., Hartikainen, P., and Riekkinen, P. J., 1992, Extrapyramidal signs in Alzheimer's disease: A 3-year follow-up study, *J. Neural Transm.* (P-D Sect.) **4**:107–119.

Stadtman, E. R., 1990, Metal ion-catalyzed oxidation of proteins, *Free Rad. Biol. Med.* **9**:315–325.

St. Clair, D., Norrman, J., Perry, R., Yates, C., Wilcock, G., and Brookes, A., 1994, Apolipoprotein E epsilon 4 allele frequency in patients with Lewy body dementia, Alzheimer's disease and age-matched controls, *Neurosci. Lett.* **176**:45-6.

Subbarao, K. V., Richardson, J. S., and Ang, L. C., 1990, Autopsy samples of Alzheimer's cortex show increased peroxidation in vitro, *J. Neurochem.* **55**(1):342–345.

Theil, E. C., 1990, The ferritin family of iron storage proteins, *Advances in Enzymology and Related Areas of Molecular Biology* **63**:421–429.

Thomas, L. O., Boyko, O. B., Anthony, D. C., and Burger, P. C., 1993, MR detection of brain iron, *Am. J. Neuroradiol.* **14**:1043–1048.

Vymazal, J., Brooks, R. A., Zak, O., McRill, C., Shen, C., and Di Chiro, G., 1992, T_1 and T_2 of ferritin at different field strengths: Effect on MRI, *Magn. Reson. Med.* **27**:368–374.

Vymazal, J., Zak, O., Bulte, J. W. M., Aisen, P., and Brooks, R. A., 1996, T_1 and T_2 of ferritin solutions: Effect of loading factor, *Magn. Reson. Med.* **36**:61–65.

Ye, F. Q., Martin, W. R. W., Hodder, J., and Allen, P. S., 1994, Brain iron imaging exploiting heterogeneous-susceptibility-enhanced proton relaxation, Proceedings of the Society on Magnetic Resonance, 2nd Annual Meeting, San Francisco, pp. 5.

Chapter 4

Neurochemical Roles of Copper as Antioxidant or Prooxidant

Joseph R. Prohaska

4.1. INTRODUCTION

The biological reactivity of copper is the basis for both its essentiality and toxicity. Prior to the advent of modern medicine, drug design, and therapeutics, there is a long historical documentation of the use of copper salts in ointments for over 2000 years. The biological activity of copper was unknown at the time but was used for a wide variety of maladies including diseases of the skin, various infections, and recoveries from neurological disorders (Deuschle and Weser, 1985). Toxic properties of copper were also recognized as copper sulfate was used as a murder weapon and a suicidal agent.

Although the presence of copper in plant and animal tissues was known for the past 200 years, it wasn't until the 1920s that the critical work of Hart and others with experimental rats demonstrated that copper was an essential transition metal for mammals. This historical background has been reviewed in much detail by Owen (1981) and Davis and Mertz (1987). These reviews discuss the background information on the essentiality of copper in a number of different animals, including humans. More recent reviews on the biochemistry and biological functions of copper are available (Linder, 1991; Prohaska, 1988). There are also reviews that discuss the distribution and functions of copper in the central nervous system (CNS) of mammals (Prohaska, 1987). Many of the biological functions of copper have been elucidated by investigation in experimental animals, domestic animals, and mutants that have abnormal copper metabolism. The biochemical changes that occur following a deficiency of copper have been summarized elsewhere (Prohaska, 1990). There have also been extensive reviews

Joseph R. Prohaska   Department of Biochemistry and Molecular Biology, School of Medicine, University of Minnesota–Duluth, Duluth, Minnesota 55812.

Metals and Oxidative Damage in Neurological Disorders, edited by Connor. Plenum Press, New York, 1997.

summarizing the changes in the brain of mammals following copper deficiency. The reader is referred elsewhere for this information (O'Dell and Prohaska, 1983).

The issue of copper, reactive oxygen species (ROS), and the CNS is highly dependent on a number of extenuating variables. Although the total copper concentration in the CNS is estimated at approximately 10^{-5} mole/liter, this is a rather simplistic way of looking at the relative amount of copper available for neurochemical reactions. This is because most of the copper present as inorganic copper; Cu^{2+}, cupric copper; and Cu^+, cuprous copper is not free in solution but rather complexed to various organic ligands. The vast majority of these organic ligands are specific proteins that bind stoichiometric amounts of copper and carry out highly regulated biochemical reactions. The Cu^{2+} ion normally associates with sulfur and nitrogen ligands in these proteins and assumes a distorted tetragonal configuration. This asymmetric metal-chelated center in these proteins is usually associated with three major geometries, the so-called Type 1 copper, Type 2 copper, and Type 3 copper. The biochemical characteristics of the copper in these particular centers are reviewed elsewhere (Deuschle and Weser, 1985). The redox center in these Cu^{2+} complexes is positive relative to the redox center in simple copper aqueous solutions by approximately 0.3 V, making this copper center very reactive. Because Cu^{2+} can be easily reduced to Cu^+ and, under certain circumstances Cu^+ can be oxidized to Cu^{2+}, this rather facile one-electron reaction is known to occur in these copper proteins. This places copper ions, like iron ions, in a unique situation to interact with molecular oxygen. Molecular oxygen prefers to participate in single-electron reactions due to spin restrictions in its molecular structure. Many of the known copper proteins catalyze functions that depend upon interaction with oxygen or ROS. These functions are referred to as antioxidant functions since they are hypothesized to be important in protecting tissues against some of the aberrant reactions known to take place when ROS become abnormally elevated. Because copper ions can also participate in electron transfer, they may be involved in the production of ROS and, therefore, could be considered to be prooxidants. Superoxide production by mitochondria is an example. It is this double-edged property that makes copper and other transition metals unique in a discussion of neurotoxicity. This review will address some of the evidence that copper can act both as an antioxidant or a prooxidant under certain specified conditions. Others have reviewed the evidence that ROS play an important role in CNS neurotoxicity when transition metal ions like copper and iron are available (Halliwell, 1992). Others in this contribution will address issues of whether or not metals play roles in the neurodegenerative processes involving oxygen in conditions such as Parkinson's disease, Alzheimer's disease, and others. Hypotheses regarding these disorders and trace metals have been studied for a long period of time and are reviewed elsewhere (Dexter *et al.,* 1991).

4.2. COPPER DISTRIBUTION

Concentration of copper in the brain is relatively high compared to other mammalian organs. Distribution is dependent on a number of factors including brain region, subcellular location, age, species, and environmental and genetic factors. Changes in the concentration and/or distribution of copper are undoubtedly antecedents of potential roles of copper acting as an antioxidant and/or prooxidant. Extensive data analysis on human and rat brain for a number of regions is available in the literature (Prohaska, 1987).

Regional analysis indicates that the concentration of copper is higher in gray matter than in white matter. In human brain, copper is especially enriched in the locus ceruleus and substantia nigra. In rat brain, the hypothalamus has a slightly higher concentration of copper compared to other regions. The concentration of copper in human brain is about twice that of rodent brain. Within the cell, copper is found in most subcellular organelles. Relative to protein, copper is enriched in the cytosolic fraction and mitochondria. When the cytoplasm of cells is further fractionated, several distinct peaks containing copper can be distinguished. Further details on this distribution can be obtained from recent reviews (Linder, 1991; Prohaska, 1987).

Of the many factors that influence the concentration of copper, and thus its potential to act as an antioxidant or prooxidant, environmental copper is especially important. The concentration of copper in the brain is largely determined during the developmental period, and changes in the availability of copper during periods of gestation have a major impact on the concentration in the offspring (Prohaska and Bailey, 1993a). When the availability of copper is not sufficient during early development, deficits in the concentration of copper are regionally dependent (Prohaska and Bailey, 1994). In a study in rats it was shown that the deficit in copper following nutritional copper deficiency was less in the hypothalamus compared to other brain regions. Changes that occur in copper levels during early development may persist for a lifetime. Recent experiments that followed recovery after deficiency in both mice and rats indicate that the concentration of copper is very resistant to restoration (Prohaska and Bailey, 1995a; Prohaska and Bailey, 1993b). Although not precisely known, these changes in early development may persist because the turnover of copper in brain is extremely slow (Levenson and Janghorbani, 1994). A recent study in rats indicated that the biological half-life of brain copper is more than one year and that the turnover decreases 15-fold when copper is limiting in the diet. Changes in the concentration of copper early on in development may have long-term consequences.

Most of the copper in the CNS of mammals is accumulated during early development (Prohaska and Wells, 1974). It is known from previous experiments that the concentration of copper does increase with age. Morita et al. (1994a) recently emphasized this fact again in some experiments in mice. This same group (Morita et al., 1994b) also showed that the accumulation of copper with age was dependent on genetic factors when they investigated the concentration of copper in different stains of mice. In their experiments they showed that copper accumulates more rapidly in the senescence-accelerated mouse. These observations suggest that the concentration of copper is dependent upon genetics, age, and a supply from the environment. It is also possible that within the cell the distribution of copper is dependent upon similar factors. A summary of changes in the subcellular distribution of copper in liver by Owen et al. (1977) indicated that the copper subcellular distribution was, in fact, dependent upon whether there was a copper-overload or a copper-deficient state. If the same is true for the CNS, then the absolute level of copper in the cell may partition differently depending upon the level present.

Other factors can influence the pool size of copper in brain of otherwise healthy subjects. For example, it is well known that certain drugs accelerate the uptake of copper from the blood to the CNS. For example, diethyldithiocarbamate (DDC), a metabolite of the drug disulfiram (Antabuse), used for treatment of alcoholism, is known to chelate copper. DDC treatment also is known to increase the concentration of copper in the CNS, presumably by transport across the blood–brain barrier. The drug also is known to have an impact on the activity of known copper enzymes, therefore, this drug or other drugs that may influence the uptake of copper and/or distribution within the tissue can be a modulating factor in the role of copper as an antioxidant or prooxidant.

4.3. NEUROCHEMICAL FUNCTIONS OF COPPER

4.3.1. Cuproenzymes

Dogma holds that the majority of intracellular and extracellular copper is associated with specific proteins. In many cases these high-affinity ligands hold the copper in specific geometries and facilitate their participation in biological reactions. These biological reactions constitute the functions that we know for copper. These binding proteins and cuproenzymes are either intracellular proteins within the CNS or are present in the extracellular fluid bathing the blood–brain barrier and interstitium of the brain (Table 4.1).

Plasma contains amine oxidases, a group of proteins similar in physical properties, which are dimers of molecular weight of 150,000–200,000, containing one Type 2 copper per subunit. These enzymes are involved in oxidative deamination of mono- and di-amines, producing hydrogen peroxide as a product. Thus, these enzymes may have a role in producing outside of the cell precursors for ROS. These amine oxidases

Table 4.1
Cuproenzymes and Copper-Binding Proteins

Enzyme	Function
Amine oxidases (EC 1.4.3.6)	Oxidative deamination
Ceruloplasmin (EC 1.16.3.1)	Fe^{2+} oxidation
Cytochrome c oxidase (EC 1.9.3.1)	Mitochondrial electron transfer
Dopamine-β-monooxygenase (EC 1.14.17.1)	Norepinephrine synthesis
Lysyl oxidase (EC 1.4.3.13)	Elastin and collagen crosslinking
Metallothionein I, II, III	Cu^+ storage, antioxidant
Neurocuprein	Unknown
Peptidylglycine α-amidating monooxygenase (EC 1.14.17.3)	Neuropeptide modification
Cu,Zn-superoxide dismutase (EC 1.15.1.1)	Disproportionation of superoxide
Extracellular superoxide dismutase	Disproportionation of superoxide
Tyrosinase (EC 1.14.18.1)	Melanin synthesis

are different from the flavoprotein mitochondrial outer membrane monoamine oxidases involved in catecholamine metabolism within the CNS. In addition to copper these proteins contain an organic cofactor, probably 6-hydroxydopa (Janes *et al.*, 1990).

Also bathing the cerebral vasculature is a large concentration of the blue glycoprotein, ceruloplasmin. This protein exhibits ferroxidase activity, that is, the ability to rapidly oxidize ferrous iron to ferric iron. It is this function that classifies ceruloplasmin as an antioxidant protein. Ceruloplasmin is a 132,000 dalton, single polypeptide that is highly glycosylated and secreted into the blood plasma. It contains six moles of copper in each of the three types of geometric centers. There is also good evidence that ceruloplasmin may be found within the brain (Linder & Moor, 1977). Although recent evidence has questioned the function of ceruloplasmin as a transport protein for copper from blood into brain, this particular function remains to be investigated further. What does seem to be clear is that ceruloplasmin, by view of its ferroxidase activity, is an important antioxidant protein. Its distribution both outside and within the CNS places this protein among a group of enzymes involved in ROS.

One of the most abundant copper proteins in the central nervous system is the inner mitochondrial membrane complex, cytochrome *c* oxidase. This multisubunit polypeptide contains two subunits that each bind a copper atom and a molecule of heme *a*. This protein is involved in the transfer of electrons to molecular oxygen, producing water. Thus, it is important to maintain levels of cytochrome *c* oxidase for there not to be a leakage of electron flow into univalent reduction products of oxygen within the mitochondria. As much as one-fourth of the brain copper may be associated with this protein. It is known that copper is important for the enzymatic activity of this complex. However, copper is also known to be important for the production of cytochrome *a* (Prohaska and Wells, 1975). Recent studies have indicated that copper may also be involved in the expression of the cytochrome oxidase subunits. This appears to be true for several mammalian species (Liao *et al.*, 1995; Sparaco *et al.*, 1993). These observations all imply that copper status is important for maintenance of cytochrome *c* oxidase in the CNS. Recent work has also indicated that alteration of copper status during early development may have long-term consequences to the restoration of cytochrome *c* oxidase activity in the brain (Prohaska and Bailey, 1995a).

Another important intracellular cuproenzyme is dopamine-β-monooxygenase (DBM). DBM is a glycoprotein tetramer of molecular weight 290,000 that contains up to 8 moles of copper (Prohaska, 1987). DBM is highly localized in its distribution, unlike the constitutively and ubiquitously distributed cytochrome *c* oxidase molecule. The function of DBM is in noradrenergic neurons converting dopamine to norepinephrine. The high concentration of copper in the locus ceruleus, for example, is probably related to its association with DBM in that brain region. When the supply of environmental copper is limiting, the activity of the DBM is altered. In the CNS this was first demonstrated in copper-deficient rats in a reduction in the concentration of norepinephrine (Prohaska and Wells, 1974). Regional analysis of norepinephrine and dopamine in mouse and rat brain indicates, in general, that accompanying copper deficiency in early development there is a significant decrease in the concentration of norepinephrine in most brain regions with, again, the hypothalamus being somewhat spared, and a corresponding elevation in the precursor, dopamine, in most regions

(Prohaska and Bailey, 1994, 1993b). Measurement of the enzyme activity of DBM is somewhat complicated due to the possible presence of increased amounts of apoprotein (Prohaska and Smith, 1982).

Another rather curious consequence of copper deficiency during early development appears to be a selective loss of dopaminergic neurons in the corpus striatum (Miller and O'Dell, 1987; Feller and O'Dell, 1980). The explanation for this selective loss of dopamine in the corpus striatum is not currently known. However, the elevation of dopamine in most other brain regions may have some consequences, as an increased substrate for oxidation of catechols, which could lead to tissue damage. This point will be elaborated on later.

Lysyl oxidase is a protein similar in structure to the plasma amine oxidases and contains, in addition to copper, an organic cofactor. Its function is in oxidative deamination of peptidyl lysine in elastin and collagen. Lysyl oxidase is thus important in the laying down of basement membranes and in the integrity of the blood–brain barrier. Its activity is also known to be sensitive to copper status.

Brain, as well as other tissues, contains a group of related proteins collectively referred to as metallothioneins. These low-molecular-weight, high-cysteine-containing proteins bind a variety of different metals, including copper. Function of these metallothioneins is still somewhat disputed (Bremner and Beattie, 1990). There are three different classes of metallothioneins that are expressed in the CNS. Metallothionein I and II consist of a series of different gene products that seem to be regulated by a number of different inducers. MT-I is higher in glial compared to neuronal cells and is induced by Zn^{2+}, Cu^{2+}, and glucocorticoids (Hidalgo et al., 1994). Metallothionein III is a related protein whose distribution is rather unique to the CNS. This protein may be deficient in Alzheimer's disease brain and has a biological function in inhibiting the growth of neuronal cells (Uchida et al., 1991). Some suggest that metallothionein also is involved as an antioxidant protein, and thus, is another example of a copper-binding protein involved in protection against ROS. Metallothionein is a small protein, molecular weight approximately 6,000, that has the potential to bind multiple atoms of copper per mole of protein but, in fact, may be a zinc-binding protein in vivo.

Another low molecular weight protein isolated from brain with a molecular weight of 9,000 is called neurocuprein. Properties of this acidic protein, its discovery, and putative functions have been described elsewhere (Prohaska, 1987).

Peptidylglycine-α-amidating monooxygenase (PAM) is a monooxygenase dependent upon both ascorbate and copper for enzyme activity (Eipper and Mains, 1988). PAM is responsible for posttranslational modification of a number of precursor neuropeptides that contain glycine at the C-terminus. PAM converts these peptides into corresponding α-amides with release of glycine carbons as glyoxylate (Bradbury and Smyth, 1991). Notable α-amidated peptides include cholecystokinin, oxytocin, vasopressin, neuropeptide Y, vasoactive intestinal peptide, substance P, and calcitonin. If changes in PAM activity limited the synthesis of these important peptides, major changes in physiological function would be predicted to occur. The catalytic domain of PAM is very similar to DBM. The copper, which seems to exchange rather rapidly, is involved in the α-hydroxylation of the precursor peptide. A second domain on PAM is responsible for splitting of the peptide and a release of the two carbons. This bifunctional protein, PAM, is widely expressed in the central nervous system of both rats and sheep, with highest activities found in the hypothalamus and very low activities in the cerebellum (Lew *et al.*, 1992; Schafer *et al.*, 1992). Alteration of copper status changes the PAM enzyme activity in experimental rodent models (Prohaska and Bailey, 1995b; Prohaska *et al.*, 1995; Mains *et al.*, 1985). Consequences of these changes in enzyme activity, however, have not been documented. It is also not clear whether or not changes in PAM activity may play a role in antioxidant or prooxidant functions of copper.

Neuronal and glial cells express two distinct proteins that catalyze the disproportionation of superoxide (O_2^-), the univalent reduction product of dioxygen. A mitochondrial matrix protein utilizes manganese as a cofactor. A cytoplasmic form is a homodimer of 32,000 molecular weight with each subunit containing a mole of copper and zinc. Cu,Zn-superoxide dismutase (SOD) has been widely studied in regards to its antioxidant function in the CNS. The discovery of the enzymatic activity of this protein, originally isolated in 1957 by Porter and Folch, and of some of the associations of this enzyme with various neurological diseases, has been reviewed previously (Prohaska, 1987). This historical information will not be reviewed here, but some more recent research on SOD and antioxidant function will be described later. Changes in copper status are known to alter the enzyme activity of SOD in the CNS (Prohaska and Wells, 1974). Many other factors, including genetic expression, also have an impact on SOD levels and thus antioxidant homeostasis in the brain. This is especially true for Down's syndrome, in which the overexpression of SOD, because of the three copies of chromosome 21, is responsible for a 50% increase in the SOD activity of cells. Another genetic disease, amyotrophic lateral sclerosis (ALS), sometimes called Lou Gehrig's disease, is associated with a mutation of the human SOD gene. Therefore, there is much interest in SOD as a candidate molecule for explaining some of the neuropathology associated with these two diseases (Chapter 15).

Anchorage-dependent cells secrete in the extracellular space a protein referred to as extracellular superoxide dismutase. This protein is also copper binding, having dismutase activity immunologically distinct from the intracellular SOD (Marklund, 1990). Extracellular SOD may have a function as an antioxidant in the milieu outside the cell. However, other studies indicate that the extracellular superoxide protects against oxygen-dependent CNS damage from nitric oxide (Oury *et al.*, 1992).

Brain also has pigment-containing cells that likely contain the copper protein tyrosinase, a Type III copper enzyme. These pigment-forming cells enriched in the substantia nigra probably account for the high copper concentration in this brain region as well as the presence of pigment. It is not known if tyrosinase has any role in antioxidant or prooxidant functions of copper.

4.3.2. Copper Complexes

Although many of the functions of copper can be ascribed to its catalytic role in the cuproenzymes mentioned above, there may be other functions of copper in the brain that are dependent upon complexes of copper released from intracellular stores. Barnea *et al.* (1989) have investigated, over the years, the hypothesis that copper is released during exocytosis, and that this released copper has a physiological function in

the subsequent stimulation of specific hormone-releasing hormones from the hypothalamus. This is an example of one type of copper function that is not directly
correlated with a known cuproenzyme. There is a substantial amount of copper in the
brain that is not yet accounted for in the cuproenzymes that were described above,
undoubtedly accounting for other functions of copper yet to be discovered. Because
copper has such a high affinity for amino acids like histidine and cysteine, it is unlikely
that free copper ion exists to any extent uncomplexed to either amino acids, small
peptides, or specific proteins. However, these complexes may have important biological activity. An example is the βA4 amyloid precursor protein, which has a high
binding capacity for copper (Hesse *et al.*, 1994).

4.3.3. Prooxidant Properties

The role of transition metals like copper in oxygen-radical reactions has been
thoroughly reviewed by multiple authors. An excellent review discussing the reactions
involved was published by Aust *et al.* (1985). If copper ions are to have a role in the
generation of ROS, two criteria must be met. First, the copper ion must be reduced.
This reducing power could come from reductants such as ascorbate or superoxide anion
itself (Rowley and Halliwell, 1983). The reduction of cupric ion by superoxide anion is
very favorable thermodynamically (see Eq. 3, Table 4.2). The copper ion that is
liberated secondly must be stabilized or solubilized, thus, low-molecular-weight peptides, or amino acids like histidine, forming soluble complexes of copper are necessary
for this type of chemistry to occur (Simpson *et al.*, 1988). Although the reduction of
ferric iron by superoxide anion is also a favorable process thermodynamically (Table
4.2), the production of the highly-reactive hydroxyl radical is actually thermodynamically more favorable with cuprous ion as an electron donor than ferrous iron
(compare Eqs. 1 and 2, Table 4.2). Thus, copper ions, like iron ions, can participate in
one-electron-redox chemistry under conditions that exist in the cell with an adequate
supply of oxygen. Traces of ionic copper can have quite an impact on physiological
processes. For example, experiments done with isolated perfused hearts showed that
traces of copper in the buffer had a major impact on the mechanical properties of that

Table 4.2
Thermodynamic Values

Reaction	E'° (V)	$\Delta G'^{\circ}$ (kJ/mole)	Eq.
$Cu^{2+} + e^{-} \rightarrow Cu^{+}$	+0.16		
$Fe^{3+} + e^{-} \rightarrow Fe^{2+}$	+0.77		
$O_2 + e^{-} \rightarrow O_2^{-}$	−0.33		
$H_2O_2 + H^{+} + e^{-} \rightarrow HO\cdot + H_2O$	+0.38		
$H_2O_2 + H^{+} + Cu^{+} \rightleftarrows HO\cdot + H_2O + Cu^{2+}$		−21.2	(1)
$H_2O_2 + H^{+} + Fe^{2+} \rightleftarrows HO\cdot + H_2O + Fe^{3+}$		+37.6	(2)
$Cu^{2+} + O_2^{-} \rightleftarrows Cu^{+} + O_2$		−47.3	(3)
$Fe^{3+} + O_2^{-} \rightleftarrows Fe^{2+} + O_2$		−106	(4)

organ (Powell and Wapnir, 1994). One would predict that if free ionic copper were available in the CNS it would act as a very powerful prooxidant and lead to neurological damage through generation of reactive oxygen species, such as the hydroxyl radical. The question is whether or not free copper ion could be released from storage proteins or cuproenzymes under certain conditions since binding of copper to these proteins and complexes is so tight. Second, it is not known what type of redox center might exist in these complexes of copper as compared to the aqueous reduction potential (Table 4.2). Nevertheless, evidence points to a role for copper ion as a prooxidant under certain neurological conditions.

4.3.4. Antioxidant Roles

Evidence that copper has an important antioxidant function has been thoroughly reviewed recently by Johnson *et al.* (1992). The hypotheses that discuss the generation of hydroxyl radical usually involve a reaction in which ferrous iron reacts with hydrogen peroxide (Table 4.2, Eq. 2). Superoxide anion is believed to play a role in these processes since it is an excellent reducing agent (Table 4.2). If the concentration of superoxide anion were exceedingly small, it would not be available to perform this function. This is believed to be the antioxidant role of SOD. Also, if ferrous iron were not available to catalyze the Fenton reaction, the generation of hydroxyl radical would also be limited. Ceruloplasmin (ferroxidase) is believed to provide this antioxidant function by quickly oxidizing any ferrous iron to ferric iron (without formation of ROS). Because of the abundant concentration of cysteine in the metallothionein molecule, it is believed to act, perhaps as an antioxidant, through stabilization of free radicals. These three copper-binding proteins, namely SOD, ceruloplasmin, and metallothionein, are examples of potential copper-dependent antioxidant proteins. Whether or not copper can secondarily influence the expression and activity of other antioxidant enzymes is somewhat questionable. This would include manganese-dependent SOD, catalase, and the selenoenzyme, glutathione peroxidase. The latter three proteins all play a role in antioxidant defense in tissues, but whether copper has any direct influence on these proteins in the central nervous system is not known.

Another factor that potentially has an impact on the antioxidant properties would be the redox environment of the copper-binding protein centers. If the copper in the centers is more difficult to reduce, it may not be able to accept an electron from the ascorbate or superoxide electron donor to generate reduced forms of copper, thus, depending on the nature of the copper ligand versus low-molecular-weight or free copper ions, the transition metal can act as a prooxidant or antioxidant.

4.4. COPPER DEFICIENCY

4.4.1. Dietary Copper Deficiency

The most compelling evidence that dietary copper deficiency is associated with altered antioxidant function are studies in experimental rats that demonstrated that copper-deficient rats produced more ethane, a gas formed following lipid peroxidation of (n-3) fatty acids (Saari *et al.,* 1990; Lawrence and Jenkinson, 1987). When rats where further challenged by administration of carbon tetrachloride, the copper-deficient animals had a much greater ethane evolution, indicating that lipid peroxidation induced by carbon tetrachloride was increased in the copper deficiency (Lawrence and Jenkinson, 1987). The question of whether the CNS of copper-deficient rats or other mammals have lipid peroxidation or oxidative damage occurring when copper status is limiting is much more difficult to document. Prohaska and Wells (1975) found no evidence for enhanced lipid peroxidation in the brains of copper-deficient rats. The brains of these same animals had a decreased concentration of myelin; however, this was likely due to a delay in developmental synthesis rather than an enhanced breakdown. Analysis of the fatty acids from phospholipids of brains of copper-deficient rats similarly did not support rampant lipid peroxidation (Sun and O'Dell, 1992a). In these experiments an increase in linoleic acid (18:2 n-6) was observed. Myelin concentration was again significantly decreased. There has been somewhat of a controversy in brains of copper-deficient mammals as to whether the myelin changes are due to a reduction in biosynthesis (hypomyelination) or an enhanced breakdown (demyelination). Studies by Komoly *et al.* (1992) and others have shown that young mice treated with the copper chelator, cuprizone, produce a pronounced demyelination in the CNS. Therefore, the issue of copper complexation versus decreased copper concentration in the CNS and myelin needs further study. Focal loss of neurons in the copper-deficient brain is known to occur accompanying dietary copper deficiency during perinatal development (Sun and O'Dell, 1992b). The cause of the loss of these neurons in the corpus striatum is not known but could be related to altered antioxidant status and enhanced degradation of these sensitive dopaminergic neurons.

Limitation of copper to the CNS might result in altered antioxidant status through secondary mechanisms. In the mouse, copper deficiency is associated with decreased concentrations of ascorbate in the brain. This phenomena is rapidly reversed by injection of cupric chloride but not sodium chloride (Prohaska and Cox, 1983). The observation that ascorbic acid is lower in the brain of copper-deficient mice was not confirmed in a recent study of copper-deficient rats (Kubat and Prohaska, 1996). Therefore, it is not known whether copper status affects brain ascorbate levels across species.

Another important metabolite that could have an impact on antioxidant status is the tripeptide, glutathione. In copper-deficient liver there is a pronounced elevation of reduced glutathione (Allen *et al.,* 1988). It is not known if brain glutathione changes accompanying dietary copper deficiency. Another indirect way in which dietary copper deficiency might impact antioxidant status is that the elevation of dopamine in most brain regions might provide a reactant that could autooxidize. Transition metals like iron, manganese, and copper can accelerate catechol-mediated protein crosslinking known to be associated with Lewy bodies (Montine *et al.,* 1995).

4.4.2. Genetic Copper Deficiency

In humans, the genetic condition Menkes' disease is associated with neuropathological lesions, which are similar to experimental animals with copper deficiency (Danks *et al.,* 1972). Defective efflux mechanisms, normally catalyzed by a copper

ATPase, results in the accumulation of copper in certain tissues and deficiency of copper in the CNS. Brains are characterized by decreases in cytochrome c oxidase and DBM activity. There is also evidence for enhanced lipid peroxidation (French et al., 1972). This might be due to an impairment in Cu,Zn-SOD activity, but this remains to be established. Mutations in mice that are analogous to humans have also been used as models for Menkes' disease. Brains of brindled mice are also characterized by copper deficiency. Many researchers have contributed to this field. Some of that information is summarized elsewhere (Prohaska, 1987). Although evidence for less myelination is present, brains of these mice do not appear to have evidence of lipid peroxidation. However, there is one report in heterozygous females with this mutation that suggests an accumulation of lipofusion granules in certain brain regions (Tanaka et al., 1990).

Whether changes in the CNS following dietary copper deficiency in rats and in piglets (Pletcher and Banting, 1983), which include spongy myelopathy, and changes observed in ataxic sheep born to copper-deficient ewes, are due to changes in antioxidant status is yet to be established. Evidence of whole animal lipid peroxidation occurring during dietary copper deficiency is solid.

4.5. COPPER TOXICITY

4.5.1. CNS Toxicity

Acute mortality is observed when copper sulfate is injected into rats that are deficient in vitamin E and selenium (Dougherty and Hoekstra, 1982). Ethane evolution increased sixfold. This suggests enhanced lipid peroxidation due to copper toxicity in this animal paradigm. Vitamin E also has a modifying affect on the toxicity of excess dietary copper (Sokol et al., 1990). Evidence of in vivo peroxidation of membrane lipids in mitochondria in these copper-overloaded rats was evident. It is possible that if copper was elevated to the same extent in brain as in the liver, increased lipid peroxidation due to copper toxicity would be observed. Not only would there be evidence of lipid peroxidation in these situations but evidence of site-specific oxygen radical damage to DNA catalyzed by copper complexes (Gutteridge, 1984). Thus, copper toxicity might induce a number of radical-based reactions if the tissue copper levels were sufficiently high. Some of the increased copper in the brain might likely induce metallothionein and sequester some of the incoming copper flux. It is known that brain metallothionein I and II are induced by copper and other factors (Gasull et al., 1994). Chronic copper toxicity in sheep is characterized by glial and astrocytic vacuole formation (Howell et al., 1974). In sheep, liver accumulation of copper chronically occurs until it spills into the blood in a hemolytic crisis. The toxicity associated with this toxicity to the CNS may be due directly to the accumulation of copper in the late stages of the disease or secondarily due to humoral factors as a result of the liver poisoning. From the classic work of Peters it is known that one of the lesions of ionic copper on nervous tissue is an inhibition of the Na^+,K^+-ATPase (Donaldson et al., 1971). Thus, the toxicity of copper to the central nervous system may or may not involve an oxygen-dependent scenario. The chelator, diethyldithiocarbamate, when injected into rats, is associated with CNS neurotoxicity and an increased concentration of copper in the brain (Allain and Krari, 1991). In follow-up studies, increased malondialdehyde levels were measured in hippocampus and cerebellum of DDC-intoxicated rats and the correlation of increased copper and increased malondialdehyde production was evident (Delmaestro and Trombetta, 1995). This suggests that if ionic copper, either in a free or chelated form, reaches the CNS, lipid peroxidation can be enhanced. Although this situation may be rare, it nevertheless is potentially deleterious to neurological function.

4.5.2. Genetic Diseases

The most well-known genetic disease involving copper excess is the autosomal-inherited Wilson's disease. Recently it has been shown that a mutation of chromosome 13 in humans is responsible for defective production of a gene product that has high homology with the Menkes' disease gene and is involved in intracellular copper transport (Tanzi *et al.,* 1993). Wilson's disease is characterized by a chronic accumulation of hepatic copper. When this organ becomes saturated, copper is released into the blood and can accumulate in other tissues, such as the cornea and the brain, leading to neurological Wilson's disease. There is a marked elevation in brain copper concentration (Walshe and Gibbs, 1987). There is also a significant increase in copper concentration in cerebral spinal fluid (Kodama *et al.,* 1988). In the brain, copper accumulates predominantly in basal ganglia and subthalmic nuclei. The corpus striatum is pigmented and microscopically there is cavitation and loss of neurons. Copper does accumulate in other areas but the focal lesions appear to be in these regions (Alt *et al.,* 1990). Neuropathological descriptions of cases of Wilson's disease vary greatly, thus, it is difficult to summarize on particular lesions. For example, there are some cases where there are major cavitation of the cerebral white matter with loss of myelin (Ishino *et al.,* 1972). Others have described a type of astrocytic hyperplasia similar to certain types of Alzheimer's disease (Anzil *et al.,* 1974). It seems clear in the liver that the excess copper that accumulates results in oxidative damage in a free radical-type mechanism to the mitochondria (Sokol *et al.,* 1994). Whether oxidative damage is also involved in the neuropathology of Wilson's disease is not clearly established. The increased copper concentration in the brains of those afflicted by Wilson's disease may not reach a critical level in order to accelerate Fenton-type chemical reactions.

Altered brain antioxidative mechanisms may be involved in other genetically-related diseases in humans. Two such diseases, Alzheimer's disease and Parkinson's disease, have been widely described in the literature. The basal ganglia of Parkinsonian patients appear to have altered iron levels but may, in fact, have decreased copper content (Dexter *et al.,* 1991). Whether the decrease in copper is associated with an altered antioxidant status is not known. The cerebral spinal fluid (CSF) of Parkinson's disease patients was assayed for the presence of ceruloplasmin and found to be unaltered compared to controls (Loeffler *et al.,* 1994). The ceruloplasmin levels may correlate with brain copper levels and, therefore, suggest no major changes in Parkinsonian disease patients. The same paper detected an increase in ceruloplasmin in the CNS of Alzheimer's disease patients compared to aged, normal subjects. This is in contrast to the report by Connor *et al.* (1993) that suggested a loss of one-third of the ceruloplasmin in both the gray and white matter of the superior temporal gyrus in Alzheimer's disease brains compared to controls. Further research will be required to determine whether or not brain ceruloplasmin has any role in these neurodegenerative processes.

Recently, papers have been published describing mutations in the ceruloplasmin gene resulting in a condition in humans in which ceruloplasmin protein is not produced. The consequences of this gene mutation appear to be primarily related to alterations in iron rather than in copper metabolism, raising the question of the role of ceruloplasmin in copper transport. For example, liver copper levels are approximately normal as are brain copper levels in patients with this particular genetic defect. Patients with this disease do develop a severe hemosiderosis (Morita *et al.*, 1995). Analysis from biopsy material of the copper concentrations in the liver indicate an approximate threefold increase, whereas the iron increase in the same specimens is 1000-fold elevated. In brain the iron accumulates primarily in basal ganglia and cerebellum. Although there is widespread iron deposition, the necrosis is primarily located in the putamen and caudate nucleus. There are also retinal lesions in these patients similar to those described in Hallervorden–Spatz disease. These patients demonstrate the essentiality for ceruloplasmin in humans and identifies at least one function for this protein in normal iron metabolism. Indirectly, a mutation in a copper-binding protein has an impact on neurological disease.

There are a number of mutations in animals that also involve accumulations of excess copper in tissues. One mutation, the LEC rat, is analogous to the mutation in Wilson's disease, involving the same gene product (Yamaguchi *et al.*, 1994). Analogous to Wilson's disease there is a large hepatic accumulation of copper later on in the disease process. Brain copper, which is initially low, increases in seven of eight different brain regions studied (Sugawara *et al.*, 1992). In confirmation of these results, Sato *et al.* (1994) have shown decreased copper in most of the regions compared to normal rats. Therefore, it may be that the Wilson's disease gene product is involved in normal copper metabolism in the brain, and when mutated decreased copper levels are present early on. After the liver is saturated with copper, neurotoxicity is associated with the LEC mutant. Expression of the LEC mutant is associated with tissue damage involving oxidative metabolism (Yamamoto *et al.*, 1993). In these experiments, 8-hydroxy-deoxy-guanosine levels in DNA were elevated in brain, kidney, and especially liver, indicating that oxygen radicals in the presence of excess copper can damage DNA. This is strong evidence that copper toxicity can generate oxygen radicals when tissue levels are very high.

There are some other inherited disorders involving copper toxicity. Bedlington terriers accumulate large amounts of copper in their livers, however, there does not appear to be a major spillover to the CNS (Su *et al.*, 1982). This phenotype appears to be different from the LEC rat or Wilson's disease. Although there is a 30-fold increase in the concentration of copper in liver, there are no changes in copper in the cerebellum or cerebrum. The mutation in mice, toxic-milk, involves accumulation of liver copper in the homozygous animals of approximately 50-fold (Howell and Mercer, 1994). In the same animals the concentration of copper in brain did not double. Therefore, neurological problems associated with the toxic-milk mouse are not particularly evident. There have been other mutations in the mouse ascribed to alterations in copper homeostasis, namely, crinkled and quaking; however, these results have been questioned and are discussed elsewhere (Prohaska, 1987). Undoubtedly other mutations exist in which copper accumulates in various animals. There have been reported instances of abnormal copper storage in teleost fish and in the wild mute swan. Neither of these reportedly has CNS abnormalities and will not be discussed further.

4.6. Cu,Zn-SUPEROXIDE DISMUTASE

Many neurological diseases of humans have been implicated to changes in antioxidant status in particular, emphasizing changes in Cu,Zn-SOD. These include Down's syndrome, ALS, Parkinson's disease, and Alzheimer's disease. In the case of ALS, a subclass of this disease has been shown to be related to a gene defect in the Cu,Zn-SOD molecule (Rosen *et al.*, 1993). Down's syndrome overexpresses Cu,Zn-SOD due to an extra copy of chromosome 21. Parkinson's disease and Alzheimer's disease have implicated metal centers and abnormal antioxidant status in their etiology as well. Recently, Furuta *et al.* (1995) have shown abnormal distribution in Alzheimer's disease and Down's syndrome of both copper-zinc and manganese SOD in brain. Superoxide dismutases have also been implicated in the etiology of damage following anoxia and ischemia. Oxygen radical mechanisms appear to be involved in the brain injury following these conditions (Traystman *et al.*, 1991). Neuronal cells may even respond to oxidative stress and ischemic by inducing more superoxide dismutase to protect themselves (Ohtsuki *et al.*, 1993). These implicated roles for SOD as an antioxidant or prooxidant in the case of Down's syndrome has led scientists to investigate this hypothesis further.

The strategy utilizes transgenic mice. Transgenic mice overexpressing Cu,Zn-SOD increase two- to three-fold the amount of this enzyme across the various brain regions. Modest changes in some other antioxidant enzymes also are known to occur (Przedborski *et al.*, 1992). Mice overexpressing this human gene were resistant to neurotoxicity from glutamate as well as to reperfusion injury after cerebral ischemia (Yang *et al.*, 1994; Chan *et al.*, 1990). These transgenic experiments suggest that Cu,Zn-SOD plays an important role as an antioxidant in nervous tissue and confirms speculations from other experiments and other genetic diseases. However, others have shown that brains of transgenic mice overexpressing SOD have enhanced basal lipid peroxidation (Ceballos-Picot *et al.*, 1992).

4.7. SUMMARY

Copper has many known neurochemical functions. These functions generally involve an association with a specific protein ligand and participation in some sort of enzyme-catalyzed reaction. Copper may also have a role in nonenzyme-mediated release of various humoral factors from the CNS. Some of these copper-dependent enzymes catalyze reactions that utilize molecular oxygen or oxygen metabolites as substrates, therefore, these copper-dependent enzymes can be considered antioxidant proteins. However, traces of inorganic copper can readily react with oxygen radicals to generate other reactive species that can cause focal necrosis or DNA damage. Evidence for altered antioxidant status due to dietary copper deficiency is apparent from both dietary and genetic experiments. There is also evidence for excess copper accumulation either from environmental exposure or due to genetic factors such as Wilson's disease and the LEC rat. Experiments with transgenic animals indicates an important role for Cu,Zn-superoxide dismutase as an antioxidant protein in the CNS. Future research will be required to elucidate further roles for copper as a protective agent for the brain or as a potential prooxidant in tissue damage.

ACKNOWLEDGMENT.   The skillful editorial and secretarial assistance of Doreen Fleetwood is appreciated.

4.8.   REFERENCES

Allain, P., and Krari, N., 1991, Diethyldithiocarbamate, copper and neurological disorders, *Life Sci.* **48**:291–299.

Allen, K. G. D., Arthur, J. R., Morrice, P. C., Nicol, F., and Mills, C. F., 1988, Copper deficiency and tissue glutathione concentration in the rat, *Proc. Soc. Exp. Biol. Med.* **187**:38–43.

Alt, E. R., Sternlieb, I., and Goldfischer, S., 1990, The cytopathology of metal overload, *Int. Rev. Exp. Pathol.* **31**:165–188.

Anzil, A. P., Herrlinger, H., Blinzinger, K., and Heldrich, A., 1974, Ultrastructure of brain and nerve biopsy tissue in Wilson disease, *Arch. Neurol.* **31**:94–100.

Aust, S. D., Morehouse, L. A., and Thomas, C. E., 1985, Role of metals in oxygen radical reactions, *J. Free Radicals Biol. Med.* **1**:3–25.

Barnea, A., Hartter, D. E., and Cho, G., 1989, High-affinity uptake of ^{67}Cu into a veratridine-releasable pool in brain tissue, *Am. J. Physiol.* **257**:C315–C322.

Bradbury, A. F., and Smyth, D. G., 1991, Peptide amidation, *Trends Biochem. Sci.* **16**:112–115.

Bremner, I., and Beattie, J. H., 1990, Metallothionein and the trace minerals, *Ann. Rev. Nutr.* **10**:63–83.

Chan, P. H., Chu, L., Chen, S. F., Carlson, E. J., and Epstein, C. J., 1990, Reduced neurotoxicity in transgenic mice overexpressing human copper-zinc-superoxide dismutase, *Stroke* **2**:III-80–III-82.

Ceballos-Picot, I., Nicole, A., Clément, M., Bourre, J.-M., and Sinet, P.-M., 1992, Age-related changes in antioxidant enzymes and lipid peroxidation in brains of control and transgenic mice overexpressing copper-zinc superoxide dismutase, *Mut. Res.* **275**:281–293.

Connor, J. R., Tucker, P., Johnson, M., and Snyder, B., 1993, Ceruloplasmin levels in the human superior temporal gyrus in aging and Alzheimer's disease, *Neurosci. Lett.* **159**(1–2):88–90.

Danks, D. M., Campbell, P. E., Stevens, B. J., Mayne, V., and Cartwright, E., 1972, Menkes' kinky hair syndrome, *Pediatrics* **50**:188–201.

Davis, G. K., and Mertz, W., 1987, Copper, in: *Trace Elements in Human and Animal Nutrition,* Volume 1 (W. Mertz, ed.), Academic Press, New York, pp. 301–364.

Delmaestro, E., and Trombetta, L. D., 1995, The effects of disulfiram on the hippocampus and cerebellum of the rat brain: A study on oxidative stress, *Toxicol. Lett.* **75**:235–243.

Deuschle, U., and Weser, U., 1985, Copper and inflammation, in: *Progress in Clinical Biochemistry and Medicine,* Volume 2 (E. Baulieu, D. T. Forman, L. Jaenicke, J. A. Kellen, Y. Nagai, G. F. Springer, L. Träger, L. Will-Shahab, and J. L. Wittliff, eds.), Springer-Verlag, New York, pp. 97–130.

Dexter, D. T., Carayon, A., Javoy-Agid, F., Agid, Y., Wells, F. R., Daniel, S. E., Lees, A. J., Jenner, P., and Marsden, C. D., 1991, Alterations in the levels of iron, ferritin and other trace metals in Parkinson's disease and other neurodegenerative diseases affecting the basal ganglia, *Brain* **114**:1953–1975.

Donaldson, J., St-Pierre, T., Minnich, J., and Barbeau, A., 1971, Seizures in rats associated with divalent cation inhibition of Na⁺-K⁺-ATP'ase, *Can. J. Biochem.* **49:**1217–1224.

Dougherty, J. J., and Hoekstra, W. G., 1982, Effects of vitamin E and selenium on copper-induced lipid peroxidation *in vivo* and on acute copper toxicity, *Proc. Soc. Exp. Biol. Med.* **169:**201–208.

Eipper, B. A., and Mains, R. E., 1988, Peptide α-amidation, *Ann. Rev. Physiol.* **50:**333–344.

Feller, D. J., and O'Dell, B. L., 1980, Dopamine and norepinephrine in discrete areas of the copper-deficient rat brain, *J. Neurochem.* **34:**1259–1263.

French, J. H., Sherard, E. S., Lubell, H., Brotz, M., and Moore, C. L., 1972, Trichopoliodystrophy. Report of a case and biochemical studies, *Arch. Neurol.* **26:**229–244.

Furuta, A., Price, D. L., Pardo, C. A., Troncoso, J. C., Xu, Z. S., Taniguchi, N., and Martin, L. J., 1995, Localization of superoxide dismutases in Alzheimer's disease and Down's syndrome neocortex and hippocampus, *Am. J. Pathol.* **2:**357–367.

Gasull, T., Giralt, M., Hernandez, J., Martinez, P., Bremner, I., and Hidalgo, J., 1994, Regulation of metallothionein concentrations in rat brain: Effect of glucocorticoids, zinc, copper, and endotoxin, *Am. J. Physiol.* **266:**E760–E767.

Gutteridge, J. M. C., 1984, Copper-phenanthroline-induced site-specific oxygen-radical damage to DNA, *Biochem. J.* **218:**983–985.

Halliwell, B., 1992, Reactive oxygen species and the central nervous system, *J. Neurochem.* **59:**1609–1623.

Hesse, L., Beher, D., Masters, C. L., and Multhaup, G., 1994, The βA4 amyloid precursor protein binding to copper, *FEBS Lett.* **349:**109–116.

Hidalgo, J., García, A., Oliva, A. M., Giralt, M., Gasull, T., González, B., Milnerowicz, H., Wood, A., and Bremner, I., 1994, Effect of zinc, copper and glucocorticoids on metallothionein levels of cultured neurons and astrocytes from rat brain, *Chem. Biol. Interact.* **93:**197–219.

Howell, J. McC., and Mercer, J. F. B., 1994, The pathology and trace element status of the toxic milk mutant mouse, *J. Comp. Pathol.* **110:**37–47.

Howell, J. McC., Blakemore, W. F., Gopinath, C., Hall, G. A., and Parker, J. H., 1974, Chronic copper poisoning and changes in the central nervous system of sheep, *Acta Neuropath.* **29:**9–24.

Ishino, H., Mii, T., Hayashi, Y., Saito, A., and Otsuki, S., 1972, A case of Wilson's disease with enormous cavity formation of cerebral white matter, *Neurology* **22:**905–909.

Janes, S. M., Mu, D., Wemmer, D., Smith, A. J., Kaur, S., Maltby, D., Burlingame, A. L., and Klinman, J. P., 1990, A new redox cofactor in eukaryotic enzymes: 6-hydroxydopa at the active site of bovine serum amine oxidase, *Science* **248:**981–987.

Johnson, M. A., Fischer, J. G., and Kays, S. E., 1992, Is copper an antioxidant nutrient?, *Crit. Rev. Food Sci. Nutr.* **32:**1–31.

Kaler, S. G., Goldstein, D. S., Holmes, C., Salerno, J. A., and Gahl, W. A., 1993, Plasma and cerebrospinal fluid neurochemical pattern in Menkes disease, *Ann. Neurol.* **33:**171–175.

Kodama, H., Okabe, I., Yanagisawa, M., Nomiyama, H., Nomiyama, K., Nose, O., and Kamoshita, S., 1988, Does CSF copper level in Wilson disease reflect copper accumulation in the brain?, *Pediatr. Neurol.* **4:**35–37.

Komoly, S., Hudson, L. D., Webster, H.DeF., and Bondy, C. A., 1992, Insulin-like growth factor I gene expression is induced in astrocytes during experimental demyelination, *Proc. Natl. Acad. Sci. USA* **89:**1894–1898.

Kubat, W. D., and Prohaska, J. R., 1996, Copper status and ascorbic acid concentrations in rats, *Nutr. Res.* **16:**237–243.

Lawrence, R. A., and Jenkinson, S. G., 1987, Effects of copper deficiency on carbon tetrachloride-induced lipid peroxidation, *J. Lab. Clin. Med.* **109:**134–140.

Levenson, C. W., and Janghorbani, M., 1994, Long-term measurement of organ copper turnover in rats by continuous feeding of a stable isotope, *Anal. Biochem.* **221:**243–249.

Lew, R. A., Clarke, I. J., and Smith, A. I., 1992, Distribution and characterization of peptidylglycine α-amidating monooxygenase activity in the ovine brain and hypothalamo-pituitary axis, *Endocrinology* **130:**994–1000.

Liao, Z., Medeiros, D. M., McCune, S. A., and Prochaska, L. J., 1995, Cardiac levels of fibronectin, laminin, isomyosins, and cytochrome c oxidase on weanling rats are more vulnerable to copper deficiency than those of postweanling rats, *J. Nutr. Biochem.* **6:**385–391.

Linder, M. C., 1991, *Biochemistry of Copper,* Plenum Press, New York.

Linder, M. C., and Moor, J. R., 1977, Plasma ceruloplasmin: Evidence for its presence in and uptake by heart and other organs of the rat, *Biochim. Biophys. Acta* **499:**329–336.

Loeffler, D. A., DeMaggio, A. J., Juneau, P. L., Brickman, C. M., Mashour, G. A., Finkelman, J. H., Pomara, N., and LeWitt, P. A., 1994, Ceruloplasmin is increased in cerebrospinal fluid in Alzheimer's disease but not Parkinson's disease, *Alzheimer Disease and Associated Disorders* **8:**190–197.

Mains, R. E., Myers, A. C., and Eipper, B. A., 1985, Hormonal, drug, and dietary factors affecting peptidyl glycine α-amidating monooxygenase activity in various tissues of the adult male rat, *Endocrinology* **116:**2505–2515.

Marklund, S. L., 1990, Expression of extracellular superoxide dismutase by human cell lines, *Biochem. J.* **266:**213–219.

Miller, D. S., and O'Dell, B. L., 1987, Milk and casein-based diets for the study of brain catecholamines in copper-deficient rats, *J. Nutr.* **117:**1890–1897.

Montine, T. J., Farris, D. B., and Graham, D. G., 1995, Covalent crosslinking of neurofilament proteins by oxidized catechols as a potential mechanism of Lewy body formation, *J. Neuropathol. Exp. Neurol.* **54:**311–319.

Morita, A., Kimura, M., and Itokawa, Y., 1994a, Changes with age in the mineral status in brain of female SAMP1 and SAMR1, in: *The SAM Model of Senescence* (T. Takeda, ed.), Elsevier Science, New York, pp. 317–320.

Morita, A., Kimura, M., and Itokawa, Y., 1994b, The effect of aging on the mineral status of female mice, *Biol. Trace Elem. Res.* **42:**165–177.

Morita, H., Ikeda, S., Yamamoto, K., Morita, S., Yoshida, K., Nomoto, S., Kato, M., and Yanagisawa, N., 1995, Hereditary ceruloplasmin deficiency with hemosiderosis: A clinicopathological study of a Japanese family, *Ann. Neurol.* **37:**646–656.

O'Dell, B. L., and Prohaska, J. R., 1983, Biochemical aspects of copper deficiency in the nervous system, in: *Neurobiology of the Trace Elements,* Volume 1 (I. E. Dreosti, and R. M. Smith, eds.), Humana Press, Clifton, New Jersey, pp. 41–81.

Ohtsuki, T., Matsumoto, M., Suzuki, K., Taniguchi, N., and Kamada, T., 1993, Effect of transient forebrain ischemia on superoxide dismutases in gerbil hippocampus, *Brain Res.* **620:**305–309.

Oury, T. D., Ho, Y.-S., Piantadosi, C. A., and Crapo, J. D., 1992, Extracellular superoxide dismutase, nitric oxide, and central nervous system O_2 toxicity, *Proc. Natl. Acad. Sci. USA* **89:**9715–9719.

Owen, Jr., C. A., 1981, *Copper Deficiency and Toxicity,* Noyes Publications, Park Ridge, New Jersey.

Owen, Jr., C. A., Dickson, E. R., Goldstein, N. P., Baggenstoss, A. H., and McCall, J. T., 1977, Hepatic subcellular distribution of copper in primary biliary cirrhosis, *Mayo Clin. Proc.* **52:**73–80.

Pletcher, J. M., and Banting, L. F., 1983, Copper deficiency in piglets characterized by spongy myelopathy and degenerative lesions in the great blood vessels, *J. South Afr. Vet. Assoc.* **54:**43–46.

Powell, S. R., and Wapnir, R. A., 1994, Adventitious redox-active metals in Krebs–Henseleit buffer can contribute to Langendorff heart experimental results, *J. Mol. Cell. Cardiol.* **26:**769–778.

Prohaska, J. R., 1987, Functions of trace elements in brain metabolism, *Physiol. Rev.* **67:**858–901.

Prohaska, J. R., 1988, Biochemical functions of copper in animals, in: *Essential and Toxic Trace Elements in Human Health and Disease* (A. S. Prasad, ed.), Alan R. Liss, New York, pp. 105–124.

Prohaska, J. R., 1990, Biochemical changes in copper deficiency, *J. Nutr. Biochem.* **1:**452–461.

Prohaska, J. R., and Bailey, W. R., 1993a, Copper deficiency during neonatal development alters mouse brain catecholamine levels, *Nutr. Res.* **13:**331–338.

Prohaska, J. R., and Bailey, W. R., 1993b, Persistent regional changes in brain copper, cuproenzymes and catecholamines following perinatal copper deficiency in mice, *J. Nutr.* **123:**1226–1234.

Prohaska, J. R., and Bailey, W. R., 1994, Regional specificity in alterations of rat brain copper and catecholamines following perinatal copper deficiency, *J. Neurochem.* **63:**1551–1557.

Prohaska, J. R., and Bailey, W. R., 1995a, Persistent neurochemical changes following perinatal copper deficiency in rats, *J. Nutr. Biochem.* **6:**275–280.

Prohaska, J. R., and Bailey, W. R., 1995b, Alterations of rat brain peptidylglycine α-amidating monooxygenase and other cuproenzyme activities following perinatal copper deficiency, *Proc. Soc. Exp. Biol. Med.* **210:**107–116.

Prohaska, J. R., and Cox, D. A., 1983, Decreased brain ascorbate levels in copper-deficient mice and in brindled mice, *J. Nutr.* **113:**2623–2629.

Prohaska, J. R., and Smith, T. L., 1982, Effect of dietary or genetic copper deficiency on brain catecholamines, trace metals and enzymes in mice and rats, *J. Nutr.* **112:**1706–1717.

Prohaska, J. R., and Wells, W. W., 1974, Copper deficiency in the developing rat brain: A possible model for Menkes' steely-hair disease, *J. Neurochem.* **23:**91–98.

Prohaska, J. R., and Wells, W. W., 1975, Copper deficiency in the developing rat brain: Evidence for abnormal mitochondria, *J. Neurochem.* **25:**221–228.

Prohaska, J. R., Bailey, W. R., and Lear, P. M., 1995, Copper deficiency alters rat peptidylglycine α-amidating monooxygenase activity, *J. Nutr.* **125:**1447–1454.

Przedborski, S., Jackson-Lewis, V., Kostic, V., Carlson, E., Epstein, C. J., and Cadet, J. L., 1992, Superoxide dismutase, catalase, and glutathione peroxidase activities in copper/zinc-superoxide dismutase transgenic mice, *J. Neurochem.* **58:**1760–1767.

Rosen, D. R., Siddique, T., Patterson, D., Figlewicz, D. A., Sapp, P., Hentati, A., Donaldson, D., Goto, J., O'Regan, J. P., Deng, H.-X., Rahmani, Z., Krizus, A., McKenna-Yasek, D., Cayabyab, A., Gaston, S. M., Berger, R., Tanzi, R. E., Halperin, J. J., Herzfeldt, B., Van den Bergh, R., Hung, W.-Y., Bird, T., Deng, G., Mulder, D. W., Smyth, C., Laing, N. G., Soriano, E., Pericak-Vance, M. A., Haines, J., Rouleau, G. A., Gusella, J. S., Horvitz, H. R., and Brown, Jr., R. H., 1993, Mutations in Cu/Zn superoxide dismutase gene are associated with familial amyotrophic lateral sclerosis, *Nature* **362:**59–62.

Rowley, D. A., and Halliwell, B., 1983, Superoxide-dependent and ascorbate-dependent formation of hydroxyl radicals in the presence of copper salts: A physiologically significant reaction? *Arch. Biochem. Biophys.* **225**(1):279–284.

Saari, J. T., Dickerson, F. D., and Habib, M. P., 1990, Ethane production in copper-deficient rats, *Proc. Soc. Exp. Biol. Med.* **195:**30–33.

Sato, M., Sugiyama, T., Daimon, T., and Iijima, K., 1994, Histochemical evidence for abnormal copper distribution in the central nervous system of LEC mutant rat, *Neurosci. Lett.* **17:**97–100.

Schafer, M. K.-H., Stoffers, D. A., Eipper, B. A., and Watson, S. J., 1992, Expression of peptidylglycine α-amidating monooxygenase (EC 1.14.17.3) in the rat central nervous system, *J. Neurosci.* **12:**222–234.

Simpson, J. A., Cheeseman, K. H., Smith, S. E., and Dean, R. T., 1988, Free-radical generation by copper ions and hydrogen peroxide, *Biochem. J.* **254:**519–523.

Sokol, R. J., Devereaux, M., Mierau, G. W., Hambidge, K. M., and Shikes, R. H., 1990, Oxidant injury to hepatic mitochondrial lipids in rats with dietary copper overload, *Gastroenterology* **99:**1061–1071.

Sokol, R. J., Twedt, D., McKim, Jr., J. M., Devereaux, M. W., Kårrer, F. M., Kam, I., von Steigman, G., Narkewicz, M. R., Bacon, B. R., Britton, R. S., and Neuschwander-Tetri, B. A., 1994, Oxidant injury to hepatic mitochondria in patients with Wilson's disease and Bedlington terriers with copper toxicosis, *Gastroenterology* **107:**1788–1798.

Sparaco, M., Hirano, A., Hirano, M., DiMauro, S., and Bonilla, E., 1993, Cytochrome c oxidase deficiency and neuronal involvement in Menkes' kinky hair disease: Immunohistochemical study, *Brain Pathol.* **3:**349–354.

Su, L.-C., Ravanshad, S., Owen, Jr., C. A., McCall, J. T., Zollman, P. E., and Hardy, R. M., 1982, A comparison of copper-loading disease in Bedlington terriers and Wilson's disease in humans, *Am. J. Physiol.* **243:**G226–G230.

Sugawara, N., Ikeda, T., Sugawara, C., Kohgo, Y., Kato, J., and Takeichi, N., 1992, Regional distribution of copper, zinc and iron in the brain in Long-Evans Cinnamon (LEC) rats with a new mutation causing hereditary hepatitis, *Brain Res.* **588:**287–290.

Sun, S. H.-H., and O'Dell, B. L., 1992a, Low copper status of rats affects polyunsaturated fatty acid composition of brain phospholipids, unrelated to neuropathology, *J. Nutr.* **122:**65–73.

Sun, S. H.-H., and O'Dell, B. L., 1992b, Elevated striatal levels of glial fibrillary acidic protein associated with neuropathology in copper-deficient rats, *J. Nutr. Biochem.* **3:**503–509.

Tanaka, H., Kasama, T., Inomata, K., and Nasu, F., 1990, Abnormal movements in brindled mutant mouse heterozygotes: As related to the development of their offspring—biochemical and morphological studies, *Brain Dev.* **12:**284–292.

Tanzi, R. E., Petrukhin, K., Chernov, I., Pellequer, J. L., Wasco, W., Ross, B., Romano, D. M., Parano, E., Pavone, L., Brzustowicz, L. M., Devoto, M., Peppercorn, J., Bush, A. I., Sternlieb, I., Pirastu, M., Gusella, J. F., Evgrafov, O., Penchaszadeh, G. K., Honig, B., Edelman, I. S., Soares, M. B., Scheinberg, I. H., and Gilliam, T. C., 1993, The Wilson disease gene is a copper transporting ATPase with homology to the Menkes disease gene, *Nature Genet.* **5:**344–350.

Traystman, R. J., Kirsch, J. R., and Koehler, R. C., 1991, Oxygen radical mechanisms of brain injury following ischemia and reperfusion, *J. Appl. Physiol.* **71:**1185–1195.

Uchida, Y., Takio, K., Titani, K., Ihara, Y., and Tomonaga, M., 1991, The growth inhibitory factor that is deficient in the Alzheimer's disease brain is a 68 amino acid metallothionein-like protein, *Neuron* **7:**337–347.

Walshe, J. M., and Gibbs, K. R., 1987, Brain copper in Wilson's disease, *Lancet* (**II**):1030.

Yamaguchi, Y., Heiny, M. E., Shimizu, N., Aoki, T., and Gitlin, J. D., 1994, Expression of the Wilson disease gene is deficient in the Long-Evans Cinnamon rat, *Biochem. J.* **301:**1–4.

Yamamoto, F., Kasai, H., Togashi, Y., Takeichi, N., Hori, T., and Nishimura, S., 1993, Elevated level of 8-hydroxydeoxyguanosine in DNA of liver, kidneys, and brain of Long-Evans Cinnamon rats, *Jpn. J. Cancer Res.* **84**(5):508–511.

Yang, G., Chan, P. H., Chen, J., Carlson, E., Chen, S. F., Weinstein, P., Epstein, C. J., and Kamii, H., 1994, Human copper-zinc superoxide dismutase transgenic mice are highly resistant to reperfusion injury after focal cerebral ischemia, *Stroke* **25:**165–170.

Chapter 5

Manganese Neurotoxicity and Oxidative Damage

Michael Aschner

5.1. INTRODUCTION

Manganese (Mn) is of critical importance to enzyme and membrane transport systems. Both deficiencies and excess body-burdens of Mn, whether genetic or acquired, can seriously impair vital physiological and biochemical processes. Although one of the least toxic of the heavy metals, Mn toxicity occasionally occurs in miners and ore milling plant workers exposed to prolonged inhalation of manganese dioxide (MnO_2) (Chia *et al.*, 1993). The impact of Mn on human health and disease upon low-level chronic exposure to environmental Mn is uncertain. While it remains largely speculative as to whether excessive Mn exposure plays a role in the pathogenesis of neurodegenerative disorders, distinct similarities between Parkinson's disease and manganism have led to an intense pursuit of its possible role in Parkinson's disease, as well as in a number of other neurodegenerative disorders of, as yet, uncertain etiologies (Calne *et al.*, 1994).

Mn is an abundant and essential metal; it is especially critical during development where it serves as a necessary constituent of some metalloproteins, including the mitochondrial enzymes, superoxide dismutase (SOD) and pyruvate carboxylase (PC), as well as the glial-specific enzyme, glutamine synthetase (GS) (review, Prohaska, 1987; Wedler, 1993). Mn readily crosses the blood–brain barrier in both the adult and the developing fetus (Aschner and Aschner, 1990; Mena *et al.*, 1974). The brain normally contains only a small amount of Mn (Cotzias *et al.*, 1968), but is critically

Michael Aschner   Department of Physiology and Pharmacology, Bowman Gray School of Medicine, Winston–Salem, North Carolina 27157.
Metals and Oxidative Damage in Neurological Disorders, edited by Connor. Plenum Press, New York, 1997.

affected by both deficiency and excessive exposure to Mn. Deficiency of Mn causes seizure activity, probably due to decreased Mn-SOD and GS activities (Critchfield *et al.,* 1993). Excessive exposure to Mn is associated with an irreversible brain disease with prominent psychological and neurological disturbances, characterized initially by a psychiatric disorder (*locura manganica*) that closely resembles schizophrenia. Symptoms include compulsive or violent behavior and emotional instability, as well as hallucinations. Subsequently, ataxia ensues, followed by an extrapyramidal syndrome, resembling several clinical disorders collectively described as "extrapyramidal motor system dysfunction," and in particular, Parkinson's disease and dystonia (for a review, see Barbeau *et al.,* 1976; Barbeau, 1984; Calne *et al.,* 1994). This condition is associated with elevated brain levels of Mn, primarily in those areas known to contain high concentrations of non-heme iron (Fe), including the caudate-putamen, globus pallidus, substantia nigra, and subthalamic nuclei (Eriksson *et al.,* 1992; Komura and Sakamoto, 1993), as well as decreased regional cerebral blood flow in the caudate and thalamus (Lill *et al.,* 1994). At the morphological level, manganism is associated with depletion of striatal dopamine, the degeneration of dopaminergic nerve endings, and massive cell loss in the internal segment of the globus pallidus, as well as gliosis and edema in the posterior limb of the internal capsule (Bernheimer *et al.,* 1973; Eriksson *et al.,* 1984, 1987a, 1992; Heilbronn *et al.,* 1982; Komura and Sakamoto, 1994; Hauser *et al.,* 1994).

The primary objective of this review is to focus on the disparate effects of Mn within the CNS and its role in the etiology of extrapyramidal disorders, with particular emphasis on oxidative stress as a triggering and sustaining mechanism for its pathophysiology. It begins with a brief synopsis of the role of Mn in brain physiology and pathology, focusing on homeostatic mechanisms associated with Mn transport and distribution in the CNS. This is followed by a discussion of potential mechanisms associated with the cytotoxicity of Mn, and the small amount of information currently available on the ability of Mn to attenuate neurotoxic injuries. Both pros and cons for the prooxidative and antioxidative properties of Mn are provided, with a critical evaluation of their respective roles in facilitating or protecting the CNS from Mn-induced oxidative damage.

For additional information on the biological significance of Mn in mammalian systems the reader is referred to a comprehensive review by the late Dr. Frederick Wedler (Wedler, 1993) to whom I wish to dedicate this chapter. His pursuit of mechanisms of Mn homeostasis, and his invaluable contributions to our current understanding of the requirement for Mn in enzyme function and activation, particularly astrocytic glutamine synthetase (GS) function (Wedler *et al.,* 1982, 1994; Wedler and Denman, 1984; Wedler and Ley, 1994), make it imperative that his lifelong effort in unraveling the role Mn in neurodegenerative disorders continue to be vigorously pursued.

5.2.   MANGANESE TRANSPORT IN THE CNS

Mn exists in a number of physical and chemical forms in the earth's crust, in water, and in the atmosphere's particulate matter. Because its outer electron shell can donate up to 7 electrons (review, US EPA, 1984), Mn can assume 11 different oxidation states. Of environmental importance are Mn^{2+}, Mn^{4+}, and Mn^{7+}. In living tissue, Mn has been found as Mn^{2+}, Mn^{3+}, and Mn^{4+} (Archibald and Tyree, 1987). Mn^{5+}, Mn^{6+}, Mn^{7+}, and other complexes of Mn at higher oxidation states, are generally unrecognized in biological materials (review, Keen, 1995). The ability of Mn to assume valence states ranging all the way from $3-$ to $7+$ in combination with other elements is extremely important, emphasizing the potential for Mn to act either as a prooxidant or antioxidant in biological medium. Typically found in compounds with a coordination number of 6 and lacking octahedral coordination complexes, Mn tends to form very tight complexes with other substances (review, Keen, 1995). As a result, its free plasma and tissue concentrations tend to be extremely low (Cotzias et al., 1968).

Mn enters the brain from the blood either across the cerebral capillaries and/or the cerebrospinal fluid (CSF). At normal plasma concentrations, Mn appears to enter into the CNS primarily across the capillary endothelium, whereas at high plasma concentrations, transport across the choroid plexus appears to predominate (Murphy et al., 1991; Rabin et al., 1993), consistent with observations on the rapid appearance and persistent elevation of Mn in this organ (London et al., 1989). Radioactive Mn injected into the blood stream is concentrated in the choroid plexus within 1 hour after injection and 3 days post-injection, Mn is localized to the dentate gyrus and CA3 of the hippocampus (Takeda et al., 1994).

How and in what chemical form Mn is transported across the blood–brain barrier remains controversial. It appears that facilitated diffusion (Rabin et al., 1993), active transport (Murphy et al., 1991; Aschner and Gannon, 1994; Rabin et al., 1993) as well as transferrin (Tf)-dependent transport (Aschner and Gannon, 1994) mechanisms are all involved in shuttling Mn across the blood–brain barrier. Although non-protein-bound Mn enters the brain more rapidly than Tf-bound Mn (Murphy et al., 1991; Rabin et al., 1993), the question remains as to which form represents the predominant mechanism of transport in situ. Analyses of transport mechanisms that are based on tracer techniques employing a bolus injection of Mn into the circulation, while a very sensitive technique for quantifying Mn transport, is not easily interpretable because the information derived from such studies does not necessarily reflect the chemically active or functional forms in which Mn exists and is transported in vivo.

There is much theoretical, and some experimental, evidence to suggest that Tf, the principal Fe-carrying protein of the plasma, functions prominently in Mn transport across the blood–brain barrier. In the absence of Fe, the binding sites of Tf can accommodate a number of other metals raising the possibility that Tf functions in vivo as a transport agent for many of these metals. Mn binding to Tf is time-dependent (Keefer et al., 1970; Scheuhammer and Cherian, 1985; Aschner and Aschner, 1990). When complexed with Tf, Mn is exclusively present in the trivalent oxidation state, with 2 metal ions tightly bound to each Tf molecule (Aisen et al., 1969). At normal plasma Fe concentrations (0.9–2.8 μg/ml), normal iron binding capacity (2.5–4 μg/ml), and at normal Tf concentration in plasma (3 mg/ml), with 2 metal-ion-binding sites per molecule (M_r 77000) of which only 30% are occupied by Fe^{+3}, Tf has available 50 μmole of unoccupied Mn^{3+} binding sites per liter.

Since Tf receptors are present on the surface of the cerebral capillaries (Fishman *et al.,* 1985; Jeffries *et al.,* 1984; Partridge *et al.,* 1987) and the endocytosis of Tf is known to occur in these capillaries (Partridge *et al.,* 1987), it has been suggested that Mn (in the trivalent oxidation state) enters the endothelial cells complexed with Tf. Mn is then released from the complex in the endothelial cell interior by endosomal acidification and the apo-Tf Tf complex is returned to the luminal surface (Morris *et al.,* 1992a, 1992b). Mn released within the endothelial cells is subsequently transferred to the abluminal cell surface for release into the extracellular fluid. The endothelial Mn is delivered to brain-derived Tf for extracellular transport and subsequently taken up by neurons, oligodendrocytes, and astrocytes for usage and storage. Support for receptor-mediated endocytosis of a Mn-Tf complex in cultured neuroblastoma cells (SHSY5Y) was recently demonstrated by Suarez and Eriksson (1993). Sloot and Gramsbergen (1994) have demonstrated anterograde axonal transport of [54]Mn in both nigrostriatal and striatonigral pathways. Furthermore, *in vivo,* intravenous administration of ferric-hydroxide dextran complex significantly inhibits Mn brain uptake, and high Fe intake reduces CNS Mn concentrations, corroborating a relationship between Fe and Mn transport (Aschner and Aschner, 1990; Diez-Ewald *et al.,* 1968).

The distribution of Tf receptors in relationship to CNS Mn accumulation is noteworthy. Pallidum, thalamic nuclei, and substantia nigra contain the highest Mn concentrations (Barbeau *et al.,* 1976). Interestingly, Fe concentrations in these structures are the highest as well (Hill and Switzer, 1984). Although the areas with dense Tf distribution (Hill *et al.,* 1985) do not correspond to the distribution of Mn (or Fe), the fact that Mn-accumulating areas are efferent to areas of high Tf receptor density suggests that these sites may accumulate Mn through neuronal transport (Sloot and Gramsbergen, 1994). For example, the Mn-rich areas of the ventral-pallidum, globus pallidus, and substantia nigra receive input from the nucleus accumbens and the caudate-putamen (Walaas and Fonnum, 1979; Nagy *et al.,* 1978)—two areas abundantly rich in Tf receptors.

5.3. MANGANESE NEUROTOXICITY: WHAT POSSIBLE MECHANISMS?

5.3.1. Manganese Neurotoxicity—A Dual Spectrum

Both deficiency and excess exposure to Mn impair brain function. Although rare, Mn deficiency and the ensuing low blood Mn concentration are known to be associated with epilepsy in both humans and experimental animal models (Carl *et al.,* 1989, 1993). The associated epilepsy has been attributed by some directly to the low blood Mn concentration, whereas others have proposed that the low Mn concentration may be secondary to genetic mechanisms underlying the epilepsy (i.e., decreased GS and Mn-SOD activity). Recent studies in genetically epilepsy-prone rats (GEPRs) suggest abnormalities in Mn-dependent enzymes that are apparently independent of seizure activity. GEPRs were confirmed to have low whole brain GS activity (Critchfield *et al.,* 1993); it would thus appear that the seizure activity in the GEPR does not stem from an increased nutritional or metabolic need for Mn (i.e., is not acquired) but rather is attributable to a genetic susceptibility. The biochemical basis for the epileptic seizure activity is that low Mn^{2+} concentrations in glial cells, and in astrocytes in particular, lowers GS activity, which, in turn, leads to elevated extracellular levels of the excitatory amino acid, glutamate, within the extracellular fluid (review, Wedler, 1993). Notably, glutamate has been implicated in the toxicity of excess brain Mn concentration (for further discussion see below).

Neurotoxicity resulting from excessive accumulation of Mn because of environmental or occupational exposure is well documented (review, Yanagihara, 1982; Chia et al., 1993; review, Florence, 1995). During the Japanese occupation of World War II (1941–1944), Guamanian Chamorros were forced to work in Mn mines (Yanagihara, 1982). A subsequent epidemiological study on the incidence of amyotrophic lateral sclerosis and Parkinson's disease in this mining population suggested an increased incidence of both amyotrophic lateral sclerosis and Parkinson's disease in miners with prolonged exposure to Mn (Yase, 1972). Unlike Parkinsonism, however, manganism also produces dystonia, a neurological sign associated with damage to the globus pallidus (Calne et al., 1994). A comprehensive survey of patients afflicted by Parkinson's disease or manganism concludes that although similar in many respects, there are distinct differences between the two neurological disorders. Similarities between Parkinson's disease and manganism include the presence of generalized bradykinesia and widespread rigidity. Dissimilarities between Parkinson's disease and manganism were also recognized, notably the following in manganism: (1) a less frequent resting tremor, (2) more frequent dystonia, (3) a particular propensity to fall backward, (4) failure to achieve a sustained therapeutic response to levodopa, and (5) failure to detect a reduction in fluorodopa uptake by positron emission tomography (PET; for further details see Calne et al., 1994).

5.3.2. Mechanisms Associated with Manganese-Induced Neurotoxicity

As alluded to above, there is compelling evidence to suggest that excessive exposure to Mn produces region-specific neurotoxicity, especially within the basal ganglia. The precise biochemical trigger remains unknown and a plethora of theories have been advanced to explain this region-specific effect (review, Archibald and Tyree, 1987). These include: (1) a direct toxic effect of Mn in its divalent oxidation state (or perhaps Mn in a higher oxidation state; Donaldson et al., 1980; Archibald and Tyree, 1987) to dopamine-containing cells (Parenti et al., 1988); (2) a Mn-induced decrease in the content of peroxidase and catalase within the substantia nigra (Ambani et al., 1975); (3) the production of superoxide (SO; $O_2^{\cdot-}$), hydrogen peroxide (H_2O_2), or hydroxyl free ($\cdot OH$) radicals by Mn, which, in turn, "attack" dopamine, dopaminergic cells, and dopamine receptors (Cohen et al., 1974; Cohen, 1984; Marinho and Manso, 1993; Sun et al., 1993); (4) the production of 6-hydroxydopamine or other toxic catecholamines by Mn^{2+} and a decrease in protective thiols (Cohen et al., 1974; Graham et al., 1978; Graham, 1984; Liccione and Maines, 1988; Perry et al., 1982), (5) the autooxidation of dopamine, leading to formation of toxic (semi) quinones, concomitantly depleting tissue dopamine (Donaldson et al., 1980, 1982; Graham, 1984; Garner and Nachtman, 1989; Millar et al., 1990; Roy et al., 1994), and most recently, (6) an excitotoxic mechanism in which the activation of glutamate-gated cation channels contributes to neuronal degeneration (Brouillet *et al.*, 1993). Although experimental evidence in support of each of these mechanisms is available, their relative contributions and the primary biochemical event associated with Mn neurotoxicity remain largely speculative. In accordance with the theme of this book, the next section concentrates on the prooxidant activity of Mn.

5.3.2.1. The Production of Superoxide (SO; $O_2^{\cdot-}$), H_2O_2, or Hydroxyl (·OH) Free Radicals by Mn

A free radical is defined as any atom or molecule containing an unpaired electron in the outermost orbit. This electrical configuration is unstable, necessitating that the free radical either take or donate an electron to an adjacent compound in order to restore its own orbital stability. The ensuing electron transfer not only disrupts the structure of the adjacent compound, but it may impair its function (Zhang *et al.*, 1994). Oxygen-free radicals are constantly produced in the CNS, where they may have beneficial, but also deleterious, effects. Figure 5.1 depicts a number of reactions that may lead to the generation of oxygen radicals. Oxygen-based free radicals have been implicated in a number of disorders and may have been one of the more pervasive driving forces in the course of evolution (Halliwell and Gutteridge, 1984). The highly reactive superoxide (SO; $O_2^{\cdot-}$) free radical is formed during many normal biochemical events (Coyle and Puttfarcken, 1993). Its hyperproduction, however, can lead to neuronal degeneration via its ability to mutate DNA and initiate a lipid peroxidation chain reaction. Arrayed against such damage is an extensive system of free radical defenses capable of scavenging and transforming oxygen-free radicals into non-toxic species (see below).

A number of underlying causes can lead to the ubiquitous formation of reactive oxygen species (review, Bondy and LeBel, 1992). These include: (1) enhanced phospholipase activity and the ensuing release of arachidonic acid (AA), (2) cytosolic acidity, (3) oxidative phosphorylation, (4) the presence of metal ions with multivalence potential, and (5) the induction of various oxidases. Enhanced phospholipase (PL) activity is associated with the release of AA. AA contains four ethylenic bonds and is readily autooxidizable. Its enzymatic conversion to bioactive prostaglandins, leukotrienes, and thromboxanes by cyclooxygenases and lipooxygenases leads to considerable oxygen radical generation (Freeman and Crapo, 1982). Lowered pH resulting from

$$\text{1. Xanthine} \quad + \quad O_2 \xrightarrow{\overset{\text{Xanthine}}{\text{Oxidase}}} \text{Uric Acid} \quad + \quad O_2^{\cdot-} \quad + \quad H_2O_2$$

$$\text{2. Substrate} \quad + \quad O_2 \xrightarrow{\overset{\text{Mixed Function}}{\text{Oxidase}}} \text{Product} \quad + \quad H_2O_2$$

$$H_2O_2 \quad + \quad Fe^{2+} \longrightarrow OH^- \quad + \quad \cdot OH \quad + \quad Fe^3$$

$$\text{3. NO} \quad + \quad O_2^{\cdot-} + \quad H^+ \longrightarrow ONOOH \longrightarrow \cdot OH \quad + \quad NO_2$$

FIGURE 5.1. Reactions leading to the production of oxygen radicals.

excess glycolytic activity would be expected not only to accelerate the process of liberating protein-bound Fe, but also to lead to an impairment of oxidative ATP generation and to the appearance of the prooxidant protonated superoxide (Siesjo, 1988). Although enhanced phospholipase activity and reduced cellular pH may potentially be involved in Mn-induced reactive oxygen formation, experimental data on their relative contribution to Mn neurotoxicity are presently lacking.

Experimental evidence in support of the ability of Mn to induce free radical formation by the remaining three underlying causes is compelling, and is briefly discussed below:

5.3.2.1a. Oxidative Phosphorylation (Mitochondrial Connection).

A disproportionate amount of the body's oxygen is consumed by the CNS, as it derives its energy almost exclusively from oxidative metabolism of the mitochondrial respiratory chain. Approximately 2% of the molecular oxygen consumed by mitochondria is incompletely reduced, thus generating oxygen radicals. This proportion may be further increased when the mitochondrial electron transport system is compromised. Neuronal oxidative phosphorylation, which generates ATP while reducing O_2 to H_2O by the sequential addition of 4 electrons and 4 H^+, varies in proportion to neuronal firing. "Leakage" of high energy electrons along the mitochondrial electron transport chain causes the formation of both O_2^- and H_2O_2 (Lehninger, 1972).

Mn is known to have a special affinity for mitochondria. As early as 1955 Maynard and Cotzias found that radiolabeled ^{54}Mn readily concentrates in rat liver mitochondria within 15 min. after exposure. Liccione and Maines (1988) found elevated Mn concentrations in brain mitochondria of rats subchronically exposed to the metal, and Mn^{2+} has been found to bind after uptake to the inner mitochondrial membrane (Gunter et al., 1975), where the electron transport is located. Other studies on the mitochondrial uniport transporter (Chance, 1965; Lehninger, 1972) established the kinetic profile of both the uptake and efflux of Mn^{2+}, and its ability to competitively inhibit Ca^{2+} efflux from mitochondria (Gavin et al., 1990). Mn^{2+} accumulation was related to a combination of Ca^{2+} enhanced Mn^{2+} influx, as well as extremely slow Mn^{2+} efflux from brain mitochondria (Gavin et al., 1990). With succinate as substrate, Mn^{2+} inhibited oxygen consumption by suspensions of rat liver mitochondria after the addition ADP but not after the addition of uncoupler. Mn^{2+} also inhibited ADP-stimulated respiration as well as uncoupler-stimulated respiration with glutamate/malate as a substrate. Whereas state 3 respiration was inhibited, state 4 respiration appeared unchanged, indicative of the retention of the inner mitochondrial membrane to proton impermeability, and providing direct support for the interference of Mn^{2+} with oxidative phosphorylation, most likely by binding to the F1 ATPase (Gavin et al., 1990).

Whether the affinity of Mn^{2+} for mitochondria and its inhibition of oxidative phosphorylation is related the distinctive pathology of excessive Mn exposure and the unique morphological damage to the globus pallidus and the substantia nigra remains largely speculative. The brain consumes a full 20% of the total oxygen required by the human at rest (Sokoloff, 1974) and in synaptically active brain tissue approximately 40% of the total energy production is utilized to maintain ion gradients (Bradford, 1986). The striatum, which preferentially accumulates Mn, is among the most synaptically active areas of brain and thus, may be highly vulnerable to its effects (Gavin et *al.*, 1990). Should Mn-loaded neurons be put under strenuous metabolic conditions, necessitating an increase in their firing rates, their compromised mitochondria might be unable to supply them with needed additional ATP. Within this context it is noteworthy that an increase in the tonic activity of both globus pallidus and substantia nigra has been measured after depletion of striatal dopamine (Miller and DeLong, 1986; Mitchell *et al.*, 1986). When the increased activity is accompanied by elevated intramitochondrial Mn^{2+}, the rate of ATP production declines just as the demand for energy is soaring. In addition to a direct effect on oxidative phosphorylation, Mn can affect mitochondrial function indirectly by strongly inhibiting Ca^{2+} efflux from mitochondria (Gavin *et al.*, 1990). Elevated intramitochondrial Ca^{2+} triggers loss of membrane potential (Crompton, 1990; Gunter and Pfeiffer, 1990), which would be expected to further compromise neuronal ATP levels.

5.3.2.1b. The Presence of Metal Ions with Multivalence Potential. Liberation of intracellular protein-bound iron can occur by enhanced degradation of important iron-binding proteins such as ferritin and transferrin. A small increase in levels of free iron within cells can dramatically accelerate rates of oxygen radical production.

It has been demonstrated that both Mn^{2+} and Fe^{2+} catalyze Fenton-like reactions with amino acids, involving inner sphere bi-dentate coordination of the amino acid, 2 HCO_3^- moieties per Mn^{2+}, and H_2O_2, leading to radical extraction of the alpha-H as a primary step, and proceeding through an AA-NHO intermediate to yield an α-keto acid, an aldehyde, or carboxylic acid plus carbon dioxide. Mn^{2+}- and Fe^{2+}-promoted oxidative damage to specific residues in proteins, as precursor steps that trigger proteolytic degradation and protein turnover, have been identified and studied in detail for glutamine synthetase (GS) (review, Wedler, 1993).

It has been recently also suggested that CNS Fe stores may actively modulate the neurotoxic effects of Mn in its divalent oxidation state. Fe storage pathways and Fe transport (Aschner and Aschner, 1990, 1991; Hill and Switzer, 1984; Hill, 1990; Morris *et al.*, 1992b) are believed to exert regulatory control over the selective accumulation of Mn^{2+} in the basal ganglia (Eriksson *et al.*, 1987b; London *et al.*, 1989; Newland *et al.*, 1989) and its transport in nigrostriatal and stratonigral neurons (Sloot and Gramsbergen, 1994). As alluded to earlier in this review, Mn in its +3 oxidation state is known to bind to transferrin, a predominantly Fe transport protein, and to its receptor. Accordingly, by directly disturbing Fe homeostasis, particularly within the mitochondria, Mn may liberate endogenous Fe, which in turn, can catalyze the Haber Weiss reaction (Koster and Sluiter, 1994) forming ·OH. Alternatively, ·OH may be generated independently from Fe by the decomposition of peroxynitrite, which is a reaction product of nitric oxide (NO) and O_2^- (Beckman *et al.*, 1990; Ben Sachar *et al.*, 1991; Hammer *et al.*, 1993).

5.3.2.1c. Oxidases. Chemical induction of cytochrome P-450 containing mixed function monooxidases (MFO) can increase the rate of detoxification reactions. At the cellular level most of the cerebral MFO are mitochondrial rather than microsomal. Whereas in the microsomal fraction only cytochrome P-450 activity is altered by Mn^{2+}, the concentration of all mitochondrial hemeproteins, i.e., cytochrome P-450, and respiratory hemeproteins a, b, cl, and c appear to be susceptible to Mn^{2+} treatment (Liccione and Maines, 1989). When the mitochondrial fraction of whole brain is compared with that of the striatum, the latter appears to be particularly vulnerable to the ability of Mn^{2+} to alter drug biotransformation activities. In accordance with the site-specific damage induced by excessive Mn exposure, the mitochondrial fraction within the striatum appears to be more responsive to Mn^{2+} than the microsomal fraction, consistent with the mitochondria as a target organelle for Mn^{2+}. The mechanism by which Mn affects the cellular concentration of various hemeproteins in the brain is presently unknown. However, the finding that mitochondrial cytochrome P-450 concentration is elevated in the face of a decreased activity of δ-aminolevulinic acid (ALA) synthetase and a decrease in respiratory hemeprotein concentrations is likely to reflect one of the following factors: (1) the preferential utilization of heme for production of cytochrome P-450, (2) stabilization of cytochrome in the presence of Mn^{2+}, or (3) inhibition of its degradation by mitochondrial heme degrading systems (Kutty and Maines, 1987).

5.3.2.2. The Autooxidation of Dopamine, Leading to Formation of Quinones and Concomitantly Depleting Tissue Dopamine

Oxidant stress elicited by reactive oxygen species generated by metal-catalyzed dopamine autooxidation may be the common neurodegenerative process involved in selective nigrostriatal degeneration in Parkinson's disease produced by environmental neurotoxins such as Mn. Quinone autooxidation products of L-DOPA and dopamine have been shown to exert toxicity on C1300 neuroblastoma cells *in vitro* via their neutrophilic reactivity (Graham *et al.*, 1978; review, Archibald and Tyree, 1987). It has also been observed that the substantia nigra has detectable levels of H_2O_2 and high levels of monoamine oxidase (MAO), able to generate H_2O_2 (Ambani *et al.*, 1975). Donaldson *et al.* (1980) have presented evidence that Mn^{2+} *in vitro* increases dopamine autooxidation, proposing that Mn, in a higher oxidation state, may perpetrate oxidative damage. A similar hypothesis, invoking Mn in its trivalent oxidation state as the neurotoxic culprit, was postulated by Archibald and Tyree (1987). Whether originally entering in the $+2$, $+3$, or $+4$ oxidation state, Mn will, via spontaneous oxidation and dismutation, peroxidatic activity, or oxygen-radical-mediated oxidation, give rise to Mn^{3+}, which in a simple complex, perhaps with catecholamines themselves, will oxidatively destroy dopamine, epinephrine, norepinephrine, and their precursor DOPA in an expedient manner (Archibald and Tyree, 1987). If the resulting quinones generate oxygen radicals, additional Mn^{3+} forms and the process is further accelerated. If the quinones are cytotoxic, as the dopaminergic cells die off and the level of tissue dopamine decreases, a higher and higher proportion of the remaining dopamine would be destroyed by a constant concentration of striatal Mn (Chiueh *et al.*, 1993). In theory, metal-catalyzed dopamine autooxidation may lead to formation of not only semiquinone anion radicals but also reactive oxygen species. It has been also suggested that Mn in its divalent oxidation state may form a complex with dopamine, and that this complex undergoes oxidation to generate a trivalent Mn–dopamine complex (review, Florence, 1995), liberating $O_2^{\cdot -}$ in the process. The $O_2^{\cdot -}$ radical, in turn, reacts with the trivalent Mn–dopamine complex to generate a semiquinone free radical, liberating H_2O_2. The semiquinone free radical is subsequently oxidized into dopaquinone-manganese. This sequence of reactions, in addition to effectively depleting striatal dopamine, replaces the latter with two compounds that are of known toxic potency.

It has been extremely difficult to assess the generation of short-lived oxygen free radicals and quinones in the CNS *in vivo*. Nevertheless, it is likely that dopamine plays a role in Mn-induced neurotoxicity. This is supported by studies showing that pretreatment with the dopamine synthesis blockers α-methyltyrosine and lisuride effectively attenuate Mn^{2+} neurotoxicity (Parenti *et al.*, 1988), whereas the monoamine oxidase inhibitor pargyline, as well as L-DOPA, potentiate its neurotoxicity (Parenti *et al.*, 1986). Consistent with the concept of quinone production by Mn (Roy *et al.*, 1994) is the observation on the attenuation of Mn^{2+}-induced dopamine depletion by vitamin E treatment (Puppo and Halliwell, 1988).

5.3.2.3.   An Excitotoxic Mechanism in Which the Activation of Glutamate-Gated Cation Channels Contributes to Neuronal Degeneration

The attractive feature of the oxidative damage theory described above is that it can account for cumulative damage associated with exposure to Mn. Recent evidence also points to the potentiation of oxidative stress damage within the CNS by sequential activation of glutamate-gated cation channels, synergistically reinforcing the neurodegenerative damage (Coyle and Puttfarcken, 1993). Accordingly, attention is directed here to the possibility that activation of glutamate-gated cation channels contribute directly or indirectly to Mn-induced neurotoxicity.

In view of marked changes in the striata one week after injection of $MnCl_2$, consistent with a *N*-methyl-D-aspartate (NMDA) excitotoxicity, Brouillet *et al.* (1994) have recently suggested that Mn exerts an indirect excitotoxic process secondary to its ability to impair oxidative energy metabolism. Prior removal of cortico-striatal glutamatergic input or treatment with the noncompetitive NMDA antagonist MK-801 blocked the Mn-induced effect, suggesting an NMDA receptor-mediated process similar to that of other mitochondrial toxins, such as aminooxyacetic acid and 1-methy-4-phenylpyridinium. Chiueh *et al.* (1993) proposed that striatal damage associated with exposure to Mn may be the outcome of a lifetime accumulation of ·OH-induced oxidant stress elicited by dopamine, oxygen, and iron. ·OH, in addition to attacking membrane lipid bilayers, induces Ca^{2+} conductances, leading to intracellular Ca^{2+} overload and the activation of proteases, which, in turn, cause K^+ overflow and release of excitatory amino acids such as glutamate. Elevated intraneuronal Ca^{2+} is also known to activate peptidases, such as calpain I, which can catalyze the enzymatic conversion of xanthine dehydrogenase to xanthine oxidase. The catabolism of purine bases by xanthine oxidase yields $O_2^{·-}$. This reaction may become quite prominent since NMDA receptor agonists, such as glutamate, cause depletion of ATP and elevation of AMP. Furthermore, as lactic acid increases, the acidic conditions favor the liberation of cellular stores of Fe^{2+}, which promotes the Fenton reaction to yield ·OH from H_2O_2 (review, Coyle and Puttfarcken, 1993).

While it is reasonable to accept that acute brain insult causing massive release of glutamate could damage neurons expressing NMDA receptors, it is more difficult to link the neurotoxic action of glutamate to the long latent period associated with low-level Mn exposure and the slow progression of neurodegenerative disorders. There is a mismatch between the millisecond dynamics of glutamate-gated ion kinetics and the years, perhaps decades, involved in the progressive neuronal injury associated with these disorders. Environmental levels insufficient to cause acute, overt CNS toxicity, but associated with chronic low level elevation of glutamate, *could* nevertheless accelerate the processes of excitotoxic neurodegeneration associated with Mn neurotoxicity. Evidence is now emerging that activation of glutamate-gated cation channels may be an important source of oxidative stress (reviewed by Coyle and Puttfarcken, 1993), referring to the cytotoxic consequences of oxygen radicals (superoxide anion, hydroxy radical, and hydrogen peroxide), which are generated as byproducts of normal and aberrant metabolic processes that use molecular O_2. Thus the convergence of sequential and interactive mechanisms (glutamate-gated ion channel activation and oxidative stress) may provide a link between chronic low-level environmental Mn exposure and glutamate (or other EAA)-induced neuronal degeneration. From this perspective, glutamate would be acting as a long-term "pulsing" stimulus, and the neuronal damage associated with long-term exposure to Mn would then be measurable as the sum total of the lifetime glutamate linked events.

5.3.3. Manganese as an Antioxidant and Brain Protectant

The CNS, like all other organs, shares an extensive system of free radical defenses by scavenging and transforming oxygen-free radicals into non-toxic species (H_2O_2; review, Coyle and Puttfarcken, 1993; Bondy and LeBel, 1992). Included in these defensive systems is the enzymatic conversion of SO anions to hydrogen peroxide (H_2O_2) and subsequently to water (H_2O). Superoxide dismutase (SOD) plays a vital protective role by catalyzing the transformation of SO anion radicals into H_2O_2. However, in the presence of unbound ions of iron, H_2O_2 is catalytically converted into potent and long-lasting hydroxyl ($\cdot OH$) radicals that can cause peroxidative damage to proteins, lipids, and DNA.

SOD is a family of enzymes responsible for dismutation of $O_2{}^-$ to H_2O_2, thereby reducing the risk of $\cdot OH$ formation (Hassan and Schrum, 1994). The reaction catalyzed by SOD is shown in Figure 5.2. Three forms of SOD, encoded by three separate genes, are expressed in eukaryotic cells (review, Coyle and Puttfarcken, 1993). Cu- and Zn-dependent SOD (CuZn-SOD) is localized to the cytoplasm. An extracellular CuZn-SOD is expressed in low concentrations in extracellular fluids, and a Mn-dependent SOD (Mn-SOD) is preferentially localized to the inner membranes of mitochondria. Given that production of oxygen free radicals through respiratory chain reactions is especially high in mitochondria, Mn-SOD would appear be of particular importance in an antioxidant role. Mn-SOD exists as a dimeric or tetrameric enzyme with a molecular

$$O_2{}^{\cdot -} + O_2{}^- + 2\,H^+ \xrightarrow{\text{SOD}} H_2O_2 + O_2$$

FIGURE 5.2. The superoxide dismutase (SOD)-catalyzed dismutation of superoxide (SO; $O_2{}^{\cdot -}$) to hydrogen peroxide (H_2O_2) reduces the risk of hydroxyl ($\cdot OH$) formation.

weight of 45 kD and 90 kD, respectively. Mn-SOD activity can be induced by various stimuli, such as interleukin I (IL-I), tumor necrosis factor (TNF), as well as molecular oxygen or oxygen-free radicals. Mn-SOD is hetereogeniously distributed within the human brain and is abundant within the substantia nigra (Zhang *et al.,* 1994). Since the concentration of Mn-SOD is thought to correlate with free radical levels within the normal brain, its high content within the substantia nigra may render this region particularly vulnerable to free-radical-induced damage should defensive mechanisms be compromised. Elevation in free radical levels is likely associated with activation of defense mechanisms against oxidative stress. In line with this concept, increased Mn-SOD activity has been noted in several conditions in which oxidative stress is suspected to occur, namely, (1) during normal aging in the rat brain, (2) after facial nerve transection, and (3) in the substantia nigra of Parkinson's disease tissue, corroborating an important role for Mn-SOD in neuronal defense mechanisms (Zhang *et al.,* 1994).

In addition to the ability of Mn-SOD to afford cellular protection, some Mn chelates can act as a true SODs by dismutating 2 O_2^- radicals to water and H_2O_2. Examples include Mn^{2+}-pyrophosphate and Mn^{2+}-tartarate, which can block lipid peroxidation mediated by O_2^- and ionizing radiation (Cavallini *et al.,* 1984; Donaldson *et al.,* 1982; reviewed by Archibald and Tyree, 1987). Free Mn^{2+} in solution, while it appears appear to have little or no ability to scavenge ·OH directly can scavenge radicals produced either by metabolic reactions, especially those involving NADPH, chemotherapeutic agents, or ionizing radiation (the Fe^{3+}-EDTA-catalyzed production of ·OH by γ-irradiation; reviewed by Wedler, 1993). Furthermore, a number of simple Mn^{3+} complexes are relatively stable and can selectively oxidize a variety of specific biological molecules. Several manganic porphyrins (stable Mn^{3+} complexes), with substitutes on their methine bridge carbons were shown to catalyze the dismutation of superoxide radicals *in vivo* and to mimic the function of SOD *in vitro* (Faulkner *et al.,* 1994). Antioxidant properties associated with Mn (II) in phospholipid peroxidation have also been recently detailed (Tampo and Yonaha, 1992).

5.4. SUMMARY AND FUTURE DIRECTIONS

A wealth of data indicates that Mn can lead to profound effects on brain functions. Currently, interference with neuronal circuit and network models dominate our views on the way by which Mn exerts its neurotoxic effect. A heterogeneous distribution of Mn and lack of firm understanding of transport mechanisms of Mn from the blood into the CNS limits the interpretation of molecular events associated with Mn neurodegeneration. As illustrated in the present review, the possibilities abound, and one or a combination of them may act as the biochemical trigger for Mn-induced CNS degeneration. As several countries have replaced, or are considering the replacement of, lead in gasoline with a Mn antiknock compound (Wedler, 1993), methylcylopentadienyl manganese tricarbonyl (MMT), it is imperative that further attention be directed at the potential of Mn to elicit long-term neurodegenerative disorders. Thus, the need arises for a more complete understanding of Mn functions both in health and disease, and the as yet unresolved mechanisms of its neurotoxicity.

ACKNOWLEDGMENTS.   This work was supported in part by grants NIEHS 05223 and USEPA R-824087.

5.5. REFERENCES

Aisen, P., Aasa, R., and Redfield, A. G., 1969, The chromium, manganese, and cobalt complexes of transferrin, *J. Biol. Chem.* **244:**4628–4633.

Ambani, L. M., Vanwoert, M. H., and Murphy, S., 1975, Brain peroxidase and catalase in Parkinson's disease, *Arch. Neurol.* **32:**114–118.

Archibald, F. S., and Tyree, C., 1987, Manganese poisoning and the attack of trivalent manganese upon catecholamines, *Arch. Biochem. Biophys.* **256:**638–650.

Aschner, M., and Aschner, J. L., 1990, Manganese transport across the blood–brain barrier: Relationship to iron homeostasis, *Brain Res. Bull.* **24:**857–860.

Aschner, M., and Aschner, J. L., 1991, Manganese neurotoxicity: Cellular effects and blood–brain barrier transport mechanisms, *Neurosci. Biobehav. Rev.* **15:**333–340.

Aschner, M., and Gannon, M., 1994, Manganese (Mn) transport across the blood–brain barrier: Saturable and transferrin-dependent transport mechanisms, *Brain Res. Bull.* **33:**345–349.

Barbeau, A., 1984, Manganese and extrapyramidal disorders, *Neurotoxicol.* **5:**13–36.

Barbeau, A., Inoué, N., and Cloutier, T., 1976, Role of manganese in dystonia, *Adv. Neurol.* **14:**339–352.

Beckman, J. S., Beckman, T. W., Chen, J., Marshall, P. M., and Freeman, B. A., 1990, Apparent hydroxyl radical production from peroxynitrite: Implications for endothelial injury by nitric oxide and superoxide, *Proc. Natl. Acad. Sci. USA* **87:**1620–1624.

Ben Sachar, D., Eshel, G., Finberg, J. P. M., and Youdim, M. B. H., 1991, The iron chelator desferrioxamine (Desferal) retards 6-hydroxydopamine-induced degeneration of nigrostriatal dopamine neurons, *J. Neurochem.* **56:**1441–1444.

Bernheimer, H., Birkmayer, W., Hornykiewicz, O., Jellinger, K., and Seitelberger, F., 1973, Brain dopamine and the syndromes of Parkinson and Huntington-clinical, morphological and neurochemical correlations, *J. Neurol. Sci.* **20:**415–425.

Bondy, S. C., and LeBel, C. P., 1992, Formation of excess reactive oxygen species within the brain, in: *The Vulnerable Brain and Environmental Risks,* Volume 2 (R. L. Isaacson and K. F. Jensen, eds.), Plenum Press, New York, pp. 255–272.

Bradford, H. F., (ed.), 1986, in: *Chemical Neurobiology,* Freeman, New York.

Brouillet, E. P., Shinobu, L., McGarvey, U., Hochberg, F., and Beal, M. F., 1993, Manganese injection into the rat striatum produces excitotoxic lesions by impairing energy metabolism, *Experim. Neurol.* **120:**89–94.

Calne, D. B., Chu, N. S., Huang, C. C., Lu, C. S., and Olanow, W., 1994, Manganism and idiopathic parkinsonism: similarities and difference, *Neurology* **44:**1583–1586.

Carl, G. F., Critchfield, J. W., Thompson, J. L., McGinnis, L. S., Wheeler, G. A., Gallagher, B. B., Holmes, G. L., Hurley, L. S., and Keen, C. L., 1989, Effect of kainate-induced seizures on tissue trace element concentrations in the rat, *Neuroscience* **33:**223–227.

Carl, G. F., Blackwell, L. K., Barnett, F. C., Thompson, L. A., Rissinger, C. J., Olin, K. L., Critchfield, J. W., Keen, C. L., and Gallagher, B. B., 1993, Manganese and epilepsy: Brain glutamine synthetase and liver arginase activities in genetically epilepsy prone and chronically seizured rats, *Epilepsia* **34:**441–446.

Cavallini, L., Valente, M., and Bindoli, A., 1984, On the mechanism of inhibition of lipidperoxidation by manganese, *Inorg. Chim. Acta* **91:**117–120.

Chance, B., 1965, The energy-linked reaction of calcium with mitochondria, *J. Biol. Chem.* **240:**2729–2748.

Chia, S. E., Foo, S. C., Gan, S. L., Jeyaratnam, J., and Tian, C. S., 1993, Neurobehavioral functions among workers exposed to manganese ore, *Scand. J. Work* **19:**264–270.

Chiueh, C. C., Murphy, D. L., Miyake, H., Lang, K., Tulsi, P. K., and Huang, S.-J., 1993, Hydroxyl free radical (·OH) formation reflected by salicylate hydroxylation and neuromelanin. In vivo markers for oxidant injury of nigral neurons, *Ann. NY Acad. Sci.* **679:**370–375.

Cohen, G., 1984, Oxy-radical toxicity in catecholamine neurons, *Neurotoxicol.* **5:**77–82.

Cohen, G., and Heikkila, R. E., 1974, The generation of hydrogen peroxide, superoxide radical, and hydroxyl radical by 6-hydroxydopamine, dialuric acid, and related cytotoxic agents, *J. Biol. Chem.* **249:**2447–2452.

Cotzias, G. C., Horiuchi, K., Fuenzalida, S., and Mena, I., 1968, Chronic manganese poisoning: Clearance of tissue manganese concentrations with persistence of the neurological picture, *Neurology* **18:**376–382.

Coyle, J. T., and Puttfarcken, 1993, Oxidative stress, glutamate, and neurodegenerative disorders, *Science* **262:**689–695.

Critchfield, J. W., Carl, G. F., and Keen, C. L., 1993, The influence of manganese supplementation on seizure onset and severity, and brain monoamines in the genetically epilepsy prone rat, *Epilepsy Res.* **14:**3–10.

Crompton, M., 1990, The role of Ca^{2+} in the function and dysfunction of heart mitochondria, in: *Calcium and the Heart* (G. A. Langer, ed.), Raven Press, New York, pp. 167–197.

Diez-Ewald, M., Weintraub, L. R., and Crosby, W. H., 1968, Inter relationship of iron and manganese metabolism, *Proc. Soc. Exp. Biol. Med.* **129:**448–151.

Donaldson, J., 1987, The physiopathologic significance of manganese in brain: Its relation to schizophrenia and neurodegenerative disorders, *Neurotoxicol.* **8:**451–462.

Donaldson, J., and Barbeau, A., 1985, in: *Metal Ions in Neurology and Psychiatry,* Alan R. Liss, New York, pp. 259–285.

Donaldson, J., Labella, F. S., and Gesser, D., 1980, Enhanced autoxidation of dopamine as a possible basis of manganese neurotoxicity, *Neurotoxicol.* **2:**53–64.

Donaldson, J., McGregor, D., and Labella, F. S., 1982, Manganese neurotoxicity: A model for free radical mediated neurodegeneration? *Can. J. Physiol. Pharmacol.* **60:**1398–1405.

Eriksson, H., Morath, C., and Heilbronn, E., 1984, Effects of manganese on the nervous system, *Acta Neurol. Scand.* **70:**89–93.

Eriksson, H., Lenngren, S., and Geilbronn, E., 1987a, Effect of long-term administration of Mn on biogenic amine levels in discrete striatal regions of rat brain, *Arch. Toxicol.* **59:**426–431.

Eriksson, H., Magista, K., Plantin, L.-O., Fonnum, E., Hedstrom, K.-G., Theodorsson-Norheim, E., Kristensson, K., Stalberg, E., and Heilbronn, E., 1987b, Effects of manganese oxide on monkeys as revealed by a combined neurochemical, histological and neurophysiological evaluation, *Arch. Toxicol.* **61:**46–52.

Eriksson, H., Tedroff, J., Thuomas, K. A., Aquilonius, S. M., Hartvig, P., Fasth, K. J., Bjurling, P., Langstrom, B., Hedstrom, K. G., and Heilbronn, E., 1992, Manganese induced brain lesions in Macaca fascicularis as revealed by positron emission tomography and magnetic resonance imaging, *Arch. Toxicol.* **66:**403–407.

Faulkner, K. M., Liochev, S. I., and Fridovich, I., 1994, Stable Mn(III) porphyrins mimic superoxide dismutase in vitro and substitute for it in vivo, *J. Biol. Chem.* **269:**23471–23476.

Fishman, J. B., Handrahan, J. B., Rubir, J. B., Connor, J. R., and Fine, R. E., 1985, Receptor-mediated trancytosis of transferrin across the blood–brain barrier, *J. Cell. Biol.* **101:**423A.

Florence, M., Environmental exposure to Mn in Groote Eyland, Australia, 1995, in: *Proceedings of the Workshop on the Bioavailability and Oral Toxicity of Manganese,* (S. Velazquez, EPA Liaison), US EPA, Environmental Criteria and Assessment Office, pp. 83–94.

Freeman, B., and Crapo, J. D., 1982, Biology of disease: Free radicals and tissue injury, *Lab. Invest.* **47:**412–426.

Garner, C. D., and Nachtman, J. P., 1989, Manganese catalyzed auto-oxidation of dopamine to 6-hydroxydopamine in vitro, *Chem. Biol. Interactions* **69:**345–351.

Gavin, C. E., Gunter, K. K., and Gunter, T. E., 1990, Manganese and calcium efflux kinetics in brain mitochondria: Relevance to manganese toxicity, *Biochem. J.* **266:**329–334.

Gavin, C. E., Gunter, K. K., and Gunter, T. E., 1992, Mn^{2+} sequestration by mitochondria and inhibition of oxidative phosphorylation, *Toxicol. Appl. Pharmacol.* **115:**1–5.

Graham, D. G., 1984, Catecholamine toxicity: a proposal for the molecular pathogenesis of manganese neurotoxicity and Parkinson's disease, *Neurotoxicol.* **5:**83–96.

Graham, D. G., Tiffany, S. M., Bell, W. R., Jr., and Gutknecht, W. F., 1978, Autooxidation versus covalent binding of quinones as the mechanism of toxicity of dopamine, 6-hydroxydopamine, and related compounds toward C 1300 neuroblastoma cells in vitro, *Molec. Pharmacol.* **14:**644–653.

Gunter, T. E., and Pfeiffer, D., 1990, Mechanisms by which mitochondria transport calcium, *Am. J. Physiol.* **258:**C755–C786.

Gunter, T. E., Puskin, J. S., and Russell, P. R., 1975, Quantitative magnetic resonance studies of manganese uptake by mitochondria, *Biophys. J.* **15:**319–333.

Halliwell, B., 1984, Manganese ions, oxidation reactions and the superoxide radical, *Neurotoxicol.* **5:**113–118.

Halliwell, B. J., and Gutteridge, J. M. C., 1984, Oxygen toxicity, oxygen radicals, transition metals and disease, *Biochem. J.* **219:**1–14.

Hammer, B., Parker, W. D., Jr., and Bennett, J. P., Jr., 1993, NMDA receptors increase OH radicals in vivo by using nitric oxide synthase and protein kinase C, *Neuroreport* **5:**72–74.

Hassan, H. M., and Schrum, L. W., 1994, Roles of manganese and iron in the regulation of the biosynthesis of manganese-superoxide dismutase in *Escherichia coli, FEMS Microbiol. Rev.* **14:**315–323.

Hauser, R. A., Zesiewicz, T. A., and Rosemurgy, A. S., 1994, Manganese intoxication and chronic liver failure, *Annal. Neurol.* **36:**871–875.

Heilbronn, E., Eriksson, H., and Haggblad, J., 1982, Neurotoxic effects of manganese: Studies on cell cultures, tissue homogenates and intact animals, *Neurobehav. Toxicol. Teratol.* **4:**655–658.

Hill, J. M., 1990, Iron and proteins of iron metabolism in the central nervous system, in: *Iron Transport and Storage* (P. Ponka, H. M. Schulman, and R. C. Woodworth, eds.), CRC Press, Boca Raton, Florida, pp. 315–330.

Hill, J. M., and Switzer, R. C., III, 1984, The regional distribution and cellular localization of iron in the rat brain, *Neuroscience* **11:**595–603.

Hill, J. M., Ruff, M. R., and Weber, R. J., 1985, Transferrin receptors in rat brain: Neuropeptide-like pattern and relationship to iron distribution, *Proc. Natl. Acad. Sci. USA* **82:**4553–4557.

Jeffries, W. A., Brandon, M. R., Hunt, S. V., Williams, A. F., and Mason, D. Y., 1984, Transferrin receptor on endothelium of brain capillaries, *Nature* **132:**162–163.

Keefer, R. C., Barak, A. J., and Boyett, J. D., 1970, Binding of manganese and transferrin in rat serum, *Biochim. Biophys. Acta* **221:**390–393.

Keen, C. L., 1995, Overview of manganese toxicity, in: *Proceedings of the Workshop on the Bioavailability and Oral Toxicity of Manganese,* (S. Velazquez, EPA Liaison), US EPA, Environmental Criteria and Assessment Office, pp. 3–11.

Komura, J., and Sakamoto, M., 1994, Chronic oral administration of methylcyclopentadienyl manganese tricarbonyl altered brain biogenic amines in the mouse: comparison with inorganic manganese, *Toxicol. Lett.* **73:**65–73.

Koster, J. F., and Sluiter, W., 1994, Physiological relevance of free radicals and their relation to iron, in: *Free Radicals in the Environment, Medicine and Toxicology* (N. Nohl, H. Esterbauer, and C. Rice-Evans, eds.), Richelieu Press, London, pp. 409–427.

Kutty, R. K., and M. D. Maines, 1987, Characterization of an NADH-dependent haem-degrading system in ox heart mitochondria, *Biochem. J.* **246:**467–474.

Lehninger, A. L., 1972, The coupling of Ca transport to electron transport in mitochondria, in: *Molecular Basis of Electron Transport* (J. Schultz and B. F. Cameron, eds.), Academic Press, New York, pp. 133–151.

Liccione, J. J., and Maines, D. M., 1988, Selective vulnerability of glutathione metabolism and cellular defense mechanisms in rat striatum to manganese, *J. Pharmacol. Exp. Therap.* **247:**157–161.

Liccione, J. J., and Maines, D. M., 1989, Manganese-mediated increase in the rat brain mitochondrial cytochrome P-450 and drug metabolism activity: Susceptibility of the striatum, *J. Pharmacol. Exp. Therap.* **248:**222–228.

Lill, D. W., Mountz, J. M., and Darji, J. T., 1994, Technetium-99m-HMPAO brain SPECT evaluation of neurotoxicity due to manganese toxicity, *J. Nuclear Med.* **35:**863–866.

London, R. E., Toney, G., Gabel, S. A., and Funk, A., 1989, Magnetic resonance imaging studies of the brains of anesthetized rats treated with manganese chloride, *Brain Res. Bull.* **23:**229–235.

Marinho, C. R., and Manso, C. F., 1993, O_2 generation during neuromelanin synthesis. The action of manganese, *Acta Med. Portug.* **6:**547–554.

Maynard, L. S., and Cotzias, G. C., 1955, Partition of manganese among organs and intracellular organelles of the rat, *J. Biochem. Chem.* **214:**489–495.

Mena, I., Horiuchi, K., and Lopez, G., 1974, Factors enhancing entrance of manganese into brain: iron deficiency and age, *J. Nuc. Med.* **15:**516.

Millar, D. M., Buttner, G. R., and Aust, S. D., 1990, Transition metals as catalysts of "autooxidation" reactions, *Free Rad. Biol. Med.* **8:**95–108.

Miller, W. C., and DeLong, M., 1986, Altered tonic activity of neurons in the globus pallidus and subthalamic nucleus in the primate MPTP model of parkinsonism, in: *The Basal Ganglia,* Volume 2 (M. B. Carpenter, and A. Jayarman, eds.), Plenum Press, New York, pp. 415–427.

Mitchell, I. J., Cross, A. J., Sambrook, M. A., and Crossman, A. R., 1986, Neural mechanisms mediating 1-methyl-4-phenyl-1,2,3,4-tetrahydropyridine-induced Parkinsonism in the monkey: Relative contributions of the striatopallidal and striatonigral pathways as suggested by 2-deoxyglucose uptake, *Neurosci. Lett.* **63:**61–66.

Morris, C. M., Keith, A. B., Edwardson, J. A., and Pullen, R. G. L., 1992a, Uptake and distribution of iron and transferrin in the adult brain, *J. Neurochem.* **59:**300–306.

Morris, C. M., Candy, J. M., Keith, A. B., Oakley, A., Taylor, G., Pullen, R. G. L., C. A. Bloxham, Gocht, A., and Edwardson, J. A., 1992b, Brain iron homeostasis, *J. Inorganic Biochem.* **47:**257–265.

Murphy, V. A., Wadhwani, K. C., Smith, Q. R., and Rapoport, S. I., 1991, Saturable transport of manganese (II) across the rat blood–brain barrier, *J. Neurochem.* **57:**948–954.

Nagy, J. I., Carter, D. A., and Fibiger, H. C., 1978, Evidence for a GABA-containing projection from the enopenduncular nucleus to the lateral habenula in the rat, *Brain Res.* **145:**360–364.

Newland, M. C., Ceckler, T. L., Kordower, J. H., and Weiss, B., 1989, Visualizing manganese in the basal ganglia with magnetic resonance imaging, *Exp. Neurol.* **106:**251–258.

Parenti, M., Flauto, C., Parati, E., Vescovi, A., and Groppetti, A., 1986, Manganese neurotoxicity: Effect of L-DOPA and pargyline treatments, *Brain Res.* **367:**8–13.

Parenti, M., Rusconi, L., Cappabianca, V., Parati, E., and Groppetti, A., 1988, Role of dopamine in manganese neurotoxicity, *Brain Res.* **473:**236–240.

Partridge, W. M., Eisenberg, J., and Yang, J., 1987, Human blood–brain barrier transferrin receptor, *Metabolism* **36:**892–895.

Perry, T. L., Godin, D. V., and Hansen, S., 1982, Parkinson's disease: a disease due to nigral glutathione deficiency?, *Neurosci. Lett.* **33:**305–310.

Prohaska, J. R., 1987, Function of trace elements in brain metabolism, *Physiol. Rev.* **67:**858–901.

Puppo, A., and Halliwell, B., 1988, Formation of hydroxyl radicals from hydrogen peroxide in the presence of iron. Is haemoglobin a biological Fenton reagent? *Biochem. J.* **249:**185–190.

Rabin, O., Hegedus, L., Bourre, J. M., and Smith, Q. R., 1993, Rapid brain uptake of manganese(II) across the blood-brain barrier, *J. Neurochem.* **61:**509–517.

Roy, B. P., Paice, M. G., Archibald, F. S., Misra, S. K., and Misiak, L. E., 1994, Creation of metal-complexing agents, reduction of manganese dioxide, and promotion of manganese peroxidase-mediated Mn(III) production by cellobiose:quinone oxidoreductase from *Trametes versicolor, J. Biol. Chem.* **269:**19745–19750.

Scheuhammer, A. M., and Cherian, M. G., 1985, Binding of manganese in human and rat plasma, *Biochim. Biophys. Acta* **840:**163–169.

Siesjo, B. K., 1988, Acidosis and ischemic brain damage, *Neurochem. Pathol.* **9:**31–88.

Sloot, W. N., and Gramsbergen, J. B., 1994, Axonal transport of manganese and its relevance to selective neurotoxicity in the rat basal ganglia, *Brain Res.* **657:**124–132.

Sokoloff, L., 1974, Changes in enzyme activities in neural tissues with maturation and development of the nervous system, in: *The Neurosciences: Third Study Program* (F. O. Schmitt and F. G. Worden, eds.), MIT Press, Cambridge, pp. 885–898.

Suarez, N., and Eriksson, H., 1993, Receptor-mediated endocytosis of a manganese complex of transferrin into neuroblastoma (SHSY5Y) cells in culture, *J. Neurochem.* **61:**127–31, 1993.

Sun, A. Y., Yang, W. L., and Kim, H. D., 1993, Free radical and lipid peroxidation in manganese-induced neuronal cell injury, *Annal. NY Acad. Sci.* **679:**358–363.

Takeda, A., Akiyama, T., Sawashita, J., and Okada, S., 1994, Brain uptake of trace metals, zinc and manganese, in rats, *Brain Res.* **640:**341–344.

Tampo, Y., and Yonaha, M., 1992, Antioxidant mechanism of Mn(II) in phospholipid peroxidation, *Free Rad. Biol. Med.* **13:**115–120.

USEPA, 1984, in: *Health Assessment Document for Manganese,* United States Environmental Protection Agency, EPA 600/8-83-013F.

Walaas, I., and Fonnum, F., 1979, The distribution and origin of glutamate decarboxylase and choline acetyltransferase in ventral pallidum and other basal forebrain regions, *Brain Res.* **177:**325–336.

Wedler, F. C., 1993, Biological significance of manganese in mammalian systems, in: *Progress in Medicinal Chemistry,* Volume 30 (G. P. Ellis, and D. K. Luscombe, eds.), Elsevier Science P., Amsterdam, pp. 89–133.

Wedler, F. C., and Denman, R. B., 1984, Glutamine synthetase: the major Mn (II) enzyme in mammalian brain, *Curr. Top. Cell. Regul.* **24:**153–169.

Wedler, F. C., Denman, R. B., and Roby, W. G., 1982, Glutamine synthetase from bovine brain is a manganese (II) enzyme, *Biochemistry* **21:**6389–6396.

Wedler, F. C., and Ley, B., 1994, Kinetic, ESR, and trapping evidence for in vivo binding of Mn (II) to glutamine synthetase in brain cells, *Neurochem. Res.* **19:**139–144.

Wedler, F. C., Vichnin, M. C., Ley, B. W., Tholey, G., Ledig, M., and Copin, J. C., 1994, Effects of Ca(II) ions on Mn(II) dynamics in chick glia and rat astrocytes: Potential regulation of glutamine synthetase, *Neurochem. Res.* **19:**145–151.

Yanagihara, R., 1982, Heavy metals and essential minerals in motor neuron disease, in: *Human Motor Neuron Diseases* (L. P. Rowland, ed.), Raven Press, New York, 233–247.

Yase, Y., 1972, The pathogenesis of amyotrophic lateral sclerosis, *Lancet* **2:**292–296.

Zhang, P., Anglade, P., Hirsch, E. C., Javoy-Agid, F., and Agid, Y., 1994, Distribution of manganese-dependent superoxide dismutase in the human brain, *Neuroscience* **61:**317–330.

Chapter 6

The Role of Zinc in Brain and Nerve Functions

Ananda S. Prasad

6.1. INTRODUCTION

Although the role of zinc in microorganisms, plants, and animals has been known for many years, its role in humans was recognized only in the early 1960s (Prasad *et al.,* 1963). Severe growth retardation, hypogonadism in males, hepatosplenomegaly, rough and dry skin, mental lethargy, and susceptibility to infections were reported in zinc-deficient humans from the Middle East in 1963 (Prasad *et al.,* 1963).

During the past three decades, nutritional deficiency of zinc has been reported from many parts of the world, including the United States (Prasad, 1993). A conditioned deficiency of zinc has been observed to occur in patients with liver disease, sickle cell disease, malabsorption syndrome, renal disease, drug-induced disorders, and following excessive alcohol ingestion. Neurosensory changes such as abnormal taste, abnormal dark adaptation, and delayed wound healing have been related to zinc deficiency in humans. Recent studies show that zinc plays an important role in cell-mediated immune functions.

At present, we know of approximately 300 enzymes that require zinc for their activities (Prasad, 1993). Equally important, we now know that there may be a large number of nucleoproteins containing zinc that are involved in gene expression of various proteins. In this review, the role of zinc in brain and nerve functions will be summarized.

Ananda S. Prasad   Division of Hematology–Oncology, Department of Internal Medicine, Wayne State University School of Medicine, and Harper Hospital, Detroit, Michigan 48201.
Metals and Oxidative Damage in Neurological Disorders, edited by Connor. Plenum Press, New York, 1997.

6.2. ZINC AND CELL DIVISION

Zinc is required for the development of primitive neural tubule. In the offspring of zinc-deficient female rats, agenesis and dysmorphogenesis of the brain, spinal cord, eyes, and olfactory tract have been observed. Hydrocephalus caused by closure of the aqueducts of sylvius has been reported as a result of zinc deficiency. The pattern of early brain malformations are consistent with impaired mitosis during embryonic development, and suggests the involvement of zinc in DNA synthesis and cell division. It has been suggested that the developing brain is more sensitive to zinc deficiency with respect to cell division than are other organs (Eckert and Hurley, 1977).

Zinc is required for several enzymes in transcription and translation, but the primary role of zinc in cell division appears to be its effect on the enzyme thymidine kinase (TK), which is widely recognized to represent a rate-limiting step in DNA synthesis (Prasad and Oberleas, 1974; Dreosti, 1984).

The teratogenicity of zinc is widely ascribed to impaired DNA synthesis during embryonic development and to the resulting asynchrony in histogenesis and organogenesis, which would distort the differential rate of growth necessary for normal morphogenesis (Swenerton et al., 1969; Dreosti et al., 1972).

Dreosti (1993) reported that the activity of TK fell more in the brain (53%) than in the liver (34%) of zinc-deficient 20 day-old fetuses compared to restricted fed controls. The TK pathway for DNA synthesis represents a mechanism for salvaging existing thymidine, while the other pathway utilizes the enzyme thymidylate synthetase and permits synthesis of the metabolite de novo. It appears that the brain has a greater reliance on the supply of preformed nucleotides for DNA synthesis than does the liver (Dreosti, 1984).

It has been proposed that widespread zinc deficiency in the Middle East may account for the greater incidence of teratology in that region (Sever and Emanuel, 1973; Cavdar et al., 1980). Zinc deficiency also is known to be teratogenic in chicks and rats (Blamberg et al., 1960; Hurley and Swenerton, 1966).

Zinc plays a role in the polymerization of tubulin into microtubules. Microtubule reassembly is reduced in brain extracts from zinc-deficient animals because of impairment in tubulin repolymerization (de la Torre et al., 1981), and this leads to defective neural folds.

6.3. TRANSPORT OF ZINC

Studies utilizing ^{65}Zn suggest that zinc enters the brain via transport sites that are not anatomically well defined (Kasarskis, 1984). The net transport of zinc is increased in response to a decrease in plasma zinc concentration. Most likely zinc is taken up by neurons (and glia as well), and a proportion of zinc is transported distally by slow axonal transport (Knull and Wells, 1975). Inasmuch as zinc is involved in the assembly of microtubules, it is possible that zinc may be transported via tubulin. Zinc may be incorporated into the metalloenzymes or other ligands either in the neuronal stroma, the axon, or the synapse. Excess zinc is most likely eliminated via cerebrospinal fluid pathways and probably re-enters plasma by transport throughout the choroid plexus. The possibility that zinc may egress through other barrier tissues, such as cerebral capillaries, cannot be ruled out.

6.4. BIOCHEMICAL ROLE OF ZINC

Decreased activities of myelin marker enzyme $2'$, $3'$ cyclic nucleotide phosphohydratase and brain alkaline phosphatase as a result of zinc deficiency have been reported, and it is believed that zinc may play a role in myelination and brain maturation via these enzymes (Dreosti, 1984). In one infant with acrodermatitis enteropathica, an improvement in myelination was observed following zinc therapy (Dreosti, 1984). Zinc deficiency may have an important effect on the metabolism of the neurotransmitter glutamic acid dehydrogenase and on glutamergic neural systems (Dreosti, 1984).

Wallwork and Sandstead (1993) observed that the levels of norepinephrine and dopamine were higher in the brains of zinc-deficient weanling rats, and they suggested that behavioral changes in zinc-deficient animals may be partly related to catecholamine alterations. Further studies are needed to confirm these observations.

Of all brain regions, the hippocampus accumulates the most zinc, and the hippocampal mossy fibers are among the most densely stained using heavy metal histochemical methods and dynorphin antisera (McGinty et al., 1984). The mere colocalization of zinc and opioid peptides in mossy fibers has suggested that there may be a functional interaction between them, and it has been suggested that zinc ions may be important regulators of opioid action in brain. Thiol-reducing agents restore the binding capacities of $ZnCl_2$-treated membranes, suggesting that oxidation of opioid receptor SH groups by zinc ions may be the mechanism through which zinc blocks opioid binding.

The following evidence suggests interactions between zinc and opioid peptides in the hippocampus: 1) zinc- and enkephalin-containing opioid peptides are selectively localized in hippocampal mossy fibers, 2) zinc blocks opioid binding, 3) zinc complexes with enkephalins in vitro, and 4) zinc and opioid peptides administered intraventricularly induce limbic seizure. These findings suggest that zinc may interact with opioids in the physiological regulation of hippocampal excitability. These endogenous metabolites may act synergistically with an excitatory amino acid during hippocampal synaptic transmission. Together they could regulate normal excitability and seizure threshold in limbic circuits, possibly by altering the degree of GABA (γ-aminobutyric acid) ergic inhibition in the hippocampus.

It has been suggested that zinc may act to downregulate binding of enkephalins to an opiate receptor site (Stengaard-Pedersen et al., 1981). Other studies have also shown an effect of zinc on the receptor binding of other neuroactive substances, such as γ-aminobutyric acid, acetyl choline and benzodiazepine (Baraldi et al., 1984; Ebadi and Pfeiffer, 1984; Slevin and Kasarskis, 1985).

Anorexia is a major symptom of zinc deficiency in animals. An important role for endogenous opiate peptides in appetite regulation has been suggested. Dynorphin, a leucine-enkephalin-containing opiate peptide, is a potent inducer of spontaneous feeding, and zinc-deficient animals were found to be resistant to dynorphin-induced feeding (Essatara *et al.,* 1984a, 1984b). The effects of zinc deficiency on endogenous opiate action may include alterations in receptor affinity, a postreception defect, and/or alterations in synthesis and/or release of dynorphin.

Zinc is concentrated in the hippocampus, and it accumulates in the intrahippocampal mossy fiber pathway during the first 2–4 weeks postnatal, a period when most development of the region is known to occur (Crawford and Connor, 1972). In chronic zinc deficiency, hippocampal electrophysiology in the rat has been shown to be disturbed, suggesting that the presence of zinc in the mossy fiber boutons may be important for synaptic transmission (Dreosti, 1984).

The role of zinc in the mossy fibers may be associated with the metabolism of the neurotransmitter glutamic acid dehydrogenase, or as a stable zinc glutamate storage complex in the giant boutons (Crawford and Connor, 1972). The levels of glutamic acid and the activity of the enzyme glutamic acid dehydrogenase are both higher in the region of the horn of Ammon containing the giant boutons than in other hippocampal areas. Glutamic acid appears to be released following electrical stimulation of the mossy fibers, and binding sites with a high affinity for glutamic acid have been found on the membranes of the giant mossy fiber boutons. These findings are pertinent as they may provide a neurochemical basis for the disturbed hippocampal function in zinc-deficient rats.

Axon terminals of glutaminergic neurons account for a large proportion of synaptic contacts made in brain, and these terminals are high in zinc. Zinc appears to be released during synaptic transmission and to decrease the effect of glutamate at N-methyl-D-aspartate (NMDA) receptors, the receptor subclass most associated with excito-toxic damage. Magnesium ions also have a role in modulating these receptors. Dietary deficiency of zinc is shown to alter neurotransmission at certain glutaminergic synapses.

In the brain there are two types of GABA receptors, $GABA_A$ and $GABA_B$, the former coupled with chloride channels and the latter linked with divalent cation channels. Zinc is known to regulate the activity of glutamic acid decarboxylase (GAD), the target enzyme of GABA synthesis in nerve terminals. At low levels zinc stimulates GAD, while at higher concentrations it inhibits the activity of this enzyme. Intracerebroventricular administration of zinc may cause convulsions in rats, which has been attributed to an inhibition of GAD activity and to an inhibition of Na^+, K^+-ATPase in hippocampus and hypothalamus (Baraldi *et al.,* 1984). On the other hand, Baraldi *et al.* (1984) have reported a decrease in the amount of zinc in several brain areas associated with changes in [^{3}H] GABA binding characteristics in cases of experimentally induced hepatic coma. These observations indicate that both an increase and a decrease in zinc content of the brain may be associated with pathological states.

GABA is synthesized by α-decarboxylation of L-glutamate under the catalytic activity of GAD, which has an absolute requirement for pyridoxal phosphate (PLP). A positive correlation has been observed between the activity of GAD and the concentration of GABA. Conversion of pyridoxal to pyridoxal phosphate requires a zinc-dependent enzyme, pyridoxal kinase.

When GAD activity was assayed in the presence of 0.2 mM PLP, no differences were noted in tissues obtained from saline-treated or zinc-sulfate-treated animals. On the other hand, in the absence of PLP, GAD activity was significantly reduced in the zinc-treated sample only, in the hippocampus area (Itoh and Ebadi, 1982). Interestingly, the hippocampus had the highest concentration of zinc and the lowest level of PLP. In another experiment, zinc did not reduce the concentration of PLP. Therefore, it was suggested that zinc may have inhibited the binding of PLP to Gad apoenzyme in the hippocampus (Miller *et al.,* 1978).

In zinc-deficient weanling rat brains, norepinephrine was consistently increased, and in one experiment brain serotonin was increased while 5-hydroxyindole acetic acid was decreased (Essatara *et al.,* 1984a, 1984b). These changes were reversed upon zinc supplementation. Interestingly, the zinc level in the brain was not affected by zinc deficiency; however, the copper level was increased.

Prenatal zinc deprivation from the 14th through the 20th days of gestation resulted in depressed fetal weight, brain weight, and total brain DNA (McKenzie *et al.,* 1975). In postnatally zinc-deprived pups, the levels of brain DNA, RNA, and protein were lower at all ages compared with control pups (Fosmire *et al.,* 1975). Effects of zinc deficiency on morphology of the brain were characteristic of immaturity in general (Sandstead, 1984).

6.5. ZINC AND FREE RADICALS

The enzyme cupro-zinc superoxide dismutase in brain may protect against damage by superoxide radicals. The activity of this enzyme and that of manganese-dependent superoxide dismutase increases dramatically in the first 2 months postnatal in rat brain, and in aging rat brains the activity of superoxide dismutase declines. This is accompanied by membrane injury. Superoxide dismutase may protect catecholamines from being oxidized by superoxide, and indeed during development its activity is much higher in catecholamine-rich areas of the brain. Zinc deficiency may alter the activity of the enzyme copper-zinc superoxide dismutase (Cu-Zn SOD), resulting in an excess of free radicals, which is damaging to the cell membranes. In the rat brain, cytosolic Cu-Zn-SOD is known to decrease with aging (Vanella *et al.,* 1982).

Excess free radicals may produce a deleterious effect on membrane lipids. The cellular membranes of brain are rich in polyunsaturated fatty acids. These polyunsaturated fatty acids control the fluidity of membranes, and hence their enzymatic activities and electrophysiological properties. Peroxidation products affect the membrane asymmetry, which is essential for the proper functioning of membranes (Bourre, 1988). Moreover, inasmuch as neuronal renewal is impossible, the integrity of the nerve tissue depends largely on its ability to protect itself.

Excess free radicals are major cellular linkage agents. Cross linkage of both DNA and other macrolmolecules may be a causative factor in error accumulation, and this may be responsible for a cascade of errors in protein synthesis.

Zinc is known to induce metallothionein (MT) synthesis, primarily by increasing the rate of MT gene transcription (Prasad, 1993). The molecular weight of MT is 6000, and it contains 20 cysteine (CYS) residues. Inasmuch as MT contains a high number of CYS residues, this protein serves as an excellent scavenger of OH^- ions. Thus, zinc may play an important role in suppression of free radicals via its effect on MT synthesis.

In zinc-deficient animals, peroxidation of tissue lipids has been noted to occur (Burke and Fenton, 1985; Coppen et al., 1988). The effect of zinc may be also related to its function as an inhibitor of NADPH-dependent lipid peroxidation. NADPH oxidation is crucial to generation of H_2O_2. Malondialdehyde formation is increased three- to four-fold as a result of zinc deficiency, suggesting that lipid peroxidation is enhanced by zinc deficiency.

The in situ hybridization study revealed that MT-1 mRNA was located in several areas of brain, with the highest concentrations found in the cerebellum, hippocampus, and ventricles, and the administration of zinc dramatically enhanced MT-1 mRNA in the brain in the rat (Hao et al., 1994). MT may participate in the growth and development of brain. The occurrence of MT-1 mRNA in the endothelial cells not only may signify the presence of a zinc transport mechanism but also the availability of a specific mechanism to safeguard against toxic metals such as cadmium. The concentrations of Zn, Zn MT, and MT mRNA are high in the hippocampus, a component of the limbic system that has been linked to various psychological processes, including internal inhibition, response inhibition, attentional shift, attentional "turn out," recognition memory, long-term memory selection, contextual retrieval, spatial memory, and working memory (Hao et al., 1994). The precise mechanism of zinc action, however, needs to be delineated.

6.6. ZINC AND BRAIN FUNCTION IN ANIMALS AND HUMANS

Behavioral changes in zinc-deficient animals include impaired long-term memory, some reduction in short-term memory and learning ability, coupled with decreased activity, increased emotional lability, and greater susceptibility to stress (Caldwell et al., 1970, 1973; Sandstead, 1984). Reduced cell division and impaired synaptogenesis and myelination have been suggested as possible explanations responsible for behavioral changes seen in zinc deficiency (Sandstead, 1984; Duerre et al., 1977; Dreosti, 1993).

Caldwell et al., (1970) were the first to report behavioral changes in zinc-deficient rats. Learning and memory deficits and increased aggression resulting from zinc deficiency have now been demonstrated in rats, mice, and monkeys (Caldwell et al., 1973; Sandstead, 1984). Behavioral changes in rats have been observed both as a result of zinc deficiency occurring during prenatal or early postnatal development. In one study, impaired working memory and persistent injury across all eight regions of the hippocampus as a result of mild zinc deficiency was observed in the rats (Dreosti, 1984). These deficits included impaired learning and/or working memory, and significant trends toward reduced areas of neuropil and neuronal density in the hippocampal CA3 stratum pyramidale and the dentate stratum granulosum (Dreosti, 1984).

In humans, a direct relationship between zinc deficiency and brain dysmorphogenesis has not been established. Nonetheless, current evidence suggests that the developing human fetus is no less vulnerable to zinc depletion than are the offsprings of other species.

Mental lethargy has been reported to be a feature of the zinc-deficiency syndrome reported from the Middle East (Prasad *et al.*, 1961), and psychological disturbances occur regularly as a zinc-responsive symptom associated with acrodermatitis enteropathica. In cases of iatrogenically induced severe zinc deficiency in infants and adults, psychological disturbances include jitteriness, impaired concentration, depression, and mood lability. Patients with cirrhosis of the liver and chronic renal disease often have low plasma zinc levels and accompanying taste disorders, anorexia, and mental disturbances (Lindeman *et al.*, 1978).

Chronic alcoholics are known to be zinc deficient. Zinc concentrations are reduced in the hippocampus of alcoholic patients, and the granular cell layer of the dentate gyrus is abnormally thin, suggesting that zinc deficiency may be pathogenically involved in alcohol-related mental deterioration. In adult rats, chronic ethanol consumption led to a loss of hippocampal pyramidal cells and granular cells in the dentate gyrus (Walker *et al.*, 1980). The organization of mossy fibers in the hippocampus was significantly altered in rats exposed to alcohol prenatally (West *et al.*, 1981). One of the manifestations of the alcohol abstinence syndrome in chronic alcoholics is grand mal convulsive seizures which sometimes precede the tremulousness. Whether or not zinc deficiency plays a biochemical role in alcoholic seizures remains to be established.

6.7.   ZINC AND EPILEPSY

Trace elements analysis of hair in epileptic subjects indicated that, whereas magnesium was increased, the zinc concentration was low (Shrestha and Oswaldo, 1987). The significance of this observation is not known.

High levels of dietary zinc are known to protect against the toxic effect of lead, probably due to its inhibitory effect on lead absorption. Lead intoxication in humans causes seizures. In addition, subconvulsive doses of lead are known to enhance the convulsant activity of picrotoxin, isoniazid, mercaptopropionic acid, and strychnine. Chronic lead exposure produces cholinergic hypoactivity and catecholaminergic hyperactivity (Ebadi and Pfeiffer, 1984). Also, lead-induced seizures may be induced by the inhibition of GABAergic transmission.

In rats, the intraventricular administration of zinc or copper produced epileptic seizures, and it was proposed that the divalent ions inhibited Na^+, K^+, ATPase and hence had ouabain-like actions (Donaldson *et al.*, 1971). The intracerebroventricular administration of GABA (0.4 μmol in 10 μl) prevented the zinc-induced epileptic seizures, suggesting that zinc may interfere with GABAergic transmission.

Intravenously or intraperitoneally administered zinc, even in large doses (up to 100 mg/kg), did not cause convulsive seizures, in contrast to intracerebroventricular administration of zinc, which induced epileptic seizures in rats (Itoh and Ebadi, 1982).

It was assumed that the peripherally administered zinc became bound to various plasma proteins and was not available to the central nervous system. Although under normal conditions the level of zinc in the hippocampus is high, it is possible that the free zinc concentration is low and, therefore, GAD activity is not affected adversely. The existence of three zinc binding proteins with apparent estimated molecular weights of 15,000, 25,000, and 210,000 in the brain has been documented (Itoh and Ebadi, 1982; Itoh et al., 1983). It was shown that only intracerebroventricular, but not the intraperitoneal, administration of zinc stimulates the synthesis of zinc binding proteins in the brain.

Acute zinc deficiency has been implicated in a syndrome of unexplained neonatal convulsions (Pryor et al., 1981; Goldberg and Sheehy, 1982) which occur 4–6 days after birth. These seizures are self limited. The concentration of zinc in cerebrospinal fluid is considerably reduced in such cases. Low serum zinc concentrations have been demonstrated in untreated male epileptics (Palm and Hallmans, 1982). Phenytoin, phenobarbital, or carbamazepine, individually or in combination, have been shown to increase the plasma zinc concentration.

6.8. ZINC AND NERVE FUNCTIONS

Growing guinea pigs and chicks seem to be most prone to stiffness, abnormal gait, and hypersensitivity to touch as a result of zinc deficiency (O'Dell et al., 1989). The neuromuscular symptoms in guinea pigs and chicks were observed after approximately 4 weeks on the zinc-deficient diet, and severe signs were seen after 5–6 weeks. A single intraperitoneal dose of Zn SO_4 (50 μ mol/kg) caused remission of signs within 4–5 days, following which all signs regressed within 7 days. Zinc level decreased only in the plasma and bone in severely deficient guinea pigs, and major soft tissues such as muscle, brain, liver, and skin showed no changes. These findings suggest that major soft tissues do not serve as mobilizable stores for zinc for other critical metabolic functions. Although the bone zinc slowly mobilized during zinc deficiency, the rate of mobilization was insufficient to maintain health or even life of the animals. The neuromuscular signs related to zinc deficiency may arise from a defect in muscle, nerve, the neuromuscular junction, or a combination of these sites. The low pain threshold most likely resides in the abnormal nerve metabolism, the biochemical basis of which is unknown. Alkaline phosphatase and 2',3'-cyclic nucleotide 3'-phosphohydrotase are involved primarily with axonal myelination, while glutamate dehydrogenase, dopamine-β-hydroxylase and phenylethanolamine-N-methyl transferase function in relation to the metabolism of neurotransmitters (Dreosti, 1993). Glutamate decarboxylase appears to be needed for the synthesis of the inhibitory neurotransmitter γ-aminobutyric acid and is inhibited by intraventricularly injected excess zinc.

The critical role of zinc for normal brain development is dramatically illustrated in the severe neural teratology associated with zinc deficiency in vitro. It also appears that zinc may be equally important in regulating the release of neurotransmitters (Sloviter, 1985) and in binding of neuroactive substances to neuroreceptors, thereby modulating postsynaptic activity, possibly as an inhibitor of excitatory synapses at physiological levels and of inhibitory synapses at supranormal levels (Slevin and Kasarskis, 1985; Ebadi and Pfeiffer, 1984). These effects of zinc may account for suboptimal mental alertness and performance in chronically zinc-deficient subjects.

Patients with Alzheimer's disease (AD) exhibit progressive dementia and the gradual deposition of extracellular neuritic plaques, particularly in the hippocampus area. Neuritic plaques contain β-amyloid (Aβ), a set of oligopeptides of about 40–43 acids. β-amyloid is derived from a much larger β-amyloid precursor protein (β-APP), the product of a single gene.

Autosomal dominant forms of AD (familial AD) result in the relatively early onset of the disease. Three genes account for all the known cases of early onset FAD. The first of these genes is that for β-APP itself, located on chromosome 21, in which several point mutations within or close to the Aβ region are associated with FAD in different families (Dewji and Singer, 1996). Mutations in the gene for β-APP account for only a small fraction (2–3%) of cases of early onset FAD. However, the remaining cases of FAD are associated with mutations in two other genes, one on chromosome 14 that encodes the protein S 182, a seven-transmembrane-spanning integral protein, and the other on chromosome 1, which encodes the protein STM 2, another seven-trans-membrane integral protein. STM 2 has 67% homology in amino acid sequence with S 182. Mutations in chromosome 14 gene accounts for 70–80% of familial cases.

β-APP-like proteins are found throughout the body, but normal functions of β-APP, and for that matter functions of S 182 and STM 2, are not known. It is unclear as to what mechanistic connection may exist among these three proteins, the mutations of which collectively account for all known cases of early onset FAD.

Dewji and Singer (1996) have proposed that one or more forms of the β-APP and S 182 (or alternatively, the closely homologous STM 2) proteins, may be normal components of an intercellular signaling system. The β-APP protein on the surfaces of appropriate neurons, and the S 182 (or alternatively STM 2) proteins on the surfaces of neighboring auxiliary cells in the hippocampus and other regions of the brain would bind to one another specifically through their amino terminal extracellular domains protruding from the cell membranes. This binding would then lead to cell–cell adhesion and transmit a signal into the neuron that ultimately would cause certain normal responses. This normal signaling function, however, would not directly result in Aβ production. Instead, as a byproduct of the intercellular interaction of β-APP and S 182 (or STM 2), perhaps by the process of mutual capping of the two proteins into the membrane regions of cell–cell contact, vesicles would be pinched off the cell surfaces and incorporated into the neuronal cell. These vesicles would then fuse with multi-vesicular bodies inside the neuronal cell, where the β-APP would be proteolyzed by enzymes in the multivesicular bodies, Aβ being a product of this proteolysis. The usual intracellular traffic between the lysosomal compartment and the plasma membrane would then release the Aβ from the neuronal cell, resulting, ultimately, in the formation of the extracellular neuritic plaques containing Aβ.

Recent studies indicate that in AD and other amyloid diseases, the formation of a β-sheet structure in key proteins can dictate the assembly of the insoluble fibers that form plaques. Partial denaturation of a soluble amyloid protein, such as acidic conditions in lysosomes, could denature the protein and allow it to adopt a β-sheet conformation. Accessory proteins, including APo E, may accelerate or prevent plaque formation by binding to amyloid-β and stabilizing its β-sheet structures. It has been reported that people who carry a particular APo E allele (APo E 4) have an increased risk of developing AD. Whether APo E 4 accelerates amyloid-fiber formation more than APo E 2 or APo E 3, or whether it prevents amyloid-β aggregation less efficiently than the other two ISO forms, needs to be settled. Knowledge of the structure of amyloid peptides and the dynamics of their aggregation may lead to the development of drugs that slow plaque formation by bonding to appropriate peptides in the early states of aggregation.

Involvement of chromosome 19 in late onset AD has been confirmed by the finding of an association between AD and the apolipoprotein E locus (APo E) on chromosome 19 (Corder *et al.*, 1993). APo E has three alleles: APo E-E2, Apo E-E3, and APo E-4. A total of 80% of familial and 64% of sporadic AD late onset cases have at least one APo E-4 compared to 31% of control subjects. Corder *et al.* (1993) have concluded that APo E-4 gene dose is a major risk factor for late onset AD and, in these families, homozygosity for APo E-4 was virtually sufficient to cause AD by age 80.

It has been observed that extracts prepared from AD brains could increase the survival of rat cortical neurons *in vitro* (Erickson *et al.*, 1994). This enhanced neurotrophic activity of AD brain was due to reduction of a growth inhibitory factor (GIF) that was later shown to be a new member of the metallothionein (MT) gene family (MT-III). Erickson *et al.* (1994) have shown that AD extracts stimulated the survival of approximately twofold more rat cortical neurons than control extracts, demonstrating that AD brain possesses elevated neurotrophic activity. MT-III but not MT-1, had an inhibitory effect on neuron survival, confirming that MI-III is a specific inhibitory factor in this assay. However, in contrast to previous reports, neither MT-III, mRNA nor MT-III protein levels were significantly decreased in the AD group. Thus, the difference in neurotrophic activity between the AD and control brain samples examined in this study is probably not mediated by MT-III.

Burnet (1982) was the first to hypothesize a possible role of zinc deficiency in the pathogenesis of Alzheimer's dementia. The cerebrospinal fluid zinc has been reported to be decreased in Alzheimer's patients, although the decrease was not statistically significant (Kapaki, 1989). In general, the zinc content of different areas of brain remains relatively steady throughout the adult life (Markesbury *et al.*, 1984). However, in the hippocampus, it gradually declines from 17.8 mg/kg at 40–59 years of age to 15.2 mg/kg in the 80–99 age group.

The two most characteristic histopathological brain lesions of Alzheimer's disease are neurofibrillary tangles (NFT) and the senile plaques (SP) (Mann, 1985). NFT and SP are found mainly in the hippocampus, and their density strongly correlates with the degree of dementia (Ball, 1977). The formation of senile plaques precedes the neurofibrillar lesions (Constantinidis, 1991a, 1991b).

Endogenous zinc may serve a protective function, preventing normal levels of excitatory synaptic activity from becoming neurotoxic. Peters *et al.* (1987) demonstrated that zinc selectively blocks the action of NMDA on critical neurons. Thus, a reduction in zinc may lead to increased NMDA receptor-mediated neuronal death.

Rapid aging is one of the characteristics of Down's syndrome (trisomy 21), and the brain lesions are similar to that observed in Alzheimer's disease (Burger and Vogel, 1973; Lai and Williams, 1989). The Cu-Zn-SOD gene is present in chromosome 21, and brain zinc content is known to be decreased in patients with Down's syndrome. In Alzheimer's disease, brain cells exhibit an age-associated increase in the content of lipofuscin, which accumulates preferentially in cortex and hippocampus. The accumulation of lipofuscin is dependent on the rate of formation of free radicals. The lipofuscin is composed of polymerized products for malondialdehyde (formed by peroxydation of polyenoic acids), amine-containing lipids, proteins, and other components (Bourre, 1988).

Neurofibrillary tangles (NFT) in human encephalopathies of various etiologies may result from a functional decrease of zinc leading to an abnormal neuronal DNA and synthesis of pathological proteins such as NFT (Constantinidis, 1991a, 1991b). In encephalopathia saturnica, zinc decreases in the hippocampus displaced by lead; in Guam's encephalopathy, calcium deficiency permits the increase of toxic metals in the brain which displace zinc; in Boxer's dementia and some viral encephalitides, blood–brain barrier (BBB) is altered, and abnormal metals may reach the brain; in Down's syndrome and Alzheimer's disease, precapillary and capillary amyloidosis disturbs the BBB, and metals such as iron and aluminum increase in the brain, whereas zinc decreases, especially in the hippocampus. A deficiency of zinc enzymes of neuronal detoxification, of glutamate catabolism, and of some neurotransmitters metabolism may also contribute in the neuronal dysfunction of these encephalopathies. A non-toxic zinc compound that crosses the BBB may be useful in the treatment of these encephalopathies, and especially for AD.

Constantinidis (1991a, 1991b) has proposed the following mechanism for amyloid (AM)-induced zinc deficiency: The AM is formed within the walls of capillaries (senile plaques), disturbs the BBB, and allows the toxic metals to enter the cerebral cortex, displacing zinc in some enzymes. Zinc deficiency may affect zinc enzymes, resulting in 1) abnormal DNA metabolism, leading to abnormal protein synthesis such as paired helical filaments (PHF) and neurofibrillary tangles (NFT); 2) an excitotoxic increase of GLU; and 3) impaired neuronal detoxification due to an adverse effect of zinc deficiency on enzymes such as superoxide dismutase (SOD), carbonic anhydrase, and lactate dehydrogenase. During the window between AM formation and PHF-NFT production (14–52 months), a zinc complex crossing the BBB may be useful to prevent AM from generating PHF-NFT and normalize neuronal detoxification. Thus, treatment with zinc may prevent onset of clinical dementia in AD patients.

Zinc concentration in hippocampus is known to be decreased, whereas iron, aluminum, mercury, and silicon may be increased in AM-SP in AD patients (Constandinidis, 1991a, 1991b). CSF silicon was reported to be increased in patients with AD; however, other trace elements such as aluminum, arsenic, lead, or manganese were not altered (Hershey et al., 1983).

Bush et al. (1994) recently published data from an in vitro study of an interaction of aqueous Zn ions with β-amyloid protein (Aβ) and observed that Aβ 1–40 in Tris-saline was significantly enhanced by Zn cations. The accuracy and reproducibility of their measurements, and the extrapolation of results obtained at higher concentrations of peptide and metal to the much lower ones found physiologically, were questioned by other investigators on two grounds (Maggio *et al.,* 1995). First, their method of measurement of peptide concentration and aggregation was incorrect, and second, their extrapolation from data gathered at from 1–25 μM, (one-thousandfold lower) was unsubstantiated (Maggio *et al.,* 1995). According to Maggio *et al.* (1995), significant aggregation is induced by Zn^{++}, but at concentrations one-hundredfold higher than those reported by Bush *et al.* (1994).

Bush *et al.* (1994) demonstrated that *in vitro* concentrations of free Zn^{++} above 300 nM rapidly destabilized human Aβ I-40, a major component of Alzheimer's disease cerebral amyloid. They also cite anecdotal observation (unpublished) that administration of zinc to Alzheimer's patients resulted in a marked deterioration in a matter of four days. Based on their *in vitro* data, they speculated that zinc supplements, even in physiological amounts, may be detrimental to Alzheimer's patients.

The translation of their *in vitro* observation to *in vivo* clinical relevance in Alzheimer's disease is fallacious. Oral zinc first enters the plasma pool, then zinc is transferred to various organs in the body. The concentration of total zinc in the plasma is approximately 15μM. In the plasma, zinc is bound to albumin, α-2 macroglobulin, transferrin, haptoglobin, ceruloplasmin, and IgG (Prasad, 1993). In our earlier studies, we determined that approximately 1.5% of total plasma zinc was ultrafiltrable. Thus, approximately 225 nM total zinc is present in the ultrafiltrate. Our studies also showed that several amino acids such as histidine, glutamine, threonine, cystine, and lysine, as well as small peptides are present in the ultrafiltrate, which would bind zinc (Prasad, 1993). Metallothionein (MT, MW 6000–7000) is also present in the ultrafiltrate, and this protein has the capability to bind 7 atoms of zinc per mole of MT. Other zinc ligands in the ultrafiltrate include citrate, lactate, and phosphates. Thus, the concentration of free zinc ions in plasma ultrafiltrate is extremely low. It is truly inconceivable that the concentration of free zinc ions (Zn^{++}) in extracellular fluids *in vivo* would approximate 300 nM or higher. In the brain itself, zinc is bound to various proteins, and virtually no free zinc ions are present.

Zinc concentration of cerebrospinal fluid (CSF), when measured by careful techniques, ranges from 150–450 nM (Prasad, 1993). These values are similar to zinc in the plasma ultrafiltrate and, as mentioned earlier, inasmuch as most of the zinc in the ultrafiltrate is bound to various organic molecules, the concentration of free zinc ions (Zn^{++}) in CSF is also most probably exceedingly low. Furthermore, various reports do not show any increase in the CSF zinc in Alzheimer's disease patients (Prasad, 1993), contrary to what has been quoted by Bush *et al.* (1994). In one study (Kapaki *et al.,* 1989), zinc in CSF was assayed in six autopsy-verified Alzheimer's disease patients. In five cases zinc concentration ranged from 150–375 nM, and in one case it was 1320 nM. This patient died of an acute hemorrhagic cerebral infarction, and in this sample, presence of cells that were rich in zinc may have accounted for a high level of zinc. Thus, there is no evidence that CSF zinc is increased in Alzheimer's disease patients.

In a recent study, by the use of inductively coupled plasma source mass spectrometry (ICP-MS), increased concentrations of aluminum and silicon, and reduced concentrations of zinc and selenium in hippocampal tissues were observed (Corrigan *et al.,* 1993). The authors postulated that displacement of hippocampal zinc by heavy metals may have played an important role in producing clinical memory disturbance. We are not aware of any study that has shown an increase in zinc in hippocampal tissues of Alzheimer's disease patients.

Bush *et al.* (1994) observed a marked deterioration of cognitive functions in Alzheimer's disease patients within four days of zinc administration. We are not told the amount and mode of zinc administration, nor do we know the clinical condition of these subjects prior to zinc therapy. In view of our clinical experience with zinc for the past three decades, their observation is simply incredible. We have used zinc 50 to 150 mg/d orally therapeutically for treatment of Wilson's disease and sickle cell disease, but to date we have not observed any adverse effects on mental functions.

6.9. SUMMARY

Zinc appears to play an important role in neurobiology. Zinc deficiency during embryogenesis leads to brain malformations in fetuses. The primary role of zinc in cell division appears to be its effect on the enzyme thymidine kinase, which is known to be involved in DNA synthesis. Zinc plays a role in myelination, and zinc deficiency affects the metabolism of the neurotransmitter glutamic acid dehydrogenase. Zinc ions may be important regulators of opioid action in brain.

The enzyme cupro-zinc superoxide dismutase in brain may protect against damage by superoxide radicals. Zinc deficiency may alter the activity of the enzyme superoxide dismutase, resulting in an excess of free radicals, which is harmful to the integrity of cell membranes. Zinc is known to induce metallothionein (MT) synthesis. Inasmuch as MT contains a high number of CYS residues, this protein serves as an excellent scavenger of OH-ions. Thus zinc may play an important protective role against damage caused by free radicals.

Behavioral changes in zinc-deficient animals include impaired long-term memory, some reduction in short-term memory and learning ability, coupled with decreased activity, increased emotional lability, and greater susceptibility to stress. In zinc-deficient rats persistent injury across all eight regions of the hippocampus was observed. Mental lethargy and psychological disturbances such as impaired concentration, depression, mood lability, and learning impairment have been observed in zinc-deficient humans.

Zinc is concentrated in the hippocampus and accumulates in the intrahippocampal mossy fiber pathway during the first 2–4 weeks postnatal. In chronic zinc deficiency, hippocampal electrophysiology in the rat is disturbed, suggesting that the presence of zinc in the mossy fiber boutons may be important for synaptic transmission.

Growing guinea pigs and chicks appear to be most prone to stiffness, abnormal gait, and hypersensitivity to touch as a result of zinc deficiency. The neuromuscular signs related to zinc deficiency may arise from a defect in muscle, nerve, neuromuscular junction, or a combination of these sites.

Burnet (1982) was the first to hypothesize a possible role of zinc deficiency in the pathogenesis of Alzheimer's disease (AD). The cerebrospinal fluid zinc has been reported to be decreased in AD patients and zinc level in the hippocampus has been observed to decline with advancing age.

It has been proposed that neuro-fibrillary tangles (NFT) in human encephalopathies of various etiologies may result from a functional decrease of zinc leading to an abnormal neuronal DNA and synthesis of pathological proteins such as NFT. The other possibility is that zinc deficiency is induced by alteration in blood–brain barrier caused by amyloid deposits in the walls of the capillaries leading to the accumulation of toxic metals, which displace zinc from enzymes. Zinc deficiency in turn causes abnormal DNA metabolism leading to synthesis of abnormal proteins such as NFT. Zinc concentration of hippocampus is known to be decreased, whereas iron, aluminum, mercury, and silicon may be increased in AD patients. Further studies are required for delineating the role of zinc in AD and whether zinc supplementation is beneficial to patients with AD.

ACKNOWLEDGMENTS.   Supported in part by National Institutes of Health/National Institute of Diabetes and Digestive and Kidney Diseases Grant No. DK-31401, Food and Drug Administration Grant No. FDA-U-000457, NIH/National Cancer Institute Grant No. CA 43838, and Labcatal Laboratories.

6.10.   REFERENCES

Ball, M. J., 1977, Neuronal loss, neurofibrillary tangles and granulvacuolar degeneration in hippocampal cortex of aging and demented patients: A quantitative study, *Acta. Neuropath.* **37**:111–118.

Baraldi, M., Caselgrandi, E., and Santi, M., 1984, Effect of zinc on specific binding of GABA to rat brain membranes, in: *Neurobiology of Zinc, Part A* (C. J. Fredrickson, G. A. Howell, and E. J. Kasarskis, eds.), Wiley-Liss, New York, pp. 59–71.

Blamberg, D. L., Blackwood, U. B., Supplee, W. C., and Combs, G. F., 1960, Effect of zinc deficiency in hens on hatchability and embryonic development, *Proc. Soc. Biol. Med.* **104**:217–220.

Bourre, J. M., 1988, The effect of dietary lipids on the central nervous system in aging and disease: Importance of protection against free radicals and peroxydation, in: *Importance of Protection against Free Radicals and Peroxydation* (M. Bergener, M. Ermini, and H. B. Stahelin, eds.), Academic Press, London, pp. 141–167.

Burger, P. C., and Vogel, F. S., 1973, The development of the pathologic changes of Alzheimer's disease and senile dementia in patients with Down's syndrome, *Am. J. Pathol.* **73**:457–476.

Burke, J. P., and Fenton, M. R., 1985, Effect of zinc deficient diet on lipid peroxidation in liver and tumor subcellular membranes, *Proc. Soc. Exp. Biol. Med.* **179**:187–191.

Burnet, F. M., 1982, New horizons in the role of zinc in cellular function, in: *Clinical Applications of Recent Advances in Zinc Metabolism,* (A. S. Prasad, I. E. Dreosti, and B. S. Hetzel, eds.), Alan R. Liss, New York, pp. 181–192.

Bush, A. I., Pettingel, W. H., Multhaup, G., Paradis, M., Vonsattel, J. P., Gusella, J. F., Beyreuther, K., Masters, C. L., and Tanzi, R. E., 1994, Rapid induction of Alzheimer Aβ amyloid formation by zinc, *Science* **265**:1464–1467.

Caldwell, D. F., Oberleas, D., Clancy, J. J., and Prasad, A. S., 1970, Behavioral impairment in adult rats following acute zinc deficiency, *Proc. Soc. Exp. Biol. Med.* **133**:1417–1421.

Caldwell, D. F., Oberleas, D., and Prasad, A. S., 1973, Reproductive performance of chronic mildly zinc deficient rats and the effects on behavior of their offspring, *Nutr. Rep. Int.* **7**:309–319.

Cavdar, A. O., Arcasoy, A., Baycu, T., and Himmetoglu, O., 1980, Zinc deficiency and anencephaly in Turkey, *Teratology* **23**:141 (letter).

Constantinidis, J., 1991a, Hypothesis regarding amyloid and zinc in the pathogenesis of Alzheimer's disease: Potential for preventive intervention, *Alzheimer Disease and Associated Disorders* **5**:31–35.

Constantinidis, J., 1991b, The hypothesis of zinc deficiency in the pathogenesis of neurofibrillary tangles, *Med. Hypotheses.* **35**:319–323.

Coppen, D. E., Richardson, D. E., and Cousins, R. J., 1988, Zinc suppression of free radicals induced in cultures of rat hepatocytes by iron, T-butyl hydroperoxide, and 3 methylindole, *Proc. Soc. Exp. Biol. Med.* **189**:100–109.

Corder, E. H., Saunders, A. M., Strittmalter, W. J., Schmechel, D. E., Gaskell, P. C., Small, G. W., Roses, A. D., Haines, J. L., and Pericak-Vance, M. A., 1993, Gene dose of apolipoprotein E Type 4 allele and the risk of Alzheimer's disease in late onset families, *Science* **261**:828–829.

Corrigan, F. M., Reynolds, G. P., and Ward, N. I., 1993, Hippocampal tin, aluminum and zinc in Alzheimer's disease, *Bio. Metals.* **6**:149–154.

Crawford, I. L., and Connor, J. D., 1972, Zinc in maturing rat brain: Hippocampal concentration and localization, *J. Neurochem.* **19**:1451–1458.

de la Torre, J., Villasante, A., Corral, J., and Avila, J., 1981, Factors implicated in determining the structure of zinc tubulin-sheets: Lateral tubulin-tubulin interaction is promoted by the presence of zinc, *J. Supramol. Struct. Cell Biochem.* **17**:183–196.

Dewji, N. N., and Singer, S. J., 1996, Genetic clues to Alzheimer's disease, *Science* **271**:159–160.

Donaldson, J., St. Pierre, T., Minnich, J., and Barbeau, A., 1971, Seizures in rats associated with divalent cation inhibition of Na^+K^+ ATPase, *Can. J. Biochem.* **49**:1217–1224.

Dreosti, I. E., 1984, Zinc in the central nervous system: The emerging interactions, in *The Neurobiology of Zinc, Part A* (C. J. Fredrickson, G. A. Howell, and E. J. Kasarskis, eds.), Alan R. Liss, New York, pp. 1–26.

Dreosti, I. E., 1993, Zinc in brain development and function, in: *Essential and Toxic Trace Elements in Human Health and Disease: An Update,* (A. S. Prasad, ed.), Wiley-Liss, New York, pp. 81–90.

Dreosti, I. E., Grey, P. C., and Wilkins, P. J., 1972, Deoxyribonucleic acid synthesis, protein synthesis and teratogenesis in zinc deficient rats, *S. Afr. Med J.* **46**:1585–1588.

Duerre, J. A., Ford, K. M., and Sandstead, H. H., 1977, Effect of zinc deficiency on protein synthesis in brain and liver of suckling rats, *J. Nutr.* **107**:1082–1093.

Ebadi, M., and Pfeiffer, R. F., 1984, Zinc in neurological disorders and in experimentally induced epileptiform seizures, in *The Neurobiology of Zinc, Part B* (C. J. Fredrickson, G. A. Howell, and E. J. Kasarskis, eds.), Alan R. Liss, New York, pp. 307–324.

Eckert, C. D., and Hurley, L. S., 1977, Reduced DNA synthesis in zinc deficiency: Regional differences in embryonic rats, *J. Nutr.* **107**:855–861.

Erickson, J. C., Sewell, A. K., Jenson, L. T., Winge, D. R., and Palmiter, R. D., 1994, Enhanced neurotrophic activity in Alzheimer's disease cortex is not associated with down-regulation of metallothionein-III (GIF), *Brain Res.* **649**:297–304.

Essatara, M. B., McClain, C. J., Levine, A. S., and Morley, J. E., 1984a, Zinc deficiency and anorexia in rats: The effect of central administration of norepinephrine, muscimol and bromergocryptine, *Physiol. Behav.* **32**:479–482.

Essatara, M. B., Morley, J. E., Levine, A. S., Elson, M. K., Shafer, R. B., and McClain, C. J., 1984b, The role of endogenous opiates in zinc anorexia, *Physiol. Behav.* **32**:475–478.

Fitzgerald, D. J., 1995, Zinc and Alzheimer's disease, *Science* **268**:1920.

Fosmire, G. J., al-Ubaidi, Y. Y., and Sandstead, H. H., 1975, Some effects of postnatal zinc deficiency on developing rat brain, *Pediatr. Res.* **9**:89–93.

Goldberg, H. J., and Sheehy, E. M., 1982, Fifth day fits: An acute zinc deficiency syndrome? *Arch. Dis. Child.* **57**:632–635.

Halsted, J. A., Ronaghy, H. A., Abadi, P., Haghshenass, M., Amirhakemi, G. H., Barakat, R. M., and Reinhold, J. G., 1972, Zinc deficiency in man: Shiraz experiment, *Am. J. Med.* **53**:277–284.

Hao, R., Cerutis, D. R., Blaxall, H. S., Rodriguez-Sierra, J. F., Pfeiffer, R. F., and Ebadi, M., 1994, Distribution of zinc metallothionein I mRNA in rat brain, *Neurochem. Res.* **19**:761–767.

Hershey, C. O., Hershey, L. A., Varnes, A., Vibhakar, S. D., Lavin, P., and Strain, W. H., 1983, Cerebrospinal fluid trace element content in dementia: Clinical radiologic, and pathologic correlations, *Neurology* **33**:1350–1353.

Hurley, L. S., and Swenerton, H., 1966, Congenital malformations resulting from zinc deficiency in rats, *Proc. Soc. Exp. Biol. Med.* **123:**692–696.

Itoh, M., and Ebadi, M., 1982, The selective inhibition of hippocampal glutamic acid decarboxylase in zinc-induced epileptic seizures, *Neurochem. Res.* **7:**1287–1298.

Itoh, M., Ebadi, M., and Swanson, S., 1983, The presence of zinc binding proteins in brain, *J. Neurochem.* **41:**823–829.

Kapaki, E., Segditsa, J., and Papageorgiou, C., 1989, Zinc, copper and magnesium concentrations in serum and CSF of patients with neurological disorders, *Acta. Neurol. Scand.* **79:**373–378.

Kasarskis, E. J., 1984, Regulation of zinc homeostasis in rat brain, in: *Neurobiology of Zinc, Part A* (C. J. Fredrickson, G. A. Howell, and E. J. Kasarskis, eds.) Alan R. Liss, New York, pp. 27–37.

Knull, H. R., and Wells, W. W., 1975, Axonal transport of cations in the chick optic system, *Brain Res.* **100:**121–124.

Lai, F., and Williams, R. S., 1989, A prospective study of Alzheimer disease in Down syndrome, *Arch. Neurol.* **46:**849–853.

Lindeman, R. D., Baxter, D. J., Yunice, A. A., and Kraikitpanitch, S., 1978, Serum concentration and urinary excretions of zinc in cirrhosis, nephrotic syndrome and renal insufficiency, *Am. J. Med. Sci.* **275:**17–31.

Maggio, J. E., Esler, W. P., Stemson, E. R., Jennings, J. M., Ghilardi, J. R., and Mantyh, P. W., 1995, Zinc and Alzheimer's disease, *Science* **268:**1920–1921.

Mann, D. M. A., 1985, The neuropathology of Alzheimer's disease: A review with pathogenic, etiological and therapeutic considerations, *Mech. Ageing Dev.* **31:**213–255.

Markesbury, W. R., Ehmann, W. D., Alauddin, M., and Hossain, T. I. M., 1984, Brain trace element concentrations in aging, *Neurobiol. Aging* **5:**19–28.

McGinty, J. F., Henriksen, S. J., and Chavkin, C., 1984, Is there an interaction between zinc and opioid peptides in hippocampal neurons? in: *Neurobiology of Zinc, Part A* (C. J. Fredrickson, G. A. Howell, and W. J. Kasarskis, eds.), Alan R. Liss, New York, pp. 73–89.

McKenzie, J. M., Fosmire, G. J., and Sandstead, H. H., 1975, Zinc deficiency during the latter third of pregnancy: Effects on fetal rat brain, liver and placenta, *J. Nutr.* **105:**1466–1475.

Miller, L. P., Martin, D. L., Mazumdar, A., and Walters, J. P., 1978, Studies on the regulation of GABA synthesis: Substrate-promoted dissociation of pyridoxal-5'-phosphate from GAD, *J. Neurochem.* **30:**361–369.

O'Dell, B. L., Becker, J. K., Emery, M. P., and Browning, J. D., 1989, Production and reversal of the neuromuscular pathology and related signs of zinc deficiency in guinea pigs, *J. Nutr.* **119:**196–201.

Palm, R., and Hallmans, G., 1982, Zinc and copper metabolism in phenytoin therapy, *Epilepsia* **23:**453–461.

Peters, S., Koh, J., and Choi, D. W., 1987, Zinc selectively blocks the action of N-methyl-D-aspartate on cortical neurons, *Science* **236:**589–593.

Prasad, A. S., 1993, *Biochemistry of Zinc*, Plenum Press, New York, pp. 149–164.

Prasad, A. S., and Oberleas, D., 1974, Thymidine kinase activity and incorporation of thymidine into DNA in zinc deficient tissue, *J. Lab. Clin. Med.* **83:**634–639.

Prasad, A. S., Halsted, J. A., and Nadimi, M., 1961, Syndrome of iron deficiency anemia, hepato-splenomegaly, hypogonadism, dwarfism and geophagia, *Am. J. Med.* **31:**532–546.

Prasad, A. S., Miale, A., Farid, Z., Schulert, A., and Sandstead, H. H., 1963, Zinc metabolism in patients with the syndrome of iron deficiency anemia, hypogonadism, and dwarfism, *J. Lab. Clin. Med.* **61:**537–549.

Pryor, D. S., Don, N., and Macourt, D. C., 1981, Fifth day fits: A syndrome of neonatal convulsions, *Arch. Dis. Child.* **56:**753–758.

Sandstead, H. H., 1984, Neurobiology of zinc, in: *Neurobiology of Zinc, Part B* (C. J. Fredrickson, G. A. Howell, and E. J. Kasarskis, eds.), Alan R. Liss, New York, pp. 1–16.

Sandstead, H. H., Fosmire, G. J., McKenzie, J. M., and Halas, E. S., 1975, Zinc deficiency and brain development in the rat, *Fed. Proc.* **34:**86–88.

Sever, L. E., and Emanuel, I., 1973, Is there a connection between maternal zinc deficiency and congenital malformations of the central nervous system? *Teratology* **7:**117.

Shrestha, K. P., and Oswaldo, A., 1987, Trace elements in hair of epileptic and normal subjects, *Sci. Total Environ.* **67:**215–225.

Slevin, J. I., and Kasarskis, E. J., 1985, Effects of zinc on markers of glutamate and aspartate neurotransmission in rat hippocampus, *Brain Res.* **334:**281–286.

Sloviter, R. S., 1985, A selective loss of hippocampal mossy fiber Timm stain accompanies granule cell seizure activity induced by perforans path stimulation, *Brain Res.* **330:**150–153.

Stengaard-Pederson, K., Fredens, K., and Larsson, L. I., 1981, Enkephalin and zinc in the mossy fiber system, *Brain Res.* **212:**230–233.

Swenerton, H., Shrader, R. E., and Hurley, R. L., 1969, Zinc deficient embryos: reduced thymidine incorporation, *Science* **166:**1014–1015.

Vanella, A., Geremia, E., D'Urso, G., Tiriolo, P., Di Silvestro, K., Grimaldi, R., and Pinturo, R., 1982, Superoxide dismutase activities in aging rat brain, *Gerontology* **28:**108–113.

Walker, D. W., Barnes, D. E., Zormetzer, S. F., Hunter, B. E., and Kubanis, P., 1980, Neuronal loss in hippocampus induced by prolonged ethanol consumption in rats, *Science* **209:**711–713.

Wallwork, J. C., and Sandstead, H. H., 1993, Zinc and brain function in: *Essential and Toxic Trace Elements in Human Health and Disease: An Update* (A. S. Prasad, ed.) Wiley-Liss, New York, pp. 65–80.

West, J. R., Hodges, C. A., and Black, A. C., 1981, Prenatal exposure to ethanol alter the organization of the hippocampal mossy fibres, *Science* **211:**957–959.

Chapter 7

Delivery of Metals to Brain and the Role of the Blood–Brain Barrier

Quentin R. Smith, Olivier Rabin, and Elsbeth G. Chikhale

7.1. INTRODUCTION

Metals serve critical roles in brain as essential cofactors, catalysts, second messengers, and modulators of gene, enzyme, and receptor activity. Currently, eight metals, including calcium, magnesium, iron, copper, zinc, manganese, cobalt, and molybdenum, are known to be required for the normal development and function of the brian (Prohaska, 1987). Each must be supplied at specific levels to avoid signs of deficiency or toxic excess. Others, such as lead, aluminum, and mercury, are not essential, but are toxic if allowed to accumulate in the nervous system. Several metals, including iron, stimulate free radical formation and have been linked to oxidative damage in neurological disorders, including ischemia, stroke, and Parkinson's disease (Riederer *et al.,* 1989; Dexter *et al.,* 1989; Griffiths and Crossman, 1993).

The ability of metals to both nurture and harm has lead to the development of a wide range of mechanisms to regulate metal uptake and distribution in tissues. Nowhere is this more important than in brain, where cellular regeneration is limited and the toxic effects of metals can be both catastrophic and irreversible.

The first step in metal uptake into brain is passage across the blood–brain barrier. The blood–brain barrier is formed principally by the brain capillary endothelium and choroid plexus epithelium (Figure 7.1), which together restrict and regulate solute exchange between blood and brain interstitial fluid. This regulation is mediated at the

Quentin R. Smith, Olivier Rabin, and Elsbeth G. Chikhale   Neurochemistry and Brain Transport Section, Laboratory of Neurosciences, National Institute on Aging, National Institutes of Health, Bethesda, Maryland 20892; *current address of QRS:* Department of Pharmaceutical Sciences, Texas Tech University HSC, Amarillo, Texas 79106.
Metals and Oxidative Damage in Neurological Disorders, edited by Connor. Plenum Press, New York, 1997.

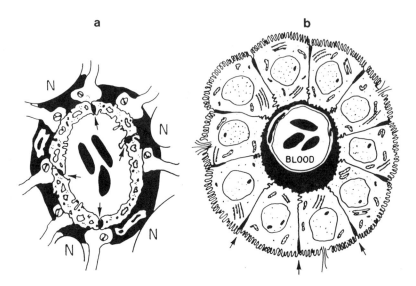

FIGURE 7.1. Schematic representation of structure of the blood–brain barrier at the cerebral capillary (a) and the choroid plexus epithelium (b). Arrows point to tight junctions. The disc-shaped elements in the core of each diagram represent erythrocytes; the darkened areas around the blood vessels represent tissue interstitial space. The abluminal surface of the brain capillary endothelium is covered with astrocytic foot processes, with the neurons (N) slightly removed. The capillaries of the choroid plexus are leaky and allow ready passage of solutes into the choroid interstitial space.

barrier by a series of active or receptor-mediated transport systems that together serve to preserve and control the environment of the nervous system.

Ten years ago, very little was known about metal uptake and regulation in the central nervous system, with the exception of the monovalent ions, Na^+, K^+, and Cl^- (Bradbury, 1979). In the past decade, however, significant progress has been made on the rates and mechanisms by which the common divalent and trivalent metals gain access to the brain. This chapter reviews current knowledge concerning metal transport systems at the blood–brain barrier and how they change with disease and age. These barrier mechanisms, together with neuronal and glial transport systems, determine the access and distribution of metals in the nervous system.

7.2.   CIRCULATION IN SERUM AND BLOOD

The first step in metal uptake into brain is delivery via the circulation. Most metals gain access to the brain from the blood, and thus their uptake is critically influenced by the form, speciation, and concentration of the metal in the circulation.

Table 7.1 summarizes typical concentrations and binding species for nine metals in serum or blood. Normally, most metals circulate in serum bound heavily to proteins (70–99.9%) (Table 7.1), including transferrin (Fe^{3+}, Mn^{3+}, Co^{3+}, Cr^{3+}, Sc^{3+}, Al^{3+}), albumin (Zn^{2+}), ceruloplasmin (Cu^{2+}), and α_2-macroglobulin (Mn^{2+}) (Smith, 1990).

For some, binding is quite tight (i.e., Fe-transferrin and Cu-ceruloplasmin), so that rates of dissociation and exchange are slow. Many metals also form low-molecular-weight complexes with organic acids (e.g., citrate), amino acids (e.g., histidine and cysteine), and inorganic ions (e.g., hydroxyl, chloride, bicarbonate ions) in serum (May *et al.,* 1977). Though usually present in low concentrations ($<10\%$ of the serum metal), these complexes often are good substrates for pore or carrier transport systems because of their small size (for review, see Dawson and Ballatori, 1995). With few exceptions, the serum concentration of free, uncomplexed metal ions is quite low (10^{-24}–10^{-8} M) and represents only a minute fraction ($<<1\%$) of the total concentration (May *et al.,* 1977). Two exceptions to this are calcium and magnesium, which circulate predominantly (50–70%) in the free, uncomplexed form (Smith, 1990).

The dominant species for brain uptake varies from metal-to-metal, with some thought to be transported primarily in the protein-bound form (e.g., Fe-transferrin) via receptor-mediated transcytosis (Fishman *et al.,* 1987; Morris *et al.,* 1992; Ueda *et al.,* 1993), and others gaining access as low-molecular-weight complexes or ions (Kerper *et al.,* 1992; Bradbury and Deane, 1993; Rabin *et al.,* 1993; Buxani-Rice *et al.,* 1994). However, the dominant mechanism should not be taken as a constant, as the speciation may change under different conditions in response to serum concentrations of metals, proteins, or low-molecular-weight binding compounds.

7.3. BLOOD–BRAIN BARRIER

Once a metal is delivered to brain via the circulation, its ability to gain access to brain interstitial fluid is determined by the transport properties of the blood–brain barrier. The blood–brain barrier, as shown in Figure 7.1, is formed at both the brain

Table 7.1
Metal Concentrations in Serum or Blood and Principal Binding Species[a]

Metal	Physiological status	Concentration (mM)	Primary serum binding proteins	Small MW complexes
Calcium	Essential	2100–2500 (S)	Albumin (40%)	Bicarbonate/ Citrate ($<10\%$)
Magnesium	Essential	800–1100 (S)	Albumin (30%)	
Iron	Essential	12–27 (S)	Transferrin ($>99\%$)	Citrate (?)
Zinc	Essential	11–20 (S)	Albumin (66%) α_2-Macroglobulin (32%)	Histidine Cysteine
Copper	Essential	15–20 (S)	Ceruloplasmin (90%) Albumin (6%)	Histidine
Manganese	Essential	0.01–0.1 (S)	Transferrin ($>95\%$)	Citrate (?)
Aluminum	Toxic	0.1–0.2 (S)	Transferrin	Citrate (?)
Lead	Toxic	0.5–0.7 (B)	Albumin	Cysteine
Methyl mercury	Toxic	0.02–0.09 (B)		Cysteine

[a]Modified from Smith (1990). S, serum; B, blood

capillary endothelium and choroid plexus epithelium by a single layer of cells joined by tight junctions (Brightman and Tao-Cheng, 1993). These junctions, with their associated ZO-1, ZO-2, and cingulin proteins (Citi *et al.*, 1988; Watson *et al.*, 1991; Dermietzel and Krause, 1991; Petrov *et al.*, 1994), closely link adjacent cell membranes, thereby sealing off the paracellular cleft to aqueous diffusion. Therefore, to cross the barrier, most solutes must either 1) dissolve in and diffuse across the lipoid endothelial cell membrane by passive diffusion (Rapoport *et al.*, 1979; Takasato *et al.*, 1984) or 2) be transported across by specific carrier, channel, or receptor mechanisms. Over 20 separate transporters are present at the barrier that facilitate and regulate the influx of organic nutrients, hormones, vitamins, and peptides into brain (Smith, 1993). The barrier also contains ion transporters (e.g., Na^+-K^+ ATPase, Ca^{2+} ATPase) and channels that aid in the secretion of cerebrospinal fluid and contribute to the regulation of the ionic homeostasis of brain extracellular fluid (Schielke and Betz, 1992).

Cerebrospinal fluid is formed at the choroid plexus epithelium and flows from the brain ventricles into the subarachnoid space where it subsequently exits the CNS at the vascular sinuses and spinal roots. The CSF cushions the brain and acts like a "neural lymph" as there is free exchange of solutes between CSF and brain interstitial fluid. Some solutes gain access to brain indirectly from the choroid plexus via the CSF pathway (Smith and Rapoport, 1986; Murphy *et al.*, 1988b). In addition, some solutes can also gain access to brain from the small areas of the brain that lack a blood–brain barrier (the "circumventricular organs"), or via retrograde transport from areas that are outside the barrier, such as the olfactory receptor cells and the terminal endings of peripheral nerves.

The restrictive transport properties of the blood–brain barrier are considerable, with permeability values 100–1000-times less than that of peripheral capillaries for many polar, hydrophilic compounds (Smith, 1990). For such compounds, half-times for uptake and interstitial fluid equilibration range from hours to days in brain, as compared to seconds to minutes in most peripheral tissues. The electrical resistance of the brain capillaries, a measure of the tightness of the barrier, is also quite large ($\sim$1900 ohm·cm²), comparable to that of the tightest epithelial tissues and on the same order of magnitude of some cell membranes (Crone and Olesen, 1982). The combination of the low passive permeability, together with the facilitated and active transport, creates a highly regulated environment for optimal brain function. The protective and regulatory functions of the barrier are apparently necessary for higher neural function as a barrier system is found in all vertebrates and most advanced invertebrates (Abbott, 1992), and starts to form during the first trimester of fetal human life.

7.4. TRANSPORT METHODS AND KINETIC ANALYSIS

Most studies to examine blood–brain barrier metal transport use one of two techniques. In the first, a metal radionuclide is injected or infused intravenously into animals and the time course of tracer accumulation in brain is examined relative to the time course of plasma concentration. This method provides the most physiologic estimates of brain metal uptake and allows the effects of metal concentration and inhibitors to be examined under normal *in vivo* conditions (Bradbury and Deane, 1986; Murphy *et al.*, 1991a; Pullen *et al.*, 1991; Morris *et al.*, 1992; Ueda *et al.*, 1993). However, the inherent difficulty of these *in vivo* methods, have led to development of short-term perfusion methods where the brain circulation is briefly taken over with artificial fluid containing the metal radionuclide of interest (Takasato *et al.*, 1984; Fishman *et al.*, 1987; Deane and Bradbury, 1990; Rabin *et al.*, 1993). The brain perfusion approach allows complete control over fluid composition, so that the levels of serum proteins, ions, metal competitors, and organic ligands can be varied at will. The results obtained with these whole animal methods can be compared to those using isolated brain microvessels or cultured cerebrovascular endothelium to further define and isolate transport mechanisms (Pardridge *et al.*, 1987). Ultimately, when transport systems are identified, their expression and regulation can be studied with antibodies and probes of mRNA (Jefferies *et al.*, 1984; Bloch *et al.*, 1987; Ueda *et al.*, 1993).

The kinetics of metal uptake into brain have been analyzed using simple "compartmental" models that allow estimation of permeability coefficients and influx rates. For a given metal, the total influx of the metal into brain or CSF (J_{tot}) is the sum of the contributions (ΣJ_i) of the individual species ($i = 1, 2, 3, \ldots n$),

$$J_{tot} = \Sigma J_i = \Sigma P_i A C_i \tag{1}$$

where influx for each species (J_i) is defined as the product of barrier permeability (P_i), surface area (A) and serum concentration (C_i). Because each species makes up a certain fraction (f_i) of the total serum concentration of the metal (C_{tot}) ($f_i = C_i/C_{tot}$), Eq. 1 can be written alternatively as,

$$J_{tot} = \Sigma J_i = \Sigma P_i A f_i C_{tot} \tag{2}$$

In most *in vivo* or perfusion experiments, the metal radionuclide is delivered via the circulation and the time course of tracer accumulation in brain or CSF is determined in relation to serum exposure. If the time of exposure is limited so that uptake is unidirectional (Rapoport *et al.*, 1979; Smith, 1989), then the rate of tracer influx (J^*_{tot}) can be expressed as,

$$J^*_{tot} = dq^*_{br}/dt = \Sigma P_i A C^*_i \tag{3}$$

where q^*_{br} is the quantity of tracer in brain, corrected for vascular isotope, at the time of measurement, t is the circulation time, and C^*_i is the serum tracer concentration for each species of interest. Eq. 3 can be integrated from the time of tracer injection ($t = 0$) to the time of brain measurement (T) to provide,

$$q^*_{br} = \Sigma P_i A \int_0^T C^*_i dt \tag{4}$$

If only one species of radiotracer is present in serum that contributes to brain uptake, the blood–brain barrier PA for that species can be obtained explicitly by rearranging Eq. 4 and solving for $P_i A$,

$$P_i A = q_{br}^* / \int_0^T C_i^* dt \tag{5}$$

Alternatively, if multiple species are present that contribute to brain uptake, a general index of the ability of the metal to cross the blood–brain barrier can be obtained as,

$$K_{in} = q_{br}^* / \int_0^T C_{tot}^* dt \tag{6}$$

where K_{in} is a blood-to-brain transfer constant and $C_{tot}^* = \Sigma C_i^*$. If all the relevant species of the metal in serum are labeled equally (i.e., the specific activity is the same) so that the fractional distribution of label among the species ($f_i^* = C_i^*/C_{tot}^*$) is the same as that of the metal ($f_i^* = f_i$), then K_{in} is the true transfer constant for brain uptake of the metal and $J_{tot} = K_{in} C_{tot}$ is the total brain influx rate. If not all the species are labeled equally ($f_i^* \neq f_i$), but the fractional distribution of label among species remains effectively constant over the time course of the experiment, then the measured K_{in} will not represent the true transfer coefficient, but instead will represent a composite term (K_{in}^*) defined as,

$$K_{in}^* = \Sigma f_i^* P_i A. \tag{7}$$

This latter case is common for metals with different oxidation states (i.e., Fe^{3+} vs. Fe^{2+} and Mn^{2+} vs. Mn^{3+}) where it is difficult to get all the serum species labeled to the same specific activity (Murphy *et al.*, 1991a; Morris *et al.*, 1992; Ueda *et al.*, 1993).

Unidirectionality of influx is assessed by examining the time course of tracer uptake into brain and ensuring that accumulation is linear and extrapolates to zero as predicted by Eq. 4. An example of this is shown in Figure 7.2 for $^{54}Mn^{2+}$. The brain concentration at each time point is divided by the corresponding serum concentration (i.e., q_{br}^*/C_{tot}^*) to obtain an effective "uptake space."

$$\text{Uptake Space} = q_{br}^*/C_{tot}^* = (K_{in} \text{ or } K_{in}^*)T' \tag{8}$$

where $T' = [\int C_{tot}^* dt]/C_{tot}^*$. If the serum tracer concentration is maintained constant, then $T' = T$ and $\int C_{tot}^* dt = C_{tot}^* \times T$. Linear regression provides K_{in} or K_{in}^* as

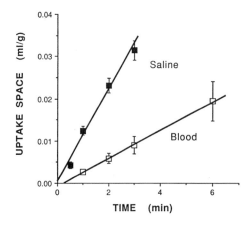

FIGURE 7.2. Time course of $^{54}Mn^{2+}$ uptake into rat cerebral cortex during perfusion with rat whole blood or physiologic saline containing $^{54}Mn^{2+}$ and [^{14}C]sucrose. The uptake space was calculated by dividing the quantity of ^{54}Mn per gram brain at the end of the experiment by the concentration of ^{54}Mn in serum or saline perfusate. Values were corrected for intravascular tracer by subtracting the uptake space for [^{14}C]sucrose from that for ^{54}Mn. Values represent means ± SE for $n = 3–8$ animals. Best fit lines were obtained by linear regression.

the slope of q^*_{br}/C^*_{tot} vs. T. Brain tracer that is trapped or bound to the walls of brain blood vessels, and that has not crossed the blood-brain barrier, can be removed at the end of the experiment by brief vascular perfusion with tracer-free saline, as described by Bradbury and Deane (1986) and Rabin et al. (1993). Alternatively, the vascular tracer can be incorporated into the analysis by plotting q^*_{tot}/C^*_{tot} vs. T' using the equation,

$$q^*_{tot}/C^*_{tot} = K^*_{in}T' + (q^*_{vas}/C^*_{tot}) \tag{9}$$

where q^*_{tot} is the total brain tracer content and q^*_{vas} is the amount bound to the blood vessel wall or trapped in residual blood (Smith, 1989). Care should also be taken to distinguish tracer that has actually crossed the blood–brain barrier from that which is simply trapped in the barrier, either in brain endothelial cells or choroid plexus epithelium. This can be performed using autoradiography or by fractionation of barrier components from whole brain (Fishman et al., 1987; Rabin et al., 1993).

7.5.   RATES OF METAL TRANSPORT INTO BRAIN AND CSF

Over the past 10 years, a number of studies have used the intravenous infusion and injection methods to determine in vivo rates of metal transfer rates into brain under normal conditions. Figure 7.3 summarizes blood–brain barrier K^*_{in} values for 13

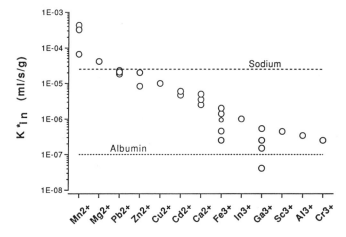

FIGURE 7.3. Blood–brain barrier K^*_{in} values for brain uptake of divalent and trivalent metals after intravenous infusion or injection in the rat. Each point represents the mean value from a separate study. Data were compiled from: Mn^{2+} (Murphy et al., 1991a; Bradbury, 1992; Rabin et al., 1993), Mg^{2+} (Smith, 1990), Pb^{2+} (Bradbury and Deane, 1986; Murphy et al., 1988a; Bradbury, 1992), Zn^{2+} (Smith, 1990; Pullen et al., 1991), Cu^{2+} (Smith, 1990), Cd^{2+} (Murphy et al., 1988a, 1989a), Ca^{2+} (Tai et al., 1986; Murphy et al., 1988b, Smith, 1990), Fe^{3+} (Banks et al., 1988; Murphy and Rapoport, 1992; Morris et al., 1992; Ueda et al., 1993; Radunovic, 1994), In^{3+} (Smith, 1990), Ga^{3+} (Murphy et al., 1988a; Pullen et al., 1990; Murphy and Rapoport, 1992; Radunovic, 1994), Sc^{3+} (Murphy and Rapoport, 1992), Al^{3+} (Radunovic, 1994), and Cr^{3+} (Smith, 1990). Values for sodium (Smith and Rapoport, 1986) and albumin (Murphy and Rapoport, 1992) are shown for comparison.

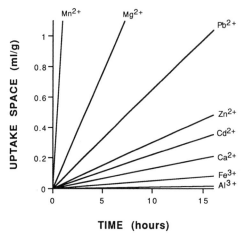

FIGURE 7.4. Rates of metal radiotracer uptake into brain at constant level of tracer in serum. Lines were generated based on K_{in}^* values reported in the literature for the rat (see Figure 7.3 legend). Uptake space (ml/g) was defined as the metal tracer content in brain, corrected for vascular activity, divided by the tracer concentration in serum. Uptake was assumed to be unidirectional because, for most metals, the steady-state concentration of unlabeled metal in brain greatly exceeds the plasma concentration (Smith, 1990).

divalent and trivalent metals that have been reported for the rat. For comparison, predicted time courses of metal uptake are illustrated in Figure 7.4.

Metals differ tremendously in their ability to gain access to brain from the circulation. For the 13 metals listed in Figure 7.3, K_{in}^* varies over 3–4 orders of magnitude with Mn^{2+} as the most permeant ($K_{in}^* = 0.5$–4.7×10^{-4} ml/s/g;) (Murphy *et al.*, 1991a; Bradbury, 1992; Rabin *et al.*, 1993), and Ga^{3+} as the least permeant ($K_{in}^* = 4.1$–53×10^{-8} ml/s/g) (Pullen *et al.*, 1990; Murphy and Rapoport, 1992; Radunovic, 1994).

The K_{in}^* values can be used to predict half-times for metal uptake and equilibration in brain, as

$$t_{1/2} = \ln 2 / K_{in}^* \tag{10}$$

assuming constant serum tracer metal concentration and a final brain/serum distribution ratio of 1.0 (Smith, 1989). Predicted half-times range from 0.5–10 hr for the more permeant metals (Mn^{2+}, Mg^{2+}, and Pb^{2+}) to 8–194 days for the less permeant metals (Ga^{3+}, Sc^{3+}, Cr^{3+}, and Al^{3+}). All the $t_{1/2}$ values are fairly large, indicating restricted uptake into the nervous system.

Metal transport rates into brain are orders of magnitude less than into other tissues, including the choroid plexus epithelium (Kasarskis, 1984; Bradbury and Deane, 1986; Murphy *et al.*, 1989a; Ueda *et al.*, 1993; Radunovic, 1994). K_{in}^* for Ca^{2+}, Pb^{2+}, and Mn^{2+} into choroid plexus epithelial cells *in vivo* exceed those into brain by ~100-fold (Murphy *et al.*, 1991a, 1991c; Rabin *et al.*, 1993). However, this cannot be taken as evidence that the choroid plexus is the primary site for metal uptake into the nervous system. Based on autoradiographic evidence and analysis of regional brain and CSF K_{in}^* values, most metals appear to gain access to brain predominantly from the cerebral blood vessels (brain $K_{in}^* \geq$ CSF K_{in}^*). Only for Ca^{2+}, which has a K_{in}^* into CSF (2–3×10^{-4} ml/s/g) ~100 times that into brain (2–5×10^{-6} ml/s/g), is the choroid plexus/CSF the principal route of entry (Tai *et al.*, 1986; Murphy *et al.*, 1989b, 1991c).

Metal transfer rates into brain also depend critically on the form and species of the metal that is presented to the brain. Serum protein binding reduces uptake rates for most metals, as shown in Figure 7.5, where P_iA values, determined in the absence of serum proteins or small molecular weight organic solutes using the perfusion method, are compared to K^*_{in} constants obtained in normal animals *in vivo*. For the four metals, the relative enhancement of uptake in the absence of protein was 5–500-fold. The P_iA values also exceed those for free calcium and sodium by 8–300-fold, suggesting that the observed transport cannot be explained by nonselective diffusion through paracellular channels or pores (i.e., tight junctions) (Crone, 1984). The rapid uptake rates of the nonprotein-bound metals are more consistent with facilitated transport through special blood–brain barrier carriers or pores. Below follows a synopsis of what is currently known concerning the blood–brain barrier transport systems for iron, manganese, zinc, and calcium, as well as the mechanisms used by the toxic metals, aluminum, lead, and methylmercury, to gain access to the nervous system.

7.6. TRANSPORT MECHANISMS

7.6.1. Iron

The brain has a large requirement for iron as a cofactor in brain energy metabolism and neurotransmitter synthesis. This requirement is met in large part by receptor mediated transport of iron into brain at the blood–brain barrier.

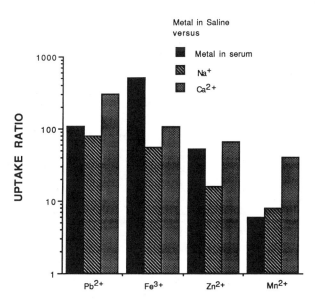

FIGURE 7.5. Comparison of metal uptake rate (P_iA) obtained from saline perfusate in the absence (or reduced levels) of protein and organic binding agents to K^*_{in} from serum and P_iA for free Na^+ or Ca^{2+}. Data for P_iA are from: Pb^{2+} (Deane and Bradbury, 1990); Fe^{3+} (Bradbury, 1994); Zn^{2+} (Buxani-Rice *et al.*, 1994); and Mn^{2+} (Rabin *et al.*, 1993). K^*_{in} values are from: Pb^{2+} (Bradbury and Deane, 1986); Fe^{3+} (Ueda *et al.*, 1993); Zn^{2+} (Pullen *et al.*, 1991); and Mn^{2+} (Rabin *et al.*, 1993). P_iA values for free Na^+ and Ca^{2+} are from Smith and Rapoport (1986) and Tai *et al.* (1986).

Iron circulates in plasma almost completely in the Fe^{3+} form bound to transferrin, an ~ 80 kD glycoprotein that serves to shuttle iron from sites of absorption and storage to tissues throughout the body (Huebers and Finch, 1987). Serum transferrin is synthesized predominantly by the liver and expression is influenced by a number of factors, including iron deficiency or excess, glucocorticoids, or estrogen (Crichton, 1991). Transferrin has two independent iron binding domains with very high affinities for Fe^{3+} of 10^{22} M^{-1} (Martin *et al.*, 1987). At normal plasma concentrations (25–40 μM), transferrin is only $\sim 30\%$ saturated with iron, and thus significant excess binding capacity exists. Other trivalent metal cations also bind tightly to transferrin, including Mn^{3+}, Co^{3+}, Cr^{3+}, and Al^{3+} (Aisen *et al.*, 1969; Martin *et al.*, 1987).

Transferrin uptake into cells is mediated by the transferrin receptor, which is a disulfide-linked, integral membrane glycoprotein with a molecular weight of $\sim 180,000$ Ds (Huebers and Finch, 1987). The receptor exhibits highest affinity for diferric-transferrin (K_d 2–7 nM), with the affinity for apotransferrin some two orders of magnitude lower. The first indication that this receptor may have an important role in brain iron uptake came from the study of Jefferies *et al.* (1984), using monoclonal antibodies, which demonstrated that the transferrin receptor was present at high levels at the luminal membrane of brain capillaries. Subsequently, Pardridge *et al.* (1987) showed that isolated brain capillaries selectively bind and endocytose diferric-^{125}I-transferrin *in vitro* with an affinity (K_d 5 nM) comparable to transferrin receptors in other tissues.

The first *in vivo* demonstration of receptor-mediated transcytosis of iron-^{125}I-transferrin across the blood–brain barrier was made by Fishman *et al.* (1987) using a rat brain perfusion system. Intact diferric-^{125}I-transferrin was found to be taken up into brain within 12–30 min by a mechanism that could be blocked by addition of excess unlabeled transferrin, consistent with saturable transport. Tissue fractionation demonstrated uptake of ^{125}I-transferrin first into capillaries, with subsequent transfer to the brain parenchyma. Experiments were performed at tracer transferrin concentration to optimize signal. Uptake of diferric-^{125}I-transferrin at normal serum transferrin concentrations (20–40 μM) (Crichton, 1991) would be expected to be much less because the K_d of the receptor (5 nM) is three orders of magnitude less than the circulating serum transferrin concentration and thus the receptor is heavily saturated ($>99\%$) *in vivo*. Competitive brain uptake of diferric-^{125}I-transferrin has recently been confirmed by Skarlatos *et al.* (1995) using a similar perfusion preparation. A role for transferrin is also supported by the fact that brain ^{59}Fe uptake: 1) is influenced by the species source of the injected transferrin (Taylor and Morgan, 1991) and 2) can be blocked by monoclonal antibodies to the transferrin receptor (Ueda *et al.*, 1993).

While the evidence for receptor-mediated iron uptake into brain via transferrin is quite strong, some controversy has arisen concerning the handling of the iron-transferrin complex once it has been endocytosed into the capillary endothelium. In an electron microscope study using horseradish peroxidase-transferrin conjugate, Roberts *et al.* (1992, 1993) found ready binding and uptake of the transferrin complex via clathrin-coated vesicles at the capillary luminal membrane, but little transcytosis to the brain interstitial space via the capillary abluminal membrane. Approximately 90% of the transferrin complex appeared to be recycled to the capillary luminal membrane. This finding is consistent with tracer studies showing more rapid entry of ^{59}Fe into brain than ^{125}I-transferrin (Banks *et al.*, 1988; Morris *et al.*, 1992). This observation has lead to the suggestion that iron is released from transferrin in acidic endosomal vesicles, similar to that of other cells (Heubers and Finch, 1987) and is then transported across the capillary abluminal membrane by an unknown mechanism (Roberts *et al.*, 1992, 1993), possibly involving reduction to Fe^{2+} (Ueda *et al.*, 1993). Bradbury (1994) has shown that Fe^{2+} is readily taken up across the capillary endothelium by a nontransferrin-dependent mechanism. It is also possible that a small fraction of brain iron gains access as Fe^{3+} bound to a low-molecular-weight complex (Murphy and Rapoport, 1992). Work involving hypotransferrinemic mice (Ueda *et al.*, 1993) has shown that iron can readily gain access to brain in the absence of transferrin, and thus nontransferrin-dependent mechanisms are present that can contribute if substrate is available.

A model illustrating the iron transport pathways across the brain capillary endothelium is illustrated in Figure 7.6. Similar pathways may be used by other metals that bind tightly to transferrin (see below).

Once within brain interstitial fluid, iron may bind to intracerebral transferrin for delivery to brain cell membranes (i.e., neurons and glia). The brain has been shown to synthesize its own transferrin (Idzerda *et al.*, 1986) and the abnormally high CSF/plasma transferrin concentration ratio is consistent with this hypothesis. However, prelimi-

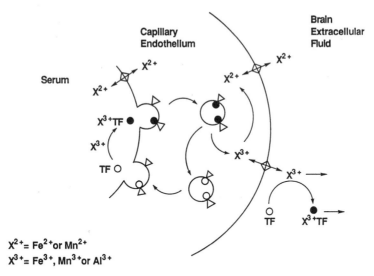

FIGURE 7.6. Model of Fe, Mn, and Al transport from blood across the brain capillary endothelium. Fe^{3+}, Mn^{3+}, and Al^{3+} bind to transferrin (○) in serum and are taken up into the capillary endothelium via transferrin receptors (△) at the capillary luminal membrane by receptor-mediated endocytosis. The internalized vesicle is processed into an endosome and the intracellular fluid acidified (<pH 5.5) to dissociate metal from transferrin. Apotransferrin is then recycled back to the luminal membrane and released into the circulation. The intracellular metal is either transported out of the cell into the brain interstitial fluid in the 3+ state or is reduced to 2+ prior to transport. Uptake can also occur via transport of Mn^{2+} and Fe^{2+}. The model is based on work of Jefferies *et al.*, 1984; Fishman *et al.*, 1987; Roskams and Connor, 1990; Murphy *et al.*, 1991a; Roberts *et al.*, 1993; and Ueda *et al.*, 1993.

nary data suggesting a critical role of the choroid plexus epithelium in this process appear valid only for the rat (Aldred *et al.*, 1987; Bloch *et al.*, 1987). Most other species exhibit only low levels of choroid plexus transferrin production (Schreiber and Aldred, 1993). Immunohistochemical studies have demonstrated transferrin receptors on cortical neurons, oligodendrocytes and choroid plexus.

7.6.2. Manganese

Manganese, like iron, is an essential metal that is required as a cofactor for enzymatic reactions. It is also thought to circulate in plasma predominantly ($>$95%) in the Mn^{3+} form bound to transferrin (Aisen *et al.*, 1969; Davidsson *et al.*, 1989), though a small fraction may also be present as Mn^{2+}, free in serum or bound weakly to albumin, transferrin, or α_2-macroglobulin (Rabin *et al.*, 1993).

Manganese likely gains access to brain in part via a transferrin-dependent mechanism (Figure 7.6). Receptor-mediated uptake of Mn^{3+}-transferrin has been shown into neuroblastoma cells in culture (Morris *et al.*, 1987; Suárez and Eriksson, 1993) and evidence in favor of a transferrin/blood–brain barrier mechanisms has been presented (Aschner and Aschner, 1990b; Aschner and Gannon, 1994). However, the slow rate at which the transferrin-dependent mechanism operates at the *in vivo* blood–brain barrier (K^*_{in} for Fe^{3+}-transferrin $= 2.5$–20×10^{-7} ml/s/g; Banks *et al.*, 1988; Murphy and Rapoport, 1992; Morris *et al.*, 1992; Ueda *et al.*, 1993; Radunovic, 1994), may allow contributions from other Mn species if the transport rates are sufficiently rapid.

Murphy *et al.* (1991a) determined the *in vivo* blood–brain barrier transfer rate for Mn^{2+} in the rat and obtained a value of $K^*_{in} = 3.2 \times 10^{-4}$ ml/s/g in the parietal cortex. This value is 3–4 orders of magnitude more rapid than Fe^{3+}-transferrin, and may suggest a significant role of Mn^{2+} in brain Mn uptake, even if Mn^{2+} represents only 0.01–1% of total serum Mn. Rabin *et al.* (1993) confirmed the rapid brain uptake of Mn^{2+} with the brain perfusion technique and demonstrated that transfer is most likely as the free ion, as addition of transferrin, albumin, or α_2-macroglobulin to the perfusate at physiologic concentrations significantly reduced uptake. Mn^{2+} transport into brain is saturable with a K_m of 1 μM and a V_{max} of 23 pmol/min/g (Murphy *et al.*, 1991a). Uptake may be mediated through an endogenous blood–brain barrier Mn^{2+} transporter, or via a channel or transporter for some other ion, as Mn^{2+} is known to be able to substitute for Ca^{2+} at Ca channels and the Na/Ca exchanger (Narita *et al.*, 1990; Frame and Milanick, 1991).

7.6.3. Zinc

Zinc enters brain at a fairly significant rate (Figure 7.3, $K^*_{in} = 0.8$–2.0×10^{-5} ml/s/g; Smith, 1990; Pullen *et al.*, 1991) and has important roles in a number of brain functions, including synaptic transmission. Recently, Buxani-Rice *et al.* (1994) examined $^{65}Zn^{2+}$ uptake with the brain perfusion technique and found that transfer (P_iA) of free, ionic zinc was quite rapid and exceeded that of *in vivo* $^{65}Zn^{2+}$ uptake (K^*_{in}) by some 50-fold (Figure 7.5). Transfer was further enhanced by addition of histidine to the perfusate. However, the effect was not stereospecific and could not be blocked by addition of large concentrations of competing amino acids, suggesting that the enhanced uptake was not mediated via an amino acid transporter. Buxani-Rice *et al.* (1994) proposed that histidine, which binds Zn^{2+}, may help deliver Zn^{2+} across the capillary glycocalyx as a $[ZnHis^+]$ complex to an endothelial zinc transporter (Figure 7.7). Zn influx into brain follows Michaelis–Menten saturation kinetics with a K_m of 16 nM and a V_{max} of 44 pmol/min/g. Most zinc in serum circulates bound to albumin and α_2-macroglobulin (Smith, 1990).

7.6.4. Calcium

Calcium uptake into CSF is mediated by saturable and nonsaturable uptake at the choroid plexus epithelium (Murphy *et al.*, 1988b, 1989b, 1991c). The K_m of the saturable system is low ($K_m \leq 0.5$ mM) and it is likely that the saturable system operates at near maximal capacity at normal plasma ionized calcium concentrations (1.2–1.6 mM). The V_{max} of the saturable Ca^{2+} influx is quite large (16–44 nmol/min/g) relative to other blood–brain barrier metal transport systems and exceeds that for Zn^{2+} by over 1000-fold.

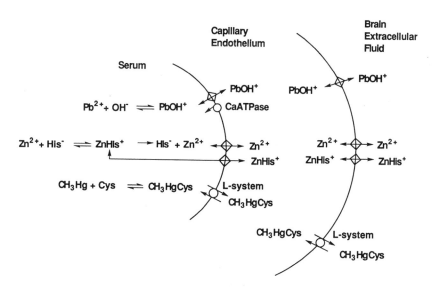

FIGURE 7.7. Summary of proposed means by which Pb^{2+}, Zn^{2+}, and CH_3Hg gain access to the brain from the circulation. Lead is taken up into brain capillary cells as $PbOH^+$, as proposed by Deane and Bradbury (1990). Intracellular Pb^{2+} may be transported back out of the cell into blood via the capillary Ca^{2+} ATPase pump. Zn^{2+} uptake into brain is facilitated by binding to histidine (His). Transport may be via a Zn^{2+} carrier (Buxani-Rice *et al.*, 1994). Methylmercury is transported into brain, bound to L-cysteine, by the cerebrovascular neutral amino acid transporter (System L) (Kerper *et al.*, 1992).

7.6.5. Toxic Metals (Aluminum, Lead, Mercury)

Aluminum, like iron, is thought to be taken up into brain in part by a transferrin-dependent mechanism (Radunovic, 1994). The binding of Al^{3+} to transferrin is not as tight as that of Fe^{3+}, and some aluminum may circulate in a small molecular weight form bound to citrate (Martin *et al.,* 1987). Roskams and Connor (1990) demonstrated that Al-transferrin binds to the transferrin receptor with a K_d (13 nM) only slightly higher than that for Fe-transferrin. The transfer of Al^{3+} into brain is quite slow ($K^*_{in} = 3 \times 10^{-7}$ ml/s/g; Radunovic, 1994), consistent with a transferrin mechanism, though it is also possible that small-molecular weight complexes (Al-citrate) may be important. Allen *et al.* (1995) have presented evidence for active efflux of Al from the central nervous system using a rat microdialysis system. The lack of an inexpensive, easily measured Al radiotracer has impeded studies of Al's uptake and handling by the central nervous system.

Lead uptake into brain proceeds at an appreciable rate ($K^*_{in} = 2 \times 10^{-5}$ ml/s/g; Bradbury and Deane, 1986; Murphy *et al.,* 1988a) consistent with its action as a potent CNS toxin. Using a perfusion approach, Deane and Bradbury (1990) have found that Pb^{2+} enters brain as a free ion, not complexed to protein or a small organic solute. Uptake was enhanced at slightly alkaline pH, and correlated with the presence of $PbOH^+$ (Figure 7.7). Unlike uptake in red cells and adrenal medullary cells (Dawson and Ballatori, 1995), Pb^{2+} transport at the brain capillary endothelium was not mediated via a Ca channel or the Cl^-/HCO_3^- exchanger, as appropriate blockers of both these systems did not affect brain Pb^{2+} uptake. Addition of metabolic inhibitors to the perfusate greatly enhanced brain Pb^{2+} influx, suggesting a role of Ca^{2+}-ATPase in back transport of lead into blood. The specific mechanism that mediates lead uptake into brain was not identified but was proposed to involve passive uptake of $PbOH^+$ (Deane and Bradbury, 1990).

Finally, methylmercury has been shown to gain access to brain bound to L-cysteine via the L System neutral amino acid transporter (Aschner and Aschner, 1990a; Kerper *et al.,* 1992). Uptake is stereospecific and can be blocked by selective ligands of the L System, including 2-aminobicyclo[2.2.1]heptane-2-carboxylic acid. Kerper *et al.* (1992) estimated a K_m for saturable methyl[203]Hg-L-cysteine uptake of 0.39 mM with a V_{max} of 33 nmol/min/g. The rapid uptake of methylmercury into brain is consistent with methylmercury's role as a potent CNS neurotoxin.

7.7. CHANGES WITH DIET, DISEASE, AND DEVELOPMENT

Because the quantitative aspects of blood–brain barrier metal transport have only begun to be examined in the past 10 years, the literature on alterations with diet, disease, and development is not extensive. Metal uptake into brain is known to be critically influenced by diet and to be enhanced for Fe, Ca, and Zn under conditions of respective metal deficiency (Kasarskis, 1984; Idzerda *et al.,* 1986; Murphy *et al.,* 1986, 1988b; Taylor *et al.,* 1991). Metals can also cross-interact so that a deficiency or excess in one metal can influence the uptake and concentration of other metals (Murphy *et al.,* 1991b, 1991d). Developmental changes have been reported in iron uptake into brain with the highest level at 15 days after birth in the rat (Taylor and Morgan, 1990). Although iron and aluminum have been suggested to have roles in the etiology of Alzheimer's disease, Kalaria *et al.* (1992) did not find statistically significant changes in the level of transferrin receptor in brain capillaries in the Alzheimer brain. The eventual identification and cloning of the barrier metal transport systems will allow much more insightful studies into the mechanisms that regulate metal access and availability to brain. Some work in this regard has been started with iron (Idzerda *et al.,* 1986).

7.8.   CONCLUSIONS

Significant progress has been made in the past 10 years on the mechanisms that regulate metal uptake into brain at the blood-brain barrier. For at least 13 metals, quantitative information is available on rates of uptake into brain, and of those, preliminary data exists on transport mechanisms for at least seven (i.e., iron, manganese, zinc, calcium, aluminum, lead, and methylmercury). The results collected thus far suggest that the blood–brain barrier has a critical role in regulating metal uptake and distribution in the central nervous system. The eventual cloning of the genes for the barrier transport systems will allow more insightful studies into the regulation of metal transport into brain and changes with diet and disease. Some work in this regard has been started with iron (Idzerda *et al.,* 1986). In the future it may be possible to selectively upregulate or downregulate barrier transporters to aid in treatment of metal toxicity or disease.

7.9.   REFERENCES

Abbott, N. J., 1992, Comparative physiology of the blood–brain barrier, in: *Physiology and Pharmacology of the Blood-Brain Barrier, Handbook of Experimental Pharmacology,* Volume 103, (M. W. B. Bradbury, ed.), Springer-Verlag, Berlin, pp. 371–396.

Aisen, P., Aasa, R., and Redfield, A. G., 1969, The chromium, manganese, and cobalt complexes of transferrin, *J. Biol. Chem.* **244:**4628–4633.

Aldred, A. R., Dickson, P. W., Marley, P. D., and Schreiber, G., 1987, Distribution of transferrin synthesis in brain and other tissues in the rat, *J. Biol. Chem.* **262:**5293–5297.

Allen, D. D., Orvig, C., and Yokel, R. A., 1995, Evidence for energy-dependent transport of aluminum out of brain extracellular fluid, *Toxicology* **98:**31–39.

Aschner, M., and Aschner, J. L., 1990a, Mercury neurotoxicity: Mechanisms of blood–brain barrier transport, *Neurosci. Biobehav. Rev.* **14:**169–176.

Aschner, M., and Aschner, J. L., 1990b, Manganese transport across the blood–brain barrier: Relationship to iron homeostasis, *Brain Res. Bull.* **24:**857–860.

Aschner, M., and Gannon, M., 1994, Manganese (Mn) transport across the rat blood–brain barrier: Saturable and transferrin-dependent transport mechanisms, *Brain Res. Bull.* **33:**345–349.

Banks, W. A., Kastin, A. J., Fasold, M. B., Barrera, C. M., and Augereau, G., 1988, Studies of the slow bidirectional transport of iron and transferrin across the blood–brain barrier, *Brain Res. Bull.* **21:**881–885.

Bloch, B., Popovici, T., Chouham, S., Levin, M. J., Tuil, D., and Kahn, A., 1987, Transferrin gene expression in choroid plexus of the adult rat brain, *Brain Res. Bull.* **18:**573–576.

Bradbury, M. W. B., 1979, *The Concept of a Blood-Brain Barrier,* Wiley & Sons, Chichester.

Bradbury, M. W. B., 1992, Trace metal transport at the blood-brain barrier, in: *Physiology and Pharmacology of the blood-brain barrier, Handbook of Experimental Pharmacology,* Volume 103, (M. W. B. Bradbury, ed.), Springer-Verlag, Berlin, pp. 263–278.

Bradbury, M. W. B., 1994, Transport of Fe^{2+} into brain during cerebrovascular perfusion in the anaesthetized rat, *J. Physiol.* **479:**37P.

Bradbury, M. W. B., and Deane, R., 1986, Rate of uptake of lead-203 into brain and other soft tissues of the rat at constant radiotracer levels in plasma, *Ann. NY Acad. Sci.* **481:**142–160.

Bradbury, M. W. B., and Deane, R., 1993, Permeability of the blood-brain barrier to lead, *Neurotoxicology* **14:**131–136.

Brightman, M. W., and Tao-Cheng, J. H., 1993, Tight junctions of brain endothelium and epithelium, in: *The Blood-Brain Barrier* (W. M. Pardridge, ed.), Raven Press, New York.

Buxani-Rice, S., Ueda, F., and Bradbury, M. W. B., 1994, Transport of zinc-65 at the blood-brain barrier during short cerebrovascular perfusion in the rat: Its enhancement by histidine, *J. Neurochem.* **62:**665–672.

Citi, S., Sabanay, H., Jakes, R., Geiger, B., and Kendrick-Jones, J., 1988, Cingulin: a new peripheral component of tight junctions, *Nature* **333:**272–276.

Crichton, R. R., ed., 1991, *Inorganic Biochemistry of Iron Metabolism,* Ellis Horwood, New York.

Crone, C., 1984, Lack of selectivity to small ions in paracellular pathways in cerebral and muscle capillaries of the frog. *J. Physiol (Lond)* **353:**317–337.

Crone, C., and Olesen, S. P., 1982, Electrical resistance of brain microvascular endothelium, *Brain Res.* **241:**49–55.

Davidsson, L., Lönnerdal, B., Sandström, B., Kunz, C., and Keen, C. L., 1989, Identification of transferrin as the major plasma carrier protein for manganese introduced orally or intravenously or after in vitro addition in the rat, *J. Nutr.* **119:**1461–1464.

Dawson, D. C., and Ballatori, N., 1995, Membrane transporters as sites of action and routes of entry for toxic metals, in: *Toxicology of Metals: Biochemical Aspects, Handbook of Experimental Pharmacology,* Volume 115, (R. A. Goyer and M. G. Cherian, eds.), Springer-Verlag, Berlin, pp. 53–76.

Deane, R., and Bradbury, M. W. B., 1990, Transport of lead-203 at the blood–brain barrier during short cerebrovascular perfusion with saline in the rat, *J. Neurochem.* **54:**905–914.

Dermietzel, R., and Krause, D., 1991, Molecular anatomy of the blood–brain barrier as defined by immunocytochemistry, *Int. Rev. Cytol.* **127:**57–109.

Dexter, D. T., Wells, F. R., Lees, A. J., Agid, F., Agid, Y., Jenner, P., and Marsden, C. D., 1989, Increased nigral iron content and alterations in other metal ions occurring in brain in Parkinson's disease, *J. Neurochem.* **52:**1830–1836.

Fishman, J. B., Rubin, J. B., Handrahan, J. V., Connor, J. R., and Fine, R. E., 1987, Receptor-mediated transcytosis of transferrin across the blood–brain barrier, *J. Neurosci. Res.* **18:**299–304.

Frame, M. D. S., and Milanick, 1991, Mn and Cd transport by the Na–Ca exchanger of ferret red blood cells, *Am. J. Physiol.* **261:**C467–C475.

Griffiths, P. D., and Crossman, A. R., 1993, Distribution of iron in the basal ganglia and neocortex in postmortem tissue in Parkinson's disease and Alzheimer's disease, *Dementia* **4:**61–65.

Huebers, H. A., and Finch, C. A., 1987, The physiology of transferrin and transferrin receptors, *Physiol. Rev.* **67:**520–582.

Idzerda, R. L., Huebers, H., Finch, C. A., and McKnight, G. S., 1986, Rat transferrin gene expression: Tissue-specific regulation by iron deficiency, *Proc. Natl. Acad. Sci. USA* **83:**3723–3727.

Jefferies, W. A., Brandon, M. R., Hunt, S. V., Williams, A. F., Gatter, K. C., and Mason, D. Y., 1984, Transferrin receptor on endothelium of brain capillaries, *Nature* **312:**162–163.

Kalaria, R. N., Sromek, S. M., Grahovac, I., and Harik, S. I., 1992, Transferrin receptors of rat and human brain and cerebral microvessels and their status in Alzheimer's disease, *Brain Res.* **585:**87–93.

Kasarskis, E. J., 1984, Zinc metabolism in normal and zinc-deficient rat brain, *Exp. Neurology* **85:**114–127.

Kerper, L. E., Ballatori, N., and Clarkson, T. W., 1992, Methylmercury transport across the blood–brain barrier by an amino acid carrier, *Am. J. Physiol.* **262:**R761–R765.

Martin, R. B., Savory, J., Brown, S., Bertholf, R. L., and Wills, M. R., 1987, Transferrin binding of Al^{3+} and Fe^{3+}, *Clin. Chem.* **33**:405–407.

May, P. M., Linder, P. W., and Williams, D. R., 1977, Computer simulation of metal-ion equilibria in biofluids: Models for the low-molecular-weight complex distribution of calcium (II), magnesium (II), manganese (II), iron (III), copper (II), and lead (II) ions in human blood plasma, *J. Chem. Soc. Dalton* 588–595.

Morris, C. M., Candy, J. M., Court, J. A., Whitford, C. A., and Edwardson, J. A., 1987, The role of transferrin in the uptake of aluminium and manganese by the IMR 32 neuroblastoma cell line, *Biochem Soc. Transactions* **15**:498–499.

Morris, C. M., Keith, A. B., Edwardson, J. A., and Pullen, R. G. L., 1992, Uptake and distribution of iron and transferrin in the adult rat brain, *J. Neurochem.* **59**:300–306.

Murphy, V. A., and Rapoport, S. I., 1992, Brain transfer coefficients for ^{67}Ga: Comparison to ^{59}Fe and effect of calcium deficiency, *J. Neurochem.* **58**:898–902.

Murphy, V. A., Smith, Q. R., and Rapoport, S. I., 1986, Homeostasis of brain and cerebrospinal fluid calcium concentrations during chronic hypo- and hypercalcemia, *J. Neurochem.* **47**:1735–1741.

Murphy, V. A., Smith, Q. R., and Rapoport, S. I., 1988a, Transfer coefficients for uptake of Ga-67, Cd-109, and Pb-203 into brain and cerebrospinal fluid (CSF), *Abstr. Soc. Neurosci.* **14**:1037.

Murphy, V. A., Smith, Q. R., and Rapoport, S. I., 1988b, Regulation of brain and cerebrospinal fluid calcium by brain barrier membranes following vitamin D-related chronic hypo- and hypercalcemia in rats, *J. Neurochem.* **51**:1777–1782.

Murphy, V. A., Smith, Q. R., and Rapoport, S. I., 1989a, Rates of tracer cadmium uptake into various tissues, *The Pharmacologist* **31**:137.

Murphy, V. A., Smith, Q. R., and Rapoport, S. I., 1989b, Uptake and concentration of calcium in rat choroid plexus during chronic hypo- and hypercalcemia, *Brain Res.* **484**:65–70.

Murphy, V. A., Wadhwani, K. C., Smith, Q. R., and Rapoport, S. I., 1991a, Saturable transport of manganese (II) across the rat blood-brain barrier, *J. Neurochem.* **57**:948–954.

Murphy, V. A., Rosenberg, J. M., Smith, Q. R., and Rapoport, S. I., 1991b, Elevation of brain manganese in calcium-deficient rats, *Neurotoxicology* **12**:255–264.

Murphy, V. A., Smith, Q. R., and Rapoport, S. I., 1991c, Saturable transport of Ca into CSF in chronic hypo- and hypercalcemia, *J. Neurosci. Res.* **30**:421–426.

Murphy, V. A., Embrey, E. C., Rosenberg, J. M., Smith, Q. R., and Rapoport, S. I., 1991d, Calcium deficiency enhances cadmium accumulation in the central nervous system, *Brain Res.* **557**:280–284.

Narita, K., Kawasaki, F., and Kita, H., 1990, Mn and Mg influxes through Ca channels of motor nerve terminals are prevented by verapamil in frogs, *Brain Res.* **510**:289–295.

Pardridge, W. M., Eisenberg, J., and Yang, J., 1987, Human blood–brain barrier transferrin receptor, *Metabolism* **36**:892–895.

Petrov, T., Howarth, A. G., Krukoff, T. L., and Stevenson, B. R., 1994, Distribution of tight junction-associated protein ZO-1 in circumventricular organs of the CNS, *Mol. Brain Res.* **21**:235–246.

Prohaska, J. R., 1987, Functions of trace elements in brain metabolism, *Physiol. Rev.* **67**:858–901.

Pullen, R. G. L., Candy, J. M., Morris, C. M., Taylor, G., Keith, A. B., and Edwardson, J. A., 1990, Gallium-67 as a potential marker for aluminium transport in rat brain: Implications for Alzheimer's disease, *J. Neurochem.* **55**:251–259.

Pullen, R. G. L., Franklin, P. A., and Hall, G. H., 1991, ^{65}Zn uptake from blood into brain in the rat, *J. Neurochem.* **56**:485–489.

Rabin, O., Hegedus, L., Bourre, J. M., and Smith, Q. R., 1993, Rapid brain uptake of manganese (II) across the blood-brain barrier, *J. Neurochem.* **61**:509–517.

Radunovic, A., 1994, Transport of aluminium, compared with 67-Ga and 59-Fe, into brain and other tissues of the young adult rat, Ph.D. thesis, University of London.

Rapoport, S. I., Ohno, K., and Pettigrew, K. D., 1979, Drug entry into the brain, *Brain Res.* **172**:354–359.

Riederer, P., Sofic, E., Rausch, W. D., Schmidt, B., Reynolds, G. P., Jellinger, K., and Youdim, M. B. H., 1989, Transition metals, ferritin, glutathione, and ascorbic acid in Parkinsonian brains, *J. Neurochem.* **52**:515–520.

Roberts, R., Sandra, A., Siek, G. C., Lucas, J. J., and Fine, R. E., 1992, Studies of the mechanism of iron transport across the blood-brain barrier, *Ann. Neurol.* **32:**S43–S50.

Roberts, R. L., Fine, R. E., and Sandra, A., 1993, Receptor-mediated endocytosis of transferrin at the blood-brain barrier, *J. Cell Sci.* **104:**521–532.

Roskams, A. J., and Connor, J. R., 1990, Aluminum access to the brain: A role for transferrin and its receptor, *Proc. Natl. Acad. Sci. USA* **87:**9024–9027.

Schielke, G. P., and Betz, A. L., 1992, Electrolyte transport, in: *Physiology and Pharmacology of the Blood–Brain Barrier, Handbook of Experimental Pharmacology,* Volume 103, (M. W. B. Bradbury, ed.), Springer-Verlag, Berlin pp. 221–243.

Schreiber, G., and Aldred, A. R., 1993, Molecular cloning of choroid plexus-specific transport proteins, in *The Blood–Brain Barrier* (W. M. Pardridge, ed.), Raven Press, New York, pp. 441–459.

Skarlatos, S., Yoshikawa, T., and Pardridge, W. M., 1995, Transport of [^{125}I]transferrin through the rat blood–brain barrier, *Brain Res.* **683:**164–171.

Smith, Q. R., 1989, Quantitation of blood–brain barrier permeability, in: *Implications of the Blood–Brain Barrier and Its Manipulation,* Volume 1, (E. A. Neuwelt, ed.), Plenum Press, New York, pp. 85–118.

Smith, Q. R., 1990, Regulation of metal uptake and distribution within brain, in: *Nutrition and the Brain,* Volume 8 (R. J. Wurtman and J. J. Wurtman, eds.), Raven Press, New York, pp. 25–74.

Smith, Q. R., 1993, Drug Delivery to brain and the role of carrier-mediated transport, in: *Frontiers in Cerebral Vascular Biology: Transport and Its Regulation,* (L. R. Drewes, and A. L. Betz, eds.), Plenum Press, New York, pp. 83–93.

Smith, Q. R., and Rapoport, S. I., 1986, Cerebrovascular permeability coefficients to sodium, potassium, and chloride, *J. Neurochem.* **46:**1732–1742.

Suárez, N., and Eriksson, H., 1993, Receptor-mediated endocytosis of a manganese complex of transferrin into neuroblastoma (SHSY5Y) cells in culture, *J. Neurochem.* **61:**127–131.

Tai, C., Smith, Q. R., and Rapoport, S. I., 1986, Calcium influxes into brain and cerebrospinal fluid are linearly related to plasma ionized calcium concentration, *Brain Res.* **385:**227–236.

Takasato, Y., Rapoport, S. I., and Smith, Q. R., 1984, An in situ brain perfusion technique to study cerebrovascular transport in the rat, *Am. J. Physiol.* **247:**H484–H493.

Taylor, E. M., and Morgan, E. H., 1990, Developmental changes in transferrin and iron uptake by the brain in the rat, *Dev. Brain Res.* **55:**35–42.

Taylor, E. M., and Morgan, E. H., 1991, Role of transferrin in iron uptake by the brain: a comparative study, *J. Comp. Physiol. B* **161:**521–524.

Taylor, E. M., Crowe, A., and Morgan, E. H., 1991, Transferrin and iron uptake by the brain: Effects of altered iron status, *J. Neurochem.* **57:**1584–1592.

Ueda, F., Raja, K. B., Simpson, R. J., Trowbridge, I. S., and Bradbury, M. W. B., 1993, Rate of ^{59}Fe uptake into brain and cerebrospinal fluid and the influence thereon of antibodies against the transferrin receptor, *J. Neurochem.* **60:**106–113.

Watson, P. M., Anderson, J. M., Vanltaille, C. M., and Doctorow, S. R., 1991, The tight junction-specific protein ZO-1 is a component of the human and rat blood–brain barriers, *Neurosci. Lett.* **129:**6–10.

Chapter 8

Ferritin

Intracellular Regulator of Metal Availability

J. G. Joshi

8.1. INTRODUCTION

Toxicity of cations results from excessive intake or from their presence at high concentrations in an unsequestered form. The toxic level and the expression of toxicity vary with the cation. For example, in experimental animals, exposure to Pb^{2+} causes, among other ill-effects, inhibition of specific enzymes involved in heme synthesis. In addition, Pb^{2+} causes replacement of iron in the heme by Zn^{2+}, to generate a nonfunctional zinc protporphyrine (Jeslow, 1980). In some instances, living systems respond to toxic levels of metal ions such as Cd^{2+}, Zn^{2+}, or Cu^{2+} by synthesizing a family of proteins, metallothioneins, to sequester the detoxicant (Kojima and Kagi, 1978).

Iron is essential to all forms of life. It exists predominantly in two oxidation states, Fe^{2+} and Fe^{3+}. At physiological pH, Fe^{2+} is highly soluble and Fe^{3+} is practically insoluble. The redox potential ($E'o$) of the Fe^{3+}/Fe^{2+} couple is $+0.77V$, but when bound to proteins and other chelators it covers an entire range, from $-0.34v$ for $NADP^+/NADPH$ to $+0.81v$ for $1/2O_2/H_2O$. Because of this versatility, the concentration of iron is the highest among all known transition metal ions found in the living systems.

During normal protein turnover the iron is released, sequestered by chelators of different molecular weights, and recycled rather than excreted. The amount of iron released is far in excess of the combined capacity of these chelators to bind it. Either form of unsequestered iron react with H_2O_2, a normal metabolite, and generate highly

J. G. Joshi   Department of Biochemistry, University of Tennessee, Knoxville, Tennessee 37996.

Metals and Oxidative Damage in Neurological Disorders, edited by Connor. Plenum Press, New York, 1997.

reactive free radicals: O_2^-, OH^-, and HO_2, which react with numerous biomolecules and in many instances adversely affect their function (Fridovich, 1989). For all forms of life a mechanism is essential to store increasing amounts of excess iron. This is provided by transferrin, which binds the circulating iron and deposits it via a specific receptor-mediated system into ferritin for storage.

Ferritin (MW $\approx$ 480,000) is composed of 24 subunits, H (heavy, MW $\approx$ 21,000) and L (light, MW $\approx$ 19,000) H chains are more acidic than L chains. Isoferritins varying in the proportions of H and L chains can be resolved by isoelectrofocusing. The subunits form a shell with an inside diameter of 80A° that can store up to 4,500 molecules of Fe^{3+} as ferrihydrite of an approximate composition of $[(Fe\ OOH)_8$ $(FeO:OPO_3H_2)]$ (Granick and Hahn, 1944). Sucrose density centrifugation resolves native ferritin varying in iron content. Thus all ferritin molecules are neither fully nor equally saturated with iron and very little, if any, apoferritin (ferritin devoid of iron) can be isolated from a native ferritin pool.

Fe^{2+} enters the protein's shell, where it is oxidized to Fe^{3+} on the ferroxidase center of H chain. L chains are more efficient in core formation. Understandably, ferritin composed of exclusively H chains (H_{24}) or L chains (L_{24}) have not been detected in any tissue. As required, iron is released as Fe^{2+}. Thus, alternate oxidation and reduction are essential steps for the storage and release of iron (Aisen and Listowsky, 1980). Ferritins rich in H chains tend to contain less iron and are found in tissues, such as heart and brain, that require protection against oxidative damage, whereas storage organs such as liver and spleen contain L-chain-rich ferritins (Aisen and Listowsky, 1980).

Regulation of Iron Metabolism

Cellular levels of iron regulate its own concentration. This is accomplished by the interactions between a stretch of a 28 nucleotide sequence, IRE (iron regulatory element), present in the 5' untranslated region of the mRNAs for H and L chains, and its several repeats in the 3' untranslated region for the transferrin receptors and IRE-binding protein (IREBP). At low levels of iron, the IREBP binds to IRE. This prevents translation of ferritin chains and stabilizes the receptor mRNA, thus increasing the concentration of the receptors as well as iron uptake. At high levels of iron, the metal complexes with the IREBP rather than to IRE, the complex leaves mRNA, and enhances the synthesis of ferritin chains. Two IREBPs have been isolated. One has been identified as the cytosolic aconitase and is termed IREBP. The details of this elegant mechanism of the posttranscriptional regulation of the bioavailability of iron have appeared in recent articles (Harford and Klausner, 1990; Klausner et al., 1993; Munro, 1993). The second IREBP is termed IRF_B (iron regulatory factor B). Its role is yet to be established.

The major function of ferritin, that of the detoxification, storage, and transport of iron, was recognized first, while its other functions (Table 8.1) are related to this major function and were recognized later (Joshi and Clauberg, 1988; Joshi and Zimmerman, 1988).

This discussion is restricted to the role of ferritin as a regulator of metal availability.

Table 8.1
Multifunctions of Ferritin

Function	Reference
1. Detoxification, storage, and transport of iron	Aisen and Listowsky, 1980
2. (i) Detoxicant for non-ferrous metal ions	Joshi *et al.*, 1984; Price and Joshi, 1984; Lindenschmidt *et al.*, 1986; Noda *et al.*, 1991; Muller *et al.*, 1988; Suarez and Eriksson, 1993
(ii) Zinc and copper ion donor	Price and Joshi, 1982
3. A source of oxygenated free radical	Mazur *et al.*, 1958; Aust, 1988; Reif, 1992
4. A "limited" dephosphorylating agent	Grant and Taborsky, 1966; Deshpande and Joshi, 1985
5. A probable contributor in energy production	Theil, 1987

8.2. FORMATION OF HOLOFERRITIN AND LOADING OF IRON INTO APOFERRITIN

In vitro the stoichiometry of iron loading depends upon the amount Fe^{2+} presented to the apoferritin. Thus, Eq. 1 below

$$Fe^{2+} + \tfrac{1}{2} O_2 + 2H_2O \rightarrow FeOOH \text{ core} + \tfrac{1}{2} H_2O_2 + 2H^+ \qquad (1)$$

changes to

$$2Fe^{2+} + O_2 + 4H_2O \rightarrow 2 FeOOH_{core} + H_2O_2 + 4H^+ \qquad (2)$$

Oxidation occurs on the ferroxidase centers of H chains when apoferritin is presented with small (24 Fe^{2+}/proteins) increments of Fe^{2+}. If, however larger amounts of Fe^{2+} are made available (240–960 Fe^{2+}/protein), four Fe^{2+} are oxidized per O_2 reduced, without the production of H_2O_2.

$$4Fe^{2+} + O_2 + 6H_2O \rightarrow 4FeOOH_{core} + 8H^+ \qquad (3)$$

Here the oxidation also occurs on the surface of the iron core. Several divalent metal ions competitively inhibit ferroxidase activity and iron uptake. (Sun and Chasteen, 1992, 1994).

8.3. REDUCTIVE RELEASE OF IRON

Interconversion of Fe^{2+} and Fe^{3+} requires transfer of a single electron. Such univalent transfer of electrons loads Fe^{2+} into ferritin as Fe^{3+}, and also reduces it to Fe^{2+} for release. *In vitro,* this reductive release can be accomplished by reductants such as ascorbate, thiols, $FMNH_2$, and superoxides, or a superoxide-generating enzyme-mediated system such as xanthine + xanthine oxidase and O_2. Superoxide anions are also formed by radiolysis, metal-catalyzed oxidation, redox cycling of xenobiotics, and single-electron redox reactions involving paraquat, adriamycin (Reif, 1992).

$$\text{Ferritin-Fe}_n^{3+} + O_2^- \rightleftarrows [\text{Ferritin-Fe}_n^{3+} \cdot O_2^-] \rightleftarrows \text{Ferritin-Fe}_{n-1}^{3+} + Fe^{2+} + O_2 \qquad (4)$$

As seen, the release of iron would continue, provided Fe^{2+} is not reoxidized to Fe^{3+}. The role of the xanthine-xanthine oxidase system in releasing the iron from ferritin was first suggested by Green and Mazur (1957), Mazur et al., (1958), and Mazur and Carleton (1965), and expanded by Aust et al., (1988). Because the rate of O_2^- production by this system is sensitive to pO_2, the reduction of Fe^{2+} will be most effective when the supply of oxygen and, therefore, of O_2^-, is not limited. Since Reaction 4 is reversible, the products of the reaction are a potential source of free radicals.

$$O_2^- + O_2^- + 2H^+ \rightarrow H_2O_2 + O_2 \qquad (5)$$

$$Fe^{2+} + H_2O_2 \rightarrow OH^- + OH\cdot + Fe^{3+} \qquad (6)$$

$$Fe^{3+} + H_2O_2 \text{ ---(very slow)}\rightarrow HO_2\cdot + H^+ + Fe^{2+} \qquad (7)$$

It is well established that while oxygenated free radicals serve a useful function in normal metabolism (Babior et al., 1988), they can modify a host of macro- and micromolecules. A defense system comprised of enzymes such as superoxide dismutases, catalase, and peroxidases scavange the free radicals and protect the organism against the ill effects. (Fridovich, 1989; McCord, 1986; Nelson et al., 1994, 1995).

Regulating the concentration of O_2^- is essential because it also affects the regulation of the synthesis of ferritin. Accordingly, O_2^- inactivates mitochondrial as well as cytosolic aconitase, (IREBP), in two ways. It not only directly interacts with IREBP to produce inactive proteins (Hansladen and Fridovich, 1994), but it also reacts with nitric oxide. The resulting peroxynitrite, ONOO, rather than nitric oxide, also rapidly inactivates aconitase (Hausladen and Fridovich, 1994; Castro et al., 1994). The combined effect could profoundly affect iron metabolism because the concentration of O_2^- and NO are not controlled by diffusion. Thus, the effect of an altered concentration of O_2^- would affect synthesis of ferritin as well as the reductive release of iron from ferritin.

8.4. FERRITIN AND BINDING OF NONFERROUS METAL IONS

Metal-binding studies have been done on ferritin as well as apoferritin. The techniques used include electron paramagnetic resonance spectroscopy (EPR), X-ray absorption fine-structure measurements, UV absorption spectroscopy, gel-filtration, centrifree technique, microcentrifuge desalting technique, and titration with ion-specific electrodes (Macara et al., 1973a, 1973b; Wardeska et al., 1986; Price and Joshi, 1983, 1984; Fleming and Joshi, 1991, Sczekan and Joshi, 1989). Studies with apoferritin have yielded less ambiguous data because the metals were bound to the sites on the protein, yielding insight into the essential microenvironment prevailing at the metal-binding sites. The binding constants varied with the metal ion and were dependent on pH and presence of the competing metal ion. Similar studies with holoferritin, the predominant species *in vivo,* showed that, compared to apoferritin, holoferritin bound substantially larger amounts of nonferrous metal ions suggesting that the iron core, as well as the phosphate in the core, participated is sequestration of metal ions. Consistent with this possibility, synthetic iron cores of a size comparable to those in ferritin bound larger quantities of metal ions than did apoferritin; the cores with phosphate bound even higher amounts, but without any change in their K_D (Sczekan and Joshi, 1989). These data suggested that ferritin may function as a general metal detoxicant (Table 8.2). The data supporting such a possibility are summarized below.

8.4.1. Ferritin and Beryllium

Beryllium is the smallest of the divalent metal ions and is also extremely toxic. The molecular mechanism of beryllium toxicity is not clear as yet. It is, however, well established that *in vivo* it has a marked affinity for cell nuclei (Witschi and Aldridge, 1968) and at micromolar concentrations specifically inhibits Na-K ATPase, alkaline phosphatase, and phosphoglucomutase (Thomas and Aldridge, 1966; Aldridge, 1966; Toda *et al.,* 1967). Extension of previous studies with phosphoglucomutase (Hashimoto *et al.,* 1967) revealed that the activity in the purified preparation was readily inhibited by micromolar concentrations of beryllium but the activity in the crude extract was only partially inhibited (Joshi *et al.,* 1984). This observation suggested the presence of a protecting factor in the crude extracts and led to the isolation from liver homogenates of a large molecular weight protein capable of binding large quantities of beryllium. The protein was identified as ferritin (Price and Joshi, 1983; Joshi *et al.,* 1984), and, as expected, *in vitro* it protected the enzymes against inhibition by beryllium and reversed the enzymes already inhibited (Price and Joshi, 1984).

In vitro, ferritin could be saturated with Cd^{2+}, Cu^{2+}, or Zn^{2+}, but not with Be^{2+}—even after 800 g atoms of Be^{2+} were bound. Significantly, Be^{2+} did not displace iron. None of the bound Be^{2+} was dialyzable at 4°C in 50 mM Tris-acetate buffer, pH 8.5, but at pH 6.5 over 80% of the bound metal was dialyzable after 72 hrs. By contrast apoferritin bound similar amounts of all four metal ions, and some of the bound metal was dialyzable. By spectrophotometric titrations at pH 6.5 of Be^{2+} with sulfosalicylic acid (SSA), $Be_{K_D^{SSA}}$ was calculated to be 5.0×10^{-6}M and by competition of SSA and ferritin for Be^{2+} $Be_{K_D^{ferritin}}$ was calculated to be 6.8×10^{-6}M (Price and Joshi, 1983).

The ability of holoferritin to bind large quantities of beryllium indicates that it seems to serve as a good protectant against beryllium toxicity *in vivo.* Indeed, rats whose ferritin synthesis in the liver was increased greater than fivefold by administration of iron salts survived otherwise toxic doses of either intravenous (Lindenschmidt *et al.,* 1986) or pulmonary (Sendelbach and Witschi, 1987) exposure to beryllium. Whether the beryllium was complexed with ferritin or hemosiderin was not established (Lindenschmidt *et al.,* 1986). Nevertheless, this protective effect of iron is compatible

Table 8.2

Metal-Binding Constants of Ferritins and Synthetic Cores[a]

	Fe content	PO_4^2 content	Zn		Cd		Be		Al	
			K_d	n	K_d	n	K_d	n	K_d	n
Core	1200[b]	186	$9.8 \cdot 10^{-6}$	8	$1.27 \cdot 10^{-7}$	2.2	$1.83 \cdot 10^{-6}$	74	$1.05 \cdot 10^{-5}$	211
Soya	1400	288	$8.01 \cdot 10^{-7}$	75	$5.62 \cdot 10^{-7}$	73	$5.55 \cdot 10^{-6}$	882	$3.06 \cdot 10^{-5}$	652
Horse[c]	1500	322	$5.25 \cdot 10^{-6}$	135	$8.01 \cdot 10^{-7}$	76	$6.8 \cdot 10^{-6}$	800	$4.52 \cdot 10^{-5}$	164

[a] Values expressed in the table are the dissociation constants (K_d) and extrapolated number of binding sites (n) for the various metals bound to either the synthetic cores or the specific protein. Number of binding sites was obtained from the Scatchard analysis x-intercept, and the relative K_d from the slope of the Scatchard plot (Sczekan and Joshi, 1989).
[b] The synthetic cores contained approximately 1200 Fe at the time of preparation. However, during dialysis, as much as 30–40% of the iron is removed prior to accretion. All values in the table for cores were derived as ratios of Fe to metal, or Fe to phosphate, and subsequently normalized to the original value.
[c] Data taken from Price and Joshi (1983).

with earlier observations where quiescent beryllium toxicity flared up in situations accompanied by heavy loss of blood such as menses, postoperative treatment, etc. (Tepper, 1972).

8.4.2. Ferritin and Zinc

Unlike beryllium, zinc is an essential metal ion; however, excess zinc is toxic. This metal ion, as well as Cu^{2+}, Cd^{2+}, and Pb^{2+}, form mercaptides, thus metallothioneins synthesized in response to excess levels of such metals chelate them and the potential toxic effects are averted. Concentrations of metal ions required to induce synthesis of metallothioneins vary with the metal ions (Kojima and Kagi, 1978) as well as other reagents injected. For example, Brown et al., (1980) studied the metabolism of Zn^{2+}, Cd^{2+}, and Cu^{2+} in pretumorous livers from mice exposed to dimethylnitrosamine, a known carcinogen. They observed that dimethylnitrosamine-exposed mice had lower concentrations of Cu^{2+} and Zn^{2+} in the cytosol of tumorous as well as pretumorous tissue, and that this reduction was due to a "heat stable, high molecular weight protein pool." Exposure to Cd^{2+} and dimethylnitrosamine resulted in the accumulation of the metal ion in the fraction that contained heat stable, high-molecular-weight proteins. One of these proteins may be ferritin, which can bind a variety of metal ions. Indeed, injection of metal salts of cadmium, zinc, and copper at levels insufficient to stimulate the synthesis of measurable amounts of metallothionein permitted sequestration of at least part of the injected metal ion into the liver ferritin (Price and Joshi, 1982). It appears therefore that a carcinogen such as dimethylnitrosamine causes a redistribution of metal ions among metallothionein and ferritin. Subsequent studies showed that the liver ferritin from rats injected with cadmium salts had more ferritin-bound Zn^{2+} than that isolated from the zinc-injected rats. Furthermore, zinc bound to ferritin activated zinc-requiring apoenzymes (Price and Joshi, 1982). Similar data have been reported for Zn^{2+} bound to metallothionein (Udom and Brady, 1980). Thus ferritin as well as metallothionein may function as zinc detoxicants as well as a zinc ion donors.

Ferritin is a constitutive protein. Its concentration in the liver is about 0.5 mg–1.0 mg per g of liver and it can bind about 50–65 moles of Zn^{2+}/mole of ferritin (Zamen and Verwilghen, 1981). In contrast, metallothionein is normally present only in nanogram quanties and it binds a maximum of 8 moles of a metal ion. Thus ferritin, because it is present at a relatively high constitutive level, is probably more suited to handle transient fluctuations in levels of divalent metal ions. In turn, binding of divalent ions to ferritin may also interfere with normal iron storage and mobilization. Indeed, deposition of iron into cellular ferritin was inhibited by Zn^{2+} in a dose-dependent manner (Zamen and Verwilghen, 1981). Similarly, Pb^{2+} bound to ferritin or ferritin-like components in the liver, spleen, and kidney of Pb-poisoned cattle suggest an altered storage of iron (Rüssell, 1970). These findings led to the suggestion that ferritin may serve as a primary defense against low levels of metal toxicity, thus sparing the organism from a biologically expensive process of synthesizing metallothionein and, beyond a threshold levels of toxic exposure, de novo synthesis of metallothionein might serve to handle excess divalent metals thus permitting ferritin to perform its primary function (Price and Joshi, 1982).

8.4.3. Ferritin and Cadmium

Studies by Noda *et al.* (1991) and Muller *et al.* (1991) have revealed different responses to cadmium toxicity in itai-itai disease and in *Xenopus laevis* cells. Itai-itai is thought to result from chronic cadmium toxicity. To determine whether this produced an imbalance in iron metabolism, undecalcified sections of iliac bones were examined by X-ray microanalysis. The results showed elevated levels of cadmium in the bones and the presence of iron at the mineralization front; however, transferrin and ferritin were not detected. Authors suggest Cd-induced deregulation of iron metabolism as a contributing factor in the disease (Noda *et al.*, 1991). The mechanism of deregulation remains to be established. More direct evidence for the role of ferritin and iron in cadmium toxicity was obtained by Muller *et al.*, (1991). These workers isolated a *Xenopus laevis* ferritin cDNA clone, XL2-17, from cadmium- or copper-poisoned XL2 cells. Sequence analysis revealed the identity of the coding region identical with that of the known H chain of the mammalian ferritin. However, it contained a stretch of 629 nucleotides at the 5' untranslated region. In particular, the IRE segment was 489 nucleotides from the 5' end compared to 54 nucleotides in the most studied H-chain mRNA (Figure 8.1). Significantly, exposure to cadmium or copper increased the mRNA levels 10–15-fold and ferritin levels 10-fold. Whether this response is due to the direct binding of cadmium or copper to the IREBP or, indirectly, via cadmium- or copper-induced elevated levels of free iron is not known. Nevertheless, this transcriptional and translational regulation of ferritin further supports the earlier suggestion that ferritin may be a primary detoxification response to heavy metals (Price and Joshi 1982; Joshi *et al.*, 1984).

8.4.4. Ferritin and Manganese

Industrial production or mining exposes living systems to several toxicants, including manganese. In humans and animals the toxicity is primarily associated with the nervous system. Although the mechanism of biochemical or behavioral alterations of manganese toxicity are unknown, studies of Suarez and Eriksson (1993) with neuroblastoma (SHSY5Y) cells in cultures have shown how the metal ion may gain entry into the brain. Accordingly, the metal ion is bound to transferrin as Mn^{3+} is internalized via endocytosis and 80% of Mn^{3+} was stored into ferritin after 24 hr exposure to the metal. Thus mechanism of transport and storage of Mn^{3+} in ferritin was analogus to that established for the detoxification of iron.

8.4.5. Ferritin and Arsenic

Arsenite is an extremely toxic nonmutagenic carcinogen. In response to exposure to arsenite, HeLa cells produce varying amounts of mRNAs. Guzzo *et al.* (1994) used a

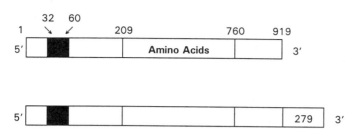

FIGURE 8.1. Schematic representation of the two mature mRNA species for the ferritin heavy chain. The numbers represent the nucleotides. The dark segment is the 28 nucleotide IRE and the shaded area is the additional 279 nucleotide sequence in the larger mRNA. In all other respects, the two mRNAs are identical. (Dhar and Joshi, 1994.)

cDNA subtraction method to compare the cDNAs grown in the absence and presence of 5×10^{-6} sodium arsenite. Three clones showed a higher hybridization signal to RNA from arsenite-exposed cells compared to control cells. Their nucleotide sequences showed that two were homologous to the human ferritin H chain and the third coded for metallothionein II. The authors suggest that arsenite exposure may lead to increased levels of oxyradicals and that elevated levels of metallothionein and ferritin may act as a defense against the free-radical damage.

8.4.6. Ferritin and Aluminum

Aluminum is practically insoluble at physiological pH. However increased bioavailability of aluminum due to acid rain has established this metal ion as toxic (Godbold *et al.* 1988). Crapper *et al.* (1976) first reported elevated levels of aluminum in the brains of patients who died of Alzheimer's disease. This observation and indeed various aspects of the role of aluminum in neurological disorders have been subjects of controversy. This was due to rapid speciation of aluminum with speciation change in the pH, methods of quantification of aluminum, varying local concentrations of aluminum in different area of a highly compartmentalized organ such as brain, and variation in the medium of intake of aluminum. It now appears that AD brains do contain high concentration of aluminum; however, the distribution is nonuniform (Lovell *et al.*, 1993).

Remarkable similarities between the coordination chemistry of aluminum and iron suggest that the two metal ions may use the same transport system (Crumbliss and Garrison, 1988). Consistent with this suggestion, aluminum binds to transferrin *in vitro* but with affinities less than that for iron (Trapp, 1983; Cochran *et al.*, 1984). Connor and coworkers suggested that the access of brain to aluminum is via transferrin (Roskams and Connor, 1990). Indeed ferritin–aluminum complex has been isolated *in vitro* and from the brains of rats exposed to 0.1 mM $AlCl_3·6H_2O$ in their drinking water for one year. Such brains also contained more ferritin, which itself also contained more iron and aluminum. Concentration of ferritin in the liver of aluminum-fed rats was also greater than the control but did not contain more iron or aluminum bound to it. Thus, the two tissues respond differently to dietary aluminum. Similar studies with human brains showed that compared to controls, more ferritin could be isolated from AD brains and such ferritin also contained more iron and aluminum (Fleming and Joshi, 1987). Cochran and Chawtar (1988) found similar binding of aluminum to ferritin and suggested that ferritin may serve as a detoxicant of nonferrous metal ions such as aluminum. Dedman *et al.* (1992a) also observed elevated levels of ferritin and iron in AD brains but did not observe increased levels of aluminum bound to ferritin from AD brains. In a separate study, when ferritin iron cores were reconstituted by adding 3000 iron atoms to apoferritin in the presence of aluminum citrate, up to 120 molecules of aluminum atoms bound per molecule (Dedman *et al.,* 1992b). This supports the suggested role of the iron core in sequestration of other metals (Sczekan and Joshi, 1989).

Finally, examination by high-energy X-ray imaging of the electron micrograph of germinating soybeans revealed the colocalization of aluminum with the crystalline arrays of ferritin in the amyloplasts (Roth *et al.,* 1987). Senile plaques and neurofibrillary tangles characteristic of AD contain aluminum (Crapper *et al.,* 1976; Perl and Brody, 1980; Good *et al.,* 1992). In view of these observations, immunohistopathological studies of Grundke-Iqbal *et al.,* (1990) and their extension by Connor *et al.,* (1992, 1995) are of particular interest. These studies establish that ferritin is a component of the neuritic senile plaque in Alzheimer's brain and suggest that colocalized iron and aluminum may indeed be contributing to the etiology of AD (Joshi, 1990, 1991; Joshi *et al.,* 1992, 1994; Clauberg and Joshi, 1993).

In vitro, aluminum reduces the rate of iron uptake by ferritin in a concentration-dependent manner but does alter the total amount of iron that could be sequestered (Fleming and Joshi, 1991). If such is the case *in vivo,* one would anticipate increased transient levels of unsequestered iron, which may generate free radicals and/or produce more ferritin to sequester it. Both of these events would depend upon the focal concentrations of the metal and the necessary machinery to cause or prevent the harmful effects (Clauberg and Joshi, 1993). Indeed San-Marina and Nicholls (1992) observed a twofold increase in ferritin synthesis after administering aluminum salts to animals. The mechanism of induction, and whether this induction represents enhanced synthesis of both H and L chains or only of the H chain as found in the cadmium-poisoned cells (Muller *et al.,* 1991) remains to be established.

Large molecular weight proteins other than ferritin with the ability to bind aluminum have been reported recently. Cochran *et al.* (1993) reported a large molecular weight protein distinct from ferritin that binds aluminum, and San-Marina and coworkers have isolated from brain, two "ferritin-like" large molecular weight proteins capable of binding aluminum and reacting with ferritin antibodies (personal communication). Significantly, these proteins bind aluminum as well as iron with affinities similar to those reported for ferritin (Sczekan and Joshi, 1989).

8.4.7. Aluminum and Free Radicals

Transition metal ions, upon reacting with H_2O_2 or free radicals O_2^-, $OH\cdot$, and $HO_2\cdot$, initiate a chain reaction that self-propagates the formation of free radicals. Although aluminum is not a transition metal, aluminum and aluminum silicates, the known components of AD brain (Candy *et al.,* 1986) seem to participate in the free-radical-mediated damage. For example, aluminum accelerates the lipid peroxidation initiated by iron. The process is more rapid at acidic pH (Gutteridge *et al.,* 1985) and Al^{3+} facilitates the oxidation of NADH by the superoxide anion. The source of O_2^- could be enzymatic or photochemical. Presumably an $O_2^- - Al^{3+}$ complex is a stronger oxidant than O_2^- alone (Kong *et al.,* 1992). Evans *et al.,* (1992) observed that compared to controls, aluminosilicate-exposed murine brain glial cells produced more free radicals. Conceivably, ferritin, hemosiderin, heme, or other iron compounds present in the cells contributed iron to generate free radicals whose effect was accentuated by aluminum. Ferritin-bound copper also stimulates decay of superoxides (Bolann and Ulvik, 1993). Thus the colocalization of aluminum and a transition metal ion would exacerbate free-radical damage.

8.5. ALUMINUM, IRON, FREE RADICALS, AND THE FORMATION OF AMYLOID PLAQUES CHARACTERISTIC FOR AD

The concentration of nonheme iron in the brain increases with age (Hallgren and Sourander, 1958; Hill, 1988; Swaiman, 1991). The brains of Alzheimer's disease patients contain plaques that are considered diagnostic for the disease (Selko, 1990). Plaques also contain β-amyloid peptide (β-AP) (Selko, 1990), α-antichymotrypsin (Nelson and Simon, 1990), and the element aluminum (Crapper-McLachlan, 1986). Indirect evidence suggests that all three are involved in the etiology of the disease (Clauberg and Joshi, 1993). The β-AP is generated by the proteolytic processing of the β-amyloid precursor proteins (Selko, 1991). Some of these contain a domain that is 60% homologous to bovine pancreatic trypsin inhibitor, which also inhibits α-chymotrypsin. At pH 6.5, aluminum doubles the enzyme activity and makes it 100-fold more resistant to the inhibitor. The inhibition by BX-9, a protease inhibitor prepared from the amyloid plaques (Sinha *et al.,* 1990), is also reduced by aluminum; so too is that by α-chymotrypsin, but to a lesser extent (Clauberg and Joshi, 1993). Iron and its storage protein ferritin are associated with the neuritic plaques of Alzheimer's disease (Grunde-Iqbal *et al.,* 1990). Acidic pH values reported in Alzheimer's disease brains (Paschen *et al.,* 1987; Munekata and Hossman, 1987; Yates *et al.,* 1990) also facilitate the release of iron from ferritin (Aisen and Listowsky, 1990) and self-aggregation of β-AP (Barrow and Zargorski, 1991). Aluminum, iron, and zinc ions promote aggregation of β-amyloid peptide. This aggregation is more rapid at acidic pH (Mantyh *et al.,* 1993). Aluminum enhances the lipid peroxidation initiated by the interaction of Fe^{2+} and O_2 (Gutteridge *et al.,* 1985). These free radicals also enhance crosslinking of β-AP (Dyrks *et al.,* 1992) and oxidize proteins, which renders them more susceptible to proteases (Stadtman and Oliver, 1991). It therefore appears that the seemingly diverse changes such as acidosis and the accumulation of iron, aluminum, and ferritin observed in AD brains contribute to plaque formation. Recently, desferrioxamine has been successfully used in slowing Alzheimer's disease-associated dementia (McLachlan *et al.,* 1991). This compound chelates aluminum as well as iron, both of which are established neurotoxins.

8.6.   THE ROLE FOR NOVEL FERRITIN H-CHAIN mRNAs

Studies designed to learn about ferritin in human brain led to the isolation of a novel ferritin mRNA for H-chain. (Dhar and Joshi, 1993; Dhar *et al.*, 1993; Chauthaiwale *et al.*, 1993). It has an extra stretch of 279 nucleotides at the 3′ untranslated region. In all other respects it is identical to the well-studied H chain mRNA (Klausner *et al.*, 1993) and is localized on chromosome 11 (Percy *et al.*, 1995). Although first discovered in the brain, it is present in other human tissues, with brain having the highest level (Dhar and Joshi, 1994). To date, a similar mRNA for the L chain has not been detected. In the absence of direct evidence one can only speculate about the physiological basis for the multiple forms of ferritin H-chain mRNAs differing only in the untranslated region.

Although all naturally occurring ferritins are mixtures of H and L chains, it is the ferroxidase center of H chain that initiates the nucleation of the iron core by oxidizing Fe^{2+} to Fe^{3+}. The presence of H chains is therefore critical for combating oxidative damage. Thus, the multiple forms of mRNA for H chain may represent a fail-safe mechanism to assure the availability of H chains. Alternatively, or in addition, the messages, although producing the same protein, may respond differentially to the metabolic stress imposed by the microenvironment. This is particularly crucial for a highly aerobic and compartmentalized tissue such as brain. In the last few years much has been learned about the role of the first identified IREBP (cytosolic aconitase) in the regulation of ferritin synthesis. However, very little is known about the second protein IRF_B (iron regulatory factor B) and its role in ferritin synthesis. Although IRF_B is devoid of aconitase activity, it binds to IRE and is immunologically distinct from aconitase. More importantly, it is present in several mammalian tissues, with brain having the highest level (Henderson *et al.*, 1993). It will be of interest to compare the translational efficiency of these mRNAs *per se* and also as affected independently and in combination by varying the concentrations of the components of the factors that affect this translation. These components include iron, other metals such as cadmium, copper, aluminum, and arsenic, and the two IRE-binding proteins.

Heavy chains of ferritin are more acidic, and acidic ferritin tends to contain less iron. Since nonferrous metal ions are also bound to the core, isoferritins would be expected to differ in their ability to bind and release other metal ions and therefore execute the multifunctions. Synthesis of ferritin H chains is also affected by factors other than iron and in a tissue-specific manner (Munro *et al.*, 1993; MacDonald *et al.*, 1994). It therefore follows that the focal concentration of the subpopulation of isoferritins within the tissue is determined by the specific needs at the loci. These variations could be crucial for the compartmentalized in the highly aerobic organ, the brain. Any alteration in them would be metabolically undesirable (Joshi and Clauberg, 1988). Indeed, Connor *et al.* (1994) observed this to be true. Their subsequent elegant studies (Connor *et al.*, 1995) showed that the distribution of isoferritins is not only area-specific but also disease-dependent. *In situ* hybridization may reveal whether the existence and the concentrations of the H-chain mRNAs parallel the observed changes in the ferritin and the concentration of ferrous, and perhaps also the nonferrous, metal ions.

ACKNOWLEDGMENTS.   Supported by The Council for Tobacco Research and The Physicians Medical Research Education Fund, University of Tennessee Medical Research Center, Knoxville, Tennessee. The help of Dr. Clauberg during the initial phase of the preparation of the manuscript is gratefully acknowledged.

8.7.   REFERENCES

Aisen, P., and Listowsky, I., 1980, Iron transport and storage proteins, *Annu. Rev. Biochem.* **49:**357–393.

Aldridge, W. N., 1966, Toxicity of beryllium, *Lab. Invet.* **15:**176–178.

Aust, S. D., 1988, Source of iron for lipid peroxidation in biological systems, in: *Oxygen Radicals and Tissue Injury,* Proceedings of the Brook Lodge Symposium (Berry Halliwell, ed.), Fed. American Soc. Exp. Biol., Augusta, Michigan, pp. 27–33.

Babior, B. M., Curnutte, J. T., and Okamura, N., 1988, The respiratory burst oxidase of the human neutrophil, in: *Oxygen Radicals and Tissue Injury.* Proceedings of the Brook Lodge Symposium (Barry Halliwell, ed.), Fed. American Soc. Exp. Biol., Augusta, Michigan, pp. 43–48.

Barrow, C. J., and Zargorski, M. G., 1991, Solution structures of beta peptide and its constituent fragments: Relation to amyloid deposition, *Science* **253:**179–182.

Bolann, B. J., and Ulvik, R. J., 1993, Stimulated decay of superoxide caused by ferritin bound to copper, *FEBS Lett.* **328:**263–267.

Brown, D. A., Chatel, K. W., Chan, A. Y., and Knight, B., 1980, Cytosolic levels and distribution of cadmium, copper and zinc in pretumorous livers from diethylnitrosamine exposed mice and in noncancerous kidneys from cancer patients, *Chem. Biol. Interact.* **32:**13–27.

Candy, J. M., Klinowki, J., Perry, R. H., Perry, E. K., Fairbrain, A., Oakley, A., Carpenter, T., Atack, J., Blessed, G., and Edwardson, J., 1986, Alumino-silicates and senile plaque formation in Alzheimer's disease, *Lancet* **1:**354–356.

Castro, L., Rodriguze, M., and Radi, R., 1994, Aconitase is readily inactivated by peroxynitrite, but not by its precursor, nitric oxide, *J. Biol. Chem.* **269:**29409–29415.

Chauthaiwale, V., Dhar, M., and Joshi, J. G., 1993, Cloning of a novel full length cDNA for ferritin heavy chain (FTH) from adult human brain, *FASEB J.* **7:**A628.

Clauberg, M., and Joshi, J. G., 1993, Regulation of serine protease activity by aluminum: Implications for Alzheimer's disease, *Proc. Natl. Acad. Sci. USA* **90:**1009–1012.

Cochran, M., and Chawtar, V., 1988, Interaction of horse-spleen ferritin with aluminum citrate, *Clin. Chim. Acta.* **178:**79–84.

Cochran, M., Coates, J., and Neoh, S., 1984, The competitive equilibrium between aluminum and ferric ions for the binding sites of transferrin, *FEBS Lett.* **176:**129–132.

Cochran, M., Goddard, G., Ramm, G., Ludwigson, N., Marshall, J., and Holliday, J., 1993, Absorbed aluminum is found with cytosolic protein fractions other than ferritin, in the rat duodenum. *Gut* **34:**643–646.

Connor, J. R., Menzies, S. L., St. Martin, S. M., and Mufson, E. L., 1992, A histochemical study of iron, transferrin and ferritin in Alzheimer's diseased brain, *J. Neurosci. Res.* **31:**75–81.

Connor, J. R., Boeshore, K. L., Benkovic, S. A., and Menzies, S. L., 1994, Isoforms of ferritin have a specific cellular distribution in the brain, *J. Neurosci. Res.* **37:**461–465.

Connor, J. R., Synder, B. S., Arosio, P., Loeffler, D. A., and Dewitt, P. A., 1995, A quantitative analysis of isoferritins in select regions of aged, parksinsonian, and Alzheimer's disease brains, *J. Neurochem.* **65:**717–724.

Crapper, D. R., Krishnan, S. S., and Quittkat, S., 1976, Aluminum, neurofibrillary degeneration and Alzheimer's disease, *Brain* **99:**67–80.

Crapper-McLachlan, D. R., 1986, Aluminum and Alzheimer's disease, *Neurobiol. Aging* **7:**525–532.

Crumbliss, A. L., and Garrison, M. A., 1988, Comparison of some aspects of coordination chemistry of aluminum (III), and iron (III), *Comm. Inorg. Chem.* **8:**1–11.

Dedman, D. J., Treffrey, A., Candy, J. M., Taylor, G. A., Morris, C. M., Bloxham, C. A., Perry, R. H., Edwardson, J. A., and Harrison, P., 1992a, Iron and aluminum in relation to brain ferritin in normal individuals and Alzheimer's disease and chronic renal-dialysis patients, *Biochem J.* **287**:509–514.

Dedman, D. J., Treffry, A., and Harrison, P. M., 1992b, Interaction of aluminum citrate with horse spleen ferritin, *Biochem. J.* **287**:515–520.

Deshpande, V. V., and Joshi, J. G., 1985, Vit C. Fe (III) induced loss of the covatently bound phosphate and enzyme activity of phosphoglucomutase, *J. Biol. Chem.* **260**:757–764.

Dhar, M., and Joshi, J. G., 1993, Differential processing of ferritin heavy chain mRNA in human liver and adult human brain, *J. Neurochem.* **61**:2140–2146.

Dhar, M., and Joshi, J. G., 1994, Detection and quantitation of a novel ferritin H-chain mRNA in human tissues, *Biofactors* **3**:147–149.

Dhar, M., Chauthaiwale, V., and Joshi, J. G., 1993, Sequence of a cDNA encoding the ferritin H-chain from an 11-week old fetal brain, *Gene* **126**:275–278.

Dyrks, T., Dyrks, E., Hartmann, Masters, C., and Beyreuther, K., 1992, Amyloidogenicity of βA_4 and βA_4 bearing amyloid precurson fragments by metal-catalyzed oxidation, *J. Biol. Chem.* **267**:18210–18127.

Evans, P. H., Peterhans, E., Bürge, T., and Klinowski, J., 1992, Aluminosilicate-induced free radical generation by murine brain glial cells in vitro. Potential significance in the aetiopathogenesis of Alzheimer's dementia, *Dementia* **3**:1–6.

Fleming, J. T., and Joshi, J. G., 1987, Ferritin: Isolation of aluminum-ferritin complex from brain, *Proc. Natl. Acad. Sci USA* **84**:7866–7870.

Fleming, J. T., and Joshi, J. G., 1991, Ferritin: The role of aluminum in ferritin function, *Neurobiol. Aging* **12**:413–418.

Fridovich, I., 1989, Superoxide dismutases: An adaptation to a paramagnetic gas, *J. Biol. Chem.* **264**:7761–7764.

Godbold, D. L., Fritz, E., and Hütermann, A., 1988, Aluminum toxicity and forest decline, *Proc. Natl. Acad. Sci. USA* **85**:3888–3892.

Good, P. F., Perl, D. P., Bierer, L. M., and Schmeidler, J., 1992, Selective accumulation of aluminum and iron in neurofibrillary tangles of Alzheimer's disease: A laser microprobe (LAMMA) study, *Ann. Neurol.* **31**:286–292.

Granick, S., and Hahn, P., 1944, Ferritin VIII—speed and uptake of iron by liver and its conversion to ferritin iron, *J. Biol. Chem.* **155**:661–669.

Grant, C. T., and Tabrosky, G., 1966, The generation of labile protein-bound phosphate by phosphoprotein oxidation linked to autooxidation of ferrous ions, *Biochemistry* **5**:544–553.

Green, S., and Mazur, A., 1957, Relation of citric acid metabolism to release iron from hepatic ferritin, *J. Biol. Chem.* **227**:653–668.

Grundke-Iqbal, I., Fleming, J. T., Tung, Y. C., Lassmann, H., Iqbal, K., and Joshi, J. G., 1990, Ferritin is a component of the neuritic (senile) plaque in Alzheimer's dementia, *Acta. Neuropathol.* **81**:105–110.

Gutteridge, J. M. C., Quinlan, G. J., Clark, I., and Halliwell, B., 1985, Aluminum salts accelerate peroxidation of membrane lipids stimulated by iron salts, *Biochem. Biophys. Acta.* **835**:441–447.

Guzzo, A., Karatzios, C., Diorio, C., and DuBow, M. S., 1994, Metallothionein-II and ferritin H mRNA levels are increased in arsenite-exposed HeLa cells, *Biochem. Biophys. Res. Commun.* **205**:590–595.

Hallgren, B., and Sourander, P., 1958, The effect of age on the nonheme iron in the human brain, *J. Neurochem.* **3**:41–51.

Harford, J. B., and Klausner, R. D., 1990, Coordinate post-transcriptional regulation of ferritin and transferrin receptor expression: the role of regulated RNA–protein interaction, *Enzyme* **44**:28–41.

Hashimoto, T., Joshi, J. G., Del Rio, C., and Handler, P., 1967, Phosphoglucomutase: IV. Inactivation by beryllium ions, *J. Biol. Chem.* **242**:1671–1679.

Hausladen, A., and Fridovich, I., 1994, Superoxide and peroxyoitrite inactivate aconitases but nitric oxide does not, *J. Biol. Chem.* **269**:29405–29408.

Hederson, B. R., Seiser, C., and Kuhn, L., 1993, Characterization of a second RNA-binding protein in rodents with specificity to iron-responsive elements, *J. Biol. Chem.* **268**:27327–27334.

Hill, J. M., 1988, The distribution of iron in the brain, in: *Brain Iron: Neurochemical and Behavioral Aspects* (M. B. H. Youdin, ed.), Taylor and Frances, London, pp. 1–24.

Jeslow, M. M., 1980, Blood zinc and lead poisoning, in: *Zinc in the Environment. Part II. Health Effects.* (J. O. Nrigau, ed.), Wiley Interscience, New York, pp. 171–181.

Joshi, J. G., 1990, Aluminum: A neurotoxin which affects diverse metabolic reactions, *Biofactors* **2:**163–169.

Joshi, J. G., 1991, Neurochemical hypothesis: Participation by aluminum in producing critical mass of colocalized errors in brain leads to neurological diseases, *Comp. Biochem. Physio.* **100C:**103–105.

Joshi, J. G., and Clauberg, M., 1988, Ferritin: An iron storage protein with diverse functions, *Biofactors* **1:**207–212.

Joshi, J. G., and Zimmerman, A., 1988, Ferritin: An expanded role in metabolic regulation, *Toxicology* **48:** 21–29.

Joshi, J. G., Price, D. J., and Fleming, J., 1984, Ferritin and metal toxicity, In *Protides of Biological Fluids,* (H. Peeters, ed.), Pergamon Press, Elmsford, New York, 183–186.

Joshi, J. G., Clauberg, and Dhar, M. S., 1992, Role of iron and aluminum in brain disorders, *Adv. Behav. Biol.* **40:**387–396.

Joshi, J. G., Clauberg, M., Dhar, M., and Chauthaiwale, V., 1994, Iron and aluminum homeostasis in neural disorders, *Environ. Health Perspect.* **102** (suppl 3):207–213.

Klausner, R. D., Roualt, T. A., and Harford, J. B., 1993, Regulating the fate of mRNA: The control of cellular iron metabolism, *Cell* **72:**19–28.

Kojima, Y., and Kagi, J. H. R., 1978, Mettallothioneins, *Trends Biochem. Sci.* **3:**90–93.

Kong, S., Liochev, S., and Fridorich, I., 1992, Aluminum (III) facilitates the oxidation of NADH by the superoxide anions, *Free Radic. Biol. Med.* **13:**79–81.

Lindenschmidt, R. C., Sendelbach, L. E., Witschi, H. P., Price, D. J., Fleming, J., and Joshi, J. G., 1986, Ferritin and *in vivo* beryllium toxicity, *Toxicol. Appl. Pharm.* **82:**344–350.

Lovell, M. A., Ehmann, W. D., and Markesbery, W. R., 1993, Laser microprobe analysis of brain aluminum in Alzheimer's disease, *Ann. Neurol.* **33:**36–42.

Macara, I. G., Hoy, T. G., and Harrison, P. M., 1973a, The formation of ferritin from apoferritin. Catalytic action of apoferritin, *Biochem. J.* **135:**343–348.

Macara, I. G., Hoy, T. G., and Harrison, P. M., 1973b, The formation of ferritin from apoferritin, inhibition of metal ion-binding studies, *Biochem. J.* **135:**785–789.

MacDoland, M. H., Cook, J. D., Epstein, M. L., and Flowers, C. H., 1994, Large amount of (apo) ferritin in the pancreatic insulin cell and its stimulation by glucose, *FASEB J.* **8:**771–781.

Mantyh, P. W., Ghilardi, J. R., Rogers, S., DeMaster, E., Allen, C. J., Stimson, E. R., and Maggion, J. E., 1993, Aluminum, iron and zinc ions promote aggregation of physiological concentrations of β-amyloid peptide, *J. Neurochem.* **61:**1167–1170.

Mazur, A., and Carleton, A., 1965, Hepatic xanthine oxidase and ferritin iron in the developing rat, *Blood* **26:**317–322.

Mazur, A., Green, S., Saha, A., and Carelton, A., 1958, Mechanism of release of ferritin iron *in vivo* by xanthine oxidase, *J. Clin. Invest.* **37:**1809–1817.

McCord, J. M., 1986, Superoxide radical: A likely link between reperfusion injury and inflammation, *Adv. Free. Rad. Biol. Med.* **2:**325–345.

McLachlan, D. R. C., Dalton, A. J., Kruck, T. P. A., Bell, M. Y., Smith, W. L., Kalow, W., Andrews, D. F., 1993, Intramuscular desferrioxamine in patients with Alzheimer's disease, *Lancet* **337:**1304–1308.

Muller, J. P., Vedel, M., Monnot, M. J., Touzet, N., Wegnez, M., 1991, Molecular cloning and expression of ferritin mRNA in heavy metal-poisoned Xenopus laevis cells, *DNA-Cell Biol.* **10:**571–579.

Munekata, K., and Hossman, K. A., 1987, Effect of five-minute ischemia on regional pH and energy state of the brain. Relation to selective vulnerability of the hippocampus, *Stroke* **18:**412–417.

Munro, H., 1993, The ferritin genes. Their response to iron status, *Nutrit. Rev.* **51:**65–73.

Nelson, R. B., and Simon, R., 1990, Clipsin, a chymotrypsin-like protease in the rat brain which is irreversibly inhibited by α-1-antichymotrypsin, *J. Biol. Chem.* **265:**3836–3843.

Nelson, S. K., Bose, S. K., and McCord, J. M., 1994, The toxicity of high dose of superoxide dismutase suggests that superoxide can both initiate and terminate lipid peroxidation in the reperfused heart, *Free Rad. Biol. Med.* **16:**195–200.

Nelson, S. K., Wong, G. H. W., and McCord, J. M., 1995, Leukemia inhibitory factor and tumor necrosis factor induce manganese superoxide dismutase and protect rabbit hearts from reperfusion injury, *J. Mol. Cell. Cardio.* **27:**223–229.

Noda, M., Yasuda, M., and Kitagawa, M., 1991, Iron as a possible aggrevating factor for osteopathy in itai-itai disease, a disease associate with chronic cadmium intoxication, *J. Bone Miner Res.* **6:**245–255.

Paschen, W., Djuricic, B., Mies, G., Schmidt, Kastner, R., and Linn, F., 1987, Lactate and pH in the brain. Association and dissociation in different pathophysiological states, *J. Neurochem.* **48:**154–159.

Percy, M. E., Bauer, S., Rainey, S., McLachlan, D. R. C., Dhar, M., and Joshi, J. G., 1995, Localization of a new ferritin heavy chain sequence present in human brain mRNA to chromosome 11, *Genome* **38:**450–457.

Perl, D. P., and Brody, A. R., 1980, Alzheimer's disease: X-ray spectrometric evidence of aluminum accumulation in neurofibrillary tangle baring neurons, *Science* **208:**297–299.

Price, D. J., and Joshi, J. G., 1982, Ferritin: A zinc detoxicant and a zinc ion donor, *Proc. Natl. Acad. Sci. USA* **57:**1482–1485.

Price, D. J., and Joshi, J. G., 1983, Ferritin: Binding of beryllium and other divalent metal ions, *J. Biol. Chem.* **258:**10873–10880.

Price, D. J., and Joshi, J. G., 1984, Ferritin: Protection of enzymatic activity against the inhibition by divalent metal ions *in vitro, Toxicology* **31:**151–163.

Reif, D. W., 1992, Ferritin as a source of iron for oxidative damage, *Free Radic. Biol. Med.* **12:**417–427.

Roskams, A. J., and Connor, J. R., 1990, Aluminum access to the brain: A role for transferrin and its receptor, *Proc. Natl. Acad. Sci. USA* **87:**9024–9028.

Roth, E. L., Dunlap, J. R., and Stacy, J., 1987, Localization of aluminum in soybean bacteriods and seeds, *Appl. Environ. Microbiol.* **53:**2548–2543.

Rüssell, H. A., 1970, Über die binding von blei an eisenhydroxidhaltige stoffe in leber, niere and milz-vergifteter rinder, *Bull. Environ. Contam. Toxicol.* **5:**115–124.

San-Marina, S., and Nicholls, D. M., 1992, Some effects of aluminum on rat brain protein synthesis, *Comp. Biochem. Physiol.* **103:**585–591.

Sczekan, S. R., and Joshi, J. G., 1989, Metal binding properties of phytoferritin and synthetic iron cores, *Biochim. Biophys. Acta* **990:**8–14.

Selko, D. J., 1990, Deciphering Alzheimer's disease: The amyloid precursor protein yields new clues, *Science* **248:**1058–1060.

Sendelbach, L. E., and Witschi, H. P., 1987, Protection against pulmonary beryllium toxicity by iron, *Toxicol. Lett.* **48:**321–325.

Sinha, S., Dovey, H. G., Seubert, P., Ward, P. J., Blacher, R. W., Blaber, M., Bradshaw, R. A., Arici, M., Mobley, W. C., and Lieberburg, I., 1990, The protease inhibitory properties of the Alzheimer's beta-amyloid precursor proteins, *J. Biol. Chem.* **265:**8983–8985.

Stadtman, E. R., and Oliver, C. N., 1991, Metal-catalysed oxidation of proteins, *J. Biol. Chem.* **266:**2005–2009.

Suarez, N., and Eriksson, H., 1993, Receptor-mediated endocytosis of a manganese complex of transferrin into neuroblastoma (SHSY5Y) cells in culture, *J. Neurochem.* **61:**127–131.

Sun, S., and Chasteen, N. D., 1992, Ferroxidase kinetics of horse spleen apoferritin, *J. Biol. Chem.* **267:**25160–25166.

Sun, S., and Chasteen, N. D., 1994, Rapid kinetics of the EPR-active species formed during initial iron uptake in horse spleen apoferritin, *Biochemistry* **33:**15095–15102.

Swaiman, K. F., 1991, Hallervorden–Spatz syndrome and brain iron metabolism, *Arch. Neurol.* **48:**1285–1292.

Tepper, L. B., 1972, Beryllium, *CRC Crit. Rev. Toxicol.* **1:**235–259.

Theil, E. C., 1987, Ferritin: Structure, gene regulation, and cellular function in animals, plants, and micro-organisms, *Annu. Rev. Biochem.* **56:**289–315.

Thomas, M., and Aldridge, W. N., 1966, Inhibition of enzymes by beryllium, *Biochem. J.* **98:**94–98.

Toda, G., Hashimoto, T., Asakura, T., and Minakami, S., 1967, Inhibition of Na-K-activated ATPase by beryllium, *Biochem. Biophys. Acta.* **135:**570–572.

Trapp, G. A., 1983, Plasma aluminum is bound to transferrin, *Life. Sci.* **33:**311–316.

Udom, A. O., and Brady, F. O., 1980, Reactivation *in vitro* of Zn-requiring apoenzymes by rat liver Zn-thionein *Biochem. J.* **187:**329–335.

Wardeska, J. G., Viglione, B., and Chasteen, N. D., 1986, Metal ion complexes of apoferritin. Evidence for initial binding in the hydrophilic channels, *J. Biol. Chem.* **261:**847–850.

Witschi, H. P., and Aldridge, W. N., 1968, Uptake, distribution, and binding of beryllium to organelles of the rat liver cell, *Biochem. J.* **106:**811–820.

Yates, C. M., Butterworth, J., Tennant, M. C., and Gordon, A., 1990, Enzyme activities in relation to pH and lactate in postmortem brain in Alzheimer type and other dementia, *J. Neurochem.* **55:**1624–1630.

Zamen, S., and Verwilghen, R. L., 1981, Influence of zinc on iron uptake by monolayer-cultures of rat hepatocytes and the hepatocellular ferritin, *Biochem. Biophys. Acta.* **675:**77–84.

Chapter 9

Ascorbate

An Antioxidant Neuroprotectant and Extracellular Neuromodulator

George V. Rebec

9.1. INTRODUCTION

Ascorbic acid, which exists primarily as the ascorbate anion at physiological pH, participates in many life-sustaining functions. It is perhaps best known as a cofactor in the synthesis of collagen, a connective-tissue protein (Englard and Seifter, 1986), but ascorbate also promotes iron absorption, regulates cholesterol synthesis and elimination, and generally seems essential for the normal operation of all bodily organs (Navas *et al.,* 1994; Meister, 1992; Tolbert, 1985). This versatility arises from a redox potential of +0.080, which allows ascorbate to donate reducing equivalents to many different compounds (Niki, 1991; Bendich *et al.,* 1986; Halliwell and Gutteridge, 1985). In fact, its role as a highly efficient reducing agent confers another critical function on ascorbate: neutralizing toxic free radicals formed during oxidative metabolism (Beyer, 1994; Rose and Bode, 1993; Sies *et al.,* 1992).

In the brain, which has the highest rate of oxidative activity of any organ (Götz *et al.,* 1994), antioxidant protection seems especially crucial. Ample evidence suggests that a wide range of neurodegenerative diseases arise, in part, from damage inflicted by oxygen-derived free radicals (Reiter, 1995; Rose and Bode, 1993; Halliwell *et al.,* 1992). Two forebrain transmitters, dopamine and glutamate, appear to play critical roles in such damage. Dopamine metabolism, which involves oxidative processes, can

George V. Rebec   Program in Neural Science, Department of Psychology, Indiana University, Bloomington, Indiana 47405.

Metals and Oxidative Damage in Neurological Disorders, edited by Connor. Plenum Press, New York, 1997.

generate free radicals, and activation of glutamate receptors promotes oxidative activity (Dugan and Choi, 1994; Coyle and Puttfarcken, 1993; Hall and Braughler, 1993; Chiueh et al., 1993). An intriguing aspect of these transmitters is that their release is closely linked to the release of ascorbate (O'Neill, 1995; Rebec and Pierce, 1994; Grünewald, 1993). In fact, forebrain areas like the basal ganglia, which are rich in dopamine and glutamate, contain levels of ascorbate that rank among the highest in the body (Oke et al., 1987). The antioxidant function of ascorbate, therefore, may provide an important line of defense against dopamine- or glutamate-induced neurodegenerative processes.

Research also suggests, however, that ascorbate can influence brain function in ways that extend beyond a role as antioxidant. Studies assessing the mechanisms by which ascorbate interacts with neural tissue, for example, suggest direct effects on neural function, including changes in electrical impulse flow (Pierce and Rebec, 1995; Gardiner et al., 1985; Ewing et al., 1983), transmitter release (Kuo et al., 1979) and receptor binding (Hadjiconstantinou and Neff, 1983). It also is interesting to note that the mechanisms controlling forebrain ascorbate release vary by region, suggesting region-specific mechanisms of action (Rebec and Pierce, 1994). In fact, ascorbate has recently gained attention as an extracellular modulator of dopaminergic and glutamatergic synaptic activity in the basal ganglia (Rebec and Pierce, 1994; Grünewald, 1993). This chapter focuses on the mechanisms by which ascorbate influences these transmitter systems, especially as they relate to oxidative processes and neuronal function.

9.2. UPTAKE INTO BRAIN TISSUE

Most mammalian species rely on enzymes in the liver to synthesize ascorbate from glucose, while others (e.g., primates and guinea pigs) obtain it through the diet as a water-soluble vitamin (vitamin C). In either case, blood-borne ascorbate enters brain tissue from the ventricular system. A selective transport mechanism in the choroid plexus insures a high concentration of ventricular ascorbate, which presumably diffuses passively into surrounding brain structures (Spector, 1989, 1981; Spector and Lorenzo, 1974). The final distribution of ascorbate in the brain, however, does not reflect a simple concentration gradient away from the ventricular system. The highest levels of ascorbate appear in cerebral cortex, amygdala, hippocampus, and basal ganglia (Oke et al., 1987). There also are anterior-posterior and dorso-ventral gradients within each structure such that most ascorbate accumulates in anterior and ventral areas (Basse-Tomusk and Rebec, 1990; Oke et al., 1987). This distribution pattern has been observed in several different species, including humans, although the human pattern appears to show the most variability (Oke et al., 1987). Humans also show strong anterior-posterior and dorso-ventral gradients throughout the entire brain, making it difficult to delineate individual structures solely on the basis of ascorbate content. Although it is not clear how ascorbate obtains its final heterogeneous distribution, both neurons and glial cells can accumulate ascorbate from the surrounding extracellular fluid via a sodium-dependent, active uptake system (Spector and Lorenzo, 1974).

Compared to blood plasma, where the concentration of ascorbate ranges between 30–50 μM, the ascorbate level is 3–5 times higher in the ventricular system and another 3–5 times higher in individual brain structures (Milbey *et al.,* 1982; Mefford *et al.,* 1981). Such concentration gradients require active transport, but extremely high plasma levels, which can be achieved with large systemic doses, may allow for passive diffusion across the blood–brain barrier. Increases in brain ascorbate can be detected within minutes after systemic injection of 1000–2000 mg/kg in rats and mice (Ewing *et al.,* 1983; Tolbert *et al.,* 1979b).

When plasma levels are low, which can occur in animals suffering an ascorbate deficiency, most organs undergo a net loss of ascorbate as outward diffusion overwhelms active transport (Spector, 1981). In the brain, however, active uptake mechanisms in the choroid plexus and at the cellular level help to maintain a high concentration of ascorbate, even under adverse conditions (Spector, 1989). In animals that die of an ascorbate deficiency, the brain retains as much as one-third of its normal ascorbate level (Hughes *et al.,* 1971). Thus, highly efficient transport mechanisms have evolved to insure a continuous supply of ascorbate to the brain, suggesting a critical role for this molecule in brain function. Only when these mechanisms are impaired, such as in meningitis and related diseases, does the brain level of ascorbate fall to that of blood plasma (Garcia-Buñel and Garcia-Buñel, 1965).

9.3. RELEASE INTO EXTRACELLULAR FLUID

The level of ascorbate in the extracellular fluid of the brain is maintained at the expense of intracellular stores. Brain slices, for example, can lose as much as 75% of their total ascorbate content in 30 min, but the level of ascorbate in the extracellular fluid remains constant over this period (Schenk *et al.,* 1982). Conversely, brain slices accumulate ascorbate when it is added in high concentrations (McIlwain *et al.,* 1956). It appears, therefore, that homeostatic mechanisms help to maintain a certain level of extracellular ascorbate in brain tissue. Under normal conditions, a strong concentration gradient drives ascorbate from intracellular stores to extracellular fluid (1–3 mM vs. 100–450 μM). Accordingly, high extracellular levels of ascorbate have been recorded in those brain structures that have high tissue levels of ascorbate: cerebral cortex, hippocampus, and basal ganglia (Basse-Tomusk and Rebec, 1991; Stamford *et al.,* 1984).

Homeostasis alone, however, cannot account for the large fluctuations in extracellular ascorbate that occur over a 24-hr period. In measurements obtained from the brains of freely moving rats, for example, ascorbate levels in extracellular fluid are lowest during quiet rest or sleep and highest during episodes of prolonged motor activity (O'Neill *et al.,* 1982). Similar results have been noted in guinea pigs (Kaufmann *et al.,* 1986). Even a brief episode of behavioral activation, such as that elicited by mild tail pinch, can elevate extracellular ascorbate by as much as 75% above resting levels (Boutelle *et al.,* 1989). Increases of comparable magnitude occur following injection of psychomotor stimulants (see Rebec and Pierce, 1994). Behavioral state, therefore, exerts an important influence on the extracellular accumulation of ascorbate.

Because extracellular ascorbate arises from intracellular stores, increases in extracellular concentration presumably result from some type of cellular release process. Insight into the neural mechanisms controlling this process emerged from studies of dopaminergic and glutamatergic systems. Most of this work has focused on the neostriatum, the major afferent structure of the basal ganglia. The neostriatum receives substantial input from dopaminergic cell bodies in the substantia nigra compacta (SNC) and glutamatergic neurons in cerebral cortex. The data on neostriatal ascorbate release, summarized in the following sections, reveal a role for both transmitter systems.

9.3.1. Modulation by Dopamine

Early indications that ascorbate release involves dopaminergic mechanisms arose from reports that amphetamine, a drug known to facilitate dopamine release, increases the extracellular level of ascorbate in the neostriatum (Zetterström et al., 1992; Pierce and Rebec, 1990; Mueller and Haskett, 1987; Louilot et al., 1985; Clemens and Phebus, 1984). This effect cannot be explained by a sympathomimetic action since para-hydroxyamphetamine, which has comparable sympathomimetic effects but does not easily cross the blood–brain barrier, fails to alter neostriatal ascorbate levels (Wilson and Wightman, 1985). It also is unlikely that amphetamine simply releases peripheral stores of ascorbate since adrenalectomized animals, which are deprived of a major source of peripheral ascorbate, show the same amphetamine-induced rise in neostriatal ascorbate as intact animals (Wilson and Wightman, 1985). Furthermore, neostriatal ascorbate release is not unique to amphetamine, but applies to a variety of dopamine agonists, including the combined administration of drugs acting selectively at either the D1 or D2 subfamily of receptors (Zetterström et al., 1992; Pierce and Rebec, 1990). Dopamine antagonists, conversely, lower the extracellular level of ascorbate in the neostriatum and also block the agonist-induced increase (Pierce and Rebec, 1992; Oh et al., 1989; Mueller and Haskett, 1987).

Surprisingly, however, amphetamine has relatively little effect on extracellular ascorbate levels in nucleus accumbens and medial prefrontal cortex (Mueller, 1990; Mueller and Kunko, 1990; Phebus et al., 1990; Louilot et al., 1985), despite the rich dopaminergic innervation of these areas (Fallon, 1988). Even multiple injections of amphetamine, which cause progressive changes in neostriatal ascorbate release, fail to alter accumbal ascorbate (Yount et al., 1991; Mueller, 1989). These results are intriguing in light of the different roles attributed to the neostriatal and accumbal dopaminergic systems in the behavioral effects of amphetamine (Seiden et al., 1993; Dunnett and Robbins, 1992; Rebec and Bashore, 1984). Conceivably, regional differences in amphetamine-induced ascorbate release contribute, at least in part, to these behavioral differences, but relatively little information is available on this point.

Another unexpected outcome of research on neostriatal ascorbate release is that the controlling dopaminergic mechanisms reside outside the neostriatum. This conclusion is based on reports that near-total destruction of neostriatal dopaminergic terminals failed to block amphetamine-induced neostriatal ascorbate release (Pierce et al., 1992; Kamata et al., 1986; Gonon et al., 1981). Such evidence also ruled out dopaminergic neurons as the immediate source of extracellular ascorbate. Moreover, direct infusion of either amphetamine or dopamine into the neostriatum does not increase ascorbate release at the infusion site and, in fact, causes a slight decline (Wilson and Wightman, 1985). When infused into the substantia nigra, however, these same compounds increase ascorbate release in the neostriatum (Wilson et al., 1986; Wilson and Wightman, 1985). Thus, although dopaminergic mechanisms play a role in neostriatal ascorbate release, they appear to operate at the level of dopaminergic cell bodies rather than terminals.

Amphetamine enhances dopamine transmission within the substantia nigra by facilitating the release of dopamine from dopaminergic dendrites in the SNC (Groves et al., 1975). Understanding how this mechanism can lead to ascorbate release in the neostriatum requires a consideration of nigral connections with thalamic and cortical projections.

9.3.2. The Nigro-Thalamo-Cortical System

As a major output nucleus of the basal ganglia, the substantia nigra reticulata (SNR) represents a critical link in the flow of information from neostriatum to cerebral cortex (Heimer et al., 1995; Alexander and Crutcher, 1990). SNR neurons receive input from a number of sources, including dopaminergic dendrites. In turn, SNR efferents exert an inhibitory influence on ventromedial thalamus (VMT) via the release of γ-aminobutyric acid (GABA). Thalamic inhibitory interneurons, which also appear to be GABAergic, then innervate neurons that comprise the thalamo-cortical projection. This pathway releases glutamate and exerts an excitatory influence on cortical neurons. A glutamate projection from cortex to neostriatum completes the circuit.

Amphetamine is known to activate SNR neurons (Kamata and Rebec, 1985; Rebec and Groves, 1975), and dopamine also may exert an excitatory effect by modulating nigral GABA transmission (Waszczak and Walters, 1983). The net result of these effects, as shown in Figure 9.1, is a disinhibition of the thalamo-cortical projection and an activation of glutamatergic afferents to the neostriatum. That this circuit is involved in amphetamine-induced neostriatal ascorbate release comes from evidence that such release is blocked by lesions placed at various sites along the circuit, including SNR (Pierce et al., 1994a), VMT (Pierce et al., 1994a; Basse-Tomusk and Rebec, 1990), or cerebral cortex (Desole et al., 1992; Basse-Tomusk and Rebec, 1990). In fact, the release caused by direct intranigral infusions of amphetamine is also blocked by VMT lesions, confirming the involvement of a multisynaptic circuit (Pierce et al., 1994a).

Cortical lesions, moreover, cause a greater than 70% decline in the basal level of neostriatal ascorbate (Basse-Tomusk and Rebec, 1990; O'Neill et al., 1983), whereas kainic acid lesions of the neostriatum, which destroy intrinsic cellular elements, have no such effect (Pierce et al., 1992). Thus, the cortico-striatal pathway appears to be a major source of extracellular ascorbate in the neostriatum. In the case of an increase in nigral dopamine transmission, which occurs in response to drugs like amphetamine or to an increase in SNC neuronal activity, the cortico-striatal pathway is activated via nigro-thalamo-cortical connections and ascorbate release is enhanced.

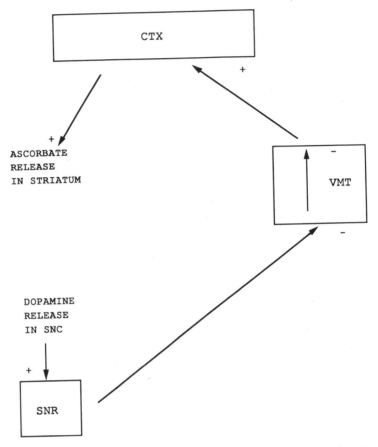

FIGURE 9.1. Schematic representation of the neural circuitry proposed to underlie amphetamine-mediated ascorbate release in striatum. The process is triggered by local dopamine release in SNC, which has a net excitatory effect on SNR neurons. Activation of SNR disinhibits VMT efferents via inhibitory interneurons resulting in cortical (CTX) activation. An increase in corticostriatal activity releases glutamate, eventually increasing extracellular ascorbate (see text).

To the extent that this model is correct, an increase in neostriatal glutamate transmission should result in an increase in ascorbate release. This indeed is the case (Pierce and Rebec, 1993; Cammack *et al.,* 1991), but rather than undergo co-release with glutamate, ascorbate appears to be released as part of the glutamate high-affinity uptake process.

9.3.3. Heteroexchange at the Glutamate Transporter

In vitro investigations of the neostriatum and other tissue known to receive glutamatergic afferents have revealed that ascorbate is not released by exocytosis but by heteroexchange with glutamate at the glutamate uptake site (see Fillenz *et al.,* 1986; O'Neill, 1995). Thus, any stimulus that activates glutamatergic neurons and causes glutamate release can be expected to activate the glutamate transporter, eventually resulting in the release of ascorbate. This heteroexchange model has been supported in subsequent work with ambulant animals (Pierce and Rebec, 1993; Ghasemzadeh *et al.,* 1991). Moreover, electrical stimulation of glutamatergic neurons has been shown to increase the level of extracellular ascorbate (O'Neill *et al.,* 1984; Cammack *et al.,* 1991), whereas destruction of cortico-striatal neurons not only lowers basal ascorbate levels but also blocks drug-induced or behavior-related ascorbate release (Basse-Tomusk and Rebec, 1990; O'Neill *et al.,* 1983).

In addition to axon terminals, the glutamate transporter is present on glial cells (e.g., Kanai *et al.,* 1994; Torp *et al.,* 1994), which may provide another source of extracellular ascorbate. Although ascorbate release via glial glutamate transporters has not been studied systematically, pharmacological blockade of glial glutamate uptake is known to attenuate, though not completely inhibit, glutamate-evoked ascorbate release (Cammack *et al.,* 1991). It seems likely, therefore, that ascorbate is released from both neuronal and glial elements associated with glutamate transport. Not all glutamate transporters, however, use ascorbate as the heteroexchange molecule (Grünewald and Fillenz, 1984), raising interesting questions about glutamate transport and the role of ascorbate in this process. It also is possible that ascorbate is released by carrier proteins not associated with glutamate transport, but little evidence is available to support this hypothesis (Fillenz *et al.,* 1986).

9.3.4. Other Considerations

To the extent that glutamatergic cortico-striatal neurons comprise the final common pathway for ascorbate release in the neostriatum, any activation of these neurons should enhance release. Thus, the nigro-thalamo-cortical circuit probably represents only one mechanism for release of neostriatal ascorbate, but it appears to be the primary mechanism by which nigral dopamine participate in this process. Other transmitter systems also could act through this circuit. For example, intranigral infusions of GABA, which inhibit reticulata efferents and thus lower the level of cortical activity, have been found to decrease neostriatal ascorbate release (Christensen and Rebec, 1994).

9.4. NEUROPROTECTANT EFFECTS

Accumulating evidence suggests that both dopamine and glutamate contribute to the neurotoxic effects of oxidative stress (Coyle and Puttfarcken, 1993). Ascorbate, which is released in conjunction with these transmitters in the neostriatum, may provide a critical element of protection. This section reviews the mechanisms by which ascorbate could perform such a function.

9.4.1. Degeneration of Dopaminergic Neurons

Dopaminergic neurons are vulnerable to degeneration because dopamine is readily oxidized, either by enzymatic or nonenzymatic reactions, to form cytotoxic hydroxyl radicals (Bonorden and Pariza, 1994). In the enzymatic metabolism of dopamine, hydroxyl radicals are formed when monoamine oxidase catalyzes the oxidative deamination of dopamine and forms hydrogen peroxide as a by-product. Dopamine also can undergo nonenzymatic autoxidation, forming neurotoxic quinones and semiquinones, along with hydroxyl radicals. Once formed, these oxidative by-products are capable of destroying vital cellular molecules, including nucleic acids, proteins, and lipids. When a free radical reacts with these molecules, other free radicals are formed and a chain reaction develops. In cellular membranes, for example, free radical chain reactions damage polyunsaturated fatty acids, resulting in lipid peroxidation and loss of membrane integrity. Antioxidants stop these chain reactions by donating hydrogen atoms, and in the process highly reactive oxygen radicals are replaced by much more stable compounds, such as the ascorbyl radical (Niki, 1991; Bendich *et al.,* 1986; Halliwell and Gutteridge, 1985).

Ascorbate is an unusual antioxidant in that it can protect against free radical chain reactions in either the aqueous or lipid phase of cells (Beyer, 1994). Because it is water soluble, ascorbate reacts directly with oxygen radicals in the aqueous phase before they can attack and destroy lipid membranes. In the lipid phase, ascorbate counteracts oxygen radical formation by reacting with the α-tocopheryl radical to regenerate α-tocopherol, a lipoprotein-associated antioxidant (vitamin E). In fact, ascorbate proved superior to α-tocopherol in protecting plasma lipids against oxidative damage (Frei, 1991). Most important for dopaminergic neurons, ascorbate inhibits dopamine oxidation (Hastings and Zigmond, 1994) and thus may provide some degree of protection against the cytotoxic effects of oxidative stress (Fornstedt and Carlsson, 1991).

Parkinson's disease is perhaps the most prominent clinical condition associated with a degeneration of dopaminergic neurons, though the same pathology also occurs following exposure to high doses of amphetamine derivatives. A persuasive case be made that the underlying cause involves inadequate protection against the neurotoxic effects of dopamine oxidation.

9.4.1.1. Parkinson's Disease

Although many different neurons show evidence of degeneration in Parkinsonian patients, the motor symptoms of this disease, such as bradykinesia and muscular rigidity, result from massive damage to the nigro-striatal dopaminergic pathway (see Duvoisin, 1991; Jellinger, 1986). In some cases exposure to toxic substances triggers the damage and the mechanism is clear: an overproduction of free radicals. Animal studies, for example, have shown that 6-hydroxydopamine (6-OHDA) and 1-methyl-4-phenyl-1,2,3,6-tetrahydropyridine (MPTP) destroy dopaminergic neurons by increasing oxidative stress (see Poirier and Thiffault, 1993; Irwin, 1986). In the case of MPTP, neuronal damage results from the conversion of MPTP to 1-methyl-4-phenylpyridinium (MPP+), which inhibits mitochondrial complex I activity and generates oxygen radicals (Langston *et al.,* 1987). This same mitochondrial abnormality has been identified in some Parkinsonian patients (Jenner, 1993; Fahn and Cohen, 1992).

In most cases of Parkinson's disease, the etiology is not so apparent, but oxidative stress appears to play a role. A consistent finding in Parkinson's disease is an elevated level of iron in the substantia nigra (Dexter *et al.,* 1992). Chelated iron catalyzes the formation of oxygen radicals and enhances lipid peroxidation (Ben-Shachar and Youdim, 1993; Adams and Odunze, 1991). In fact, the level of malondialdehyde (MDA), a product of lipid peroxidation, increases significantly in the substantia nigra of parkinsonian patients (Dexter *et al.,* 1989). These patients also suffer from a low level of nigral glutathione peroxidase, an enzyme that works in conjunction with antioxidants to oppose lipid peroxidation (Meister, 1992). Reduced glutathione, the substrate for this enzyme, is similarly low in nigral tissue (Jenner *et al.,* 1992; Riederer *et al.,* 1989).

Surprisingly, however, nigral ascorbate levels remain unchanged even in clinicopathologically severe cases of Parkinson's disease (Riederer *et al.,* 1989). The same is true of serum ascorbate levels (Fernandez-Calle *et al.,* 1993). Thus, to the extent that inadequate antioxidative capacity contributes to the pathophysiology of Parkinsonism, the problem does not appear to involve an ascorbate deficiency or an inability of ascorbate to reach the site of neuronal degeneration.

Still, ascorbate treatment, which can help restore antioxidant protection, may prove useful in combating oxidative damage. In fact, administration of antioxidants is known to protect against either 6-OHDA- or MPTP-induced neurotoxicity in animals (see Adams and Odunze, 1991). Ascorbate seems especially crucial in light of evidence that oxygen-induced membrane damage in tissue obtained from ascorbate-deficient guinea pigs persists in the presence of α-tocopherol, glutathione, and various radical-scavenging enzymes, but not after the addition of ascorbate (Chakraborty *et al.,* 1994). Dietary ascorbate also has been shown to decrease endogenous lipid peroxidation in these animals (Barja *et al.,* 1994). In an open trial of antioxidants in early Parkinson's disease, high doses of ascorbate and α-tocopherol delayed the need for standard dopamine replacement therapy with L-DOPA by 2.5 years compared to similar patients not taking antioxidants (Fahn, 1991).

Ironically, L-DOPA itself may be neurotoxic. Using data obtained from 6-OHDA-lesioned rats, Blunt *et al.* (1993) reported that L-DOPA destroys dopaminergic neurons remaining on the side of the lesion but not on the intact side. L-DOPA, moreover, is toxic for both dopaminergic and nondopaminergic neurons in culture, and ascorbate can block this effect (Pardo *et al.,* 1995). Conceivably, the surviving dopaminergic neurons in Parkinsonian patients are especially vulnerable to oxidative stress, and L-DOPA, either by undergoing autoxidation to toxic quinones or by enhancing dopamine metabolism, contributes to this effect. Further work is required to assess the toxicity of L-DOPA treatment and the protectant role of ascorbate in animal models and eventually in Parkinsonian patients.

9.4.1.2. Amphetamine Toxicity

Amphetamine and its methylated analogs release dopamine via an impulse-independent mechanism. These drugs interact with the high-affinity membrane dopamine transporter and reverse its direction of operation, increasing the extracellular concentration of dopamine (Liang and Rutledge, 1982; Raiteri *et al.,* 1979). With long-term use of these drugs, however, dopamine transmission eventually shuts down as dopaminergic terminals degenerate (Gibb *et al.,* 1994; Ricaurte *et al.,* 1994).

Although several mechanisms are likely to contribute to the neurotoxic effects of the amphetamines, dopamine itself appears to play a major role. Initial support for this view emerged from evidence that treatments that oppose the drug-induced dopamine efflux, such as administration of dopamine uptake blockers (Schmidt and Gibb, 1985) or catecholamine synthesis inhibitors (Wagner *et al.*, 1983), attenuate neurotoxicity. Protection also was conferred by treatment with dopamine antagonists, including mixed D1/D2 antagonists as well as selective D1 or D2 antagonists (O'Dell *et al.*, 1993; Sonsalla *et al.*, 1986). Although dopamine antagonists by themselves increase dopamine release, they attenuate this process when given in combination with amphetamine analogs (Marshall *et al.*, 1993). In addition, Marshall and coworkers (O'Dell *et al.*, 1991, 1993) have shown that their dose regimen for methamphetamine elicits neurotoxic effects by the fourth injection, at a time when dopamine efflux is maximal. Interestingly, after the fourth methamphetamine injection in this regimen, dopamine antagonists cause the greatest reduction in dopamine release. Thus, the magnitude of the methamphetamine-induced increase in extracellular dopamine correlates closely with dopaminergic neurotoxicity.

As in Parkinson's disease, oxidative processes appear to be responsible. Although amphetamines inhibit monoamine oxidase, the increase in extracellular dopamine caused by these drugs is likely to promote dopamine autoxidation. Consistent with this view, a single, high dose of methamphetamine has been shown to form 6-OHDA in the neostriatum (Seiden and Vosmer, 1984). In addition, pretreatment with ascorbate blocks methamphetamine-induced toxicity (Wagner *et al.*, 1985), and this protective effect is shared by other antioxidants (DeVito and Wagner, 1989). Thus, the increase in neostriatal ascorbate release that accompanies amphetamine administration may help counteract the potentially damaging effects of abnormally high levels of extracellular dopamine. Noteworthy in this regard is evidence that multiple amphetamine injections first cause a progressive enhancement of ascorbate release, but as the injections continue and neurotoxic liability increases, ascorbate release declines (Yount *et al.*, 1991).

Amphetamine analogs also facilitate the release of glutamate (Stephans and Yamamoto, 1994), an effect consistent with the glutamate heteroexchange model given that amphetamine also increases ascorbate release (see section 9.3.3). Overactivation of glutamate receptors can lead to neurodegeneration, and such receptors are found on the terminals of neostriatal dopaminergic neurons (Parent *et al.*, 1995). Thus, glutamatergic amechanisms may contribute to the neurotoxic effects of the amphetamines. Indeed, several laboratories have reported that treatment with noncompetitive antagonists of the N-methyl-D-aspartate (NMDA) glutamate receptor protects against the neurotoxic effects of methamphetamine (Layer *et al.*, 1993; Ohmori *et al.*, 1993; Sonsalla *et al.*, 1991). It is important to stress, however, that NMDA antagonists also attenuate dopamine release, raising the possibility that their neuroprotective effects result from a decline in extracellular dopamine (Marshall *et al.*, 1993; Weihmuller *et al.*, 1992). Thus, although glutamate may contribute to the neurotoxic effects of amphetamine analogs, dopamine-mediated oxidative processes appear to play a key role.

9.4.2. Glutamate-Induced Neurotoxic Effects

As a major excitatory transmitter, glutamate is found in many different brain structures, where it is stored in neuronal terminals at relatively high concentrations. Accumulation of glutamate in extracellular fluid, however, leads to cell membrane changes and death (Dugan and Choi, 1994; Götz *et al.*, 1994; Choi, 1992). High-affinity glutamate transporters normally guard against this possibility, but their operation deteriorates under anoxic conditions. When this happens, glutamate receptors become overactivated, leading to a massive influx of intracellular sodium and calcium, both of which can have neurotoxic effects (Coyle and Puttfarcken, 1993). Sodium influx forces the cell to accumulate water, which can lead to osmotic lysis. The neurotoxic effects of calcium influx, in contrast, are caused by a series of intracellular metabolic events, eventually culminating in the production of oxygen radicals. These radicals, moreover, promote glutamate release and further inhibit neuronal and glial transporters, creating a vicious cycle that can lead to large-scale neural destruction.

The effects of glutamate release can be blocked by glutamate antagonists, which offer protection against anoxic damage (Bigge and Boxer, 1994). Some endogenous protection may be provided by ascorbate. Anoxia can raise extracellular ascorbate levels in the neostriatum to approximately 1 mM, which is three times the normal concentration (Hillered *et al.*, 1988). At this level, ascorbate not only provides antioxidant protection but also blocks NMDA receptors (Majewska *et al.*, 1990). Also noteworthy is evidence that the turtle brain, which is highly resistant to anoxia, is much better than the mammalian brain at maintaining a high level of ascorbate in both intra- and extracellular compartments (Rice and Cammack, 1991; Rice and Nicholson, 1987). In fact, rats lose the ability to release ascorbate in the neostriatum within minutes after an elevation of neostriatal glutamate levels, suggesting that the protection offered by endogenous release of ascorbate is relatively short lived (Pierce and Rebec, 1993).

9.4.3. Neurotoxic Mechanisms in Schizophrenia and Tardive Dyskinesia

Neuroleptic drugs are widely used in the treatment of schizophrenia, and their ability to block dopamine receptors has been the centerpiece of evidence implicating a dopaminergic dysfunction in this disease (see Ellenbroek, 1993). Further elaborations of the dopaminergic model now include an interaction with glutamatergic or other transmitter systems as an important part of the underlying pathology (Lieberman and Koreen, 1993; Grace, 1991; Meltzer and Nash, 1991; Carlsson and Carlsson, 1990; Freed, 1989). While acknowledging a role for abnormalities in transmitter function during the early stages of schizophrenia, other theorists have focused on oxidative stress as the main mechanism underlying the progression to the defect state of the disease (Cadet and Kahler, 1994). The reason for making this distinction is that schizophrenic symptoms typically begin with hallucinations and delusions, which have a relatively good prognosis, but eventually progress to the defect condition characterized by flat affect and poor social interactions. Recovery from the defect state is rare, and neuroleptic treatment is largely ineffective. Not surprisingly, the defect state is accompanied by neuropathological changes in several forebrain areas, including frontal cortex and basal ganglia (Klausner *et al.,* 1992; Jellinger, 1986; Weinberger *et al.,* 1986).

That oxidative stress forms the basis for at least the defect state of schizophrenia is plausible given that dopaminergic excess early on in schizophrenia could easily lead to an accumulation of oxygen radicals and eventual neurodegeneration. In its current manifestation, however, this model is based entirely on indirect evidence and provides no insight into how oxidative mechanisms produce the neuropathology unique to schizophrenia. Nevertheless, it is interesting to note that schizophrenia is characterized by an ascorbate deficiency. Schizophrenic patients, for example, have a lower serum level of ascorbate and require higher amounts of dietary ascorbate to reach saturation than control subjects (Suboticanec *et al.,* 1986). Following an ascorbate load, moreover, schizophrenic patients show a lower rate of ascorbate excretion in the urine than diet-matched controls, suggesting a higher ascorbate requirement in schizophrenic patients (Suboticanec *et al.,* 1990). This interpretation is consistent with evidence from a limited number of post-mortem assessments in which the brains of schizophrenic patients showed consistently low ascorbate levels (Adams, reported in Rebec *et al.,* 1986). Whether these findings highlight mechanisms that play a direct role in the disease process or are merely secondary to the pathological and behavioral changes that accompany it remains to be determined.

When administered to schizophrenic patients, ascorbate has been reported to cause an improvement, though typically mood or motivation improves without much change in psychotic symptoms (see Rebec and Pierce, 1994). But in cases of phencyclidine (PCP) psychosis, which includes aspects of both positive and negative schizophrenic episodes, ascorbate given in conjunction with standard neuroleptics was more effective than neuroleptics alone (Giannini *et al.,* 1987). Ascorbate also appears to modulate the behavioral effects of neuroleptics in animals, though this work suggests caution in using ascorbate as a neuroleptic adjunct in human patients (see section 9.5.1).

Prolonged use of neuroleptics leads to a condition of uncontrollable motor abnormalities (tardive dyskinesia) that also has been linked to oxidative stress (Cadet and Kahler, 1994). This hypothesis is consistent with several lines of evidence, most notably the ability of neuroleptics to induce signs of lipid peroxidation in the brains of animals (Shivakumar and Ravindranath, 1992). In fact, a free-radical metabolite of haloperidol, a widely used neuroleptic, has been shown to have neurotoxic effects in rats (Rollema *et al.,* 1994). Moreover, patients treated with neuroleptics show varying degrees of neurodegeneration in the basal ganglia and other brain regions known to play a role in motor control (Dalgalarrondo and Gataz, 1994; Christensen *et al.,* 1970).

Although an up-regulation of dopamine receptors resulting from prolonged dopamine receptor antagonism has been implicated in tardive dyskinesia (Seeman, 1992; Rupniak *et al.,* 1983; Waddington *et al.,* 1983), this model fails to explain the slow onset of the disease and its persistence after neuroleptic withdrawal. Several months or years may elapse before neuroleptic treatment induces tardive dyskinesia, and the condition can become permanent (Tarsey, 1983). Up-regulation of dopamine receptors, in contrast, occurs relatively rapidly after neuroleptic administration and disappears shortly after drug withdrawal (Jeste and Wyatt, 1981). Alternative models, supported by substantial animal data and some clinical evidence, implicates a loss of neostriato-nigral GABAergic neurons (Fibiger and Lloyd, 1984) or neostriatal cholinergic inter-neurons (Miller and Chouinard, 1993) in the development of tardive dyskinesia. These suggestions are consistent with the notion of neuroleptic-induced neurodegeneration.

Neuroleptic treatment for several weeks has been shown to elevate extracellular ascorbate levels in the neostriatum of rats (Pierce *et al.,* 1994b). To the extent that neuroleptics induce free-radical production, this response could be viewed as a neuro-protective mechanism. Oxidative damage may result if this mechanism fails in the face of continued neuroleptic treatment. Whether supplemental treatment with ascorbate or other antioxidant could help counteract the neurodegenerative effects of neuroleptics remains to be established.

9.5. CLINICAL AND FUNCTIONAL IMPLICATIONS

The ease with which ascorbate both enters the brain and provides antioxidant protection appears to make this molecule an ideal prophylactic for the runaway oxida-tion underlying neurodegenerative conditions. Ascorbate also has the appeal of being a water-soluble vitamin that can be consumed in large doses with apparent safety. But until additional information becomes available on the biochemical properties of ascor-bate, caution seems to be the best clinical strategy. Under certain conditions, for example, ascorbate can promote oxidation, rather than oppose it. Fornstedt and Carls-son (1991) reported that administration of ascorbate to guinea pigs following a period of ascorbate deprivation enhances dopamine oxidation. Ascorbate also may function as a prooxidant in the presence of high iron concentrations (Montgomery, 1995; Stadt-man, 1991), a condition that characterizes the substantia nigra of Parkinsonian patients and may occur in other degenerative disorders (see section 9.4.1.1). In fact, ascorbate infusions into the substantia nigra of cats can have neurotoxic effects (Wolfarth *et al.,* 1977). But perhaps the most important reason for caution in the clinical use of ascor-bate is that apart from participation in oxidation–reduction reactions, this molecule also appears to function as an extracellular modulator of synaptic transmission (Rebec and Pierce, 1994; Hadjiconstantinou and Neff, 1983). Although extracellular modula-tors are gaining increasing attention in the research literature (Bach-y-Rita, 1993; Fuxe and Agnati, 1991), the role of ascorbate is proving extremely difficult to characterize. An emerging challenge, therefore, is to expand research on ascorbate at the level of behavioral and neuronal function. This section reviews the data currently available from this line of work as it applies to dopaminergic and glutamatergic mechanisms in the neostriatum.

9.5.1. Dopaminergic Modulation

The behavioral effects of ascorbate suggest that it can act as either a dopamine agonist or antagonist. Dopamine agonists are known to enhance open-field behavioral activity in rats and mice, the species of choice for most drug tests (Jackson and

Westlind-Danielson, 1994; Seiden *et al.*, 1993; Dunnett and Robins, 1992; Rebec and Bashore, 1984; Segal and Janowsky, 1978). This behavioral effect is often manifest as an increase in locomotion and rearing with simultaneous or independent bouts of sniffing, head movements, and occasional licking and biting. Typically, as the dose increases, behavior becomes progressively more focused and repetitive (stereotyped) as locomotion and rearing give way to prolonged sniffing as well as intense head and oral movements. In contrast, dopamine antagonists, such as haloperidol, attenuate or block most, if not all, of these behavioral changes. By themselves, dopamine antagonists decrease behavioral activity and at high doses may elicit an immobile or cataleptic posture. Ascorbate has been tested mainly in conjunction with amphetamine, though other dopamine agonists have been used with essentially similar results. Initial data, based on systemic injections of relatively high doses of ascorbate (500–2000 mg/kg), revealed that this vitamin blocked amphetamine-induced focused stereotypy (Tolbert *et al.*, 1979b) and locomotion (Heikkila *et al.*, 1981). This finding was confirmed in other behavioral tests in which ascorbate blocked responses known to require increases in forebrain dopamine transmission (Tolbert *et al.*, 1979a; Desole *et al.*, 1987).

That ascorbate could function as a dopamine antagonist was also consistent with evidence that when combined with a threshold dose of haloperidol, 1000 mg/kg ascorbate potentiated haloperidol-induced catalepsy and also enhanced the ability of haloperidol to block the behavioral activation produced by amphetamine (Rebec *et al.*, 1985). These behavioral effects also occurred when ascorbate was infused directly into the ventricular system (White *et al.*, 1988), ruling out a peripheral mechanism of action. Parallel studies also ruled out a pharmacokinetic interaction between ascorbate and amphetamine (Kiely *et al.*, 1987). Further experiments implicated the neostriatum as a critical site of ascorbate action when ascorbate infusions into this, but not surrounding structures, were found to block the behavioral effects of amphetamine (White *et al.*, 1990). Ascorbate, therefore, appears to function as a neostriatal dopamine antagonist.

This conclusion is consistent with evidence that like haloperidol, ascorbate induces behavioral supersensitivity to dopamine agonists. For example, rats treated with either substance for 21 days and then challenged the following day with apomorphine, a direct but non-selective D1/D2 agonist, showed an enhanced apomorphine-induced behavioral response (Pierce *et al.*, 1991). Only the haloperidol-treated rats, however, showed a corresponding increase in dopamine receptor binding, suggesting that to the extent ascorbate opposes dopaminergic function, it may not do so at the receptor level.

Noteworthy in this regard is biochemical evidence that although ascorbate has been shown to inhibit the binding of radiolabeled dopamine to homogenates of neostriatal tissue, this effect has not been replicated in all cases (Rebec and Pierce, 1994). Some evidence also suggests that ascorbate inhibits the binding of dopamine antagonists, though this effect appears to be due to lipid peroxidation with ascorbate acting as a prooxidant in these studies (Kimura and Sidhu, 1994; Ebersole and Molinoff, 1991). Thus, *in vitro* tests of ascorbate effects on dopamine agonist and antagonist binding have been difficult to interpret (for a discussion of this literature see Rebec and Pierce, 1994).

Even tests of ascorbate function at the level of the dopamine receptor have led to discrepant results. In neostriatal homogenates, for example, activation of the D1 dopamine receptor stimulates adenylate cyclase activity, but whereas Tolbert *et al.* (1992) reported that ascorbate blocked this effect, Schulz *et al.* (1984) failed to confirm it. These discrepancies at the receptor level are unfortunate because the behavioral data on ascorbate can be explained independently of dopamine receptor blockade. Neostriatal dopamine, for example, exerts its behavioral effects by acting on GABAergic output neurons that, in turn, are regulated by other neuronal systems and circuits, any one of which could be influenced by ascorbate. Ascorbate is known to influence glutamate transmission (see section 9.5.2), and in view of the close interaction between glutamatergic and dopaminergic terminals in the neostriatum (Parent *et al.,* 1995), an ascorbate-induced inhibition of the amphetamine behavioral response may reflect a glutamatergic change rather than a blockade of dopamine receptors (see section 9.5.2). Regardless of the mechanism of action, however, it seems clear that high doses of ascorbate inhibit dopamine-mediated behavioral effects.

Surprisingly, this conclusion does not seem to apply to low doses of ascorbate. Thus, systemic injection of 50 or 200 mg/kg ascorbate has been found to potentiate amphetamine-induced effects on open-field behavior and to antagonize the cataleptic response to haloperidol (Wambebe and Sokomba, 1986). Similarly, 100 mg/kg ascorbate both enhances amphetamine performance (Pierce *et al.,* 1995) and antagonizes haloperidol effects (Gulley and Rebec, 1995) on conditioning tasks known to be sensitive to changes in dopaminergic transmission. It appears, therefore, that the behavioral effects of ascorbate are dose dependent such that low doses enhance and high doses oppose dopaminergic function.

Ascorbate also has been tested on neostriatal neurons. Both systemic and iontophoretic applications of ascorbate have been shown to increase neuronal activity (Pierce and Rebec, 1995; Gardiner *et al.,* 1985; Ewing *et al.,* 1983). In some neurons, the increase exceeded 1000% of the rate before ascorbate administration. This effect is not simply due to oxidation because iso-ascorbate, an ascorbate isomer with the same redox potential, has relatively little effect on neostriatal neurons (Gardiner *et al.,* 1985). Simultaneous measurements of the ascorbate concentration at the tip of the iontophoresis electrode revealed that the neuronal excitation, which could be elicited by ejection currents of 10–80 nA, occurred at ascorbate levels within the normal physiological range (Gardiner *et al.,* 1985).

In awake, unrestrained rats, neostriatal neurons often are excited by spontaneous movement (Haracz *et al.,* 1993; Wang and Rebec, 1993; West *et al.,* 1987). These same cells also are excited by iontophoretic dopamine, and the dopamine-induced response is enhanced by the simultaneous application of ascorbate (Pierce and Rebec, 1995). In fact, this research has shown that the neuronal response to dopamine and ascorbate together is greater than the response to either substance alone, suggesting a supra-additive interaction. Thus, these data support the low-dose behavioral work indicating that ascorbate enhances dopaminergic effects. But, again, dose may be critical. Ascorbate iontophoresis at high ejection currents (e.g., 120–160 nA) inhibits neostriatal activity (Gardiner *et al.,* 1985; Pierce and Rebec, 1995). Although it remains to be established if high-dose ascorbate iontophoresis opposes dopaminergic function, it seems clear that ascorbate resists classification as either a dopamine agonist or antagonist. Functionally, either role may apply, depending on the concentration of ascorbate in extracellular fluid.

9.5.2. Glutamatergic Modulation

Neostriatal neurons are highly responsive to the excitatory effects of glutamate. In either anesthetized (e.g., Herrling, 1985) or freely moving rats (Kiyatkin and Rebec, 1996; Pierce and Rebec, 1995), iontophoretic application of glutamate at low ejection currents (<20 nA) induces a rapid and pronounced increase in neuronal activity in almost all cases. Interestingly, this effect is enhanced by the co-application of ascorbate. Thus, as with dopamine (see section 9.5.1), the effect of ascorbate on glutamate transmission is supra-additive (Pierce and Rebec, 1995). This glutamate–ascorbate interaction supports data from anesthetized rats that glutamate iontophoresis enhances the responsiveness of neostriatal neurons to iontophoresis of ascorbate (Gardiner et al., 1985). The basis for this interaction is unclear, but because glutamate and ascorbate undergo heteroexchange at the same membrane transport system (see section 9.3.3), it seems likely that an increase in extracellular glutamate will promote ascorbate release and vice versa. Ascorbate, therefore, may enhance the effects of glutamate on neostriatal neurons by interfering with glutamate transport, rather than by interacting with postsynaptic glutamate receptors.

Ascorbate may exert postsynaptic effects at relatively high extracellular concentrations. Perfusion of rat neocortical neurons *in vitro* with ascorbate concentrations of 1–3 mM blocked the NMDA-induced influx of depolarizing current (Majewska et al., 1990). Application of 500 μM ascorbate had no effect on this response. The NMDA receptor appears to be regulated by a redox site (Tauck, 1992), and a high level of ascorbate may be required to occupy this site and oppose glutamatergic effects (Majewska et al., 1990). *In vivo* extracellular ascorbate concentrations approach 1 mM during anoxic conditions, suggesting that this increase in ascorbate is a protective response to the massive release of glutamate (Hillered et al., 1988). An NMDA receptor blockade by high concentrations of ascorbate also could explain the inhibitory effects of ascorbate iontophoresis at high ejection currents (see section 9.5.1). As with dopamine, therefore, it appears that ascorbate exerts opposing concentration-dependent effects on glutamatergic transmission.

9.6. CONCLUSIONS

The high level of ascorbate maintained in brain tissue and the large number of stimuli that trigger its release into extracellular fluid from neural and, possibly glial, sources underscores the importance of this vitamin for normal brain function. Ample biochemical evidence indicates that, as an effective antioxidant, ascorbate offers some degree of protection against the oxidative stress associated with the normal operation of dopaminergic and glutamatergic neurons. But before ascorbate is used as treatment against the oxidative processes believed to underlie certain neurodegenerative disorders, research is required to identify the precise biochemical conditions under which this protective function of ascorbate occurs. Ascorbate also has been shown to influence dopamine-mediated behavioral effects and to potentiate the synaptic action of both dopamine and glutamate on neostriatal neurons. Thus, apart from participating in oxidation-reduction reactions, ascorbate appears to function as extracellular modulator of synaptic transmission. It now becomes important to identify the mechanisms by which ascorbate performs this role.

9.7. REFERENCES

Adams, J. D., and Odunze, I. N., 1991, Oxygen free radicals and Parkinson's disease, *Free Rad. Biol. Med.* **10**(2):161–169.

Alexander, G. E., and Crutcher, M. D., 1990, Functional architecture of basal ganglia circuits: neural substrates of parallel processing, *Tr. Neurosci.* **13**:266–271.

Bach-y-Rita, P., 1993, Neurotransmission in the brain by diffusion through the extracellular fluid—a review, *Neuroreport* **4**(4):343–350.

Barja, G., Lopeztorres, M., Perezcampo, R., Rojas, C., Cadenas, S., Prat, J., and Pamplona, R., 1994, Dietary vitamin C decreases endogenous protein oxidative damage, malondialdehyde, and lipid peroxidation and maintains fatty acid unsaturation in the guinea pig liver, *Free Rad. Biol. Med.* **17**(2):105–115.

Basse-Tomusk, A., and Rebec, G. V. (1990), Corticostriatal and thalamic regulation of amphetamine-induced ascorbate release in the neostriatum, *Pharmacol. Biochem. Behav.* **35**:55–60.

Basse-Tomusk, A., and Rebec, G. V., 1991, Regional distribution of ascorbate and 3,4-dihydroxyphenylacetic acid (DOPAC) in rat neostriatum, *Brain Res.* **538**:29–35.

Bendich, A., Machlin, L., J., Scandurra, O., Burton, G. W., and Wayner, D. D. M., 1986, The antioxidant role of vitamin C, *Adv. Free Rad. Biol. Med.* **2**:419–444.

Ben-Shachar, D., and Youdim, M. B. H., 1993, Iron, melanin and dopamine interaction—relevance to Parkinson's disease, *Prog. Neuro–Psychopharmacol. Biol. Psychiat.* **17**(1):139–150.

Beyer, R. E., 1994, The role of ascorbate in antioxidant protection of biomembranes: Interaction with vitamin E and coenzyme Q, *J. Bioenerget. Biomemb.* **26**(4):349–358.

Bigge, C. F., and Boxer, P. A., 1994, Neuronal cell death and strategies for neuroprotection, *Annu. Rep. Med. Chem.* **29**:13–22.

Blunt, S. B., Jenner, P., and Marsden, C. D., 1993, Suppressive effect of L-dopa on dopamine cells remaining in the ventral tegmental area of rats previously exposed to the neurotoxin 6-hydroxydopamine, *Mov. Disord.* **8**:129–133.

Bonorden, W. R., and Pariza, M. W., 1994, Antioxidant nutrients and protection from free radicals, in: *Nutritional Toxicology* (F. N. Kotsonis, M. Mackey, J. Hjelle, eds.), Raven Press, New York, pp. 19–48.

Boutelle, M. G., Svensson, L., and Fillenz, M., 1989, Rapid changes in striatal ascorbate in response to tailpinch monitored by constant potential voltammetry, *Neuroscience* **30**:11–17.

Cadet, J. L., and Kahler, L. A., 1994, Free radical mechanisms in schizophrenia and tardive dyskinesia, *Neurosci. Biobehav. Rev.* **18**(4):457–467.

Cammack, J., Ghasemzadeh, B., and Adams, R. N., 1991, The pharmacological profile of glutamate-evoked ascorbic acid efflux measured by in vivo electrochemistry, *Brain Res.* **565**(1):17–22.

Carlsson, M., and Carlsson, A., 1990, Interactions between glutamatergic and monoaminergic systems within the basal ganglia: implications for schizophrenia and Parkinson's disease, *Tr. Neurosci.* **13**:272–276.

Chakraborty, S., Nandi, A., Mukhopadhyay, M., Mukhopadhyay, C. K., and Chatterjee, I. B., 1994, Ascorbate protects guinea pig tissues against lipid peroxidation, *Free Rad. Biol. Med.* **16**:417–426.

Chiueh, C. C., Miyake, H., and Peng, M.-T., 1993, Role of dopamine autoxidation, hydroxyl radical generation, and calcium overload in underlying mechanisms involved in MPTP-induced parkinsonism, in: *Advances in Neurology,* Volume 60, (H. Narabayashi, T. Nagatsu, N. Yanagisawa, and Y. Mizuno, eds.), Raven Press, New York, pp. 251–258.

Choi, D. W., 1992, Excitotoxic cell death, *J. Neurobiol.* **23**:1261–1276.

Christensen, E., Moller, J. E., and Faurbye, A., 1970, Neuropathological investigation of 28 brains from patients with dyskinesia *Acta Psychiatr. Scand.* **46:**14–23.

Christensen, J. R. C., and Rebec, G. V., 1994, Further evidence for control of neostriatal ascorbate release via a nigro-thalamo-corticoneostriatal loop, *Soc. Neurosci. Abstr.* **20:**732.

Clemens, J. A., and Phebus, L. A., 1984, Brain dialysis in conscious rats confirms in vivo electrochemical evidence that dopaminergic stimulation releases ascorbate, *Life Sci.* **35:**671–677.

Coyle, J. T., and Puttfarcken, P., 1993, Oxidative stress, glutamate, and neurodegenerative disorders. *Science* **262:**689–695.

Dalgalarrondo, P., and Gattaz, W. F., 1994, Basal ganglia abnormalities in tardive dyskinesia—and possible relationships with duration of neuroleptic treatment, *Eur. Arch. Psychiatr. Clin. Neurosci.* **244:**272–277.

Desole, M. S., Anania, V., Esposito, G., Carboni, F., Senini, A., and Miele, E., 1987, Neurochemical and behavioural changes induced by ascorbic acid and d-amphetamine in the rat, *Pharmacol. Res. Commun.* **19:**441–450.

Desole, M. S., Miele, M., Enrico, P., Fresu, L., Esposito, G., Denatale, G., and Miele, E., 1992, The effects of cortical ablation on d-amphetamine-induced changes in striatal dopamine turnover and ascorbic acid catabolism in the rat, *Neurosci. Lett.* **139:**29–33.

DeVito, M. J., and Wagner, G. C., 1989, Methamphetamine-induced neuronal damage: A possible role for free radicals, *Neuropharmacology* **28**(10):1145–1150.

Dexter, D. T., Carter, C. J., Wells, F. R., Javoy-Agid, F., Agid, Y., Lees, A., Jenner, P., Marsden, C. D., 1989, Basal lipid peroxidation in substantia nigra is increased in Parkinson's disease, *J. Neurochem.* **52:**381–389.

Dexter, D. T., Jenner, P., Schapira, A. H. V., and Marsden, C. D., 1992, Alterations in levels of iron, ferritin, and other trace metals in neurodegenerative diseases affecting the basal ganglia, *Ann. Neurol.* **32**(suppl.):S94–S100.

Dugan, L. L., and Choi, D. W., 1994, Excitotoxicity, free radicals, and cell membrane changes, *Ann. Neurol.* **35:**S17–S21.

Dunnett, S. B., and Robbins, T. W., 1992, The functional role of mesotelencephalic dopamine systems, *Biol. Rev. Cambridge Philosophical Soc.* **67**(4):491–518.

Duvoisin, R. C., 1991, Diseases of the extrapyramidal system, in: *Comprehensive Neurology* (R. N. Rosenberg, ed.), Raven Press, New York, pp. 337–364.

Ebersole, B. J., and Molinoff, P. B., 1991, Inhibition of binding of <H-3>PN200-110 to membranes from rat brain and heart by ascorbate is mediated by lipid peroxidation, *J. Pharmacol. Exp. Therap.* **9**(1):337–344.

Ellenbroek, B. A., 1993, Treatment of schizophrenia—a clinical and preclinical evaluation of neuroleptic drugs, *Pharmac. Ther.* **57**(1):1–78.

Englard, S., and Seifter, S., 1986, The biochemical functions of ascorbic acid, *Annu. Rev. Nutr.* **6:**365–406.

Ewing, A. G., Alloway, K. D., Curtis, S. D., Dayton, M. A., Wightman, R. M., and Rebec, G. V., 1983, Simultaneous electrochemical and unit recording measurements: characterization of the effects of d-amphetamine and ascorbic acid on neostriatal neurons, *Brain Res.* **261:**101–108.

Fahn, S., 1991, An open trial of high-dosage antioxidants in early Parkinson's disease, *Am. J. Clin. Nutr.***53**(1):S380–S382.

Fahn, S., and Cohen, G., 1992, The oxidant stress hypothesis in Parkinson's disease–evidence supporting it, *Ann. Neurol.* **32**(6):804–812.

Fallon, J. H., 1988, Topographic organization of ascending dopaminergic projections, *Ann. NY Acad. Sci.*, **537:**1–9.

Fernandez-Calle, P., Jimenez-Jimenez, F. J., Molina, J. A., Cabreravaldivia, F., Vazquez, A., Urra, D. G., Bermejo, F., Matallana, M. C., and Codoceo, R., 1993, Serum levels of ascorbic acid (vitamin-C) in patients with Parkinson's disease, *J. Neurolog. Sci.* **118:**25–28.

Fibiger, H. C., and Lloyd K. G., 1984, Neurobiological substrates of tardive dyskinesia: The GABA hypothesis. *Tr. Neurosci.* **7:**462–464.

Fillenz, M., O'Neill, R. D., and Grunewald, R. A., 1986, Changes in extracellular brain ascorbate concentration as an index of excitatory aminoacid release, in: *Monitoring Neurotransmitter Release During Behaviour* (M. H. Joseph, M. Fillenz, I. A. MacDonald, and C. A. Marsden, eds.) Ellis Norwood, Chichester, England, pp. 144–163.

Fornstedt, B., and Carlsson, A., 1991, Vitamin-C deficiency facilitates 5-S-cysteinyldopamine formation in guinea pig striatum, *J. Neurochem.* **56:**407–414.

Freed, W. J., 1989, An hypothesis regarding the antipsychotic effect of neuroleptic drugs, *Pharmacol. Biochem. Behav.* **32:**337–345.

Frei, B., 1991, Ascorbic acid protects lipids in human plasma and low-density lipoprotein against oxidative damage, *Am. J. Clin. Nutr.* **54:**S1113–S1118.

Fuxe, K., and Agnati, L. F., 1991, Two principal modes of electrochemical communication in the brain—volume versus wiring transmission, *Vol. Transm. Brain* **1:**1–9.

Garcia-Buñuel, L., and Garcia-Buñuel, V. M., 1965, Cerebrospinal fluid levels of free myoinositol in some neurological disorders. *Neurology* **15:**348–350.

Gardiner, T. W., Armstrong-James, M., Caan, A. W., Wightman, R. M., and Rebec, G. V., 1985, Modulation of neostriatal unit activity by iontophoresis of ascorbic acid, *Brain Res.* **344:**181–185.

Ghasemzedah, B., Cammack, J., and Adams, R. N., 1991, Dynamic changes in extracellular fluid ascorbic acid monitored by in vivo electrochemistry, *Brain Res.* **547**(1):162–166.

Giannini, A. J., Loiselle, R. H., DiMarzio, L. R., and Giannini, M. C., 1987, Augmentation of haloperidol by ascorbic acid in phencyclidine intoxication. *Am. J. Psychiat.* **144:**1207–1209.

Gibb, J. W., Hanson, G. R., and Johnson, M., 1994, neurochemical mechanisms of toxicity, in: *Amphetamine and Its Analogs* (A. K. Cho, and D. S. Segal, eds.), Academic Press, San Diego, pp. 269–295.

Gonon, F., Buda, M., Cespuglio, R., Jouvet, M., and Pujol, J. F., 1981, Voltammetry in the striatum of chronic freely moving rats: detection of catechols and ascorbic acid, *Brain Res.* **223:**69–80.

Gotz, M. E., Kunig, G., Riederer, P., and Youdim, M. B. H., 1994, Oxidative stress: Free radical production in neural degeneration, *Pharmacol. Therap.* **63**(1):37–122.

Grace, A. A., 1991, Phasic versus tonic dopamine release and the modulation of dopamine system responsivity—a hypothesis for the etiology of schizophrenia, *Neuroscience* **41**(1):1–24.

Groves, P. M., Rebec, G. V., and Harvey, J. A., 1975, Alteration of the effects of (+)-amphetamine on neuronal activity in the striatum following lesions of the nigrostriatal bundle, *Neuropharmacology* **14:**369–376.

Grunewald, R. A., 1993, Ascorbic acid in the brain, *Brain Res. Rev.* **18**(1):123–133.

Grunewald, R. A., and Fillenz, M., 1984, Release of ascorbate from synaptosomal fraction of rat brain, *Neurochem. Int.* **6:**491–500.

Gulley, J. M., and Rebec, G. V., 1995, Dose-dependent effects of ascorbate on conditioned avoidance response, *Soc. Neurosci. Abstr.* **21:**2088.

Hadjiconstantinou, M., and Neff, N. H., 1983, Ascorbic acid could be hazardous to your experiments: a commentary on dopamine receptor binding studies with speculation on a role for ascorbic acid in neuronal function, *Neuropharmacology* **22:**939–943.

Hall, E. D., and Braughler, J. M., 1993, Free radicals in CNS injury, in: *Molecular and Cellular Approaches to the Treatment of Neurological Disease* (S. G. Waxman, ed.), Raven Press, New York, pp. 81–105.

Halliwell, B., and Gutteridge, J. M. C., 1985, *Free Radicals in Biology and Medicine,* Clarendon, Oxford.

Halliwell, B., Gutteridge, J. M. C., and Cross, C. E., 1992, Free radical, antioxidants, and human disease: Where are we now?, *J. Lab. Clin. Med.* **119:**598–620.

Haracz, J. L., Tschanz, J. T., Wang, Z., White, I. M., and Rebec, G. V., 1993, Striatal single-unit responses to amphetamine and neuroleptics in freely moving rats, *Neurosci. Biobehav. Rev.* **17:**1–12.

Hastings, T. G., and Zigmond, M. J., 1994, Identification of catechol-protein conjugates in neostriatal slices incubated with [H-3]dopamine: Impact of ascorbic acid and glutathione, *J. Neurochem.* **63**(3):1126–1132.

Heikkila, R. E., Cabbat, F. S., and Manzino, L., 1981, Differential inhibitory effects of ascorbic acid on the binding of dopamine agonists and antagonists to neostriatal membrane preparations: correlations with behavioral effects, *Res. Commun. Chem. Path. Pharmacol.* **34:**409–421.

Heimer, L., Zahm, D. S., and Alheid, G. F., 1995, Basal ganglia, in: *The Rat Nervous System,* 2nd Edition, (G. Paxinos, ed.), Academic Press, San Diego, pp. 579–628.

Herrling, P. L., 1985, Pharmacology of the corticocaudate excitatory postsynaptic potential in the cat: Evidence for its mediation by quisqualate- or kainate-receptors, *Neuroscience* **14**(2):417–426.

Hillered, L., Persson, L., Bolander, H. G., Hallström, Å., and Ungerstedt, U., 1988, Increased levels of ascorbate in the striatum after middle cerebral artery occlusion in the rat monitored by intracerebral microdialysis, *Neurosci. Lett.* **95**:286–290.

Hughes, R. E., Hurley, R. J., and Jones, P. R., 1971, The retention of ascorbic acid by guinea pig tissues. *Brit. J. Nutr.* **26**:433–438.

Irwin, I., 1986, The neurotoxin 1-methyl-4-phenyl-1,2,3,6-tetrahydropyridine (MPTP): A key to Parkinson's disease? *Pharmac. Res.* **3**(1):7–11.

Jackson, D. M., and Westlind-Danielsson, A., 1994, Dopamine receptors: molecular biology, biochemistry and behavioural aspects, *Pharmacol. Therap.* **64**(2):291–370.

Jellinger, K., 1986, Pathology of parkinsonism, in: *Recent Developments in Parkinson's Disease* (S. Fahn *et al.*, eds.), Raven Press, New York, pp. 33–66.

Jenner, P., 1993, Altered mitochondrial function, iron metabolism and glutathione levels in parkinson's disease, *Acta Neurol. Scand.* **87**(Suppl. 146):6–13.

Jenner, P., Dexter, D. T., Sian, J., Schapira, A. H. V., and Marsden, C. D., 1992, Oxidative stress as a cause of nigral cell death in Parkinson's disease and incidental lewy body disease, *Ann. Neurol.* **32**(Suppl.):S82–S87.

Jeste, D. V., and Wyatt, R. J., 1981, Dogma disputed: Is tardive dyskinesia due to postsynaptic dopamine receptor supersensitivity? *J. Clin. Psychiat.* **42**:455–157.

Kamata, K., and Rebec, G. V., 1985, Nigral reticulata neurons: Potentiation of responsiveness to amphetamine with long-term treatment, *Brain Res.* **332**:188–193.

Kamata, K., Wilson, R. L., Alloway, K. D., and Rebec, G. V., 1986, Multiple amphetamine injections reduce the release of ascorbic acid in the neostriatum of the rat, *Brain Res.* **362**:331–338.

Kanai, Y., Smith, C. P., and Hediger, M. A., 1994, A new family of neurotransmitter transporters: the high-affinity glutamate transporters, *FASEB J.* **8**:1450–1459.

Kaufmann, P., Wiens, W., Dirks, M., and Krehbiel, D., 1986, Changes in social behavior and brain catecholamines during the development of ascorbate deficiency in guinea pigs, *Behav. Processes* **13**:13–28.

Kiely, M. E., Lal, S., and Vasavan Nair, N. P., 1987, effect of ascorbic acid on brain amphetamine concentrations in the rat, *Prog. Neuro-Psychopharmacol. Biol. Psychiat.* **11**:287–290.

Kimura, K., and Sidhu, A., 1994, Ascorbic acid inhibits I-125-SCH 23982 binding but increases the affinity of dopamine for D-1 dopamine receptors, *J. Neurochem.* **63**(6):2093–2098.

Kiyatkin, E. A., and Rebec, G. V., 1996, Dopaminergic modulation of glutamate-induced excitations of neurons in the neostriatum and nucleus accumbens of awake, unrestrained rats, *J. Neurophy.* **75**:142–153.

Klausner, J. D., Sweeney, J. A., Deck, M. D. F., Haas, G. L., and Kelly, A. B., 1992, Clinical correlates of cerebral ventricular enlargement in schizophrenia—further evidence for frontal lobe disease, *J. Nerv. Ment. Dis.* **180**(7):407–412.

Kuo, C.-H., Hata, F., Yoshida, H., Yamatodani, A., and Wada, H., 1979, Effect of ascorbic acid on release of acetylcholine from synaptic vesicles prepared from different species of animals and release of noradrenaline from synaptic vesicles of rat brain, *Life Sci.* **24**:911–915.

Langston, J. W., Irwin, I., and Ricaurte, G. A., 1987, Neurotoxins, parkinsonism and Parkinson's disease, *Pharmac. Therap.* **32**:19–49.

Layer, R. T., Bland, L. R., and Skolnick, P., 1993, MK-801, but not drugs acting at strychnine-insensitive glycine receptors, attenuate methamphetamine nigrostriatal toxicity, *Brain Res.* **625**(1):38–44.

Liang, N. Y., and Rutledge, C. O., 1982, Comparison of the release of [3H]dopamine from isolated corpus striatum by amphetamine, fenfluramine and unlabelled dopamine, *Biochem. Pharm.* **31**:983–992.

Lieberman, J. A., and Koreen, A. R., 1993, Neurochemistry and neuroendocrinology of schizophrenia—a selective review, *Schizophrenia Bull.* **19**(2):371–429.

Louilot, A., Gonon, F., Buda, M., Simon, H., Le Moal, M., and Pujol, J. F., 1985, Effects of d- and l-amphetamine on dopamine metabolism and ascorbic acid levels in nucleus accumbens and olfactory tubercle as studied by in vivo differential pulse voltammetry, *Brain Res.* **336**:253–263.

Majewska, M. D., Bell, J. A., and London, E. D., 1990, Regulation of the NMDA receptor by redox phenomena—inhibitory role of ascorbate, *Brain Res.* **537**(1–2):328–332.

Marshall, J. F., Odell, S. J., and Weihmuller, F. B., 1993, Dopamine–glutamate interactions in methamphetamine-induced neurotoxicity, *J. Neural Trans.* **91**:(2–3):241–254.

McIlwain, H., Thomas, J., and Bell, J. L., 1956, The composition of isolated cerebral tissues: ascorbic acid and cozymase, *Biochem. J.* **64:**332–335.

Mefford, I. N., Oke, A. F., and Adams, R. N., 1981, Regional distribution of ascorbate in human brain, *Brain Res.* **212:**223–226.

Meister, A., 1992, On the antioxidant effects of ascorbic acid and glutathione, *Biochem. Pharmacol.* **44**(10):1905–1915.

Meltzer, H. Y., and Nash, J. F., 1991, Effects of antipsychotic drugs on serotonin receptors, VII, *Pharmacolog. Rev.* **43**(4):587–604.

Milby, K., Oke, A., and Adams, R. N., 1982, Detailed mapping of ascorbate distribution in rat brain, *Neurosci. Lett.* **28:**169–174.

Miller, R., and Chouinard, G., 1993, Loss of striatal cholinergic neurons as a basis for tardive and L-dopa-induced dyskinesias, neuroleptic-induced supersensitivity psychosis and refractory schizophrenia, *Biol. Psychiat.* **34:**713–738.

Montgomery, E. B., 1995, Heavy metals and the etiology of Parkinson's disease and other movement disorders, *Toxicology* **97**(1–3):3–9.

Mueller, K., 1989, Repeated administration of high doses of amphetamine increases release of ascorbic acid in caudate but not nucleus accumbens, *Brain Res.* **494:**30–35.

Mueller, K., 1990, The effects of haloperidol and amphetamine on ascorbic acid and uric acid in caudate and nucleus accumbens of rats as measured by voltammetry in vivo, *Life Sci.* **47:**735–742.

Mueller, K., and Haskett, C., 1987, Effects of haloperidol on amphetamine-induced increases in ascorbic acid as determined by voltammetry in vivo, *Pharmacol. Biochem. Behav.* **27:**231–234.

Mueller, K., and Kunko, P. M., 1990, The effects of amphetamine and pilocarpine on the release of ascorbic and uric acid in several brain areas, *Pharmacol. Biochem. Behav.* **35:**871–876.

Navas, P., Villalba, J. M., and Cordoba, F., 1994, Ascorbate function at the plasma membrane, *Biochim. Biophy. Acta Rev. Biomemb.* **1197**(1):1–13.

Niki, E., 1991, Action of ascorbic acid as a scavenger of active and stable oxygen radicals, *Am. J. Clin. Nutr.* **54**(6):S1119–S1124.

O'Dell, S. J., Weihmuller, F. B., and Marshall, J. F., 1991, Multiple methamphetamine injections induce marked increases in extracellular striatal dopamine which correlate with subsequent neurotoxicity, *Brain Res.* **564**(2):256–260.

O'Dell, S. J., Weihmuller, F. B., and Marshall, J. F., 1993, Methamphetamine-induced dopamine overflow and injury to striatal dopamine terminals—attenuation by dopamine D(1) or D(2) antagonists. *J. Neurochem.* **60**(5):1792–1799.

Oh, C., Gardiner, T. W., and Rebec, G. V., 1989, Blockade of both D1- and D2-dopamine receptors inhibits amphetamine-induced ascorbate release in the neostriatum, *Brain Res.* **480:**184–189.

Ohmori, T., Koyama, T., Muraki, A., Yamashita, I., 1993, Competitive and noncompetitive N-Methyl-D-aspartate antagonists protect dopaminergic and serotonergic neurotoxicity produced by methamphetamine in various brain regions, *J. Neural Transm.* **92**(2–3):97–106.

Oke, A. F., May, L., and Adams, R. N., 1987, Ascorbic acid distribution patterns in human brain, in: *Third Conference on Vitamin C,* Annals of the New York Academy of Sciences, Volume 498 (J. J. Burns, J. M. Rivers, and L. J. Machlin, eds.), New York Academy of Sciences, New York, pp. 1–12.

O'Neill, R., 1995, The measurement of brain ascorbate in vivo and its link with excitatory amino acid neurotransmission, in: *Voltammetric Methods in Brain Systems,* Neuromethods, Volume 27 (A. Boulton, G. Baker, R. N. Adams, eds.), Humana Press, Clifton, New Jersey, pp. 221–268.

O'Neill, R. D., Fillenz, M., and Albery, W. J., 1982, Circadian changes in homovanillic acid and ascorbate levels in the rat striatum using microprocessor-controlled voltammetry, *Neurosci. Lett.* **34:**189–193.

O'Neill, R. D., Grunewald, R. A., Fillenz, M., and Albery, W. J., 1983, The effect of unilateral cortical lesions on the circadian changes in rat striatal ascorbate and homovanillic acid levels measured in vivo using voltammetry, *Neurosci. Lett.* **42:**105–110.

O'Neill, R. D., Fillenz, M., Sundstrom, L., and Rawlins, J. N. P., 1984, Voltammetrically monitored brain ascorbate as an index of excitatory amino acid release in the unrestrained rat, *Neurosc. Lett.* **52:**227–233.

Pardo, B., Mena, M. A., Casarejos, M. J., Paino, C. L., and Deyebenes, J. G., 1995, Toxic effects of L-DOPA on mesencephalic cell cultures: Protection with antioxidants, *Brain Res.* **682**(1–2):133–143.

Parent, A., Cote, P. Y., and Lavoie, B., 1995, Chemical anatomy of primate basal ganglia, *Prog. Neurobiol.* **46**:131–197.

Phebus, L. A., Roush, M. E., and Clemens, J. A., 1990, Effect of direct and indirect dopamine agonists on brain extracellular ascorbate levels in the striatum and nucleus accumbens of awake rats, *Life Sci.* **47**:1317–1323.

Pierce, R. C., and Rebec, G. V., 1990, Stimulation of both D1 and D2 dopamine receptors increases behavioral activation and ascorbate release in the neostriatum of freely moving rats, *Eur. J. Pharmacol.* **191**:295–302.

Pierce, R. C., and Rebec, G. V., 1992, Dopamine-, NMDA-, and sigma-receptor antagonists exert differential effects on neostriatal ascorbate and DOPAC in awake, behaving rats, *Brain Res.* **579**:59–66.

Pierce, R. C., and Rebec, G. V., 1993, Intraneostriatal administration of glutamate antagonists increases behavioral activation and decreases neostriatal ascorbate via non-dopaminergic mechanisms, *J. Neurosci.* **13**:4272–4280.

Pierce, R. C., and Rebec, G. V., 1995, Iontophoresis in the neostriatum of awake, unrestrained rats: differential effects of dopamine, glutamate, and ascorbate on motor- and nonmotor-related neurons, *Neuroscience* **67**:313–324.

Pierce, R. C., Rowlett, J. K., Bardo, M. T., and Rebec, G. V., 1991, Chronic ascorbate potentiates the effects of chronic haloperidol on behavioral supersensitivity but not D2 dopamine receptor bindings, *Neuroscience* **45**:373–378.

Pierce, R. C., Miller, D. M., Reising, D., and Rebec, G. V., 1992, Unilateral neostriatal kainate, but not 6-OHDA lesions, block dopamine agonist-induced ascorbate release in the neostriatum of freely-moving rats, *Brain Res.* **597**:138–143.

Pierce, R. C., Clemens, A. J., Grabner, C. P., and Rebec, G. V., 1994a, Amphetamine promotes neostriatal ascorbate release via a nigro-thalamo-corticoneostriatal loop, *J. Neurochem.* **63**:1499–1507.

Pierce, R. C., Clemens, A. J., Shapiro, L. A., and Rebec, G. V., 1994b, Repeated treatment with ascorbate or haloperidol, but not clozapine, elevates extracellular ascorbate in the neostriatum of freely moving rats, *Psychopharmacology* **116**:103–109.

Pierce, R. C., Rowlett, J. K., Rebec, G. V., and Bardo, M. T., 1995, Ascorbate potentiates amphetamine-induced conditioned place preference and forebrain dopamine release in rats, *Brain Res.* **688**:21–26.

Poirier, J., and Thiffault, C., 1993, Are free radicals involved in the pathogenesis of idiopathic Parkinson's disease?, *Eur. Neurol.* **33**(Suppl. 1):38–43.

Raiteri, M., Cerrito, F., Cervoni, A. M., and Levi, G., 1979, Dopamine can be released by two mechanisms differentially affected by the dopamine transport inhibitor nomifensine, *J. Pharm. Exp. Ther.* **208**:195–202.

Rebec, G. V., and Bashore, T. R., 1984, Critical issues in assessing the behavioral effects of amphetamine, *Neurosci. Biobehav. Rev.* **8**:153–159.

Rebec, G. V., and Groves, P. M., 1975, Apparent feedback from the caudate nucleus to the substantia nigra following amphetamine administration, *Neuropharmacology* **14**:275–282.

Rebec, G. V., and Pierce, R. C., 1994, A vitamin as neuromodulator: Ascorbate release into the extracellular fluid of the brain regulates dopaminergic and glutamatergic transmission, *Prog. Neurobiol.* **43**:537–565.

Rebec, G. V., Centore, J. M., White, L. K., and Alloway, K. D., 1985, Ascorbic acid and the behavioral response to haloperidol: Implications for the action of antipsychotic drugs, *Science* **227**:438–440.

Rebec, G. V., Wightman, R. M., Fillenz, M., Heikkila, R. E., Gardiner, T. W., and Adams, R. N., 1986, Ascorbate: a vitamin as neuromodulator, *Soc. Neurosci. Abstr.* **12**:170.

Reiter, R. J., 1995, Oxidative processes and antioxidative defense mechanisms in the aging brain, *FASEB J.* **9**(7):526–533.

Ricaurte, G. A., Sabol, K. E., Seiden, L. S., 1994, Functional Consequences of Neurotoxic Amphetamine Exposure, in: *Amphetamine and Its Analogs* (A. K. Cho and D. S. Segal, eds.), Academic Press, San Diego, pp. 297–313.

Rice, M. E., and Cammack, J., 1991, Anoxia-resistant turtle brain maintains ascorbic acid content in vitro, *Neurosci. Lett.* **132**(2):141–145.

Rice, M. E., and Nicholson, C., 1987, Interstitial ascorbate in turtle brain is modulated by release and extracellular volume change, *J. Neurochem.* **49:**1096–1104.

Riederer, P., Sofic, E., Rausch, W.-D., Schmidt, B., Reynolds, G. P., Jellinger, K., and Youdim, M. B. H., 1989, Transition metals, ferritin, glutathione, and ascorbic acid in parkinsonian brains. *J. Neurochem.* **52:**515–520.

Rollema, H., Skolnik, M., Dengelbronner, J., Igarashi, K., Usuki, E., and Castagnoli, N., 1994, MPP(+)-like neurotoxicity of a pyridinium metabolite derived from haloperidol—in vivo microdialysis and in vitro mitochondrial studies, *J. Pharmacol. Exp. Therap.* **268**(1):380–387.

Rose, R. C., and Bode, A. M., 1993, Biology of free radical scavengers—An evaluation of ascorbate, *FASEB J.* **7**(12):1135–1142.

Rupniak, N. M. J., Jenner, P., and Marsden, C. D., 1983, The effect of chronic neuroleptic administration on cerebral dopamine receptor function, *Life Sci.* **32:**2289–2311.

Schenk, J. O., Miller, E., Gaddis, R., and Adams, R. N., 1982, Homeostatic control of ascorbate concentration in CNS extracellular fluid, *Brain Res.* **253:**353–356.

Schmidt, C., and Gibb, J., 1985, Role of the dopamine uptake carrier in the neurochemical response to methamphetamine: Effects of amfonelic acid, *Eur. J. Pharmacol.* **109:**73–80.

Schulz, D. W., Lewis, M. H., Petitto, J., and Mailman, R. B., 1984, Ascorbic acid decreases [3H]dopamine binding in striatum without inhibiting dopamine-sensitive adenylate cyclase, *Neurochem. Int.* **6:**117–121.

Seeman, P., 1992, Dopamine receptor sequences—therapeutic levels of neuroleptics occupy D2-receptors, clozapine occupies D4, *Neuropsychopharmacology* **7**(4):261–284.

Segal, D. S., and Janowsky, D. S., 1978, Psychostimulant-induced behavioral effects: Possible models of schizophrenia, in: *Psychopharmacology: A Generation of Progress* (M. A. Lipton, A. DiMascio, and K. F. Killam, eds.), Raven Press, New York, pp. 1113–1123.

Seiden, L., and Vosmer, G., 1984, Formation of 6-hydroxydopamine in caudate nucleus of the rat brain after a single large dose of methamphetamine, *Pharmacol. Biochem. Behav.* **21:**29–31.

Seiden, L. S., Sabol, K. E., Ricaurte, G. A., 1993, Amphetamine—effects on catecholamine systems and behavior, *Annu. Rev. Pharmacol. Toxicol.* **33:**639–677.

Shivakumar, B. R., and Ravindranath, V., 1992, Oxidative stress induced by administration of the neuroleptic drug haloperidol is attenuated by higher doses of haloperidol, *Brain Res.* **595:**256–262.

Sies, H., Stahl, W., and Sundquist, A. R., 1992, Antioxidant functions of vitamins—vitamin E and vitamin C, beta-carotene, and other carotenoids, in: *Beyond Deficiency,* Volume 669, Annals of the New York Academy of Sciences, (H. E. Sauberlich, and L. J. Machlin, eds.), New York Academy of Sciences, New York, pp. 7–20.

Sonsalla, P. K., Riordan, D. E., and Heikkila, R. E., 1991, Competitive and noncompetitive antagonists at N-Methyl-D-aspartate receptors protect against methamphetamine-induced dopaminergic damage in mice, *J. Pharmacol. Exp. Ther.* **256**(2):506–512.

Sonsalla, P. K., Gibb, J. W., and Hanson, G. R., 1986, Roles of D1 and D2 dopamine receptor subtypes in mediating the methamphetamine-induced changes in monoamine systems, *J. Pharmacol. Exp. Ther.* **238:**932–937.

Spector, R., 1981, Penetration of ascorbic acid from cerebrospinal fluid into brain, *Exp. Neurol.* **72:**645–653.

Spector, R., 1989, Micronutrient homeostasis in mammalian brain and cerebrospinal fluid, *J. Neurochem.* **53:**1667–1674.

Spector, R., and Lorenzo, A. V., 1974, Specificity of ascorbic acid transport system of the central nervous system, *Am. J. Physiol.* **226:**1468–1473.

Stadtman, E. R., 1991, Ascorbic acid and oxidative inactivation of proteins. *Am. J. Clin. Nutr.* **54**(6):S1125–S1128.

Stamford, J. A., Kruk, Z. L., and Millar, J., 1984, Regional differences in extracellular ascorbic levels in the rat brain determined by high speed cyclic voltammetry, *Brain Res.* **299:**289–295.

Stephans, S. E., and Yamamoto, B. K., 1994, Methamphetamine-induced neurotoxicity: Roles for glutamate and dopamine efflux, *Synapse* **17**(3):203–209.

Suboticanec, K., Folnegovic-Smalc, V., Turcin, R., Mestrovic, B., and Buzina, R., 1986, Plasma levels and urinary vitamin C excretion in schizophrenic patients, *Hum. Nutr. Clin. Nutr.* **40C:**421–428.

Suboticanec, K., Folnegovic-Smalc, V., Korbar, M., Mestrovic, B., and Buzina, R., 1990, Vitamin C status in chronic schizophrenia. *Biol. Psychiat.* **28:**959–966.

Tarsy, D., 1983, Neuroleptic-induced extrapyramidal reactions: classification, description, and diagnosis, *Clin. Neuropharmacolog.* **6:**S9–S26.

Tauck, D. L., 1992, Redox modulation of NMDA receptor-mediated synaptic activity in the hippocampus *Neuroreport* **3**(9):781–784.

Tolbert, B. M., 1985, Metabolism and function of ascorbic acid and its metabolites. *Int. J. Vit. Nutr. Res.* **27:**121–138.

Tolbert, L. C., Thomas, T. N., Middaugh, L. D., and Zemp, J. W., 1979a, Ascorbate blocks amphetamine-induced turning behavior in rats with unilateral nigro-striatal lesions, *Brain Res. Bull.* **4:**43–48.

Tolbert, L. C., Thomas, T. N., Middaugh, L. D., and Zemp, J. W., 1979b, Effect of ascorbic acid on neurochemical, behavioral, and physiological systems mediated by catecholamines. *Life Sci.* **25:**2189–2195.

Tolbert, L. C., Morris, P. E., Spollen, J. J., and Ashe, S. C., 1992, Stereospecific effects of ascorbic acid and analogues on d1 and d2 agonist binding. *Life Sci.* **51:**921–930.

Torp, R., Danbolt, N. C., Babaie, E., Bjørås, M., Seeberg, E., Storm-Mathisen, J., Ottersen, O. P., 1994, Differential expression of two glial glutamate transporters in the rat brain: An in situ hybridization study, *Eur. J. Neurosci.* **6:**936–942.

Waddington, J. L., Cross, A. J., Gamble, S. J., and Bourne, R. C., 1983, Dopamine receptor function and spontaneous orofacial dyskinesia in rats during 6-month neuroleptic treatments, in: *CNS Receptors— From Molecular Pharmacology to Behavior* (P. Mandel and F. W. DeFeudis, eds.), Raven Press, New York, pp. 299–308.

Wagner, G., Lucot, J., Schuster, C., and Seiden, L., 1983, Alpha-methyltyrosine attenuates and reserpine increases methamphetamine-induced neuronal changes, *Brain Res.* **270:**285–288.

Wagner, G C., Jarvis, M. F., and Carelli, R. M., 1985, Ascorbic acid reduces the dopamine depletion induced by MPTP, *Neuropharmacology* **24:**1261–1262.

Wambebe, C., and Sokomba, E., 1986, Some behavioral and EEG effects of ascorbic acid in rats, *Psychopharmacology* **89:**167–170.

Wang, Z., and Rebec, G. V., 1993, Neuronal and behavioral correlates of intrastriatal infusions of amphetamine in freely moving rats, *Brain Res.* **627:**79–88.

Waszczak, B. L., and Walters, J. R., 1983, Dopamine modulation of the effects of gamma-aminobutyric acid on substantia nigra pars reticulata neurons, *Science* **220:**218–221.

Weihmuller, F. B., Odell, S. J., and Marshall, J. F., 1992, MK-801 protection against methamphetamine-induced striatal dopamine terminal injury is associated with attenuated dopamine overflow, *Synapse* **11**(2):155–163.

Weinberger, D. R., Berman, K. F., and Zec, R. F., 1986, Physiological dysfunction of the dorsolateral prefrontal cortex in schizophrenia. 1: Regional cerebral blood flow evidence, *Arch. Gen. Psychiat.* **43:**114–124.

West, M. O., Michael, A. J., Knowles, S. E., Chapin, J. K., and Woodward, D. J., 1987, Striatal unit activity and the linkage between sensory and motor events, in: *Basal Ganglia and Behavior: Sensory Aspects of Motor Functioning* (J. S. Schneider, and T. I. Lidsky, eds.), Hans Huber, Toronto, pp. 27–35.

White, L. K., Carpenter, M., Block, N., Basse-Tomusk, A., Gardiner, T. W., Rebec, G. V., 1988, Ascorbate antagonizes the behavioral effects of amphetamine by a central mechanism, *Psychopharmacology* **94:**284–287.

White, L. K., Maurer, M., Kraft, M. E., Oh, C., and Rebec, G. V., 1990, Intrastriatal infusions of ascorbate antagonize the behavioral response to amphetamine, *Pharmacol. Biochem. Behav.* **36:**485–489.

Wilson, R. L., and Wightman, R. M., 1985, Systemic and nigral application of amphetamine both cause an increase in extracellular concentration of ascorbate in the caudate nucleus of the rat, *Brain Res.* **339:**219–226.

Wilson, R. L., Kamata, K., Wightman, R. M., and Rebec, G. V., 1986, Unilateral, intranigral infusions of amphetamine produce differential, bilateral changes in unit activity and extracellular levels of ascorbate in the neostriatum of the rat, *Brain Res.* **384:**342–347.

Wolfarth, S., Coelle, E.-F., Osborne, N. N., and Sontag, K.-H., 1977, Evidence for a neurotoxic effect of ascorbic acid after an intranigral injection in the cat, *Neurosci. Lett.* **185:**183–186.

Yount, S. E., Kraft, M. E., Pierce, R. C., Langley, P. E., and Rebec, G. V., 1991, Acute and long-term amphetamine treatments alter extracellular ascorbate in neostriatum but not nucleus accumbens of freely moving rats, *Life, Sci.,* **49:**1237–1244.

Zetterstrom, T., Wheeler, D. B., Boutelle, M. G., and Fillenz, M., 1992, Striatal ascorbate and its relationships to dopamine receptor stimulation and motor activity, *Eur. J. Neurosci.* **3:**940–946.

Chapter 10

Vitamin E

Neurochemical Aspects and Relevance to Nervous System Disorders

G. T. Vatassery

10.1. INTRODUCTION

The history of vitamin E dates back to 1922 when Evans and Bishop (1922) discovered it as a nutritional factor that prevented resorption of the fetus in the laboratory rat. Since then innumerable investigations have been conducted to understand the role of vitamin E in maintaining the structure and function of various tissues. A few years ago this author reviewed selected aspects of the neurochemistry of vitamin E (Vatassery, 1992). The current article is intended to describe the state of the art with respect to the neurobiology of vitamin E, with special emphasis on human neurological diseases. The last section of the chapter summarizes part of the work done in our laboratory. During the last few years many new studies have appeared on this subject. The goal of the current review is not to present a comprehensive review of the field but to highlight some of the new information that has appeared in the literature.

Vitamin E is the major, if not the only, chain-breaking antioxidant present in biological membranes. Therefore, it is not surprising that a deficiency of vitamin E affects a variety of tissues including muscle, reproductive, cardiovascular, and nervous systems. Many of the early experiments in animals raised on vitamin E-deficient diets have provided important information. Pappenheimer and Goettsch (1931) observed

G. T. Vatassery   Research Service and the Geriatric Research Education and Clinical Center (GRECC), Veterans Affairs Medical Center, Minneapolis, Minnesota 55417; and Department of Psychiatry, University of Minnesota, Minneapolis, Minnesota 55455.

Metals and Oxidative Damage in Neurological Disorders, edited by Connor. Plenum Press, New York, 1997.

that chickens on a vitamin E-deficient diet developed cerebellar encephalomalacia. According to the studies of Einarson and Telford (1960), prolonged vitamin E deficiency in the rat resulted in demyelination of axons and gliosis in the gracilis and cuneate nuclei. Pentschew and Schwartz (1962) reported that vitamin E deficiency was associated with dystrophic afferent axons, mainly in the dorsal columns, Clarke's column and the gracile and cuneate nuclei, as well as accumulation of lipopigment in neurons. These early observations on vitamin E deficiency have been confirmed by Nelson and colleagues (Nelson, 1980).

The involvement of vitamin E deficiency in human neurological disorders was reviewed by Sokol (1989), who points out that even though a dietary vitamin E deficiency is rarely seen in humans, a symptomatic vitamin E deficiency can exist in association with abetalipoproteinemia, chronic cholestatic hepatobiliary diseases, cystic fibrosis, and short bowel syndrome. The neurological syndrome is characterized by cerebellar involvement with gait and limb ataxia, peripheral neuropathy, decreased proprioception and vibration sense, and areflexia (Muller *et al.*, 1983).

Many recent reports suggest that oxidative stress from reactive oxygen species or other free radicals may be involved in the pathogenesis of neurological diseases. A role for vitamin E in symptomatic amelioration has been postulated for many of these disorders. In some cases there is clear evidence for a palliative role for vitamin E, whereas in others such a role is not well established. A few of these conditions are described below.

10.2. FAMILIAL ATAXIA WITH VITAMIN E DEFICIENCY

Recent studies on the relatively rare cases of ataxia with isolated vitamin E deficiency have yielded information on the biology of vitamin E transport in humans. Ben Hamida *et al.* (1993) found that familial isolated vitamin E deficiency is an autosomal recessive disease that closely resembles Friedrich's ataxia. The disease seems to be more frequent in the Mediterranian basin, especially Tunisia and Sicily. Detailed pedigrees and genetic data on family members of Moroccan origin who had a diagnosis of familial ataxia with vitamin E deficiency were also reported by Amiel *et al.* (1995). It has been demonstrated that this autosomal recessive form of ataxia with isolated vitamin E deficiency maps to chromosome 8q13 (Ouahchi *et al.*, 1995). The subjects with this disease were unable to incorporate alpha tocopherol into very low density lipoproteins secreted by the liver and had low vitamin E levels in the serum, even though the absorption of vitamin E was intact. Mutations were found in the gene coding for alpha tocopherol transfer protein. The latter protein has been isolated and purified from the rat liver (Sato *et al.*, 1992). The protein has two closely similar isoforms of molecular weight about 30,500, with much higher binding specificity for alpha tocopherol compared with gamma tocopherol. A similar tocopherol-binding protein was also reported by Yoshida *et al.* (1992), who found that tocopherol-binding protein immunoreactivity was present in the cytosol of rat liver and not in the cytosol of other organs such as heart, spleen, testes, and lung. The inherited vitamin E deficiency syndrome in the Tunisian families is the same as the familial isolated vitamin E deficiency that had been reported earlier. Ouahchi *et al.* term the syndrome ataxia with vitamin E deficiency. Isolated vitamin E deficiency with normal absorption of vitamin E and impaired secretion of alpha tocopherol into very low density lipoproteins has also been reported in the United States (Traber *et al.*, 1990). The occurrence of the syndrome in Japan was reported by Gotoda *et al.*, who described the genetics of its incidence on a remote Japanese island (Gotoda *et al.*, 1995). The authors tested 801 inhabitants of the island and found that 21 were heterozygous and 1 homozygous for a mutation in the tocopherol-binding protein. Serum vitamin E levels were 25% less than normal in the heterozygotes and only 10% of normal in the homozygous individual. These reports provide clear-cut examples of vitamin E deficiency that can be attributed to a genetic defect involving a protein that is crucial for the transfer of vitamin E in adult humans. It is interesting that one of the cardinal symptoms of the disorder is ataxia indicating the special vulnerability of cerebellum to deficiency of vitamin E.

10.3. PERIPHERAL NERVE DAMAGE

Reports of peripheral nerve damage resulting from vitamin E deficiency have appeared in the literature. Nelson *et al.* (1981) studied neuropathologic lesions in rhesus monkeys and observed that the principal neuropathologic alteration was loss of sensory axons in the posterior columns, sensory roots, and peripheral nerves. After studying the effect of chronic vitamin E-deficiency in the rat, Towfighi observed that the lesions in the peripheral nerves were similar to those seen in distal or dying-back type of axonapathy (Towfighi, 1981). The distal segments of the axons were affected most severely. Traber *et al.* (1987) found that alpha tocopherol is lacking in peripheral nerves of vitamin E-deficient human subjects and that the decrease in peripheral nerve alpha tocopherol levels preceded histological abnormalities. Seven children with early onset cholestasis who developed signs of peripheral neuropathy were studied by Landrieu *et al.* (1985). The neuropathy seemed to be due to neuroaxonal dystrophy. Schwann cell inclusions were also observed. The mechanism of pathogenesis of peripheral neuropathy in vitamin E deficiency is unknown at the present time. It is interesting to note that one investigation showed that the rate of rapid axonal transport is increased in vitamin E deficient rats (Wood and Boegman, 1975). In another study, however, significant reductions of both fast anterograde and retrograde transport of acetylcholinesterase were observed in vitamin E-deficient rats (Goss-Sampson *et al.*, 1988).

10.4. ABETALIPOPROTEINEMIA

This is a rare autosomal recessive disease characterized by fat malabsorption, loss of deep tendon reflexes, followed by decreased vibration and proprioception in the lower extremities and cerebellar signs such as dysmetria and ataxia. The subjects also have an atypical retinitis pigmentosa as well as malformed erythrocytes. Rader and Brewer (1993) recently reviewed this area. Kayden and Traber (1986) have postulated that a biochemical defect in this disease may be the deficiency of a microsomal transfer protein involved in the assembly and secretion of lipoprotein particles in the intestine and liver. Massive doses of vitamin E, as much as 10,000 mg per day in the adult, have been reported to be effective in retarding or preventing nervous system damage (Rader and Brewer, 1993). It should be kept in mind that deficiency of other lipids, such as essential fatty acids, could also play an important role in the pathogenesis of this disease.

10.5. AGING

A few histological features are common to vitamin E deficiency and aging. One similarity is that both result in deposition of lipofuscin or age pigment in neurons. This has been known for a number of years. Recently, Dowson *et al.* (1992) conducted a fluorescence microscopic study and confirmed that both vitamin E deficiency and aging were associated with an increase in the relatively large lipopigment regions in hippocampal neurons.

10.6. PARKINSON'S DISEASE

Recently, there has been considerable interest in the potential role of vitamin E in the treatment of age-associated neurodegenerative diseases such as Alzheimer's and Parkinson's diseases. The role of free radical damage in Parkinson's disease (PD) and the possible therapeutic role of vitamin E have been reviewed by the author (Vatassery, 1992). Parkinson's disease is characterized by tremor, rigidity, bradykinesia, and postural instability, which emerge gradually after depletion of 70%–90% of the pigmented neurons in the pars compacta of the substantia nigra (Shoulson, 1992). Oxidative damage may be a causative factor in the pathogenesis of Parkinson's disease for the following reasons: a) Basal ganglia that are affected in PD contain high concentration of dopamine, whose metabolism generates hydrogen peroxide, a source of free radicals; b) Basal ganglia contain high concentrations of iron, a metal catalyst for peroxidative reactions; c) Neuromelanin within the substantia nigra can be a source of free radicals; d) Some neurotoxins, such as 1-methyl-4-phenyl-1,2,3,6-tetrahydropyridine (MPTP), have an affinity for substantia nigra and induce peroxidative reactions through mitochondrial damage; e) Defects in enzymes of oxidative phosphorylation have been reported in PD suggesting that superoxide radical production may be increased in this disease; and f) Concentrations of some antioxidants (e.g., glutathione) have been found to be decreased in PD.

A clinical trial involving the use of pharmacological doses of vitamin E and/or L-deprenyl, a monoamine oxidase inhibitor that can attenuate the production of hydrogen peroxide, was conducted as a multicenter study and the results were published in 1993 (Parkinson's Study Group, 1993). The major conclusion was that deprenyl delayed the onset of disability associated with untreated PD, whereas vitamin E did not show this effect. Since the subjects had already lost 70–90% of neurons in the substantia nigra when the disease became manifest and the treatment was begun, it is perhaps not very surprising that vitamin E ingestion did not show any beneficial effects. Recently, we have analyzed cerebrospinal fluid from the subjects and found that vitamin E concentrations increased significantly after ingestion of the high doses (2000 I.U./day) of vitamin E (Vatassery, Fahn, and the Parkinson Study Group, unpublished). However, it is not known whether the ingestion of vitamin E resulted in an increase in vitamin E concentrations in the basal ganglia of these subjects, and what level of increases in vitamin E concentrations will be enough to have a beneficial clinical effect. Therefore, whether vitamin E treatment can prevent the incidence of human PD remains unknown.

A few experimental investigations support the idea that vitamin E deficiency is associated with abnormalities in nigrostriatal function. Dexter *et al.* (1994) studied four patients with abetalipoproteinemia with associated vitamin E deficiency and sensory ataxia. They found that in two of the most disabled patients the uptake of ^{18}F dopa was reduced in both putamen and caudate; the uptake in the putamen was similar in range to that seen in Parkinson's disease. The authors also found that vitamin E-deficient rats had reduced numbers of dopamine terminals. They conclude that vitamin E deficiency results in loss of nigrostriatal nerve terminals and may be a contributing factor in the pathogenesis of PD. Alpha tocopherol concentrations in the autopsy brains of Parkinson's patients have been examined in one study (Dexter *et al.*, 1992). No significant differences were found in the alpha tocopherol levels in the cerebellum, basal ganglia, or cerebral cortex between control subjects and those with PD. This finding cannot rule out the possibility of abnormal vitamin E status in PD since other antioxidants may have been able to bring up the steady-state concentrations of vitamin E to normal levels. Furthermore, it is also conceivable that higher than normal concentrations of vitamin E may be beneficial in PD.

10.7. ALZHEIMER'S DISEASE

A number of recent studies have suggested that reactive oxygen species may be involved in the pathogenesis of another age-associated neurodegenerative disorder, Alzheimer's disease (AD). Some evidence exists for the possibility of excessive oxidative stress in AD. a) Activities of electron transport chain enzymes were reduced in mitochondria isolated from autopsy brains of AD patients (Parker *et al.*, 1994) suggesting that more reactive oxygen species may be produced during respiration by the defective mitochondria. b) Increased lipid peroxidation has been observed in the cerebral cortex of AD patients at autopsy (Subbarao *et al.*, 1990). c) Significant increase (50%) in oxidative damage to nuclear DNA and a threefold increase in mitochondrial DNA damage have been found in AD brain tissue (Mecocci *et al.*, 1994). d) Beta amyloid, neurotoxic peptide that aggregates to form the core of amyloid plaques seen in AD, has been shown to fragment in solution to free radical peptides, suggesting that free-radical-induced reactions may be involved in plaque formation (Hensley *et al.*, 1994). e) Senile plaques and neurofibrillary tangles show increased staining with antibodies to superoxide dismutase and catalase (Pappolla *et al.*, 1992).

It is unknown whether vitamin E may have a specific role in the treatment or modulation of AD. Metcalfe *et al.* (1989) measured alpha tocopherol concentrations in samples of cortex obtained from AD patients at autopsy and found that the concentrations were not significantly different from controls. As mentioned above, this observation on the steady-state concentration of vitamin E in the AD cortex does not mean that vitamin E has no role in the incidence or treatment of AD. Studies on the effect of vitamin E on learning and memory are extremely rare. In one study it was found that neither ingestion of excess vitamin E nor vitamin E deficiency had any effect on the ability to acquire and maintain memory (Ichitani *et al.*, 1992). However, vitamin E did have an effect on learning behavior. Ichitani *et al.* (1992) used a step-through passive avoidance response task and found that the vitamin E-deficient animals showed significantly lower rates and vitamin E-supplemented rats showed significantly higher rates of passive avoidance response than did control animals. It is unlikely that the ingestion of a single nutrient could have a specific impact upon complex phenomena such as learning and memory. However, a nutrient that is important in the maintenance of the structural integrity of cells in the brain would be expected to have an impact upon all aspects of brain function.

10.8. ALTERATION OF NEUROTOXICITY BY VITAMIN E

An interaction between chemically induced toxicity and vitamin E has been reported in some cases. Pascoe *et al.* (1987) studied isolated hepatocytes and found that treatment with alpha tocopherol succinate increased cellular tocopherol levels, maintained the levels of thiols in cells, and prevented cellular injury induced by adriamycin or ethacrynic acid. Vitamin E ingestion has also been shown to modulate toxicity induced by methylmercury (Chang *et al.*, 1978). Young male hamsters exposed to 2 ppm of methylmercury showed significant neuronal necrosis in the cerebellum and calcarine cortex, which was abolished by concomitant administration of vitamin E. 6-Hydroxydopamine is a neurotoxin that accumulates in the dopaminergic terminals and destroys them. Perumal *et al.* (1992) gave rats 6-hydroxydopamine and showed that the concentrations of glutathione and superoxide dismutase decreased in the brain stem, striatum, hippocampus, and frontal cortex. Pretreatment of rats with vitamin E resulted in significant attenuation of the effects of 6-hydroxydopamine on glutathione and superoxide dismutase concentrations.

10.9. TARDIVE DYSKINESIA AND VITAMIN E

Tardive dyskinesia is an involuntary movement disorder that is caused by exposure to one or more drugs that block dopamine receptors (Jankovic, 1995). In general, the minimum drug exposure that is necessary is 3 months, and the movement abnormality lasts for at least some time even after the drug is discontinued. Orofacial movements are very common in this syndrome. Some experts also include nonrepetitive choreic and choreoathetoid movements of fingers, hands, legs, and trunk (Cadet and Kahler, 1994). It has been hypothesized that free radical-induced neuronal damage may be one of the underlying mechanisms for the production of tardive dyskinesia (reviewed by Cadet and Kahler, 1994; see also Chapter 16). Cadet and Kahler report that electron paramagnetic resonance studies suggest that phenothiazines produce free radical species. Furthermore, neuroleptic drugs are known to cause an increase in the turnover of dopamine, which can result in the production of excessive amounts of oxyradicals. Burkhardt *et al.* (1993) have shown that the neuroleptic medications haloperidol, chlorpromazine, and thiothixine inhibit complex I in the mitochondrial respiratory chain, raising the possibility of excessive production of oxyradical leak due to the inhibition. Therefore, it would be logical to assume that antioxidant action of vitamin E would be beneficial in the treatment of tardive dyskinesia. In fact, several reports on the subject have appeared in the literature. Vitamin E has been found to be beneficial by a few investigators (Adler *et al.*, 1993). Vitamin E seems to have more of an effect if the treatment is initiated as early as possible. Others have observed no effect of vitamin E on the amelioration of tardive dyskinesia (Shriqui *et al.*, 1992). It is interesting to note that in one study in rats haloperidol administration resulted in the reduction of vitamin E levels in striatum (Von Voigtlander *et al.*, 1990). Further studies are required to establish whether vitamin E is useful in the treatment of tardive dyskinesia.

10.10. VITAMIN E IN OTHER NEUROLOGICAL CONDITIONS

Neurosurgeons in Japan have proposed the use of vitamin E in a mixture of substances used to protect the brain during surgical procedures (Uenohara *et al.*, 1988). These investigators have shown that the use a solution of mannitol, vitamin E, and glucocorticoid is capable of suppressing the irreversible changes due to ischemia and prolonging the time during which the brain can withstand ischemia. This mixture has been named the Sendai cocktail and is used primarily by the Suzuki group in Japan.

The use of vitamin E for the prevention of perinatal intracranial hemorrhage has been reported by a few investigators. Fish and colleagues (1990) conducted a prospective, randomized, double-blind placebo controlled study on the effect of vitamin E supplementation upon the incidence and severity of intracranial hemorrhage in very small premature neonates. The vitamin E treatment was shown to be associated with significant reductions in the frequency of all intracranial hemorrhages.

We have examined vitamin E concentrations in red blood cells and plasma of patients with olivopontocerebellar ataxia within the Schut–Swier kindred (Vatassery and Schut, 1987). Members of this family are the descendants of an ataxic male whose two daughters migrated from the Netherlands to the United States in 1866. The concentrations of alpha tocopherol in plasma and red cells of the ataxic subjects in this family were significantly lower than those of the unaffected close relatives as well as unrelated control subjects. Total lipids, cholesterol, triglycerides, and lipoproteins in the serum of the ataxic subjects were all within normal limits, indicating that there were no defects in general lipid metabolism. This family has not been studied further with regard to vitamin E nutrition or metabolism. These reports of neurological dysfunction associated with low serum vitamin E concentrations in humans raise the possibility that such conditions may indeed be more prevalent than our current estimates.

10.11. VITAMIN E IN CELL PROLIFERATION AND GROWTH IN THE NERVOUS SYSTEM

Some work has been done in this area using cell culture models. The use of tocopherol as a component of cell or tissue culture media has been studied by a number of investigators. Halks-Miller *et al.* (1986) studied the effect of vitamin E on neuronal injury in a three-dimensional mixed neuronal and glial cell culture system derived from fetal rat prosencephalon. The central necrotic area within the aggregate culture was unaltered by treatment with vitamin E, whereas the number of viable neurons was increased and gliosis was reduced by the treatment. Nakajima *et al.* (1991) found that more than 60% of cortical neurons could survive in culture for nine days if the medium contained 0.1 to 1 mM alpha tocopherol, whereas almost all of the neurons died within two days in its absence. Prasad and coworkers (Sahu *et al.*, 1988) have conducted a series of experiments showing that alpha tocopherol succinate inhibit the growth of murine and human neuroblastoma cells and rat glioma cells *in vitro*. The mechanism of action of vitamin E in neuronal and glial growth and development is unknown at present.

10.12. NEUROBIOLOGY OF VITAMIN E: EXPERIMENTAL DATA FROM ANIMALS

One of the distinct features of the mammalian nervous system is that different neuroanatomical regions have unique cell types and biochemical characteristics. Therefore, it is of interest to examine the distribution of vitamin E in different regions of the nervous system. We found that there were no specific areas of the brain or spinal cord that were particularly rich in vitamin E (Vatassery, 1992; Vatassery *et al.*, 1984a). In general, the gray matter contained more vitamin E than white matter. Cerebellum and spinal cord had the lowest levels of vitamin E suggesting that these areas may be more susceptible to injury from oxidative stress than the other regions of the nervous system. We have also studied the uptake of radioactive vitamin E by different regions of the rat brain (Vatassery *et al.*, 1984a). It was observed that cerebellum, which had the lowest concentrations of vitamin E, had the highest level of uptake of radioactive vitamin E from blood. The importance of vitamin E for the cerebellum was further strengthened by the results of our study of vitamin E deficiency in mouse brains (Vatassery *et al.*, 1984b). In this investigation, weaning CD-1 mice were fed vitamin E-deficient or control diets for 20 weeks. Peripheral tissues such as liver, plasma, and testes were depleted of vitamin E by the sixth week of the deficiency. Brain tended to conserve vitamin E and lose it at a much slower rate than the peripheral tissues. Similar results have been found by Muller and Goss-Sampson, (1990). Concentrations of alpha tocopherol in the cerebellum sustained larger declines within six weeks of deficiency than those in cerebral hemispheres or medulla and pons. Throughout the 20 weeks of the study the vitamin E concentrations in the cerebellum remained lower than those in other regions of the brain. Thus, all our studies suggest that cerebellum is particularly active in the metabolic utilization of vitamin E. These experiments provide a biochemical rationale for the incidence of symptoms of cerebellar damage in vitamin E deficiency in both humans and animals.

Turnover rate of a nutrient is an important variable in understanding the utilization of the compound by the organism. There is some information in the literature on turnover rate of vitamin E. Most of the recent information has come from studies of Burton, Ingold, and coworkers using deuterium-labeled vitamin E and was reviewed by Burton and Traber (1990). In one study animals were maintained on 36 mg/kg alpha tocopherol in the diet for four weeks and then switched to a diet containing the same amount of deuterated alpha tocopherol. The relative amounts of deuterated and non-deuterated tocopherols in tissues were determined by GC/MS and the exchange half-lives were calculated. The turnover half-lives for vitamin E in rats were 39 and 6 days in brain and plasma, respectively. In guinea pigs the turnover rates were 107 and 4 days in brain and plasma, respectively. This low level of turnover of vitamin E in the brain makes it very difficult to study the metabolic role and interactions of vitamin E with other antioxidants. Therefore, we have used subcellular fractions to study the neuro-chemical fate of vitamin E under *in vitro* conditions.

Using the *in vitro* incubation method we compared the potencies of various oxidizing agents with respect to their ability to oxidize alpha tocopherol (Vatassery, 1993). Subcellular fractions from the cerebral hemispheres of 4-month-old, male Fischer 344 rats were used. Linoleic acid hydroperoxide was found to be a potent oxidizing agent for membrane tocopherol in all subcellular fractions. This suggests that vitamin E may play an important role in protecting biological membranes from endogenous hydroperoxides that are produced during eicosanoid biosynthesis. In addition, it is possible that vitamin E could also play a role in controlling the level of endogenous eicosanoids, such as prostaglandins, since their biosynthesis involves free radical intermediates. Another finding was that when either endogenous peroxides or synthetic compounds producing free radicals were used to induce oxidation of tocopherol, a significant portion of tocopherol remained unoxidized in all subcellular fractions. This suggests that tocopherol exists in labile and nonlabile compartments or complexes in membranes.

Oxidation of tocopherol in mitochondria and synaptosomes were compared directly in another experiment (Vatassery *et al.,* 1994). Mitochondrial alpha tocopherol was found to be much more susceptible to oxidation than synaptosomal tocopherol. Oxidation was induced by incubating the subcellular fractions at 37 C with the free radical generators 2,2′ azobis (2′-amidinopropane) dihydrochloride (ABAP) and 2,2′ azobis (2,4-dimethyl) valeronitrile (ABVN), which undergo thermal decomposition to yield free radicals. With both synaptosomes and mitochondria, tocopherol oxidations occur after short induction or lag times of 15–30 minutes. During this time other substrates are being oxidized and tocopherol is being spared. Mitochondrial alpha tocopherol began to get oxidized after a shorter lag or induction time than synaptosomal tocopherol. Therefore, the reserve of reducing compounds that are responsible for delaying tocopherol oxidation is less in mitochondria than in synaptosomes. The ease of oxidation of mitochondrial tocopherol suggests a general vulnerability of mitochondrial membranes to oxidation. Hence antioxidant protection by vitamin E may be crucial for the maintenance of tissues, such as brain, whose function is critically dependent upon efficient mitochondrial respiration and the availability of high energy phosphates. In addition, vitamin E may have a role in the pathogenesis of neuro-degenerative diseases, like Parkinson's disease and Alzheimer's disease, where mito-chondrial defects have been reported (Beal, 1995; Mecocci *et al.*, 1994).

The importance of the interactions of antioxidants such as vitamin E, vitamin C, and sulfhydryl compounds is well-known. Therefore, the rates of oxidation of these compounds in isolated mitochondria incubated with ABAP or ABVN were examined (Vatassery *et al.*, 1995). An approximate order for the *in vitro* ease of oxidation was ascorbate >> alpha-tocopherol > sulfhydryls >> cholesterol. However, it is impor-tant to note that small amounts of ascorbate were present in the mitochondria when alpha-tocopherol and sulfhydryl compounds were getting oxidized. This observation is different from those found in experiments using homogeneous biological substrates such as blood plasma or serum. The order of oxidation of the various compounds in a biological system is not solely dependent upon the redox potentials. Other factors that control the relative ease of oxidation of antioxidants are: a) concentrations of the oxidized and reduced species, b) sequestration of the compounds within different subcellular compartments, and c) enzymatic and nonenzymatic systems for the repair or regeneration of the individual antioxidants. Therefore, the fate of a specific antioxi-dant at a particular cellular location cannot be predicted with complete accuracy using the *in vitro* order for ease of oxidation shown above. The observation that one antioxi-dant (e.g., vitamin E) is oxidized prior to the total depletion of a more easily oxidized compound (vitamin C) suggests that antioxidants of different structures and redox potentials can function simultaneously in biological systems. Therefore, it is possible that novel synthetic antioxidants may find therapeutic use in human diseases by provid-ing additional antioxidant protection and/or enhancing the activities of endogenous antioxidants. Similar results were also obtained when oxidations of synaptosomal vitamin E, vitamin C and thiols were compared (Vatassery, 1995).

10.13. CONCLUSIONS

Recently, many important contributions have been made to improve our knowl-edge of the neurobiology of vitamin E. Studies on ataxia with familial vitamin E deficiency have shown that a specific genetic defect in the vitamin E binding or transfer protein can lead to pathology in the nervous system, especially in cerebellum. Investi-gations on vitamin E-deficient animals that were performed decades ago have shown that the deficiency can lead to cerebellar damage. Neurochemical studies from our laboratory and others have established that cerebellum is an especially active site of vitamin metabolism. The exact mechanism for the special vulnerability of the cerebel-lum to vitamin E deficiency is still unknown. The potential role of vitamin E in the treatment of cases of cerebellar dysfunction or xenobiotic neurotoxicity from drugs such as neuroleptics needs to be further investigated. It is imperative that the treatment with vitamin E be started as early in the progression of the disease as possible. In addition, the interactions between the various endogenous antioxidants that are known to be complex should be considered very carefully when designing clinical trials.

Studies from our laboratory suggest that different antioxidants may play protective roles simultaneously in nerve tissue. Such interactions via enzymatic and nonenzymatic mechanisms make it possible for one antioxidant to spare another. This raises the possibility that more efficient, organ-specific, novel antioxidants that are produced either via synthesis or modification of endogenous antioxidants could find new therapeutic uses.

ACKNOWLEDGMENTS.   The author would like to thank Mr. W. Ed Smith and Mr. Hung T. Quach for expert technical assistance in many of the experimental studies conducted in the author's laboratory. The scholarly review of the manuscript by Doctors Khurshed Ansari and Maurice Dysken is gratefully acknowledged. The experimental studies from the author's laboratory were supported by research funds from the Department of Veterans Affairs and partly by grant AG10528 from the National Institutes of Health.

10.14. REFERENCES

Adler, L. A., Peselow, E., Rotrosen, J., Duncan, E., Lee, M., Rosenthal, M., and Angrist, B., 1993, Vitamin E treatment of tardive dyskinesia, *Am. J. Psychiatry* **150**:1405–1407.

Amiel, J., Maziere, J. C., Beucler, I., Koenig, M., Reutenauer, L., Loux, N., Bonnefont, D., Feo, C., and Landrieux, P., 1995, Familial isolated vitamin E deficiency. Extensive study of a large family with a 5-year therapeutic follow up, *J. Inher. Metab. Dis.* **18**:333–340.

Beal, M. F., 1995, Aging, energy, and oxidative stress in neurodegenerative disease, *Ann. Neurol.* **38**:357–366.

Ben Hamida, M., Belal, S., Sirugo, G., Ben Hamida, C., Panayides, K., Ionannou, P., Beckmann, J., Mandel, J. L., Hentafi, F., Koenig, M., and Middleton, L., 1993, Friedrich's ataxia phenotype not linked to chromosome 9 and associated with selective autosomal recessive vitamin E deficiency in two inbred Tunisian families, *Neurology* **43**:2179–2183.

Burkhardt, C., Kelly, J. P., Lim, Y. H., Filley, C. M., and Parker, Jr., W. D., 1993, Neuroleptic medications inhibit complex I of the electron transport chain, *Ann. Neurol.* **33**:512–517.

Burton, G. W., and Traber, M. G., 1990, Vitamin E: antioxidant activity, biokinetics, and bioavailability, *Ann. Rev. Nutr.* **10**:357–382.

Cadet, J. L., and Kahler, L. A., 1994, Free radical mechanisms in schizophrenia and tardive dyskinesia, *Neurosci. Biobehav. Rev.* **18**:457–467.

Chang, L. W., Gilbert, M., and Sprecher, J., 1978, Modification of methylmercury neurotoxicity by vitamin E, *Environ. Res.* **17**:356–366.

Dexter, D. T., Ward, R. J., Wells, F. R., Daniel, S. E., Lees, A. J., Peters, T. J., Jenner, P., and Marsden, C. D., 1992, Alpha tocopherol levels in brain are not altered in Parkinson's disease, *Ann. Neurol.* **32**:591–593.

Dexter, D. T., Brooks, D. J., Harding, A. E., Burn, D. J., Muller, D. P. R., Goss-Sampson, M. A., Jenner, P. G., and Marsden, C. D., 1994, Nigrostriatal function in vitamin E deficiency: clinical, experimental, and positron emission tomographic studies, *Ann. Neurol.* **35**:298–303.

Dowson, J. H., Fattoretti, P., Cairns, M., James, N. T., Wilton-Cox, H., and Bertoni-Feddari, C., 1992, The effects of aging and a vitamin E-deficienct diet on the lipopigment content of rat hippocampal and Purkinje neurons, *Arch. Gerontol. Geriatr.* **14**:239–251.

Einarson, L., and Telford, I. R., 1960, Effect of vitamin E deficiency on the central nervous system in various laboratory animals, *Danske Videnskabernes Selskab* **11**:1–81.

Evans, H. M., and Bishop, K. S., 1922, On the existence of a hitherto unrecognized dietary factor essential for reproduction, *Science* **56**:650–661.

Fish, W. H., Cohen, M., Franzek, D., Williams, J. M., and Lemons, J. A., 1990, Effect of intramuscular vitamin E on mortality and intracranial hemorrhage in neonates of 1000 grams or less, *Pediatrics* **85:**578–584.

Goss-Sampson, M. A., MacEvilly, C. J., and Muller, D. P. R., 1988, Longitudinal studies of the neurobiology of vitamin E and other antioxidant systems, and neurological function in vitamin E deficient rats, *J. Neurol. Sci.* **87:**25–35.

Gotoda, T., Arita, M., Arai, H., Inoue, K., Yokota, T., Fukuo, Y., Yazaki, Y., and Yamada, N., 1995, Adult-onset spinocerebellar dysfunction caused by a mutation in the gene for the alpha tocopherol-binding protein, *New Engl. J. Med.* **333:**1313–1318.

Halks-Miller, M., Henderson, M., and Eng, L. F., 1986, Alpha tocopherol decreases lipid peroxidation, neuronal necrosis, and reactive gliosis in reaggregate cultures of fetal rat brain. *J. Neuropath. Exp. Neurol.* **45:**471–484.

Hensley, K., Carney, J. M., Mattson, M. P., Aksenova, M., Harris, M., Wu, J. F., Floyd, R. F., and Butterfield, D. A., 1994, A model for beta amyloid aggregation and neurotoxicity based on free radical generation by the peptide: relevance to Alzheimer's disease, *Proc. Natl. Acad. Sci. USA* **91:**3270–3274.

Ichitani, Y., Okaichi, H., Yoshikawa, T., and Ibata, Y., 1992, Learning behavior in chronic vitamin E-deficient and supplemented rats: radial arm maze learning and passive avoidance response, *Beh. Brain Res.* **51:**157–164.

Jankovic, J., 1995, Tardive syndromes and other drug-induced movement disorders, *Clin. Neuropharmacol.* **18:**197–214.

Kayden, H. J., and Traber, M. G., 1986, Clinical, nutritional and biochemical consequences of apolipoprotein B deficiency, *Adv. Exp. Biol. Med.* **201:**67–81.

Landrieu, P., Selva, J., Alvarez, F., Ropert, A., and Metral, S., 1985, Peripheral nerve involvement in children with chronic cholestasis and vitamin E deficiency, *Neuropediatrics* **16:**194–201.

Mecocci, P., MacGarvey, U., and Beal, M. F., 1994, Oxidative damage in mitochondrial DNA is increased in Alzheimer's disease, *Ann. Neurol.* **36:**747–751.

Metcalfe, T., Bowen, D. M., and Muller, D. P. R., 1989, Vitamin E concentrations in human brain of patients with Alzheimer's disease, fetuses with Down's syndrome, centenarians and controls. *Neurochem. Res.* **14:**1209–1212.

Muller, D. P. R., and Goss-Sampson, M. A., 1990, Neurochemical, neurophysiological, and neuropathological studies in vitamin E deficiency, *Critical Reviews in Neurobiology* **5:**239–263.

Muller, D. P. R., Lloyd, J. K., and Wolff, O. H., 1983, Vitamin E and neurological function, *Lancet* **199**(8318):225–228.

Nakajima, M., Kashiwagi, K., Hayashi, Y., Saito, M., Kawashima, T., Furukawa, S., and Kobayashi, K., 1991, Alpha tocopherol supports the survival and neurite extension of neurons cultured from various region of fetal rat brain, *Neurosci. Lett.* **133:**49–52.

Nelson, J. S., 1980, Pathology of vitamin E deficiency, in: *Vitamin E: A Comprehensive Treatise* (L. J. Machlin, ed.), Marcel Dekker, New York, pp. 391–428.

Nelson, J. S., Fitch, C. D., Fischer, V. W., Broun Jr., G. O., and Chou, A. C., 1981, Progressive neuropathologic lesions in vitamin E-deficient rhesus monkeys, *J. Neuropathol. Exp. Neurol.* **40:**166–186.

Ouahchi, K., Arita, M., Kayden, H. J., Hentati, F., Ben Hamida, M., Sokol, R., Arai, H., Inoue, K., Mandel, J. L., and Koenig, M., 1995, Ataxia with isolated vitamin E deficiency is caused by mutations in the alpha tocopherol transfer protein, *Nature Genetics* **9:**141–145.

Pappenheimer, A. M., and Goettsch M., 1931, A cerebellar disorder in chicks, apparently of nutritional origin, *J. Exp. Med.* **53:**11–26.

Pappolla, M. A., Omar, R. A., Kim, K. S., and Robakis, N. K., 1992, Immunohistochemical evidence of antioxidant stress in Alzheimer's disease, *Am. J. Pathol.* **140,**621–628.

Parker, Jr., W. D., Parks, J., Filley, C. M., and Kleinschmidt-DeMasters, B. K., 1994, Electron transport chain defects in Alzheimer's disease brain, *Neurology* **44,**1090–1096.

The Parkinson's Study Group, 1993, Effects of tocopherol and deprenyl on the progression of disability in early Parkinson's disease, *New Engl. J. Med.* **328:**176–183.

Pascoe, G. A., Olafsdottir, K., and Reed, D. J., 1987, Vitamin E protection against chemical-induced cell injury. I. Maintenance of cellular protein thiols as a cytoprotective mechanism, *Arch. Biochem. Biophys.* **256:**150–158.

Pentschew, A., and Schwarz, K., 1962, Systemic axonal dystrophy in vitamin E deficient adult rats, *Acta Neuropath.* **1:**313–334.

Perumal, A. S., Gopal, V. B., Tordzro, W. K., Cooper, T. B., and Cadet, J. L., 1992, Vitamin E attenuates the toxic effects of 6-hydroxydopamine on free radical scavenging systems in rat brain, *Brain Res. Bull.* **29:**699–701.

Rader, D. J., and Brewer, J. H. B., 1993, Abetalipoproteinemia: New insights into lipoprotein assembly and vitamin E metabolism from a rare genetic disease, *J. Am. Med. Assoc.* **270:**865–869.

Sahu, S. N., Edwards-Prasad, J., and Prasad, K. N., 1988, Effects of alpha tocopheryl succinate on adenylate cyclase activity in murine neuroblastoma cells in culture, *J. Am. Coll. Nutr.* **7:**285–293.

Sato, Y., Hagiwara, K., Arai, H., and Inoue, K., 1992, Purification and characterization of the alpha tocopherol transfer protein from rat liver. *FEBS Lett.* **288:**41–45.

Shoulson, I., 1992, Antioxidative therapeutic strategies for Parkinson's disease, *Ann. N. Y. Acad. Sci.* **648:**37–42.

Shriqui, C. L., Bradwejn, J., Annable, L., and Jones, B. D., 1992, Vitamin E in the treatment of tardive dyskinesia: a double-blind placebo-controlled study, *Am. J. Psychiatry* **140:**391–393.

Sokol, R. J., 1989, Vitamin E and neurologic function in man, *Free Rad. Biol. Med.* **6:**189–207.

Subbarao, K. V., Richardson, J. S., and Ang, L. C., 1990, Autopsy samples of Alzheimer's cortex show increased lipid peroxidation in vitro, *J. Neurochem.* **55:**342–345.

Towfighi, J., 1981, Effects of chronic vitamin E deficiency on the nervous system of the rat, *Acta Neuropathol.* **54:**261–267.

Traber, M. G., Sokol, R. J., Ringel, S. P., Neville, H. E., Thellman, C. A., and Kayden, H. J., 1987, Lack of tocopherol in peripheral nerves of vitamin E-deficient patients with peripheral neuropathy, *New. Engl. J. Med.* **317:**262–265.

Traber, M. G., Sokol, R. J., Burton, G. W., Ingold, K. U., Papas, A. M., Huffaker, J. E., and Kayden, H. J., 1990, Impaired ability of patients with familial isolated vitamin E deficiency to incorporate alpha tocopherol into lipoproteins secreted by the liver, *J. Clin. Invest.* **85:**397–407.

Uenohara, H., Imaizumi, S., Suzuki, J., and Yashimoto, T., 1988, The protective effect of mannitol, vitamin E and glucocorticoid on ischaemic brain injury: evaluation by chemiluminescence, energy metabolism and water content, *Neurosurg. Res.* **10:**73–80.

Vatassery, G. T., 1992, Vitamin E: neurochemistry and implications for Parkinson's disease, *Annals of the New York Academy of Science,* **669:**97–110.

Vatassery, G. T., 1993, Oxidation of alpha tocopherol in subcellular fractions from rat brain and its possible involvement in nerve function, *Biochem. Pharmacol.* **45:**2295–2301.

Vatassery, G. T., 1995, In vitro oxidations of vitamins C and E, cholesterol, and thiols in rat brain synaptosomes, *Lipids* **30:**1007–1013.

Vatassery, G. T., and Schut, L. J., 1987, Changes in vitamin E concentration in red blood cells and plasma of patients with olivopontocerebellar ataxia within the Schut–Swier kindred, *J. Am. Coll. Nutr.* **6:**151–156.

Vatassery, G. T., Angerhofer, C. K., Knox, C. A., and Deshmukh, D. S., 1984a, Concentrations of vitamin E in various neuroanatomical regions and subcellular fractions and the uptake of vitamin E by specific areas of rat brain, *Biochim. Biophys. Acta* **792:**118–122.

Vatassery, G. T., Angerhofer, C. K., and Peterson, F. J., 1984b, Vitamin E concentrations in the brains and some selected peripheral tissues of selenium-deficient and vitamin E-deficient mice. *J. Neurochem.* **42:**554–558.

Vatassery, G. T., Smith, W. E., and Quach, H. T., 1994, Increased suceptibility to oxidation of vitamin E in mitochondrial fractions compared with synaptosomal fractions from rat brains, *Neurochem. Int.* **24:**29–35.

Vatassery, G. T., Smith, W. E., Quach, H. T., and Lai, J. C. K., 1995, In vitro oxidation of vitamin E, vitamin C, thiols and cholesterol in rat brain mitochondria incubated with free radicals, *Neurochem. Int.* **26:**527–535.

Von Voigtlander, P. F., Burian, M. A., Althaus, J. S., and Williams, L. R., 1990, Effects of chronic haloperidol on vitamin E levels and monoamine metabolism in rats fed normal and vitamin E-deficient diets, *Res. Comm. Chem. Path. Pharmacol.* **68:**343–352.

Wood, P. L., and Boegman, R. J., 1975, Increased rate of rapid axonal transport in vitamin E deficient rats, *Brain Res.* **84:**325–328.

Yoshida, H., Yusin, M., Ren, I., Kuhlenkamp, J., Hirano, T., Stolz, A., and Kaplowitz, N., 1992, Identification, purification, and immunochemical characterization of a tocopherol-binding protein in rat liver cytosol, *J. Lipid Res,* **33:**343–350.

Chapter 11

Nitric Oxide and Oxidative Damage in the CNS

Deborah A. Dawson

11.1. NITRIC OXIDE: SYNTHESIS AND INHIBITION

11.1.1. NO Synthases: Constitutive and Inducible Enzymes

NO is synthesized by a family of enzymes known collectively as NO synthases that catalyze the reaction of L-arginine with molecular oxygen to form NO and L-citrulline. The NO synthases can be divided into 2 main groups: a) constitutive isoforms that are continuously expressed and b) inducible isoforms, the *de novo* synthesis of which is stimulated by cytokines or bacterial-derived lipopolysaccharide (LPS) acting alone or in synergy (Nathan, 1992; Moncada *et al.,* 1991). Although all NO synthases catalyze the same basic reaction, they are structurally and, to a certain extent, functionally distinct enzymes encoded by separate genes (Nathan, 1992). Several important differences exist between constitutive and inducible NO synthases in terms of requirement of Ca^{2+} for enzyme activation, cellular localization, and perhaps most importantly in the context of this chapter, the actual amount of NO generated. These differences, discussed below, have considerable implications for both NO-dependent toxicity and the development of NO synthase inhibitors as efficacious cytoprotective drugs.

Constitutive NO synthases are Ca^{2+}-calmodulin dependent enzymes that are activated by elevation of intracellular Ca^{2+} and the binding of calmodulin. These enzymes rapidly and transiently release low levels of NO in response to agonists that raise intracellular Ca^{2+} such as glutamate in neurons or acetylcholine in endothelial cells. In contrast, the activity of inducible NO synthase is independent of changes in intracellu-

Deborah A. Dawson   Stroke Branch, NINDS, National Institutes of Health, Bethesda, Maryland 20892.
Metals and Oxidative Damage in Neurological Disorders, edited by Connor. Plenum Press, New York, 1997.

lar Ca^{2+} since this enzyme retains calmodulin as a tightly bound subunit (Cho et al., 1992). Inducible NO synthase activity is regulated mainly at the transcriptional level and involves several transcription factors, of which nuclear factor-$\kappa\beta$ is of particular importance (Xie and Nathan, 1994). Once expressed, inducible NO synthases continuously produce NO over a much longer time frame (several days) and at a much higher level than the constitutive forms of the enzyme (Nathan, 1992). Consequently, the direct cytotoxic potential of NO derived from inducible NO synthase is considerably greater than that of NO derived from constitutive NO synthase. Indeed the potent cytotoxic properties of NO released from inducible NO synthase are of the upmost importance to macrophages, which rely on the production and release of high levels of NO to effect their antimicrobial actions.

11.1.2. NO Synthases: Localization within the CNS

Within the CNS, constitutive NO synthases are present in neurons, astrocytes, and endothelial cells (Tomimoto et al., 1994; Murphy et al., 1993; Bredt et al., 1990). Highest levels of neuronal NO synthase are found in the cerebellum, superior and inferior colliculi, and dentate gyrus (Bredt et al., 1990), whereas approximately only 2% of cortical neurons contain NO synthase. However these neurons possess an extensive network of NO synthase-positive fibers that permit widespread NO release throughout the entire cortex (Vincent and Kimura, 1992). NO synthase neurons also form a dense perivascular plexus over the main cerebral arteries and are found at the branch points of intracerebral arterioles (Iadecola et al., 1993; Regidor et al., 1993). Thus NO synthase neuron are strategically located for both global and more localized control of CBF.

Inducible NO synthase is predominantly located within cells of the monocyte/macrophage line and expression of inducible NO synthase within the CNS was originally thought to be limited to microglia (the brain's resident macrophages) and astrocytes, with infiltration of peripheral leukocytes following brain injury as another potential source. However, more recently it has been demonstrated that all cellular elements within the CNS have the capacity to express inducible NO synthase—following appropriate stimulation inducible NO synthase has been demonstrated in neurons (cerebellar and cortical), endothelium, and vascular smooth muscle, as well as glia cells (Bereta et al., 1994; Minc-Golomb et al., 1994; Murphy et al., 1993). This widespread ability to induce NO synthase throughout the brain increases the potential scope for NO-dependent toxicity following injury or infection.

11.1.3. NO Synthases: Inhibition of NO Production

NO synthase can be inhibited non-specifically by pharmacological agents that prevent interaction of the enzyme with it necessary cofactors (Nathan, 1992). More specific inhibition is achieved with several structural analogs of L-arginine, including N^G-monomethyl-L-arginine (L-NMMA) and N^G-nitro-L-arginine (L-NA). These compounds inhibit both constitutive and inducible NO synthases, but L-NMMA is more potent against inducible NO synthase, while L-NA is relatively selective for constitutive NO synthases (Nathan, 1992). Selective inhibitors for the neuronal and endothelial forms of constitutive NO synthase are being actively sought. One putatively selective neuronal NO synthase inhibitor is 7-nitro-indazole (7-NI) (Moore *et al.,* 1993). Selective inhibition of inducible NO synthase can be achieved with aminoguanidine (Hasan *et al.,* 1993), while corticosteroids and specific cytokines such as transforming growth factor-β represent physiological inhibitors of inducible NO synthase that act at the post-transcriptional level (Nathan, 1992; Xie and Nathan, 1994).

11.2. MECHANISMS OF NO TOXICITY

NO toxicity is mediated by several mechanisms (summarized in Figure 11.1) that include oxidative injury and inhibition of vital enzymes involved in cellular respiration, and DNA replication and repair. The relative contribution of these different mechanisms to the toxicity of NO will be dictated by the source of NO (constitutive or inducible NO synthase), the particular target and effector cells involved, and the surrounding environmental conditions (e.g., levels of other reactive oxygen species). Thus in some situations one individual mechanism may predominate while under other circumstances the toxicity of NO will be dependent on the cumulative effect of several different reactions initiated by this highly reactive molecule.

11.2.1. Oxidative Damage: Direct and by Combination with Superoxide

At relatively high concentrations the free radical form of NO can induce direct oxidative damage to DNA (Nguyen *et al.,* 1992). The high-level, sustained production

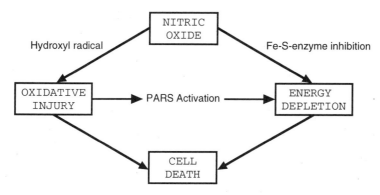

FIGURE 11.1. Nitric oxide mediates cell death by a potent combination of two major interlinked mechanisms: oxidative injury and energy depletion. Nitric oxide, itself a free radical, can also promote oxidative injury by enhancing the synthesis of hydroxyl radicals by a) reacting with superoxide and/or b) increasing free iron concentration. Nitric oxide inhibits glycolytic and aerobic ATP synthesis by binding to and inactivating vital iron-sulfur (Fe-S)-containing enzymes. The secondary activation of poly(ADP ribose) synthase (PARS) in an attempt to repair oxidative injury to DNA further depletes energy stores by the continued consumption of ATP necessary to regenerate NAD. For more details see text.

of NO from inducible NO synthase appears sufficient to induce such direct oxidative injury (Mitrovic *et al.*, 1994), but activation of constitutive NO synthase may not generate sufficient quantities of NO to induce direct DNA injury. Instead, NO derived from constitutive NO synthase mediates oxidative injury indirectly via reaction with superoxide. Under physiological conditions NO rapidly reacts with superoxide to form peroxynitrite, which subsequently decomposes into species with hydroxyl radical-like activity (Beckman *et al.*, 1990; Lipton *et al.*, 1993). Peroxynitrite and hydroxyl radical can then induce widespread oxidative injury via lipid peroxidation and oxidative damage to proteins and DNA (see Chapter 12 for more details).

The reaction of NO with superoxide is a major component of neuronal NO synthase-dependent oxidative injury in the CNS: addition of superoxide dismutase to neuronal cultures dramatically attenuates NO-induced death (Dawson *et al.*, 1993; Lipton *et al.*, 1993), despite prolonging the half-life of NO (Gryglewski *et al.*, 1986). In contrast, the neurotoxicity of inducible NO synthase is not attenuated by superoxide dismutase, and in fact is moderately increased (Boje and Arora, 1992), suggesting that inducible NO synthase releases NO in sufficient quantities to induce direct neuronal damage.

While reaction of superoxide and nitric oxide within neurons enhances neurotoxicity, the scavenging of superoxide by NO within the vasculature may be predominantly cytoprotective by preventing direct toxic actions of superoxide (Rubanyi *et al.*, 1991).

11.2.2. Oxidative Damage: Alterations in Iron Homeostasis

The interactions of NO with the mechanisms controlling iron homeostasis are complex: by displacing iron from ferritin, inhibiting ferritin synthesis, and raising extracellular free iron concentration, NO can increase the availability of reactive iron and potentiate hydroxyl radical generation (see Chapter 12).

NO, in concordance with other reactive oxygen species, induces the release of iron from the intracellular storage protein ferritin (Reif and Simmons, 1990; and see Chapter 8) and promotes iron-dependent lipid peroxidation. NO can further disrupt iron homeostasis by inhibiting ferritin synthesis via stimulation of the metalloregulatory protein iron response element-binding protein (IRE-BP). IRE-BP activity is dependent on the presence or absence of an iron-sulfur cluster—when the enzyme contains this cluster it is structurally identical to cytosolic aconitase and exhibits aconitase activity but no IRE-BP activity. Interaction of NO and/or peroxynitrite with the iron-sulfur cluster (see section 11.2.3) results in loss of aconitase activity and a concomitant rise in IRE-BP activity (Weiss, 1993). NO synthesized from either constitutive or inducible NO synthase stimulates the binding of IRE-BP to the IREs on the mRNAs for ferritin and transferrin receptor, and markedly inhibits ferritin synthesis (Weiss, 1993; Jaffrey *et al.*, 1994). The similar localizations of IRE-BP and neuronal NO synthase within the brain, and the rapid activation of IRE-BP via the NMDA receptor and NO, suggests a physiological role of the NO:IRE-BP interaction in glutamatergic signal transduction (Jaffrey *et al.*, 1994). However overactivation of this system, e.g., following the induction of inducible NO synthase, could exacerbate oxidative stress.

The displacement of iron from intracellular stores eventually causes severe cellular iron depletion (Weinberg, 1992). Loss of intracellular iron removes the ability to reactivate iron-containing enzymes inactivated by NO and contributes to the antimicrobial action of NO (Weinberg, 1992) However iron efflux also occurs from host cells and may be a major component of NO-induced autotoxicity, particularly when activation of inducible NO synthase is temporally coincident with other sources of oxidant stress and/or iron-releasing mechanism. Under such condition the host may be unable to cope with the iron overload (Weinberg, 1992) resulting in NO-dependent exacerbation of free radical injury.

11.2.3.   Inhibition of Respiration and DNA Synthesis

Proteins containing iron-sulfur or thiol groups represent some of the main reaction targets for NO within biological systems (Stamler, 1994) and NO-induced modification of the structure and function of such proteins represents an important regulatory control mechanism. However the ability of NO to react with several critical enzymes involved in respiration and DNA synthesis, and the subsequent long-lasting inhibition of these enzymes, can result in profound cytotoxicity. The potent dual inhibition of respiration and DNA synthesis/repair resulting from exposure to NO contributes significantly to the antimicrobial and antitumor effects of activated macrophages. However this powerful response to NO also has obvious autotoxic potential if misdirected against the host's own cells.

11.2.3.1.   Inhibition of Respiration

NO severely curtails ATP synthesis from both glycolysis and oxidative phosphorylation by inhibiting the enzymes glyceraldehyde-3-phosphate dehydrogenase (GADPH, involved in glycolysis), cis-aconitase (a constituent of the tricarboxylic acid cycle), and mitochondrial electron transport chain complexes I and II (Stamler, 1994). NO, or peroxynitrite derived from the combination of NO and superoxide, forms nitroso-complexes with the iron-sulfur containing enzymes (aconitases, complexes I and II), which renders them inactive. The inactivation of GADPH is more complex, involving an initial nitrosylation of a thiol group in the enzyme's active site that induces a secondary irreversible modification of the protein via the binding of NAD (Brune, 1994).

11.2.3.2.   Inhibition of DNA Synthesis and Repair

The detrimental effect of NO on DNA synthesis and repair is multifaceted. In general, any energy-dependent repair mechanisms, initiated in response to DNA damaged by direct attack from NO or other free radicals, will be impaired by NO-mediated inhibition of respiration. More specifically, NO can directly inhibit DNA replication by forming an inactive nitroso-iron-sulphur complex with the rate-limiting enzyme ribonucleotide reductase (Kwon et al., 1991). Finally the cell's already depleted energy store can be exhausted by the activation of poly (ADP-ribose) synthase (PARS) in an attempt to repair NO-dependent DNA fragmentation. Although the intention of PARS activation is the repair of damaged DNA, PARS also continuously poly ADP-ribosy-lates other nuclear components (including PARS itself) resulting in rapid depletion of NAD and the consumption of large amounts of ATP to maintain NAD levels. Thus in a situation where energy production has already been severely compromised by NO, this futile consumption of ATP hastens the cell's demise. The importance of PARS activation to NO-dependent toxicity is demonstrated by the neuroprotective action of inhibitors of poly-ADP-ribosylation, which dose-dependently inhibit NO-mediated neurotoxicity (J. Zhang *et al.*, 1993) and promote recovery of electrophysiological activity in brain slices previously exposed to NO (Wallis *et al.*, 1993).

11.3. ROLE OF NO IN CNS DISORDERS

11.3.1. NO and Glutamate Toxicity

The excitatory amino acid neurotransmitter glutamate in high concentrations is neurotoxic, and has been implicated in the pathology of several CNS disorders including stroke and Alzheimer's disease (Lipton and Rosenberg, 1994). NO is a normal component of glutamatergic signal transduction in the CNS: Ca^{2+} influx, predominantly via the NMDA receptor ion channel, activates NO synthase and stimulates NO release, which in turn activates guanylate cyclase and mediates glutamate-induced increases in cyclic GMP (Garthwaite, 1991). It is now readily apparent that NO is also a major component of glutamate-induced neurotoxicity. NO synthase inhibitors attenuate glutamate/NMDA toxicity *in vitro* (Dawson *et al.*, 1993) and reduce damage induced by infusion of glutamate into the cerebral cortex *in vivo* (Fujisawa *et al.*, 1994). The neurotoxicity induced by the glutamate-NO pathway is likely to derive predominantly from oxidative injury and facilitation of hydroxyl radical synthesis (Hammer *et al.*, 1993) by any of the mechanisms discussed in Section 11.2.

The repeated demonstration of NO-dependent glutamate toxicity now raises the possibility that NO is a significant contributor to the pathology of all CNS disorders in which excitotoxicity is proposed to be involved (Lipton and Rosenberg, 1994). In this respect much attention has focused on the pathology of stroke because of the well-documented role of excitotoxic processes in ischemic injury.

11.3.2. Stroke

Extracellular glutamate concentrations are markedly elevated in experimental focal ischemia [middle cerebral artery (MCA) occlusion] and glutamate receptor antagonists are highly efficacious in reducing ischemic damage (McCulloch *et al.*, 1991). It is now known that the synthesis of NO is also enhanced following ischemia, and NO synthase inhibitors (in low doses) have similarly proven to be effective neuroprotective agents.

Focal ischemia induces a rapid and transient activation of constitutive NO synthase in neurons and endothelium with NO reaching peak concentrations (150-fold increase over basal levels) within 5–20 min of MCA occlusion and returning to base line within 1 hour (Kader *et al.,* 1993; Malinski *et al.,* 1993). This early, transient activation of constitutive NO synthases is followed by a delayed, but more sustained, upregulation of inducible NO synthase activity, firstly within cerebromicrovessels (4–24 h post-MCA occlusion, Nagafuji, *et al.,* 1994) and subsequently in brain paren-chyma >12 h post-MCA occlusion, Iadecola *et al.,* 1995a). Chronic upregulation of endothelial constitutive NO synthase also occurs from 2 days up to 1 week post-MCA occlusion (Nagafuji *et al.,* 1994; Z. G. Zhang *et al.,* 1993). Thus there are both acute (temporally transient) and chronic (more sustained) increases in NO synthesis follow-ing stroke.

The neuroprotective efficacy of NO synthase inhibitors against focal ischemic injury has been widely investigated in the past few years. Despite conflicting early results that reported both protection and exacerbation of injury following inhibition of NO synthase (see Dawson, 1994, for review) a general consensus of opinion has now emerged. Activation of both neuronal and endothelial NO synthases occurs in acute stroke—NO derived from neuronal NO synthase is inherently neurotoxic, whereas the endothelial-derived NO is essentially neuroprotective by improving blood flow to the ischemic region. Thus administration of L-arginine or NO-donor compounds imme-diately following MCA occlusion raises CBF and reduces the volume of ischemic injury, while high doses of NO synthase inhibitors are not neuroprotective due to inhibition of NO synthesis in the vasculature and the resultant reduction in CBF. In contrast, lower doses of NO synthase inhibitors, which do not produce such pro-nounced reductions in CBF, *are* neuroprotective due to inhibition of neuronal NO synthase (Dawson, 1994).

The development of a mouse strain lacking the neuronal NO synthase gene while retaining endothelial NO synthase has confirmed the above hypothesis (Huang *et al.,* 1994). The ischemic lesion volume in these mice is significantly lower than for normal control mice, indicating the important contribution of neuronal NO synthase to the development of focal ischemic injury. Furthermore, when NO synthase inhibitors are administered to these knockout mice the lesion volume increases due to inhibition of endothelial NO synthase and the resultant detrimental effect on CBF (Huang *et al.,* 1994). To compensate for the disadvantageous effects of endothelial NO synthase inhibition, more selective neuronal NO synthase inhibitors such as 7-NI are being sought—7-NI successfully reduces ischemic damage without apparently affecting en-dothelial NO synthase (Yoshida *et al.,* 1994).

The emergence of inducible NO synthase as a potential mediator of focal ischemic injury has led to the testing of aminoguanidine (a selective inducible NO synthase inhibitor). Aminoguanidine was found to reduce infarct volume by 33%, despite a treatment delay of 24 h post-MCA occlusion (Iadecola *et al.,* 1995b). This result is of particular therapeutic importance because it raises the possibility of effective phar-macological treatment of stroke several hours after onset of symptoms. In conclusion, the role of NO derived from both constitutive and inducible NO synthases in the pathogenesis of focal ischemia is now well established.

11.3.3. Neurodegenerative Disorders

In contrast to stroke, the role of NO in neurodegenerative diseases such as Alzheimer's, Huntingdon's, and Parkinson's diseases remains to be fully elucidated. However, accumulating data suggests that NO may play a role in either the initiation and/or exacerbation of neurodegenerative processes in these disorders (for review see Simonian and Coyle, 1996), and may also contribute to the dementia associated with AIDS (Adamson et al., 1996).

Neuronal NO synthase activity equates with NADPH diaphorase activity (Matsumoto et al., 1993); it has been known for several years that neurons that exhibit NADPH diaphorase activity (reducing the dye nitroblue tetrazolium to a dark blue insoluble reaction product in the presence of NADPH) are more resistant to neurodegeneration than non-NADPH diaphorase neurons (Ferrante et al., 1985). This preferential sparing of NO synthase-containing neurons initially raised the possibility that NO is actively involved in the neurodegenerative process.

11.3.3.1. Alzheimer's Disease

NO synthase neurons (labelled with NADPH diaphorase or specific NO synthase antibodies) in specific brain regions such as the striatum (Mufson and Brandabur, 1994; Selden et al., 1994) and hippocampus (Hyman et al., 1992) are selectively preserved in Alzheimer's disease despite a marked reduction (up to 47%) in the total neuronal count, while the density of NO synthase neurons in the substantia innominata is actually increased compared with age-matched controls (Benzing and Mufson, 1995). This preferential sparing of NO-generating neurons has led to the proposal that the increased NO "load" on remaining non-NO synthase neurons may promote NO-dependent toxicity and hasten the degenerative process (Hyman et al., 1992; Benzing and Mufson, 1995). Furthermore there is a marked increase in NO synthase activity (both constitutive and inducible) in cerebral microvessels from Alzheimer's patients (Dorheim et al., 1994), suggesting that vascular derived NO also has the potential to contribute to neurotoxicity in Alzheimer's disease. Two caveats to this hypothesis must be mentioned. First, the surviving NO synthase neurons undergo profound morphological changes, including shrunken perikarya, loss of dendritic processes, and axonal distortion (Hyman et al., 1992; Mufson and Brandabur, 1994; Selden et al., 1994), which may severely compromise their functional interaction with other neurons. However, if these neurons retain the ability to release NO (which is likely since NADPH diaphorase activity is preserved) then loss of neuronal connections may be of little consequence to NO toxicity since NO is freely diffusible and does not rely on direct synaptic contacts for transfer between neurons. Second, cerebrospinal fluid (CSF) levels of the NO metabolites nitrate and nitrite are unchanged or reduced in Alzheimer's patients compared to controls (Kuiper et al., 1994; Milstien et al., 1994), suggesting reduced, rather than increase, NO production. However this reduction may reflect the generalized neuronal loss of Alzheimer's disease and therefore the concomitant loss of neurotransmitters that normally stimulate NO synthesis. Highly localized, absolute, or relative increases in NO production in discrete brain regions would be masked by sampling total CSF metabolite levels.

While NO synthase neurons are more resistant to Alzheimer's disease and therefore may indirectly promote further neurotoxicity simply by raising the relative concentration of NO surrounding remaining neurons, NO may be also more specifically
involved in Alzheimer's pathology by directly contributing to the toxicity of β-amyloid. β-amyloid protein is a major constituent of the characteristic senile plaques found
in Alzheimer's disease. β-amyloid (Aβ) and the 25–35 amino acid peptide fragment of
the protein [Aβ(25–35)] are directly neurotoxic by a mechanism involving raised
intracellular Ca^{2+} concentration. Aβ(25–35) rapidly and dose-dependently stimulates
neuronal release of NO (Hu and El-Fakahany, 1993) and potentiates NO release from
cytokine-stimulated microglia and astrocytes (Meda *et al.*, 1995; Rossi and Bianchi,
1996). Thus, activation of both constitutive and inducible NO synthases may be partially responsible for β-amyloid-mediated toxicity in Alzheimer's disease. However,
the delayed neurotoxicity associated with a relatively low concentration (25 μM) of
Aβ(25–35) in hippocampal culture was found not to be attenuated by NO synthase
inhibitors, or indeed by other free radical inhibitors (Lockhart *et al.*, 1994).

11.3.3.2. Parkinson's Disease

Parkinson's disease is characterized by degeneration of nigro-striatal dopaminergic neurons, which may be related to iron mobilization and heightened oxidative
stress (Olanow, 1992). It has been proposed (Youdim *et al.*, 1993) that NO synthesis,
stimulated by the glutamatergic projections to the striatum and substantia nigra, could
exacerbate this oxidative stress by promoting release of iron from ferritin and/or
reacting with superoxide (generated by the oxidation of dopamine) to form peroxynitrite and hydroxyl radicals. In support of this hypothesis the MPTP-dependent
generation of peroxynitrite and hydroxyl radicals observed in models of experimental
parkinsonism is significantly attenuated by NO synthase inhibitors (Schulz *et al.*, 1995;
Spencer Smith *et al.*, 1994). Furthermore, the depletion of dopaminergic neurons in
these models is markedly reduced by inhibiting NO synthesis (Schulz *et al.*, 1995;
Spencer Smith *et al.*, 1994), while mice lacking the neuronal NO synthase gene are
more resistant to MPTP toxicity (Przedborski *et al.*, 1996). Methamphetamine-depen-
dent dopaminergic toxicity is also attenuated by NO synthase inhibitors *in vitro* (Sheng
et al., 1996) and *in vivo* (Di Monte *et al.*, 1996). In clinical Parkinson's disease the
localized levels of NO surrounding dopaminergic neurons may be increased due to
preferential sparing of striatal NO synthase neurons (Mufson and Brandabur, 1994) and
release of NO from reactive microglia (Shergill *et al.*, 1996). However it must also be
noted that (similar to Alzheimer's disease) total CSF and plasma levels of NO metabolites remain unaltered or reduced in Parkinson's disease (Kuiper *et al.*, 1994; Molina *et
al.*, 1994).

11.3.3.3. Multiple Sclerosis

Perhaps the most convincing evidence for a role of NO in CNS pathology is for
multiple sclerosis (MS), an inflammatory demyelinating disorder of the CNS in which
pro-inflammatory cytokines (e.g., tumor necrosis factor-α, interferon-γ) released from

macrophages and glia are thought to be involved in the demyelination process (Hartung et al., 1992). Since these same cytokines are a major stimulus for NO synthase induction, it was postulated that NO may contribute to demyelination and oligodendrocyte cell death in MS (Sherman et al., 1992). Indeed oligodendrocytes are highly sensitive to the toxic effects of NO, and enzyme inhibition, DNA damage, and cell death in oligodendrocytes all occur at relatively low NO concentrations, to which other glial cells are resistant (Mitrovic et al., 1994).

Further evidence for a role of NO in MS pathology has come from studies utilizing experimental autoimmune encephalomyelitis (EAE) as an animal model of MS. These studies have demonstrated increases in inducible NO synthase mRNA expression and raised levels of NO metabolites in tissue derived from the CNS of affected rodents (Koprowski et al., 1993; Lin et al., 1993; MacMicking et al., 1992). Importantly, the induction of NO synthase precedes the development of overt pathology (Koprowski et al., 1993). Macrophages within and adjacent to the CNS lesion, and systemic leukocytes, have been identified as the predominant source of NO in these rodent models (Van Dam et al., 1995; MacMicking et al., 1992). Furthermore, the demonstration that NO synthase inhibitors dramatically and dose-dependently inhibit EAE (Cross et al., 1994; Ovadia et al., 1994) indicates a pivotal role for NO in rodent inflammatory demyelination.

The involvement of NO in EAE, the ease of induction of NO synthase by pro-inflammatory cytokines in rodent macrophages, and the presence of these cytokines in human MS lesions are highly suggestive of a role for NO in MS itself. However, until recently it has proven difficult to induce NO synthase in human cell lines, suggesting a differential regulation of inducible NO synthase in humans and rodents, and the possibility that NO is of less importance to the human inflammatory response. But, in repudiation of this, inducible NO synthase has now been successfully expressed in human glia (Brosnan et al., 1994; Mitrovic et al., 1994), although the optimum stimuli combinations do differ from those in rodents. Furthermore, inducible NO synthase mRNA has been shown to be markedly increased in demyelinating regions of brain from MS patients (Bo et al., 1994) and NADPH diaphorase staining has revealed intense NO synthase activity within and adjacent to MS lesions, primarily localized to reactive astrocytes (Bo et al., 1994; Brosnan et al., 1994). Circulating levels of NO in the blood and CSF of MS patients may also be increased (Johnson et al., 1995; Boullerne et al., 1995). Thus, evidence for the involvement of NO in EAE/MS is continuing to accumulate. It must be stressed however that NO should not be viewed solely as a pathological agent in EAE/MS. The anti-inflammatory actions of NO could prevent leukocyte infiltration and limit the inflammatory process in the early stages of EAE/MS. Indeed, a recent study has reported exacerbation of EAE following treatment with aminoguanidine (Zielasek et al., 1995). Thus the role of NO in MS may prove to be more complex than initially anticipated and involve both cytotoxic and cytoprotective functions—a situation analogous to the dual roles of NO uncovered in experimental stroke. Therefore, as for stroke, the potential therapeutic benefits of manipulating the NO system in MS (with either inhibitors or enhancers) may be critically dependent on choice of dose and timing of administration.

11.4. SUMMARY AND CONCLUSIONS

NO is an extremely important physiological molecule with a diverse range of functions both peripherally and within the CNS. However, overproduction of this highly reactive free radical can result in profound cytotoxicity due in large part to oxidative processes. A large body of experimental evidence (summarized in the table below) now exists to support the involvement of NO in several CNS disorders.

Experimental Evidence for Involvement of NO in the
Pathogenesis of Neurological Disorders[a]

STROKE
NO levels increased
NO synthase inhibitors reduce tissue damage
ALZHEIMER'S DISEASE
NO neurons selectively preserved—relative increase in NO levels proposed
β-amyloid stimulates NO synthesis
PARKINSON'S DISEASE
Striatal NO neurons selectively preserved—relative increase in NO levels proposed
NO synthase inhibitors attenuate MPTP toxicity
MULTIPLE SCLEROSIS
NO synthesis increased in MS lesions and animals models (EAE)
Oligodendrocytes show heightened vulnerability to NO toxicity
NO synthase inhibitors dramatically attenuate EAE pathology

[a]In all cases an absolute or relative increase in NO synthesis is postulated. While NO may have both detrimental and beneficial actions, inhibiting the production of excess NO has been demonstrated to significantly reduce injury in models of stroke, Parkinson's disease, and multiple sclerosis.

Of particular importance is the demonstration of beneficial effects of NO synthase inhibition in animal models of stroke, Parkinson's disease, and multiple sclerosis. However, animal studies of stroke have shown that broad inhibition of all NO synthesis can actually be harmful due to the wide-ranging physiological functions of NO, some of which can actively inhibit the pathological process. Thus, a clear understanding of all the functions of NO within a particular physiological system must be made before NO synthase inhibition is attempted. The continuing development of more selective inhibitors for neuronal, endothelial, and inducible NO synthases will also improve the potential clinical utility of these drugs.

In conclusion, considerable data now supports the role of NO as a mediator of oxidative damage in the CNS. Pharmacological intervention to inhibit NO synthesis in clinical disorders is now a distinct possibility.

11.5.   REFERENCES

Adamson, D. C., Wildemann, B., Sasaki, M., Glass, J. D., McArthur, J. C., Christov, V. I., Dawson, T. M., and Dawson, V. L., 1996, Immunologic NO synthase: Elevation in severe AIDS dementia and induction by HIV-1 gp41, *Science* **274**:1917–1921.

Beckman, J. S., Beckman, T. W., Chen, J., Marshall, P. A., and Freeman, B. A., 1990, Apparent hydroxyl radical production by peroxynitrite: Implications for endothelial injury from nitric oxide and superoxide, *Proc. Natl. Acad. Sci. USA* **87:**1620–1624.

Benzing, W. C., and Mufson, E. J., 1995, Increased number of NADPH-D-positive neurons within the substantia innominata in Alzheimer's disease, *Brain Res.* **670:**351–355.

Bereta, M., Bereta, J., Georgoff, I., Coffman, F. D., Cohen, S., and Cohen, M. C., 1994, Methylxanthines and calcium-mobilizing agents inhibit the expression of cytokine-inducible nitric oxide synthase and vascular cell adhesion molecule-1 in murine microvascular endothelial cells, *Exp. Cell Res.* **212:**230–242.

Bo, L., Dawson, T. M., Wesselingh, S., Mork, S., Choi, S., Kong, P. A., Hanley, D., and Trapp, B. D., 1994, Induction of nitric oxide synthase in demyelinating regions of multiple sclerosis brain, *Ann. Neurol.* **36:**778–786.

Boje, K. M., and Arora, P. K., 1992, Microglial-produced nitric oxide and reactive nitrogen oxides mediate neuronal cell death, *Brain Res.* **587:**250–256.

Boullerne, A. I., Petry, K. G., Meynard, M., and Geffard, M., 1995, Indirect evidence for nitric oxide involvement in multiple sclerosis by characterization of circulating antibodies directed against conjugated S-nitrosocysteine, *J. Neuroimmunol.* **60:**117–124.

Bredt, D. S., Hwang, P. M., and Snyder, S. H., 1990, Localization of nitric oxide synthase indicating a neural role for nitric oxide, *Nature* **347:**768–770.

Brosnan, C. F., Battistini, L., Raine, C. S., Dickson, D. W., Casadevall, A., and Lee, S. C., 1994, Reactive nitrogen intermediates in human neuropathology: An overview, *Dev. Neurosci.* **16:**152–161.

Brune, B., Dimmeler, S., Molina y Vedia, L., and Lapetina, E. G., 1994, Nitric oxide: A signal for ADP-ribosylation of proteins, *Life Sci.* **54:**61–70.

Cho, H. J., Xie, Q. W., Calaycay, J., Mumford, R. A., Swiderek, K. M., Lee, T. D., and Nathan, C., 1992, Calmodulin is a subunit of nitric oxide synthase from macrophages, *J. Exp. Med.* **176:**599–604.

Cross, A. H., Misko, T. P., Lin, R. F., Hickey, W. F., Trotter, J. L., and Tilton, R. G., 1994, Aminoguanidine, an inhibitor of inducible nitric oxide synthase, ameliorates experimental autoimmune encephalomyelitis in SJL mice, *J. Clin. Invest.* **93:**2684–2690.

Dawson, D. A., 1994, Nitric oxide and focal cerebral ischemia: Multiplicity of actions and diverse outcome, *Cerebrovasc. Brain Metabol. Rev.* **6:**299–324.

Dawson, V. L., Dawson, T. M., Bartley, D. A., Uhl, G. R., and Snyder, S. H., 1993, Mechanisms of nitric oxide-mediated neurotoxicity in primary brain cultures, *J. Neurosci.* **13:**2651–2661.

Di Monte, D. A., Royland, J. E., Jakowec, M. W., and Langston, J. W., 1996, Role of nitric oxide in methamphetamine neurotoxicity: Protection by 7-nitroindazole, an inhibitor of neuronal nitric oxide synthase, *J. Neurochem.* **67:**2443–2450.

Dorheim, M.-A., Tracey, W. R., Pollock, J. S., and Grammas, P., 1994, Nitric oxide synthase activity is elevated in brain microvessels in Alzheimer's disease, *Biochem. Biophys. Res. Comm.* **205:**659–665.

Ferrante, R. J., Kowall, N. W., Beal, M. F., Martin, J. B., Bird, E. D., and Richardson, E. P., 1987, Morphologic and histochemical characteristics of a spared subset of striatal neurons in Huntington's disease, *J. Neuropath. Exp. Neurol.* **46:**12–27.

Fujisawa, H., Dawson, D., Browne, S. E., Mackay, K. B., Bullock, R., and McCulloch, J., 1994, Pharmacological modification of glutamate neurotoxicity in vivo, *Brain Res.* **629:**73–78.

Garthwaite, J., 1991, Glutamate, nitric oxide and cell–cell signalling in the nervous system, *Trends Neurosci.* **14:**60–67.

Gryglewski, R. J., Palmer, R. M. J., and Moncada, S., 1986, Superoxide anion is involved in the breakdown of endothelium-derived vascular relaxing factor, *Nature* **320:**454–456.

Hammer, B., Davis Parker, W., and Bennett, J. P., 1993, NMDA receptors increase OH radicals in vivo by using nitric oxide synthase and protein kinase C, *Neuroreport* **5:**72–74.

Hartung, H.-P., Jung, S., and Stoll, G., 1992, Inflammatory mediators in demyelinating disorders of the CNS and PNS *J. Neuroimmunol.* **40:**197–210.

Hasan, K., Heesen, B.-J., and Corbett, J. A., 1993, Inhibition of nitric oxide formation by guanidines, *Eur. J. Pharmacol.* **249:**101–106.

Hu, J., and El-Fakahany, E. E., 1993, β-Amyloid 25–35 activates nitric oxide synthase in a neuronal clone, *Neuroreport* **4:**760–762.

Huang, Z., Huang, P. L., Panahian, N., Dalkara, T., Fishman, M., and Moskowitz, M., 1994, Effects of cerebral ischemia in mice deficient in neuronal nitric oxide synthase, *Science* **265**:1883–1885.

Hyman, B. T., Marzloff, K., Wenniger, J. J., Dawson, T. M., Bredt, D. S., and Snyder, S. H., 1992, Relative sparing of nitric oxide synthase-containing neurons in the hippocampal formation in Alzheimer's disease, *Ann. Neurol.* **32**:818–820.

Iadecola, C., Xu, X., Zhang, F., El-Fakahany, E. E., and Ross, M. R., 1995a, Marked induction of calcium-independent nitric oxide synthase activity after focal cerebral ischemia, *J. Cereb. Blood Flow Metabol.* **15**:52–59.

Iadecola, C., Zhang, F., and Xu, X., 1995b, Inhibition of inducible nitric oxide synthase ameliorate cerebral ischemic damage, *Am. J. Physiol.* **268**:R286–R292.

Iadecola, C., Beitz, A. J., Renno, W., Xu, X., Mayer, B., and Zhang, F., 1993, Nitric oxide synthase-containing neural processes on large cerebral arteries and cerebral microvessels, *Brain Res.* **606**:148–155.

Jaffrey, S. R., Cohen, N. A., Rouault, T., Klausner, R. D., and Snyder, S. H., 1994, The iron-responsive element binding protein: A target for synaptic actions of nitric oxide, *Proc. Natl. Acad. Sci. USA* **91**:12994–12998.

Johnson, A. W., Land, J. M., Thompson, E. J., Bolanos, J. P., Clark, J. B., and Heales, S. J. R., 1995, Evidence for increased nitric oxide production in multiple sclerosis, *J. Neurol. Neurosurg. Psych.* **58**:107–115.

Kader, A., Frazzinin, V. I., Solomon, R. A., and Trifiletti, R. R., 1993, Nitric oxide production during focal cerebral ischemia in rats, *Stroke* **24**:1709–1716.

Koprowski, H., Zheng, Y. M., Heber-Katz, E., Fraser, N., Rorke, L., Fu, Z. F., Hanlon, C., and Dietzschold, B., 1993, *In vivo* expression of inducible nitric oxide synthase in experimentally induced neurologic diseases, *Proc. Natl. Acad. Sci. USA* **90**:3024–3027.

Kuiper, M. A., Visser, J. J., Bergmans, P. L. M., Scheltens, P., and Wolters, E. C., 1994, Decreased cerebrospinal fluid nitrate levels in Parkinson's disease, Alzheimer's disease and multiple system atrophy patients, *J. Neurolog. Sci.* **121**:46–49.

Kwon, N. S., Stuehr, D. J., and Nathan, C. F., 1991, Inhibition of tumour cell ribonucleotide reductase by macrophage-derived nitric oxide, *J. Exp. Med.* **174**:761–768.

Lin, R. F., Lin, T.-S., Tilton, R. G., and Cross, A. H., 1993, Nitric oxide localized to spinal cords of mice with experimental allergic encephalomyelitis: an electron paramagnetic resonance study, *J. Exp. Med.* **178**:643–648.

Lipton, S. A., and Rosenberg, P. A., 1994, Excitatory amino acids as a final common pathway for neurologic disorders, *N. Engl. J. Med.* **330**:613–622.

Lipton, S. A., Choi, Y.-B., Pan, Z.-H., Lei, S. Z., Chen, H.-S., Sucher, N. J., Loscaizo, J., Singel, D. J., and Stamler, J. S., 1993, A redox-based mechanism for the neuroprotective and neurodestructive effects of nitric oxide and related nitroso-compounds, *Nature* **364**:626–632.

Lockhart, B. P., Benicourt, C., Junien, J.-L., and Privat, A., 1994, Inhibitors of free radical formation fail to attenuate direct β-amyloid25-35 peptide-mediated neurotoxicity in rat hippocampal cultures, *J. Neurosci. Res.* **39**:494–505.

MacMicking, J. D., Willenborg, D. O., Weidemann, M. J., Rockett, K. A., and Cowden, W. B., 1992, Elevated secretion of reactive nitrogen and oxygen intermediates by inflammatory leukocytes in hyper-acute experimental autoimmune encephalomyelitis: enhancement by the soluble products of encephalitogenic T cells, *J. Exp. Med.* **176**:303–307.

Malinski, T., Bailey, F., Zhang, Z. G., and Chopp, M., 1993, Nitric oxide measured by porphyrinic microsensor in rat brain after transient middle cerebral artery occlusion, *J. Cereb. Blood Flow Metabol.* **13**:355–358.

Matsumoto, T., Nakane, M., Pollock, J. S., Kuk, J. E., and Forstermann, U., 1993, A correlation between soluble brain nitric oxide synthase and NADPH-diaphorase activity is only seen after exposure of the tissue to fixative, *Neurosci. Lett.* **155**:61–64.

McCulloch, J., Bullock, R., and Teasdale, G. M., 1991, Excitatory amino acid antagonists: Opportunities for the treatment of ischaemic brain damage in man, in: *Excitatory Amino Acid Antagonists* (B. S. Meldrum, ed.), Blackwell Scientific Publications, Oxford, pp. 287–326.

Meda, L., Cassatella, M. A., Szendrei, G. I., Otvos, L., Baron, P., Villalba, M., Ferrari, D., and Rossi, F., 1995, Activation of microglial cells by β-amyloid protein and interferon-γ, *Nature* **374:**647–650.

Milstien, S., Sakai, N., Brew, B. J., Krieger, C., Vickers, J. H., Saito, K., and Heyes, M. P., 1994, Cerebrospinal fluid nitrite/nitrate levels in neurologic diseases, *J. Neurochem.* **63:**1178–1180.

Minc-Golomb, D., Tsarfaty, I., and Schwartz, J. P., 1994, Expression of inducible nitric oxide synthase by neurones following exposure to endotoxin and cytokine, *Br. J. Pharmacol.* **112:**720–722.

Mitrovic, B., Ignarro, L. J., Montestruque, S., Smoll, A., and Merrill, J. E., 1994, Nitric oxide as a potential pathological mechanisms in demyelination: Its differential effects on primary glial cells in vitro, *Neurosci.* **61:**575–585.

Molina, J. A., Jimenez-Jimenez, F. J., Navarro, J. A., Ruiz, E., Arenas, J., Cabrera-Valdivia, F., Vazquez, A., Fernandez-Calle, P., Ayuso-Peralta, L., Rabasa, M., and Bermejo, F., 1994, Plasma levels of nitrates in patients with Parkinson's disease, *J. Neurolog. Sci.* **127:**87–89.

Moncada, S., Palmer, R. M. J., and Higgs, E. A., 1991, Nitric oxide: Physiology, pathophysiology and pharmacology, *Pharmacol. Rev.* **43:**109–142.

Moore, P. K., Wallace, P., Gaffen, Z. A., Hart, S. L., Babbedge, R. C., 1993, Characterization of the novel nitric oxide synthase inhibitor 7-nitroindazole and related indazoles: Antinociceptive and cardiovascular effects, *Br. J. Pharmacol.* **110:**219–224.

Mufson, E. J., and Brandabur, M. M., 1994, Sparing of NADPH-diaphorase striatal neurons in Parkinson's and Alzheimer's diseases, *Neuroreport* **5:**705–708.

Murphy, S., Simmons, M. L., Agullo, L., Garcia, A., Feinstein, D. L., Galea, E., Reis, D. J., Minc-Golomb, D., and Schwartz, J. P., 1993, Synthesis of nitric oxide in CNS glial cells, *Trends Neurosci.* **16:**323–328.

Nagafuji, T., Sugiyama, M., and Matsui, T., 1994, Temporal profiles of Ca^{2+}/calmodulin-dependent and -independent nitric oxide synthase activity in the rat brain microvessels following cerebral ischemia, *Acta Neurochir.* **60:**285–288.

Nathan, C., 1992, Nitric oxide as a secretory product of mammalian cells, *FASEB J.* **6:**3051–3054.

Nguyen, T., Brunson, D., Crespi, C. I., Penman, B. W., Wishnok, J. S., and Tannenbaum, S. R., 1992, DNA damage and mutation in human cells exposed to nitric oxide in vitro, *Proc. Natl. Acad. Sci. USA* **89:**3030–3034.

Przedborski, S., Jackson-Lewis, V., Yokoyama, R., Shibata, T., Dawson, V. L., and Dawson, T. M., 1996, Role of neuronal nitric oxide in 1-methyl-4-phenyl-1,2,3,6-tetrahydropyridine (MPTP)-induced dopaminergic neurotoxicity, *Proc. Natl. Acad. Sci. USA* **93:**4567–4571.

Olanow, C. W., 1992, An introduction to the free radical hypothesis in Parkinson's disease, *Ann. Neurol.* **32:**S2–S9.

Regidor, J., Edvinsson, L., and Divac, I., 1993, NOS neurones lie near branchings of cortical arteriolae, *Neuroreport* **4:**112–114.

Reif, D. W., and Simmons, R. D., 1990, Nitric oxide mediates iron release from ferritin, *Arch. Biochem. Biophys.* **283:**537–541.

Rossi, F., and Bianchini, E., 1996, synergistic induction of nitric oxide by beta-amyloid and cytokines in astrocytes, *Biochem. Biophys. Res. Commun.* **225:**474–478.

Rubanyi, G. M., Ho, E. H., Cantor, E. H., Lumma, W. C., Parker Botehlo, L. H., 1991, Cytoprotective function of nitric oxide: Inactivation of superoxide radicals produced by human leukocytes, *Biochem. Biophys. Res. Comm.* **181:**1392–1397.

Schulz, J. B., Matthews, R. T., Muqit, M. M. K., Browne, S. E., and Beal, M. F., 1995, Inhibition of nitric oxide synthase by 7-nitroindazole protects against MPTP-induced neurotoxicity in mice, *J. Neurochem.* **64:**936–939.

Selden, N., Geula, C., Hersh, L., and Mesulam, M.-M., 1994, Human striatum: chemoarchitecture of the caudate nucleus, putamen and ventral striatum in health and Alzheimer's disease, *Neuroscience* **60:**621–636.

Sheng, P., Cerruti, C., Ali, S., and Cadet, J. L., 1996, Nitric oxide is a mediator of methamphetamine (METH)-induced neurotoxicity. *In vitro* evidence from primary cultures of mesencephalic cells, *Ann. N. Y. Acad. Sci.* **801:**174–186.

Shergill, J. K., Cammack, R., Cooper, C. E., Cooper, J. M., Mann, V. M., and Schapira, A. H., 1996, Detection of nitrosyl complexes in human substantia nigra, in relation to Parkinson's disease, *Biochem. Biophys. Res. Commun.* **228:**298–305.

Sherman, M. P., Griscavage, J. M., and Ignarro, L. J., 1992, Nitric oxide-mediated neuronal injury in multiple sclerosis, *Med. Hyp.* **39**:143–146.

Simonian, N. A., and Coyle, J. T., 1996, Oxidative stress in neurodegenerative diseases, *Ann. Rev. Pharmacol. Toxicol.* **36**:83–106.

Spencer Smith, T., Swerdlow, R. H., Parker, W. D., and Bennett, J. P., 1994, Reduction of MPP$^+$-induced hydroxyl radical formation and nigrostriatal MPTP toxicity by inhibiting nitric oxide synthase, *Neuroreport* **5**:2598–2600.

Stamler, J. R., 1994, Redox signalling: Nitrosylation and related target interactions of nitric oxide, *Cell* **78**:931–936.

Tomimoto, H., Nishimura, M., Suenaga, T., Nakamura, S., Akiguchi, I., Wakita, H., Kimura, J., and Mayer, B., 1994, Distribution of nitric oxide synthase in the human cerebral blood vessels and brain tissues, *J. Cereb. Blood Flow Metabol.* **14**:930–938.

Wallis, R. A., Panizzon, K. L., Henry, D., and Wasterlain, C. G., 1993, Neuroprotection against nitric oxide injury with inhibitors of ADP-ribosylation, *Neuroreport* **5**:245–248.

Weinberg, E. D., 1992, Iron depletion: A defense against intracellular infection and neoplasia, *Life Sci.* **50**:1289–1297.

Weiss, G., Goossen, B., Doppler, W., Fuchs, D., Pantopoulos, K., Werner-Felmayer, G., Wachter, H., and Hentze, M. W., 1993, Translational regulation via iron-responsive elements by the nitric oxide/NO-synthase pathway, *EMBO J.* **12**:3651–3657.

Van Dam, A.-M., Bauer, J., Man-A-Hing, W. K. H., Marquette, C., Tilders, F. J. H., and Berkenbosch, F., 1995, Appearance of inducible nitric oxide synthase in the rat central nervous system after rabies virus infection and during experimental allergic encephalomyelitis but not after peripheral administration of endotoxin, *J. Neurosci. Res.* **40**:251–260.

Vincent, S. R., and Kimura, H., 1992, Histochemical mapping of nitric oxide synthase in the rat brain, *Neurosci.* **46**:755–784.

Xie, Q., and Nathan, C., 1994, The high output nitric oxide pathway: Role and regulation, *J. Leukocyte Biol.* **56**:576–582.

Yoshida, T., Limmroth, V., Irikura, K., and Moskowitz, M. A., 1994, The NOS inhibitor, 7-nitroindazole, decreases focal infarct volume but not the response to topical acetylcholine in pial vessels, *J. Cereb. Blood Flow Metabol.* **14**:924–929.

Youdim, M. B. H., Ben-Shachar, D., Eshel, G., Finberg, J. P. M., and Riederer, P., 1993, The neurotoxicity of iron and nitric oxide. Relevance to the etiology of Parkinson's disease, *Adv. Neurol.* **60**:259–266.

Zhang, J., Dawson, V. L., Dawson, T. M., and Snyder, S. H., 1993, Nitric oxide activation of poly(ADP-ribose) synthetase in neurotoxicity, *Science* **263**:687–689.

Zhang, Z. G., Chopp, M., Zaloga, C., Pollock, J. S., and Forstermann, U., 1993, Cerebral endothelial nitric oxide synthase expression after focal cerebral ischemia in rats, *Stroke* **24**:2016–2022.

Zielasek, J., Jung, S., Gold, R., Liew, F. Y., Toyka, K. V., and Hartung, H. P., 1995, Administration of nitric oxide synthase inhibitors in experimental autoimmune neuritis and experimental autoimmune encephalomyelitis, *J. Neuroimmunol.* **58**:81–88.

Chapter 12

Iron and Oxidative Stress in Neonatal Hypoxic–Ischemic Brain Injury

Directions for Therapeutic Intervention

Charles Palmer

12.1. INTRODUCTION TO OXIDANT STRESS AND SECONDARY BRAIN INJURY

When placental or pulmonary gas exchange is compromised in the fetus or newborn infant it produces hypoxemia, hypercapnia, and metabolic acidosis. Initially during asphyxia, blood flow to the brain increases to maintain cerebral energy metabolism until cardiac depression causes hypotension, bradycardia, and cerebral ischemia (Vannucci, 1990). Unless prompt and effective resuscitation is achieved, the asphyxiated infant will die. Even when resuscitation is successful, survival may be accompanied by varying degrees of brain injury, the nature of which is related to the duration of the primary insult, the gestational age, and therapeutic interventions initiated during the post-resuscitation period.

Not until recently have we understood that brain injury evolves for hours and even days during recovery from a hypoxic–ischemic insult. Studies in human newborns as well as adult and neonatal animals show that cerebral energy metabolism (Lorek *et al.*, 1994; Azzopardi *et al.*, 1989b; Hope *et al.*, 1984) and brain injury worsens during recovery, especially after mild insults (Beilharz *et al.*, 1995; Horn and Schlote, 1992; Pulsinelli *et al.*, 1982). Recently Du *et al.* (1996) showed that after mild focal cerebral

Charles Palmer   Division of Newborn Medicine, Department of Pediatrics, Pennsylvania State University College of Medicine, M. S. Hershey Medical Center, Hershey, Pennsylvania 17033.

Metals and Oxidative Damage in Neurological Disorders, edited by Connor. Plenum Press, New York, 1997.

ischemia in mature rats, infarction developed after 3 days, whereas no infarction was present after 1 day. A prominent component of cell injury was caused by apoptosis.

The term "secondary injury" is applied to describe the brain damage that evolves during the post-resuscitation interval. Understandably there is great interest, especially amongst clinicians, in unraveling the mechanisms of secondary brain injury. Consequently new therapeutic modalities are emerging that successfully reduce ischemic brain injury even if started many hours after resuscitation. Section 12.7 is a summary of the "brain rescue strategies" effective in animal models of cerebral ischemia. The table includes neonatal animal models where possible. To emphasize the clinical relevance of such interventions during recovery, the interval (window) between resuscitation and treatment in each study is provided. The challenge is to identify those pivotal mechanisms and therapeutic strategies.

The reduction of the free-radical-mediated injury is a most effective rescue strategy, which is achieved by reducing the formation of reactive oxygen species (ROS) or scavenging them before they produce injury. It is the purpose of this chapter to review how resuscitation of the asphyxiated newborn produces damaging ROS that overwhelm the immature antioxidant defenses of the newborn and worsen brain injury. This review will emphasize how free radicals are generated by many steps during the post-resuscitation period (see Figure 12.1). Comprehensive reviews of the other biochemical mechanisms and mediators of injury can be found in references (Vannucci, 1990; Siesjo, 1988, 1992; Hara *et al.*, 1993; Traystman *et al.*, 1991; Siesjo *et al.*, 1989). This chapter will reveal how iron and nitric oxide are important collaborators and indeed conspirators in the production of brain injury (see Figure 12.2).

12.2. REACTIVE OXYGEN SPECIES (ROS): AN OVERVIEW

A free radical is a chemical species with one or more unpaired electrons in its outer orbital. This makes the species unstable, as most biologic species have their electrons arranged in pairs. Free radicals donate (reducing radical) or take (oxidizing radical) electrons from other biomolecules in an attempt to pair their electron and generate a more stable species. In this way radicals generate new radicals, forming damaging chain reactions and destroying the chemical structure and function of their target molecules. In the brain susceptible targets include DNA, protein, and most common membrane lipids. As the brain is particularly rich in polyunsaturated phospholipids, it is susceptible to free radical attack. Reduced intermediates of molecular oxygen, such as superoxide and hydrogen peroxide, are ubiquitous products of normal aerobic metabolism. Certain cells like microglia and neutrophils use superoxide and hydrogen peroxide for purposeful physiologic functions. Nitric oxide is a free radical gas produced by brain endothelial cells, some neurons, and activated microglia. It too has useful effects, but excess nitric oxide can be quite damaging. Non-protein-bound iron and nitric oxide transform mildly reactive oxygen species to more damaging free radicals (see Figure 12.1) (Reilly *et al.*, 1991; Nowicki *et al.*, 1991; Traystman *et al.*, 1991; Halliwell, 1991; Beckman *et al.*, 1990; Aruoma and Halliwell, 1987).

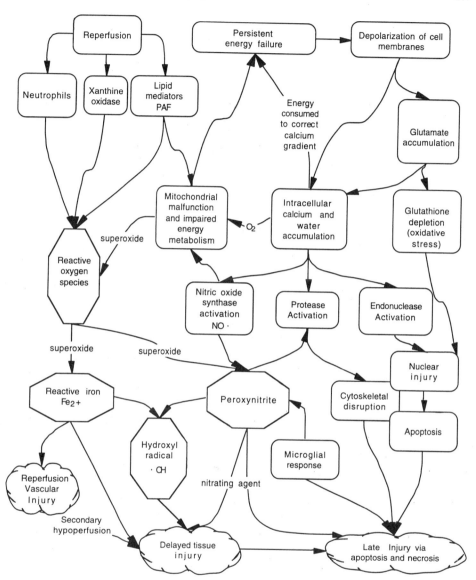

FIGURE 12.1. A flow diagram showing the interconnected mechanisms for producing free radicals and brain injury after cerebral ischemia.

The Susceptibility of the Newborn Brain to Free Radical Injury

Free radicals and reactive oxygen species (superoxide and hydrogen peroxide) are formed during normal metabolism and only cause injury when they exceed the brain's antioxidant defenses. The human newborn, especially the preterm newborn infant, may be particularly susceptible to free radical injury because of a deficiency in brain

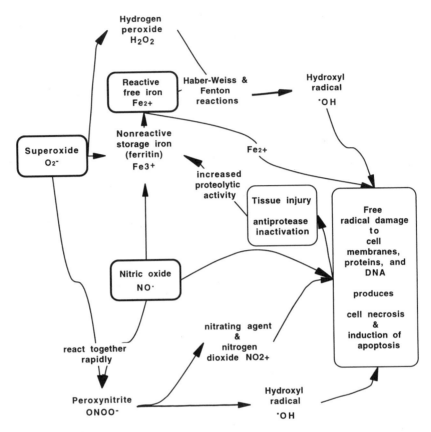

FIGURE 12.2. A flow diagram showing how superoxide, nitric oxide, and iron interact to produce brain injury following hypoxic–ischemic brain injury.

antioxidants, including the antioxidant enzymes superoxide dismutase (Nishida *et al.,* 1994; Takashima *et al.,* 1990), glutathione peroxidase (Inder *et al.,* 1994), and the soluble antioxidant, glutathione (Smith *et al.,* 1993). Newborns have a relative inability to sequester iron because of low transferrin levels (Galet and Schulman, 1976; Sullivan, 1992). Some normal newborns have detectable free iron in their cord blood (Berger *et al.,* 1995; Evans *et al.,* 1992). Diminished ferroxidase activity associated with low ceruloplasmin and high ascorbate levels may account for the presence of non-protein-bound iron (Omene *et al.,* 1981). Sullivan proposed that the premature newborns may be particularly susceptible to iron-medicated oxidative stress, because not only does the preterm newborn have limited transferrin levels, but because they receive repeated packed red blood cell transfusions, they effectively receive a rich source of iron without iron binding capacity (Sullivan, 1992).

Additionally we and others have found that the cerebral blood vessels of the newborn rat stain strongly positive for iron during the first days of life and not thereafter (Moos, 1995; Connor *et al.*, 1995). In respect to nitric oxide production, Galea *et al.* (1995) found that the inducible nitric oxide-producing enzyme is transiently expressed in vessels throughout the brain in the course of normal development: It peaks at seven days of age in the newborn rat whose brain development is similar to about a 34 week newborn human (Vannucci, 1993).

12.3. VASCULAR INJURY

12.3.1. The Primary Insult

During cerebral hypoxia–ischemia there is a disruption of energy metabolism, catabolism of ATP (Vannucci *et al.*, 1994; Palmer *et al.*, 1990), and accumulation of hypoxanthine (Hagberg *et al.*, 1987). Energy depletion leads to depolarization of cell membranes, disruption in cellular ion homeostasis, and cell swelling. These changes are aggravated by release of lipid-derived inflammatory mediators (Lindsberg *et al.*, 1990) and accumulation of excitatory neurotransmitters, such as glutamate (Rothman and Olney, 1986). Receptor-mediated excitotoxicity and energy failure disrupts cells through accumulation of intracellular calcium, sodium, and water. Endothelial cell swelling narrows the vascular lumen.

12.3.2. Reperfusion: Sources of ROS

During the first minutes to hours following reperfusion of the previously ischemic brain, animal studies show that there is a burst of superoxide radical from the surface of the brain. It arises from both the intracellular compartment (Dirnagl *et al.*, 1995) and from blood vessels. Asphyxia (Pourcyrous *et al.*, 1993) and cerebral ischemia in the piglet (Armstead *et al.*, 1988) and the cat (Nelson *et al.*, 1992) produce superoxide anion on the cortical surface during early reperfusion. Cytochemical studies show that superoxide formation happens primarily in the extracellular space associated with blood vessels and occasionally in endothelial cells (Kontos *et al.*, 1992). Perivascular sources of superoxide include xanthine oxidase, neutrophils, (Matsuo *et al.*, 1995) and fatty acid and prostaglandin metabolism (Pourcyrous *et al.*, 1993; Chan and Fishman, 1980). Figure 12.1 provides a schematic timeline for the generation of free radicals and tissue injury during recovery from cerebral ischemia.

Hypoxanthine, which accumulated during ischemia, is salvaged to regenerate ATP during reperfusion. However, some is metabolized by xanthine oxidase (located in endothelial cells), producing superoxide and hydrogen peroxide as byproducts (McCord, 1985). In the presence of transition metals like iron, oxygen-derived reactive species can produce the highly reactive and damaging hydroxyl radical (Gutteridge *et al.*, 1982). Phillis and Sen (1993) studied the temporal profile of hydroxyl radicals trapped from the cortical surface of adult rats subjected to 30 min cerebral ischemia followed by reperfusion. Hydroxyl radicals trapped in artificial cerebrospinal fluid were measured by electron spin resonance spectroscopy and showed that production peaked after 10 min of reperfusion then declined to non-detectable levels over 90 min.

Two different xanthine oxidase inhibitors, oxypurinol and ampflutizole, inhibited hydroxyl radical production, confirming the specific contribution of xanthine oxidase as a free radical generator in their model (Phillis *et al.,* 1994).

Xanthine oxidase is concentrated in endothelial cells lining cerebral blood vessels (Betz, 1985). In addition, xanthine oxidase derived from the post-ischemic liver and intestine circulates in the peripheral blood and will generate superoxide at sites distal to is release into the circulation (Tan *et al.,* 1995, 1993). Hypoxanthine, having accumulated in ischemic tissue, is an available substrate for the enzyme. *In vitro* studies indicate that xanthine oxidase inhibitors alone are not able to inhibit all free-radical-induced endothelial disturbance, suggesting that there are other sources of ROS (Watkins *et al.,* 1995).

12.3.3. Effects of ROS on the Microvasculature

Exogenous generation of oxygen radicals causes increased blood–brain barrier permeability (Schleien *et al.,* 1990; Wei *et al.,* 1986; Chan *et al.,* 1984), abnormal arteriolar reactivity (Leffler *et al.,* 1990), altered transport activity (Elliot and Schilling, 1992; Lo and Betz, 1986), and a decrease in prostacyclin production (Watkins *et al.,* 1996). Reactive oxygen species enhance neutrophil (Gasic *et al.,* 1991) and platelet adhesion to endothelium, promote phospholipase A_2 activation, platelet activating factor (PAF) production, and induce post-ischemic hypoperfusion (Rosenberg *et al.,* 1989; Thiringer *et al.,* 1987). Prostacyclin production by vascular endothelial cells is increased in the presence of superoxide scavengers. It has powerful protective effects on the blood vessel including vasodilation, inhibition of platelet aggregation, and neutrophil activation (Fantone and Kinnes, 1983).

12.3.4. Neutrophils

Neutrophils play an important role in microvascular dysfunction during ischemia by plugging capillaries and temporarily interrupting the microcirculation (Yamakawa *et al.,* 1987; Schmid-Schonbein and Lee, 1995; Del Zoppo, 1994). During reperfusion they adhere to the endothelium in response to activation of adhesion molecules on both the endothelium and neutrophil. They can obstruct blood flow and produce a host of inflammatory mediators including superoxide (Matsuo *et al.,* 1995), hydrogen peroxide, hypochlorous acid, hydroxyl radicals, protreases, and platelet activating factor (PAF) (Weiss, 1989; Candeias *et al.,* 1994, 1993). Neutrophil depletion prior to cerebral ischemia is markedly neuroprotective (Matsuo *et al.,* 1994). We found that neutropenia induced before cerebral ischemia reduces hypoxic–ischemic brain injury in neonatal rats (Palmer, 1996). More clinically relevant, is the reduction of infarct volume achieved by inhibition of neutrophil–endothelial adherence as late as 4 hrs into recovery from cerebral ischemia (Zhang *et al.,* 1995, 1994b; Jiang *et al.,* 1995). In the neonate, neutrophil function, especially migration and adhesion, is relatively impaired compared to adults so extrapolation from adult models should not be made without appropriate studies performed in immature animals (Anderson *et al.,* 1990; 1981). In the immature rat, while neutrophil depletion before cerebral ischemia is protective, we found that neutrophil depletion by 8 hrs of recovery from transient focal hypoxia–ischemia is not protective (Palmer, 1996). In addition, in the 7-day-old rat, neutrophils remained intravascular during the first 42 hrs of recovery from cerebral hypoxia–ischemia. This is in contrast to adult rats where transient cerebral ischemia is associated with marked parenchymal infiltration of neutrophils that peak at 48 hrs recovery (Zhang *et al.*, 1994a).

12.3.5. Antioxidants and Microvascular Injury

As further evidence of the pathogenic role of free radicals in microvascular injury is the neuroprotection achieved with antioxidant enzymes. Intravenous administration of superoxide dismutase and/or catalase conjugated to the macromolecule polyethylene glycol (PEG-SOD and PEG-CAT) reduce ischemic brain injury in adult animals (He *et al.*, 1993; Liu *et al.*, 1989). As these large molecules do not readily cross the blood–brain barrier, their site of action must reside within the vessel lumen or more likely within the vascular endothelium. In preliminary studies we found that PEG-SOD reduces brain swelling in 7-day-old rats at 42 hrs following a hypoxic–ischemic insult, but it does not reduce cerebral atrophy measured at 2 weeks of recovery (Palmer *et al.*, 1996). We found that 30,000U/kg PEG-SOD reduced brain swelling, but the larger dose of 100,000U/kg or a smaller dose of 10,000U/kg did not. A similar U shaped dose response with PEG-SOD has been reported by other investigators (He *et al.*, 1993).

Transgenic newborn mice that have a threefold increase in copper zinc superoxide dismutase (SOD-1) activity, have more brain damage following a transient hypoxic ischemic insult than wild-type littermates with normal enzyme activity (Ditelberg *et al.*, 1996). This is in sharp contrast to the neuroprotective effect in adult transgenic mice of the same strain (Chan *et al.*, 1995). As superoxide dismutase combines two superoxide radicals to generate hydrogen peroxide and glutathione peroxidase, the enzyme responsible for safely converting hydrogen peroxide into oxygen and water is not similarly increased; increased activity of SOD-1 could have elevated hydrogen peroxide levels. Then the availability of free iron could catalyse hydroxyl radical formation (discussed later). Iron is present in the endothelial cells of the immature brain; also, asphyxiated newborns have non-protein-bound iron circulating in the blood (van Bel *et al.*, 1994; Moison *et al.*, 1993; Evans *et al.*, 1992). In addition, peroxides derived from hydrogen peroxide or lipid peroxidation can react with heme to form reactive iron species such as perferryl and ferryl ions. This occurs when a molar excess of peroxide cause fragmentation of the cyclic tetrapyrrol rings in heme to release reactive iron (Gutteridge, 1986). Figure 12.3 shows the perivascular accumulation of histochemically detected iron (enhanced Perls stain) along a blood vessel in the cortex of the 14-day-old rat one week after cerebral hypoxia–ischemia. Figure 12.4 shows intense iron staining in the cortex and striatum of another immature rat one week after the same cerebral insult. Columns of iron reactive product extend towards the cortex in radial projections at right angles to the pial surface. In many cases, a blood vessel can be seen coursing through the center of the column (Palmer *et al.*, 1993a).

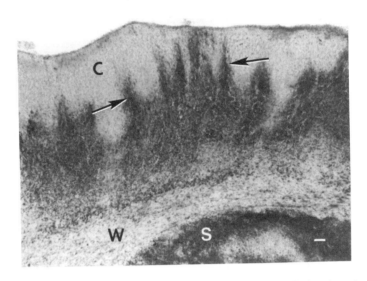

FIGURE 12.3. Photomicrograph of the cortex (C) of a 14-day-old rat stained for iron (arrowhead) with an intensified Perls stain 7 days after a cerebral hypoxic–ischemic insult. Following the insult the rats were allowed to recover for one week, then the brain was removed and fixed in neutral buffered formalin then cryoprotected in sucrose before a 50-micron-thick coronal section was stained for iron. The cortex (C) and striatum (S) stain strongly for iron. The white matter tract (W) is relatively spared. Note the finger-like projections of iron staining and tissue injury at right angles to the pial surface (arrows). Bar = 100 microns.

12.4. FREE-RADICAL-MEDIATED INJURY TO BRAIN PARENCHYMA

While free radicals have been implicated in producing microvascular injury, neurons and oligodendroglia are also susceptible to free radical attack. Until recently it has only been possible to trace free-radical-mediated injury by measuring the products of their reactions. For instance, products of lipid peroxidation are detected in the ischemia sensitive regions of the hippocampus, striatum, and cortex from 8–72 hrs post-transient cerebral ischemia (Bromont *et al.,* 1989). Fluorescent histochemical studies post-ischemia show lipid peroxidation in the same regions known to be most susceptible to injury (White *et al.,* 1993). The elusive nature of reactive free radicals, like hydroxyl radical, make them difficult to measure. However, they can be trapped by using salicylate as a target molecule and the stable hydroxylated product (2,3 or 2,5 dihydroxybenzoate) measured. Using this technique with microdialysis probes positioned within the brain parenchyma, serial measurement found elevated levels of these adducts hours after cerebral reperfusion (Kil *et al.,* 1996; Lancelot *et al.,* 1995). The same *in vivo* microdialysis technique in rat brain found that pretreatment with the amino steroid antioxidant U-74389G greatly attenuated the prolonged (60 min) increase in hydroxyl radicals generated during reperfusion (Zhang and Piantadosi,

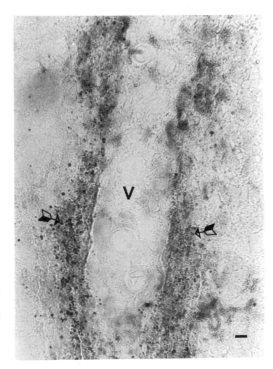

FIGURE 12.4. A high power photomicrograph of a 50-micron-thick coronal section through the damaged cortex one week after a hypoxic ischemic insult to the 7-day-old rat. The section is stained for iron (as in Figure 12.3) with intensified Perls stain. A blood vessel (V) is seen with iron staining along its outer border (arrows). Bar = 10 microns.

1994). It is likely that free radicals, including the hydroxyl radical, are continuously produced for hours after resuscitation, as free radical scavengers (PBN) administered 3 hrs after transient middle cerebral artery occlusion in adult rats reduce infarct size (Zhao *et al.*, 1994). Other protective free radical scavengers are listed in the Appendix (section 12.7).

12.4.1. Actions on Cellular Function

Free radicals impair the ability of the transmembrane enzyme Na^+K^+ ATPase, especially in cortical synaptosomal membranes (Razdan *et al.*, 1993; Mishra and Delivoria-Papadopoulos, 1988). Inhibition of the Na^+K^+ ATPase pump results in persistent membrane depolarization and excessive release of the excitatory amino acid glutamate (Schiff and Somjen, 1985). Prolonged resuscitation of the newborn pig for 2 hrs in 100% oxygen increases (Goplerud *et al.*, 1995), while Allopurinol (20mg/kg iv) diminishes, this inhibitory effect on the Na^+K^+ ATPase (Marro *et al.*, 1994).

During cerebral ischemia there is catabolism of ATP. During recovery from hypoxia–ischemia there is partial restoration of cerebral energy metabolism. This is followed some 24 hrs later by a "secondary energy failure" in human neonates, measured with P^{31} NMR spectroscopy (Wyatt, 1993; Hope *et al.*, 1984). Recent studies in human newborns using proton spectroscopy, show that during the first 24 hrs there is a persistent metabolic acidosis with increased levels of lactate especially in the infants who later have evidence of a poor prognosis (Roth *et al.*, 1992a). In newborn infants the severity of secondary energy failure is directly related to the degree of later impairment to brain growth and neurological development (Roth *et al.*, 1992b; Moorcraft *et al.*, 1991; Azzopardi *et al.*, 1989a). The underlying causes of this delayed disturbance in energy metabolism have not been clearly identified. It is probably that free radical injury contributes to the delayed secondary energy failure as early metabolic disturbances during reperfusion can be improved by metal chelation with deferoxamine (discussed later) (Hurn *et al.*, 1995).

12.4.2. Impaired Energy Metabolism and ROS

Figure 12.1 is a schematic linking many of the metabolic disturbances following cerebral ischemia. It illustrates the interconnection and interdependence of the metabolic derangements and how they are associated with production of ROS. Free radicals, including nitric oxide, cause vascular injury, secondary ischemia (hypoperfusion), mitochondrial dysfunction, and inactivation of enzymes necessary to generate ATP (Dawson, 1994; Siesjo *et al.*, 1989). At the same time that energy formation is impaired, the demands of the injured brain for energy and blood flow are not reduced, on the contrary there are many reasons why demands could be increased: seizures, hypoglycemia, spreading depression, restoration of elevated intracellular calcium, all demands on depleted energy reserves (Hossman, 1994). Persistent impairment of energy production to match the needs of the injured brain perpetuates the biochemical cascade depicted in Figure 12.1. ROS and other mediators of injury are continuously formed and, as efficient destroyers of cell membranes, essential enzymes, and even DNA, they represent a common mediator of cell necrosis.

12.4.3. Elevated Intracellular Calcium and ROS

When energy metabolism is impaired, cells depolarize and calcium enters via voltage-sensitive and receptor-mediated channels. This leads to cell swelling and death from osmotic stress. The already limited cellular energy reserves are further depleted in a futile attempt to restore the calcium gradient with the extracellular fluid (Siesjoand Bengtsson, 1989). In the immature rat, Stein and Vannucci (1988) showed that accumulation of radioactive calcium in the brain of the 7-day-old rats subjected to hypoxic–ischemic brain injury returned to normal during recovery, then after an interval of about 4 hrs radioactive calcium reaccumulated in the regions destined to become injured.

Elevated intracellular calcium acts as an intracellular signal for secondary injury mechanisms. In rats it activates endonucleases causing DNA fragmentation as early as 6 hrs after 60 min of focal cerebral ischemia (Charriaut-Marlangue *et al.*, 1995; Walker *et al.*, 1994). Calcium also activates proteases (Ostwald *et al.*, 1993), phospholipases (Al-Mehdi *et al.*, 1993; Umemura *et al.*, 1992; Rordorf *et al.*, 1991), and protein kinase

C. As elaborated on below, these reactions produce more free radicals that contribute to cell injury.

12.4.4. Excitotoxic Injury and ROS

Excitotoxic injury is also mediated by free radical injury in the following ways. First, free radicals increase extracellular glutamate accumulation by increasing the release of glutamate and inhibiting its uptake (Volterra et al., 1993; Benveniste et al., 1984). Overstimulation of glutamate receptors permits a massive influx of extracellular calcium that mediates progressive cell death (Choi, 1985). Despite the influx of calcium, neurons are relatively resistant to death in the absence of oxygen and are spared the full effects of excitotoxicity during cerebral ischemia when the oxygen supply is depleted (Dubinsky et al., 1995). It follows that oxygen is necessary for the full neurotoxic effect of intracellular calcium. N-methyl-D-aspartate (NMDA) receptor stimulation and calcium accumulation induces neuronal nitric oxide synthase and the production of nitric oxide in neurons that possess the enzyme (Northington et al., 1995). Studies performed on neuron cell cultures show that NMDA medicated cell death is mediated in part by the production of nitric oxide (Dawson et al., 1993, 1991).

12.4.5. Mitochondrial Dysfunction and ROS

Dugan et al. (1995) used an in vitro preparation of cultured murine neurons to show that in response to NMDA receptor stimulation, there was a $Ca2+$-dependent uncoupling mitochondrial electron transport in neurons and generation of reactive oxygen species. In support of this finding other investigators have shown that cerebellar granule cells produce superoxide when exposed to the excitatory amino acid N-methyl-D-aspartate (Lafon-Cazal et al., 1993a, 1993b). Mitochondria have a substantial capacity to generate superoxide and hydrogen peroxide after ischemia when the components of the respiratory chain are highly reduced and when molecular oxygen is present. Normally, mitochondria reduce oxygen completely to water in a four electron reaction. Only one or two percent of the oxygen consumed by the brain generates reactive oxygen species. When oxygen is reduced by a single electron it forms superoxide (Turrens and Boveris, 1980; Boveris and Chance, 1973).

The sites of superoxide formation in mitochondria are; complex 1 (NADH dehydrogenase) and complex III (ubiquinone-cytochrome b-c_1). The rate of superoxide production by mitochondria is increased by at least three factors that occur during cerebral ischemia/reperfusion: increased reduction of electron carriers, increase in mitochondrial $Ca2+$, and a decrease in pH. Superoxide anion produced by mitochondria rapidly dismutates to hydrogen peroxide either spontaneously or enzymatically via manganese superoxide dismutase. Hydrogen peroxide is converted to oxygen and hydrogen peroxide by glutathione peroxidase. However hydrogen peroxide may diffuse from the mitochondria to other susceptible sites within the cell. The ability to diffuse towards sources of iron is important as free iron (iron not bound to protein) acts as a catalyst to produce hydroxyl radicals (Halliwell and Gutteridge, 1992). Piantadosi demonstrated for the first time in vivo that following transient cerebral ischemia in rats, a microdiayalis probe in the hippocampus could detect a surge in hydroxyl radical production (Piantadosi and Zhang, 1996). Importantly, the production of hydroxyl radicals was directly related to complex 1 and could be inhibited by two different complex 1 inhibitors.

12.4.6.   Hydroxyl Radicals: From Iron or Nitric Oxide?

While the production of hydroxyl radicals might be related to the catalytic action of iron, a non-transition metal possibility exists: that is, from the cleavage of peroxynitrite (OHNOO⁻). Nitric oxide and superoxide when produced simultaneously react rapidly to form peroxynitrite. When peroxynitrite cleaves it produces hydroxyl radicals and another radical species, nitrogen dioxide ($NO_2\cdot$). Hydroxyl radicals are virtually unable to diffuse as they are so reactive. In contrast, peroxynitrite is relatively stable, and it can diffuse a few microns. Thus some investigators feel that peroxynitrite may be the more significant source of damaging hydroxyl radicals (Beckman, 1994; Beckman *et al.,* 1990). Rather than iron being the sole mediator of superoxide injury, nitric oxide could be its principal conspirator (Beckman *et al.,* 1991, 1990).

12.4.7.   Oxidant Stress in Young Neurons and Oligodendrocytes

In addition to neuronal toxicity, glutamate is highly toxic to oligodendroglia, the cells responsible for producing myelin in the developing brain. In the immature brain oligodendrocytes are particularly rich in iron (Connor *et al.,* 1995) and ferritin (Blissman *et al.,* 1996). Interestingly, oligodendroglial death is not produced by a glutamate receptor-mediated mechanism. Rather, glutamate enters the oligodendrocyte in exchange for cysteine. Intracellular cysteine depletion is followed by a drop in the levels of cysteine and glutathione, which is an important endogenous antioxidant and scavenger of free radicals. The cells die because they cannot protect themselves from oxidative stress (Oka *et al.,* 1993). The same mechanism of cell death occurs in embryonic cortical neurons grown in primary culture, a process that is also preventable with antioxidants (Ratan *et al.,* 1994; Murphy *et al.,* 1990).

12.4.8.   Microglia as Source of ROS

Activation and influx of reactive microglia can aggravate ischemic brain injury (Giulian *et al.,* 1994, 1993). In immature rats the microglial response to cerebral hypoxia–ischemia is obvious by 18–24 hrs after the insult and reaches a maximum after 2–3 days (Ivacko *et al.,* 1996; McRae *et al.,* 1995; Towfighi *et al.,* 1995). Delayed microglial accumulation in areas of injury can be found as long as 14 days after the insult. Microglia and astrocytes are a source of inducible nitric oxide and other ROS capable of killing neurons (Dawson *et al.,* 1994). Ohno *et al.* (1995) showed that microglia aggregated around blood vessels in the post ischemic neonatal rat brain. We have also seen a perivascular injury pattern in the neonatal rat (See Figure 12.3).

12.4.9. Apoptosis and ROS

As discussed earlier, ROS are mediators of cell necrosis as they destroy vital cellular constituents. Additionally ROS can produce cell death by triggering the process of delayed cell death (apoptosis). They achieve this by stimulating activity of the oxidative stress responsive nuclear transcription factor NF-kB (Hocking *et al.*, 1990; Buttke and Sandstrom, 1994). There is increasing evidence that programmed cell death plays a significant part in late ischemic brain injury, especially selective neuronal necrosis in adult and neonatal animals (Beilharz *et al.*, 1995; Hill *et al.*, 1995; Ferrer *et al.*, 1994; Linnik *et al.*, 1993). There is also recent *in vitro* evidence that the hydroxyl radical signals endothelial cell apoptosis in culture (Abello *et al.*, 1994). Moreover, glutathione depletion that occurs in immature cortical neurons and in oligodendroglia exposed to glutamate leads to cell death by apoptosis, a process prevented by antioxidants (Ratan *et al.*, 1994).

Thus, from within the brain parenchyma, oxygen-derived free radicals can be produced from a variety of sources including xanthine oxidase, prostaglandin synthesis, mitochondrial dysfunction, and activated inflammatory cells. Superoxide can be regarded as a ROS, which on its own does not cause injury. Transition metals and nitric oxide collaborate to produce another generation of more reactive species. Figure 12.2 is a flow diagram showing the interactions between superoxide, nitric oxide, and tissue iron as they conspire to produce brain damage.

12.5. IRON

Superoxide is comparatively innocuous on its own, but its toxicity greatly increases in the presence of a transition metal such as iron. The toxicity of iron is attributed to its ability to transfer electrons and catalyze formation of more reactive species, specifically hydroxyl radicals and other iron-oxygen compounds, like the ferryl and perferryl ions. Iron-dependent stimulation of lipid peroxidation can occur independently of hydroxyl radicals or iron–oxygen complexes as ferrous and ferric iron can initiate peroxidation by subtracting a hydrogen atom from fatty acids to form alkoxy and peroxy radicals (Halliwell, 1992, 1991).

12.5.1. Susceptibility of the Immature Brain to Iron-Mediated Injury

Brain regions with high iron contents are more susceptible to peroxidative brain injury (Subbarao and Richardson, 1990; Zaleska and Floyd, 1985). When exogenous iron is injected into the brain in the form of hemoglobin, heme, or ferric chloride, it causes lipid peroxidation with inhibition of the membrane-bound enzyme Na^+K^+ ATPase (Sadrzadeh and Eaton, 1992, 1988; Ciuffi *et al.*, 1991; Sadrzadeh *et al.*, 1987, 1984). These deleterious effects are blocked by the iron chelator deferoxamine (Sadrzadeh *et al.*, 1984). As discussed earlier, the contribution of iron-related injury may be particularly relevant to the newborn as the concentration of iron, transferrin, and ferritin in the newborn rat are highest at birth (Roskams and Connor, 1994); also, cerebral blood vessels stain prominently for iron within the first week of life, diminishing in intensity as the animal matures (Moos, 1995; Connor *et al.*, 1995). It is conceivable that the colocalization of iron in the cerebral endothelium makes the blood vessels particularly susceptible to iron mediated injury (see Figures 12.3 and 12.4).

Oligodendroglial precursors are rich in iron and ferritin. They are found concentrated in the "myelogenic foci" of the immature white matter (Blissman *et al.*, 1996; Connor *et al.*, 1995). These white matter foci are particularly susceptible to hypoxic–ischemic injury (Rice *et al.*, 1981). The greater sensitivity is correlated with glutamate-mediated glutathione depletion and lipid peroxidation in the oligodendroglial precursor (Sweeney *et al.*, 1995; Oka *et al.*, 1993). A similar mechanism for reducing intracellular glutathione occurs in immature neurons in culture (Murphy *et al.*, 1990).

The reaction of the immature brain to iron-mediated injury is relevant as cerebral hemorrhage is a common accompaniment of ischemic brain injury in the preterm infant (Volpe, 1987). Patt *et al.* (1990) demonstrated that iron depletion of chelation before cerebral ischemia reduces the extent of brain injury in gerbils. We have found that deferoxamine administered to the 7-day-old rat within 5 minutes of recovery from a hypoxic–ischemic brain insult reduces cerebral atrophy and infarction (Palmer *et al.*, 1994b) (see Figure 12.5). In humans, deferoxamine's usefulness is limited by its short half life (5–10 min) and ability to cause cardiogenic shock. This has been overcome without loss of efficacy by complexing deferoxamine to high-molecular-weight species such as dextran or hydroxyethyl starch (Mousa *et al.*, 1992; Forder *et al.*, 1990). Rosenthal *et al.* (1992a) prevented lipid peroxidation and reduced neurological injury and mortality in a rat cardiac arrest model by giving the rats a deferoxamine-hydroxyethyl starch conjugate during early resuscitation. Furthermore, iron chelation with free deferoxamine, but not the high-molecular-weight conjugate nor iron-saturated deferoxamine, reduced the secondary metabolic failure associated with severe ischemic acidosis (Hurn *et al.*, 1995). The poor recovery seen with conjugated deferoxamine indicates that the beneficial action of deferoxamine extends beyond the intravascular compartment, while the lack of effect of iron laden deferoxamine suggests that iron-mediated free radical injury is an important contributor to secondary energy failure.

12.5.2. Sources of Iron in Cerebral Ischemia

Most intracellular iron is stored tightly within ferritin in the ferric form. Circulating iron is bound to the transport protein transferrin. Mobilization of iron so it is free to react with lipids directly or catalyze formation of hydroxyl radicals requires reduction to the ferrous state. As a consequence of hypoxia–ischemia, there is an increase in those agents that can reduce ferritin-bound iron to the ferrous state. These include the enzyme xanthine oxidase in superoxide-dependent and independent mechanisms (Biemond *et al.*, 1984), Bolann and Ulvik, 1987), superoxide liberated from activated microglia (Yoshida *et al.*, 1995) or neutrophils (Matsuo *et al.*, 1995), metabolic acidosis, (Qi *et al.*, 1995; Bralet *et al.*, 1992), and nitric oxide (Reif and Simmons, 1990). It is also probable that enhanced proteolytic activity in injured tissue might release iron from storage *in vivo,* as has been found *in vitro* (Rothman, 1992). Also, cord blood samples, especially from preterm infants, contain measurable nonprotein-bound iron

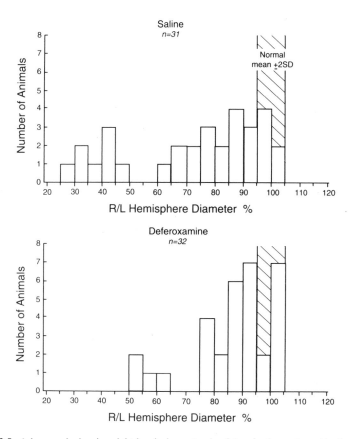

FIGURE 12.5. A bar graph showing right hemisphere atrophy determined morphometrically by the R/L hemisphere diameter ratio. The normal range (mean ± 2SD) of hemisphere asymmetry is illustrated by the striped area. Rats were treated at 5 minutes with 100 mg/kg S.C. deferoxamine mesylate (bottom) or saline (top) after a hypoxic ischemic insult that produces injury to the right hemisphere only. The deferoxamine-treated rats were less damaged than the saline-treated rats. P = .019 (Mann Whitney U test). Reprinted from Palmer *et al.*, 1994b, with permission. Copyright 1994 American Heart Association.

(Evans *et al.*, 1992; Moison *et al.*, 1993). Indeed, the same study found that the plasma of preterm infants, but not adult plasma, had a prooxidant effect of surfactant liposomes that could be neutralized by the addition of apotransferrin (Moison *et al.*, 1993).

12.5.3. Iron Accumulation in the Post-Ischemic Brain

In a preliminary report, van Bel *et al.* (1994) reported increased free iron and increased products of lipid peroxidation in the plasma of severely asphyxiated human newborns. In the immature rat model of hypoxic–ischemic brain injury, we found that there was an increase in histochemically stainable iron within the damaged brain region (see Figures 12.3 and 12.4). It still remains to be proved if iron contributes to injury or merely mineralizes dead cells. In adult rodents, ferritin staining and iron reaction product was detected weeks after ischemic brain injury (Kondo *et al.,* 1995). In the immature rat stainable iron is visible as early as 4 hrs of recovery from brain damaging hypoxia–ischemia (unpublished results). Thus in the immature rat, new iron staining occurs in the early hours of recovery and probably before brain injury is fully evolved (Vannucci *et al.,* 1993). The source of the new histochemically-detectable iron also remains to be determined.

Neutralization or elimination of transition metals like free iron is key to the development of an effective antioxidant strategy. It can be achieved with chelators like deferoxamine or by preventing the delocalization of iron from carrier proteins (Hedlund and Hallaway, 1993; Halliwell, 1991).

12.6. NITRIC OXIDE AND THE IMMATURE BRAIN

Nitric oxide (NO), also known as endothelial-derived relaxation factor, is a free radical gas that is produced by nitric oxide synthetase (NOS) from L-arginine and oxygen. When secreted by endothelial cells it produces vasodilation. During cerebral ischemia endothelial-derived NO prevents neutrophil and platelet adhesion to the endothelium. It is produced in neutrophils, astrocytes, and some neurons. In rat brain only a few select neurons have NOS activity. The brain distribution (Matsumoto *et al.,* 1993) and concentration of these cells increases over the first week of life (Tomic *et al.,* 1994). The NOS-containing neurons are particularly resistant to excitotoxic injury in immature rats (Ferriero *et al.,* 1988). Another isoform of nitric oxide synthetase is also present in microglia, neutrophils, and astrocytes. It is calcium independent, inducible by cytokines, and capable of producing large amounts of NO for days (Iadecola *et al.,* 1995b; Endoh *et al.,* 1993; Nathan, 1992). Endothelial cells lining cerebral blood vessels of the normal 7-day-old rat also stain positive for the inducible isoenzyme (Galea *et al.,* 1995).

Nitric oxide has both protective and cytotoxic effects. Beneficial effects of endothelial derived NO include vasodilation, inhibition of neutrophil and platelet aggregation, and the scavenging of superoxide and iron (Davenpeck *et al.,* 1994; Nathan, 1992). Nitrogen monoxide (NO) can exist in two redox forms, including the neurotoxic free radical form, nitric oxide (NO·) and the nitrosonium form of NO^+ (Lipton *et al.,* 1993). The nitrosonium form crosslinks sulfhydral groups on the NMDA receptor channel complex, causing persistent blockade of NMDA responses. In this form NO^+ may afford some protection from NMDA-receptor-mediated neurotoxicity (Lei *et al.,* 1992); Manzoni *et al.,* 1992).

12.6.1. NO: Injury Mechanisms

The inducible form of nitric oxide secreted by activated macrophages or produced in large quantities by stimulated endothelial cells or astrocytes can be cytotoxic. The activity of inducible NOS is enhanced at 12 hrs and peaks at 48 hrs after cerebral ischemia in adult rats (Iadecola *et al.,* 1995b). Its contribution to ischemic brain injury is important because inhibition of inducible NOS with aminoguanidine, even 24 hrs into the post-recovery period, reduces brain injury (Iadecola *et al.,* 1995c). NO impairs energy metabolism in cells by causing iron loss from enzymes essential for mitochondrial respiration (Drapier and Hibbs, 1988). NO causes DNA damage, which stimulates the enzyme poly-ADP-ribose polymerase, promoting DNA repair but also depletion of NAD and ATP (Schraufstatter *et al.,* 1986). Toxic mechanisms include nitric oxide-mediated mono ADP-ribosylation and S-nitrosylation of glyceraldehyde-3-phosphate dehydrogenase (Zhang *et al.,* 1994a). These cytotoxic effects cause depletion of energy metabolites secondary to alteration of glycolytic enzymes and depletion of energy substrates after stimulation of ADP-ribosylation (Brüne *et al.,* 1994). Inhibitors of poly-ADP ribosylation protect neurons in culture from NO-mediated injury (Wallis *et al.,* 1993).

Nitric oxide produced coincidentally with superoxide occurs during recovery from cerebral ischemia. Beckman *et al.* (1990) has shown that superoxide and nitric oxide react rapidly to form peroxynitrite. Peroxynitrite as discussed above is a toxic diffusible molecule, and can decompose to generate a hydroxyl radical and nitrogen dioxide (Radi *et al.,* 1991; Beckman *et al.,* 1991, 1990). Addition of peroxynitrite to biological fluids leads to nitration of tyrosine residues, the presence of nitrated and dysfunctional proteins. Indeed, nitrated proteins have become a marker of peroxynitrite-dependent damage *in vivo* (Royall *et al.,* 1994; Ischiropoulos *et al.,* 1992; Haddad *et al.,* 1993). Inactivation of antiprotease activity by peroxynitrite has also been demonstrated, hence peroxynitrite generation *in vivo* can facilitate both oxidative and proteolytic damage (Whiteman *et al.,* 1996; Beckman *et al.,* 1994). Whiteman *et al.* (1996) recently showed that peroxynitrite-dependent tyrosine nitration and antiprotease inactivation can be reduced by lipoic acid.

Inhibition of nitric oxide synthase with analogues of its arginine substrate provides a useful way to study nitric oxide's role in cerebral ischemia/reperfusion injury (Nathan, 1992). It has also provided new therapeutic possibilities. Experiments in primary neuronal cell culture indicate that N-methyl-D-aspartate (NMDA) neurotoxicity is mediated, at least in part, by nitric oxide (Dawson *et al.,* 1991). Izumi *et al.* (1992) used rat hippocampal slices to show that inhibitors of nitric oxide synthesis reduced neurodegeneration induced by excitatory amino acids.

12.6.2. NO and Regulation of Iron Metabolism

Not only is nitric oxide, like iron, able to turn superoxide into more damaging radicals, but it is intimately related to iron metabolism and regulation of the iron binding protein ferritin. Macrophage-derived nitric oxide represses formation of ferritin *in vitro*. Nitric oxide does this by promoting binding of the iron regulatory factor to the iron responsive element on ferritin mRNA (Weiss *et al.,* 1993; Drapier *et al.,* 1993). This binding represses translation of ferritin mRNA. This could place the cell at risk for iron mediated oxidative injury. To test this hypothesis, Juckett *et al.* (1996) showed that nitric oxide donors reduce the production of ferritin by iron compounds added to endothelial cells. This provided short-term protection against oxidative stress. Twenty-four hours after the nitric oxide donors were removed, the cells were more susceptible to oxidant injury. The investigators speculated that nitric oxide may act as a transient buffer from iron-mediated oxidative injury by sequestering free iron in nitrosyl complexes. The short-term protection afforded by NO may be offset by delayed increased susceptibility to iron-mediated oxidative injury due to inadequate ferritin synthesis (Juckett *et al.,* 1996).

12.6.3. NO Synthase Inhibition and Neuroprotection

In preliminary studies we used the immature rat model of cerebral hypoxia ischemia and gave the nonselective NOS inhibitor, L-nitroarginine-methyl-ester (L-NAME) s.c. immediately upon cessation of the insult. We used doses of 5, 10, and 20 mg/kg and found that we could significantly reduce brain swelling with the lowest dose of 5 mg/kg dose only (Palmer *et al.,* 1994a).

Huang *et al.* (1994) reported that mutant mice deficient in neuronal NOS are protected from brain injury induced by middle cerebral artery occlusion. The infarct became larger after endothelial NOS was inhibited with nitro-L-arginine. Ferriero *et al.* (1995) reported that brain injury in the immature rat can be reduced if neurons that contain nitric oxide synthase are selectively destroyed before that rat is subjected to cerebral hypoxia ischemia. These studies emphasize the importance of developing selective strategies to conserve the constitutive endothelial isoforms of NOS and inhibit the neuronal and inducible isoforms. At least in adult animals the neuroprotection achieved by late inhibition of inducible NOS (24 hrs after onset of ischemia) suggests that the therapeutic window for the appropriate antioxidant strategy is long (Iadecola *et al.,* 1995a, 1995c).

12.7. APPENDIX: RESCUE THERAPIES[a]

Strategy	Treatment	Species/model	Window	References
Hypothermia	Hypothermia	Rat, transient; Gerbil, transient global ischemia	2–12h	Coimbra and Wieloch, 1994; Colbourne and Corbett, 1994; Ginsberg et al., 1992; Busto et al., 1989; Baiping et al., 1994; Kuboyama et al., 1993; Chopp et al., 1991
	3h duration	Rat, global transient	30min	Markarian et al., 1996; Karibe et al., 1994
	3h duration	7-d rat, transient focal		Thoresen et al., 1995
	72h duration	21-d rat, transient focal		Sirimanne et al., 1996
Antioxidant / xanthine oxidase inhibition	Allopurinol (135mg/kg)	7-d rat, transient focal	15min	Palmer et al., 1993b
Radical scavengers	Oxypurinol (135mg/kg)	7-d rat, transient focal	15min	Palmer and Roberts, 1991; Phillis, 1989
	Lazeroids	7-d rat, transient focal	15min	Bagenholm et al., 1991
	U74006F			Umemura et al., 1994; Boisvert, 1991
	Thiozolidones	Rat, global ischemia	30min	Panetta and Clemens, 1994
	LY231617			
	PBN	Rat, transient focal ischemia	3hrs	Zhao et al., 1994; Cao and Phillis, 1994
	MCI-186 (3mg/kg)	Rat, transient focal		Watanabe et al., 1994
	PEG-SOD (30,000IU)	Piglet, transient global		Kirsch et al., 1993
	Nizofenone	Gerbil, transient focal	30min	Yasuda and Nakajima, 1993
Metal chelation	Deferoxamine (100mg/kg)	7-d rat, transient focal	5min	Palmer et al., 1994b; Rosenthal et al., 1992a
Nitric oxide synthase inhibition	Anti-neuronal NOS 7-Nitroindazole	Rat, MCAO	5min	Yoshida et al., 1994
	L-NAME (5mg/kg)	7d-rat, transient focal		Palmer et al., 1994a
	Anti-inducible NOS Aminoguanidine	Rat, MCAO	24h	Iadecola et al., 1995c
Platelet activating factor antagonism	BN5201	Gerbil, transient focal		Spinnewyn et al., 1987
	WEB2086			Bielenberg and Wagener, 1989; Panetta et al., 1987
Anti-neutrophil adhesion	Anti ICAM-1	Rat, transient	1h	Chopp et al., 1994
	Anti CD11b & CD18	Rat, transient	2–4h	Zhang et al., 1995
	rNIF (neutrophil inhibitory factor)	Rat, transient		Jiang et al., 1995

(continued)

12.7. APPENDIX: RESCUE THERAPIES[a] (Continued)

Strategy	Treatment	Species/model	Window	References
Phospholipase A_2 inhibition	EPC-K1	Rat, transient global		Block et al., 1995
Cytosolic calcium reduction	Nilvadipine	Rat, focal	3h	Li et al., 1994
	SNX-111 (conopeptide)	Rat, global	24h	Valentino et al., 1993
Growth factors	BDNF	Rat, transient focal		Tsukahara et al., 1994
	rh-TGF-β1	Rat, focal	2h	McNeill et al., 1994
	IGF-1	Rat, focal	2h	Guan et al., 1993
	bFGF	Rat, permanent MCAO	30min	Fisher et al., 1995
Excitotoxicity				
NMDA receptor antagonists	Dizocilpine (MK-801)	7-d rat, focal	1h	Hattori et al., 1989
	ischemia combined with 3h hypothermia	Rat, transient global	3d	Dietrich et al., 1995
	CNS-1102	Rat permanent MCA occlusion	1h	Meadows et al., 1994
	Dextromethorphan	Rabbit, transient focal	1h	Steinberg et al., 1993
Inhibition of glutamate release	Lamotrigine	Rat, permanent	1h	Smith and Meldrum, 1995
Adenosine deaminase inhibition	Deoxycoformycin	7-d rat, transient focal		Gidday et al., 1995
Adenosine transport inhibition	Propentofylline	Rat, permanent MCAO	15min	Park and Rudolphi, 1994; Parkinson et al., 1994
Gangliosides	Ganglioside GM1	Fetal sheep		Tan et al., 1994
Energy metabolism				
Substrates	Fructose-1,6 biphosphate	Rat, transient focal		Kuluz et al., 1993
Mitochondrial function	Acetyl-L-carnitine	Canine, transient global		Rosenthal et al., 1992b
Protease inhibition				
Calpain Inhibitors	AK295	Rat, permanent MCAO	1.5h	Bartus et al., 1994b
	AK275	Rat, permanent MCAO	3h	Bartus et al., 1994a

[a]The table lists 16 strategies for reducing brain damage from hypoxia–ischemia. Only studies in which treatment was started after the primary insult are included, hence the term "Rescue Therapy." In some cases permanent middle cerebral artery occlusion (MCAO) models were included where no transient ischemia studies were available. The therapeutic strategy and related interventions are listed in the far left column, the specific pharmacologic agent in the next column, and the animal model used in the middle column. The time interval after termination of a transient insult is provided under the column "window," which represents the window of opportunity for reducing damage. Where the treatment was started immediately upon reperfusion/reoxygenation no time interval is given. In the case of permanent MCAO, the time interval represents the time after the onset of arterial occlusion.

12.8. REFERENCES

Abello, P. A., Fidler, S. A., Bulkley, G. B., and Buchman, T. G., 1994, Antioxidants modulate induction of programmed endothelial cell death (apoptosis) by endotoxin, *Arch. Surg.* **129**:134–141.

Al-Mehdi, A. B., Dodia, C., Jain, M. K., and Fisher, A. B., 1993, A phospholipase A2 inhibitor decreases generation of thiobarbituric acid reactive substance during lung ischemia-reperfusion, *Biochim. Biophys. Acta Lipids Lipid Metab.* **1167**:56–62.

Anderson, D. C., Hughes, B. J., and Smith, C. W., 1981, Abnormal mobility of neonatal polymorphonuclear leukocytes, *J. Clin. Invest.* **68**:863–874.

Anderson, D. C., Rothlein, R., Marlin, S. D., Krater, S. S., and Smith, C. W., 1990, Impaired transendothelial migration by neonatal neutrophils: abnormalities of mac-1 (CD11b/CD18)-dependent adherence reactions, *Blood* **76**:2613–2621.

Armstead, W. M., Mirro, R., Busija, D. W., and Leffler, C. W., 1988, Postischemic generation of superoxide anion by newborn pig brain, *Am. J. Physiol.* **255**:H401–H403.

Aruoma, O. I., and Halliwell, B., 1987, Superoxide-dependent and ascorbate-dependent formation of hydroxyl radicals from hydrogen peroxide in the presence of iron, *Biochem. J.* **241**:273–278.

Azzopardi, D., Wyatt, J. S., Cady, E. B., Delpy, D. T., Baudin, J., Stewart, A. L., Hope, P. L., Hamilton, P. A., and Reynolds, E. O. R., 1989a, Prognosis of newborn infants with hypoxic–ischemic brain injury assessed by phosphorus magnetic resonance spectroscopy, *Pediatr. Res.* **25**:441–451.

Azzopardi, D., Wyatt, J. S., Hamilton, P. A., Cady, E. B., Delpy, D. T., Hope, P. L., and Reynolds, E. O. R., 1989b, Phosphorus metabolites and intracellular pH in the brains of normal and small for gestational age infants investigated by magnetic resonance spectroscopy, *Pediatr. Res.* **25**:440–444.

Bagenholm, R., Andine, P., and Hagberg, H., 1991, Effects of 21-aminosteroid U74006F on brain damage and edema following perinatal hypoxia–ischemia in the rat, *J. Cereb. Blood Flow Metab.* **11**:S134.

Baiping, L., Xiujuan, T., Hongwei, C., Qiming, X., and Quling, G., 1994, Effect of moderate hypothermia on lipid peroxidation in canine brain tissue after cardiac arrest and resuscitation, *Stroke* **25**:147–151.

Bartus, R. T., Baker, K. L., Heiser, A. D., Sawyer, S. D., Dean, R. L., Elliott, P. J., and Straub, J. A., 1994a, Postischemic administration of AK275, a calpain inhibitor, provides substantial protection against focal ischemic brain damage, *J. Cereb. Blood Flow Metab.* **14**:537–544.

Bartus, R. T., Hayward, N. J., Elliott, P. J., Sawyer, S. D., Baker, K. L., Dean, R. L., Akiyama, A., Straub, J. A., Harbeson, S. L., Li, Z., and Powers, J., 1994b, Calpain inhibitor AK295 protects neurons from focal brain ischemia: Effects of postocclusion intra-arterial administration, *Stroke* **25**:2265–2270.

Beckman, J. S., 1994, Peroxynitrite versus hydroxyl radical: The role of nitric oxide in superoxide-dependent cerebral injury, *Ann. NY Acad. Sci.* **738**:69–75.

Beckman, J. S., Beckman, T. W., Chen, J., Marshall, P. A., and Freeman, B. A., 1990, Apparent hydroxyl radical production by peroxynitrite: Implications for endothelial injury from nitric oxide and superoxide, *Proc. Natl. Acad. Sci. USA* **87**:1620–1624.

Beckman, J. S., Ischiropoulos, H., Chen, J., Zhu, L., and Smith, C. D., 1991, Nitric oxide as a mediator of superoxide-dependent injury, in: *Oxidative Damage and Repair: Chemical, Biological and Medical Aspects* (K. Davis, ed.), Pergamon Press, Oxford, pp. 251–255.

Beckman, J. S., Chen, J., Ischiropoulos, H., and Crow, J. P., 1994, Oxidative chemistry of peroxynitrite, *Methods Enzymol.* **233**:229–240.

Beilharz, E. J., Williams, C. E., Dragunow, M., Sirimanne, E. S., and Gluckman, P. D., 1995, Mechanisms of delayed cell death following hypoxic–ischemic injury in the immature rat: Evidence for apoptosis during selective neuronal loss, *Mol. Brain Res.* **29**:1–14.

Benveniste, H., Drejer, J., Schousboe, A., and Diemer, N. H., 1984, Elevation of the extracellular concentrations of glutamate and aspartate in rat hippocampus during transient cerebral ischemia montored by intracerebral microdialysis, *J. Neurochem.* **43**:1369–1374.

Berger, H. M., Mumby, S., and Gutteridge, J. M. C., 1995, Ferrous ions detected in iron-overloaded cord blood plasma from preterm and term babies: Implications for oxidative stress, *Free Radic. Res.* **22**:555–559.

Betz, A. L., 1985, Identification of hypoxanthine transport and xanthine oxidase activity in brain capillaries, *J. Neurochem.* **44**:574–579.

Bielenberg, G. W., and Wagener, G., 1989, Infarct reduction by PAF antagonists after MCA occlusion in the rat, *J. Cereb. Blood Flow Metabol.* **9**:(Suppl 1)S274.

Biemond, P., van Eijk, H. G., Swaak, A. J. G., and Koster, J. F., 1984, Iron mobilization from ferritin by superoxide derived from stimulated polymorphonuclear leukocytes. Possible mechanism in inflammatory diseases, *J. Clin. Invest.* **73**:(1984)1576.

Blissman, G., Menzies, S., Beard, J., Palmer, C., and Connor, J., 1996, The expression of ferritin subunits and iron in oligodendrocytes in neonatal procine brains, *Dev. Neurosci.* **18**:274–281.

Block, F., Kunkel, M., and Sontag, K. H., 1995, Posttreatment with EPC-K1, an inhibitor of lipid peroxidation and of phospholipase A2 activity, reduces functional deficits after global ischemia in rats, *Brain Res. Bull.* **36**:257–260.

Boisvert, D. P., 1991, Effectiveness of postischemic 21-aminosteroid U74006F in preventing reperfusion brain edema, *J. Cereb. Blood Flow Metab.* **11**:S135.

Bolann, B. J., and Ulvik, R. J., 1987, Release of iron from ferritin by xanthine oxidase, *Biochem. J.* **243**:55–59.

Boveris, A., and Chance, B., 1973, The mitochondrial generation of hydrogen peroxide, *Biochem. J.* **134**:707–716.

Bralet, J., Schreiber, L., and Bouvier, C., 1992, Effect of acidosis and anoxia on iron delocalization from brain homogenates, *Biochem. Pharmacol.* **43**:979–984.

Bromont, C., Marie, C., and Bralet, J., 1989, Increased lipid peroxidation in vulnerable brain regions after transient forebrain ischemia in rats, *Stroke* **20**:918–924.

Brüne, B., Dimmeler, S., Molina y Vedia, L., and Lapetina, E. G., 1994, Nitric oxide: A signal for ADP-ribosylation of proteins, *Life Sci.* **54**:61–70.

Busto, R., Dietrich, W. D., Globus, M. T., and Ginsberg, M. D., 1989, Postischemic moderate hypothermia inhibits CA1 hippocampal ischemic neuronal injury, *Neurosci. Lett.* **101**:299–304.

Buttke, T. M., and Sandstrom, P. A., 1994, Oxidative stress as a mediator of apoptosis, *Immunol. Today* **15**:7–10.

Candeias, L. P., Patel, K. B., Stratford, M., and Wardman, P., 1993, Free hydroxyl radicals are formed on reaction between the neutrophil-derived species superoxide anion and hypochlorous acid, *FEBS Lett.* **333**:151–153.

Candeias, L. P., Stratford, M., and Wardman, P., 1994, Formation of hydroxyl radicals on reaction of hypochlorous acid with ferrocyanide, a model iron(II) complex, *Free Radic. Res.* **20**:241–249.

Cao, X., and Phillis, J. W., 1994, Alpha-Phenyl-tert-butyl-nitrone reduces cortical infarct and edema in rats subjected to focal ischemia, *Brain Res.* **644**:267–272.

Chan, P. H., and Fishman, R. A., 1980, Transient formation of superoxide radicals in polyunsaturated fatty acid-induced brain swelling, *J. Neurochem.* **35**:1004–1007.

Chan, P. H., Epstein, C. J., Li, Y., Huang, T. T., Carlson, E., Kinouchi, H., Yang, G., Kamii, H., Mikawa, S., Kondo, T., Copin, J. C., Chen, S. F., Chan, T., Gafni, J., Gobbel, G., and Reola, E., 1995, Transgenic mice and knockout mutants in the study of oxidative stress in brain injury, *J. Neurotrauma* **12**:815–824.

Chan, P. H., Schmidley, J. W., Fishman, R. A., and Longar, S. M., 1984, Brain injury, edema and vascular permeability changes induced by oxygen-derived free radicals, *Neurology* **34**:315–320.

Charriaut-Marlangue, C., Margaill, I., Plotkine, M., and Ben-Ari, Y., 1995, Early endonuclease activation following reversible focal ischemia in the rat brain, *J. Cereb. Blood Flow Metab.* **15**:385–388.

Choi, D. W., 1985, Glutamate neurotoxicity in cortical cell culture is calcium dependent, *Neurosci. Lett.* **58**:293–297.

Chopp, M., Chen, H., Dereski, M. O., and Garcia, J. H., 1991, Mild hypothermic intervention after graded ischemic stress in rats, *Stroke* **22**:37–43.

Chopp, M., Zhang, R. L., Chen, H., Li, Y., Jiang, N., and Rusche, J. R., 1994, Postischemic administration of an anti-Mac-1 antibody reduces ischemic cell damage after transient middle cerebral artery occlusion in rats, *Stroke* **25**:869–875.

Ciuffi, M., Gentilini, G., Franchi-Micheli, S., and Zilletti, L., 1991, Lipid peroxidation induced "in vivo" by iron-carbohydrate complex in the rat brain cortex, *Neurochem. Res.* **16**:43–49.

Coimbra, C., and Weiloch, T., 1994, Moderate hypothermia mitigates neuronal damage in the rat brain when initiated several hours following transient cerebral ischemia, *Acta Neuropathol.* **87**:325–331.

Colbourne, F., and Corbett, D., 1994, Delayed and prolonged post-ischemic hypothermia is neuroprotective in the gerbil, *Brain Res.* **654:**265–272.

Connor, J. R., Pavlick, G., Karli, D., Menzies, S. L., and Palmer, C., 1995, A histochemical study of iron-positive cells in the developing rat brain, *J. Comp. Neurol.* **355:**111–123.

Davenpeck, K. L., Gauthier, T. W., and Lefer, A. M., 1994, Inhibition of endothelial-derived nitric oxide promotes P-selectin expression and actions in the rat microcirculation, *Gastroenterology* **107:**1050–1058.

Dawson, D. A., 1994, Nitric oxide and focal cerebral ischemia: Multiplicity of actions and diverse outcome, *Cerebrovasc. Brain Metab. Rev.* **6:**299–324.

Dawson, V. L., Dawson, T. M., London, E. D., Bredt, D. S., and Snyder, S. H., 1991, Nitric oxide mediates glutamate neurotoxicity in primary cortical cultures, *Proc. Natl. Acad. Sci. USA* **88:**6368–6371.

Dawson, V. L., Dawson, T. M., Bartley, D. A., Uhl, G. R., and Snyder, S. H., 1993, Mechanisms of nitric oxide-mediated neurotoxicity in primary brain cultures, *J. Neurosci.* **13:**2651–2661.

Dawson, V. L., Brahmbhatt, H. P., Mong, J. A., and Dawson, T. M., 1994, Expression of inducible nitric oxide synthase causes delayed neurotoxicity in primary mixed neuronal-glial cortical cultures, *Neuropharmacology* **33:**1425–1430.

Del Zoppo, G. J., 1994, Microvascular changes during cerebral ischemia and reperfusion, *Cerebrovasc. Brain Metab. Rev.* **6:**47–96.

Dietrich, W. D., Lin, B. W., Globus, M. Y. T., Green, E. J., Ginsberg, M. D., and Busto, R., 1995, Effect of delayed MK-801 (Dizocilpine) treatment with or without immediate postischemic hypothermia on chronic neuronal survival after global forebrain ischemia in rats, *J. Cereb. Blood Flow Metab.* **15:**960–968.

Dirnagl, U., Lindauer, U., Them, A., Schreiber, S., Pfister, H. W., Koedel, U., Reszka, R., Freyer, D., and Villringer, A., 1995, Global cerebral ischemia in the rat: Online monitoring of oxygen free radical production using chemiluminescence in vivo, *J. Cereb. Blood Flow Metab.* **15:**929–940.

Ditelberg, J. S., Sheldon, R. A., Epstein, C. J., and Ferriero, D. M., 1996, Brain injury after perinatal hypoxia–ischemia is exacerbated in copper/zinc superoxide dismutase transgenic mice, *Pediatr. Res.* **39:**204–208.

Drapier, J. C., and Hibbs, J., 1988, Differentiation of murine macrophages to express nonspecific cytotoxicity for tumor cells results in L-arginine-dependent inhibition of mitochondrial iron-sulfur enzymes in the macrophage effector cells, *J. Immunol.* **140:**2829–2838.

Drapier, J. C., Hirling, H., Wietzerbin, J., Kaldy, P., and Kühn, L. C., 1993, Biosynthesis of nitric oxide activates iron regulatory factor in macrophages, *EMBO J.* **12:**3643–3649.

Du, C., Hu, R., Csernansky, C. A., Hsu, C. Y., and Choi, D. W., 1996, Very delayed infarction after mild focal cerebral ischemia: A role for apoptosis? *J. Cereb. Blood Flow Metab.* 16:195–201.

Dubinsky, J. M., Kristal, B. S., and Elizondo-Fournier, M., 1995, On the probabilistic nature of excitotoxic neuronal death in hippocampal neurons, *Neuropharmacology* **34:**701–711.

Dugan, L. L., Sensi, S. L., Canzoniero, L. M. T., Handran, S. D., Rothman, S. M., Lin, T. S., Goldberg, M. P., and Choi, D. W., 1995, Mitochondrial production of reactive oxygen species in cortical neurons following exposure to N-methyl-D-aspartate, *J. Neurosci.* **15:**6377–6388.

Elliot, S. J., and Schilling, W. P., 1992, Oxidant stress alters Na^+pump and Na^+-K^+-Cl^- cotransporter activates in vascular endothelial cells, *Am. J. Physiol.* **263:**H96-H102.

Endoh, M., Maiese, K., Pulsinelli, W. A., and Wagner, J. A., 1993, Reactive astrocytes express NADPH diaphorase in vivo after transient ischemia, *Neurosci. Lett.* 154:125–128.

Evans, P. J., Evans, R., Kovar, I. Z., Holton, A. F., and Halliwell, B., 1992, Bleomycin-detectable iron in the plasma of premature and full-term neonates, *FEBS* **303:**210–212.

Fantone, J. C., and Kinnes, D. A., 1983, Prostaglandin E_1 and prostaglandin I_2 modulation of superoxide production by human neutrophils, *Biophys. Res. Commun.* **113:**129–137.

Ferrer, I., Tortosa, A., Macaya, A., Sierra, A., Moreno, D., Munell, F., Blanco, R., and Squier, W., 1994, Evidence of nuclear DNA fragmentation following hypoxia–ischemia in the infant rat brain, and transient forebrain ischemia in the adult gerbil, *Brain Pathol.* **4:**115–122.

Ferriero, D. M., Arcavi, L. J., Sagar, S. M., McIntosh, T. K., and Simon, R. P., 1988, Selective sparing of NADPH-diaphorase neurons in neonatal hypoxia–ischemia, *Ann. Neurol.* **24:**670–676.

Ferriero, D. M., Sheldon, R. A., Black, S. M., and Chuai, J., 1995, Selective destruction of nitric oxide synthase neurons with quisqualate reduces damage after hypoxia–ischemia in the neonatal rat, *Pediatr. Res.* **38:**912–918.

Fisher, M., Meadows, M. E., Do, T., Weise, J., Trubetskoy, V., Charette, M., and Finklestein, S. P., 1995, Delayed treatment with intravenous basic fibroblast growth factor reduces infarct size following permanent focal cerebral ischemia in rats, *J. Cereb. Blood Flow Metab.* **15:**953–959.

Forder, J. R., McClanahan, T. B., Gallagher, K. P., Hedlund, B. E., Hallaway, P. E., and Shlafer, M., 1990, Hemodynamic effects of intraatrial administration of deferoxamine or deferoxamine-pentafraction conjugate to conscious dogs, *J. Cardiovasc. Pharmacol.* **16:**742–749.

Galea, E., Reis, D. J., Xu, H., and Feinstein, D. L., 1995, Transient expression of calcium-independent nitric oxide synthase in blood vessels during brain development, *FASEB J.* **9:**1632–1637.

Galet, S., and Schulman, H. M., 1976, The postnatal hypotransferrinemia of early preterm newborn infants, *Pediatr. Res.* **10:**118–120.

Gasic, A. C., McGuire, G., Drater, S., Farhood, A. I., Goldstein, M. A., Smith, C. W., Entman, M. L., and Taylor, A. A., 1991, Hydrogen peroxide pretreatment of perfused canine vessels induces ICAM-1 and CD18-dependent neutrophil adherence, *Circulation* **84:**2154–2166.

Gidday, J. M., Fitzgibbons, J. C., Shah, A. R., Krujalis, M. J., and Park, T. S., 1995, Reduction in cerebral ischemic injury in the newborn rat by potentiation of endogenous adenosine, *Pediatr. Res.* **38:**1–6.

Ginsberg, M. D., Sternau, L. L., Globus, M. T., Dietrich, W. D., and Busto, R., 1992, Therapeutic modulation of brain temperature: Relevance to ischemic brain injury, *Cerebrovasc. Brain Metab. Rev.* **4:**189–225.

Giulian, D., Corpuz, M., Chapman, S., Mansouri, M., and Robertson, C., 1993, Reactive mononuclear phagocytes release neurotoxins after ischemic and traumatic injury to the central nervous system. *J. Neurosci. Res.* **36:**681–693.

Giulian, D., Li, J., Li, X., George, J., and Rutecki, P. A., 1994, The impact of microglia-derived cytokines upon gliosis in the CNS, *Dev. Neurosci.* **16:**128–136.

Goplerud, J. M., Kim, S., and Delivoria-Papadopoulos, M., 1995, The effect of post-asphyxial reoxygenation with 21% vs. 100% oxygen on Na^+,K^+-ATPase activity in striatum of newborn piglets, *Brain Res.* **696:**161–164.

Guan, J., Williams, C., Gunning, M., Mallard, C., and Gluckman, P., 1993, The effects of IGF-1 treatment after hypoxic–ischemic brain injury in adult rats, *J. Cereb. Blood Flow Metab.* **13:**609–616.

Gutteridge, J. M. C., 1986, Iron promotors of the Fenton reaction and lipid peroxidation can be released from haemoglobin by peroxides, *FEBS Lett.* **201:**291–295.

Gutteridge, J. M. C., Rowley, D. A., and Halliwell, B., 1982, Superoxide-dependent formation of hydroxyl radicals and lipid peroxidation in the presence of iron salts, *Biochem. J.* **206:**605–609.

Haddad, I. Y., Ischiropoulos, H., Holm, B. A., Beckman, J. S., Baker, J. R., and Matalon, S., 1993, Mechanisms of peroxynitrite-induced injury to pulmonary surfactants, *Amer. J. Physiol.* **265:**L555–L564.

Hagberg, H., Andersson, P., Lacarewicz, J., Jacobson, I., Butcher, S., and Sandberg, M., 1987, Extracellular adenosine, inosine, hypoxanthine, and xanthine in relation to tissue nucleotides and purines in rat striatum during transient ischemia, *J. Neurochem.* **49:**227–231.

Halliwell, B., 1991, Reactive oxygen species in living systems: Source, biochemistry and role in human disease, *Am. J. Med.* **91**(Suppl 3C):14S–22S.

Halliwell, B., 1992, Iron and damage to biomolecules, in: *Iron and Human Disease* (R. B. Lauffer, ed.), CRC Press, Boca Raton, Florida, pp. 210–230.

Halliwell, B., and Gutteridge, J. M. C., 1992, Biologically relevant metal ion-dependent hydroxyl radical generation, *FEBS* **307:**108–112.

Hara, H., Sukamoto, T., and Kogure, K., 1993, Mechanism and pathogenesis of ischemia-induced neuronal damage, *Prog. Neurobiol.* **40:**645–670.

Hattori, H., Morin, A. M., Schwartz, P. H., Fujikawa, D. G., and Wasterlain, C. G., 1989, Post-hypoxic treatment with MK-801 reduces hypoxic–ischemic damage in the neonatal rat, *Neurology* **39:**713–718.

He, Y. Y., Hsu, C. Y., Ezrin, A. M., and Miller, M. S., 1993, Polyethylene glycol-conjugated superoxide dismutase in focal cerebral ischemia-reperfusion, *Am. J. Physiol. Heart Circ. Physiol.* **265:**H252–H256.

Hedlund, B. E., and Hallaway, P. E., 1993, High-dose systemic iron chelation attenuates reperfusion injury, *Biochem. Soc. Trans.* **21:**340–343.

Hill, I. E., MacManus, J. P., Rasquinha, I., and Tuor, U. I., 1995, DNA fragmentation indicative of apoptosis following unilateral cerebral hypoxia–ischemia in the neonatal rat, *Brain Res.* **676**:398–403.

Hocking, D. C., Phillips, P. G., Ferro, T. J., and Johnson, A., 1990, Mechanisms of pulmonary edema induced by tumor necrosis factor-α, *Circ. Res.* **67**:68–77.

Hope, P. L., Costello, A. M., Cady, E. B., Delpy, D. T., Tofts, P. S., Chu, A., Hamilton, P. A., and Reynolds, E. O. R., 1984, Cerebral energy metabolism studied with phosphorus NMR spectroscopy in normal and birth-asphyxiated infants, *Lancet* **2**:366–370.

Horn, M., and Schlote, W., 1992, Delayed neuronal death and delayed neuronal recovery in the human brain following global ischemia, *Acta Neuropathol.* **85**:79–87.

Hossman, K.-A., 1994, Viability threshold and the penumbra of focal ischemia, *Ann. Neurol.* **36**:557–565.

Huang, Z., Huang, P. L., Panahian, N., Dalkara, T., Fishman, M. C., and Moskowitz, M. A., 1994, Effects of cerebral ischemia in mice deficient in neuronal nitric oxide synthase, *Science* **265**:1883–1885.

Hurn, P. D., Koehler, R. C., Blizzard, K. K., and Traystman, R. J., 1995, Deferoxamine reduces early metabolic failure associated with severe cerebral ischemic acidosis in dogs, *Stroke* **26**:688–694.

Iadecola, C., Xu, X., Zhang, F., El-Fakahany, E., and Ross, E., 1995a, Marked induction of calcium-independent nitric oxide synthase activity after focal cerebral ischemia. *J. Cereb. Blood Flow Metab.* **15**:52–59.

Iadecola, C., Zhang, F., Xu, S., Casey, R., and Ross, M. E., 1995b, Inducible nitric oxide synthase gene expression in brain following cerebral ischemia, *J. Cereb. Blood Flow Metab.* **15**:378–384.

Iadecola, C., Zhang, F., and Xu, X., 1995c, Inhibition of inducible nitric oxide synthase ameliorates cerebral ischemic damage, *Am. J. Physiol. Regul. Integr. Comp. Physiol.* **268**:R286–R292.

Inder, T. E., Graham, P., Sanderson, K., and Taylor, B. J., 1994, Lipid peroxidation as a measure of oxygen free radical damage in the very low birthweight infant, *Arch. Dis. Child Fetal Neonatal* **70**:F107–F111.

Ischiropoulos, H., Zhu, L., Chen, J., Tsai, M., Martin, J. C., Smith, C. D., and Beckman, J. S., 1992, Peroxynitrite-mediated tyrosine nitration catalyzed by superoxide dismutase, *Arch. Biochem. Biophys.* **298**:431–437.

Ivacko, J. A., Sun, R., and Silverstein, F. S., 1996, Hypoxic–ischemic brain injury induces an acute microglial reaction in perinatal rats, *Pediatr. Res.* **39**:39–47.

Izumi, Y., Benz, A. M., Clifford, D. B., Zorumski, C. F., 1992, Nitric oxide inhibitors attenuate n-methyl-d-aspartate excitotoxicity in rat hippocampus slices, *Neurosci. Lett.* **135**:227–230.

Jiang, N., Moyle, M., Soule, H. R., Rote, W. E., and Chopp, M., 1995, Neutrophil inhibitory factor is neuroprotective after focal ischemia in rats, *Ann. Neurol.* **38**:935–942.

Juckett, M. B., Weber, M., Balla, J., Jacob, H. S., and Vercellotti, G. M., 1996, Nitric oxide donors modulate ferritin and protect endothelium from oxidative injury, *Free Radic. Biol. Med.* **20**:63–73.

Karibe, H., Chen, J., Zarow, G. J., Graham, S. H., and Weinstein, P. R., 1994, Delayed induction of mild hypothermia to reduce infarct volume after temporary middle cerebral artery occlusion in rats, *J. Neurosurg.* **80**:112–119.

Kil, H. Y., Zhang, J., and Piantadosi, C. A., 1996, Brain temperature alters hydroxyl radical production during cerebral ischemia perfusion in rats, *J. Cereb. Blood Flow Metab.* **16**:100–106.

Kirsch, J. R., Helfaer, M. A., Haun, S. E., Koehler, R. C., and Traystman, R. J., 1993, Polyethylene glycol-conjugated superoxide dismutase improves recovery of post ischemic hypercapnic cerebral blood flow in pigs, *Pediatr. Res.* **34**:530–537.

Kondo, Y., Ogawa, N., Asanuma, M., Ota, Z., and Mori, A., 1995, Regional differences in late-onset iron deposition, ferritin, transferrin, astrocyte proliferation, and microglial activation after transient forebrain ischemia in rat brain, *J. Cereb. Blood Flow Metab.* **15**:216–226.

Kontos, C. D., Wei, E. P., Williams, J. I., Kontos, H. A., and Povlishock, J. T., 1992, Cytochemical detection of superoxide in cerebral inflammation and ischemia in vivo, *Am. J. Physiol. Heart Circ. Physiol.* **263**:H1234–H1242.

Kuboyama, K., Safar, P., Radovsky, A., Tisherman, S. A., Stezoski, S. W., and Alexander, H., 1993, Delay in cooling negates the beneficial effect of mild resuscitative cerebral hypothermia after cardiac arrest in dogs: A prospective, randomized study, *Crit. Care Med.* **21**:1348–1358.

Kuluz, J. W., Gregory, G. A., Han, Y., Dietrich, W. D., and Schleien, C. L., 1993, Fructose-1,6-bisphosphate reduces infarct volume after reversible middle cerebral artery occlusion in rats, *Stroke* **24**:1576–1583.

Lafon-Cazal, M., Culcasi, M., Gaven, F., Pietri, S., and Bockaert, J., 1993a, Nitric oxide, superoxide and peroxynitrite: Putative mediators of NMDA-induced cell death in cerebellar granule cells, *Neuropharmacology* **32:**1259–1266.

Lafon-Cazal, M., Pietri, S., Culacazi, M., and Bockaert, J., 1993b, NMDA-dependent superoxide production and neurotoxicity, *Nature* **364:**535–537.

Lancelot, E., Callebert, J., Revaud, M. L., Boulu, R. G., and Plotkine, M., 1995, Detection of hydroxyl radicals in rat striatum during transient focal cerebral ischemia: Possible implication in tissue damage, *Neurosci. Lett.* **197:**85–88.

Leffler, C. W., Busijia, D. W., Armstead, W. M., Shankin, D. R., Mirro, R., and Thelin, O., 1990, Activated oxygen and arachidonate effects on newborn cerebral arterioles, *Am. J. Physiol.* **259:**H1230–H1238.

Lei, S. Z., Pan, Z.-H., Aggarwal, S. K., Chen, H.-S. V., Hartman, J., Sucher, N. J., and Lipton, S. A., 1992, Effect of nitric oxide production on the redox modulatory site of the NMDA receptor-channel complex, *Neuron* **8:**1087–1099.

Li, Y., Kawamura, S., Yasui, N., Shirasawa, M., and Fukasawa, H., 1994, Therapeutic effects of nilvadipine on rat focal cerebral ischemia, *Exp. Brain Res.* **99:**1–6.

Lindsberg, P. J., Yue, T. L., Frerichs, K. U., Hallenbeck, J. M., and Feuerstein, G., 1990, Evidence for platelet-activating factor as a novel mediator in experimental stroke in rabbits, *Stroke* **21:**1452–1457.

Linnik, M. D., Zobrist, R. H., and Hatfield, M. D., 1993, Evidence supporting a role for programmed cell death in focal cerebral ischemia in rats, *Stroke* **24:**2002–2008.

Lipton, S. A., Choi, Y. B., Pan, Z. H., Lei, S. Z., Chen, H. S. V., Sucher, N. J., Loscalzo, J., Singel, D. J., and Stamler, J. S., 1993, A redox-based mechanism for the neuroprotective and neurodestructive effects of nitric oxide and related nitroso-compounds, *Nature* **364:**626–632.

Liu, T. H., Beckman, J. S., Freeman, B. A., Hogan, E. L., and Hsu, C. Y., 1989, Polyethylene glycol-conjugated superoxide dismutase and catalase reduce ischemic brain injury, *Am. J. Physiol.* **256:**H589–H593.

Lo, W. D., and Betz, A. L., 1986, Oxygen free-radical reduction of brain capillary rubidium uptake, *J. Neurochem.* **46:**394–398.

Lorek, A., Takei, Y., Cady, E. B., Wyatt, J. S., Penrice, J., Edwards, A. D. Peebles, D., Wylezinska, M., Owen-Reece, H., Kirkbridege, V., Cooper, E. E., Aldridge, R. F., Roth, S. C., Brown, G., Delpy, D. T., and Reynolds. E. O. R., 1994, Delayed ("secondary") cerebral energy failure after acute hypoxia–ischemia in the newborn piglet: Continuous 48-hour studies by phophorous magnetic resonance spectroscopy, *Pediatr. Res.* **36:**699–706.

Manzoni, O., Prezeau, L., Marin, P., Deshager, S., Bockaert, J., and Fagni, L., 1992, Nitric oxide-induced blockade of NMDA receptors, *Neuron* **8:**653–662.

Markarian, G. Z., Lee, J. H., Stein, D. J., and Hong, S. C., 1996, Mild hypothermia: Therapeutic window after experimental cerebral ischemia. *Neurosurgery* **38:**542–550.

Marro, P. J., McGowan, J. E., Razdan, B., Mishra, O. P., and Delivoria-Papadopoulos, M., 1994, Effect of allopurinol on uric acid levels and brain cell membrane Na^+,K^+-ATPase activity during hypoxia in newborn piglets, *Brain Res.* **650:**9–15.

Matsumoto, T., Pollock, J. S., Nakane, M., and Förstermann, U., 1993, Developmental changes of cytosolic and particulate nitric oxide synthase in rat brain, *Dev. Brain Res.* **73:**199–203.

Matsuo, Y., Onodera, H., Shiga, Y., Nakamura, M., Ninomiya, M., Kihara, T., and Kogure, K., 1994, Correlation between myeloperoxidase-quantified neutrophil accumulation and ischemic brain injury in the rat: Effects of neutrophil depletion, *Stroke* **25:**1469–1475.

Matsuo, Y., Kihara, T., Ikeda, M., Ninomiya, M., Onodera, H., and Kogure, K., 1995, Role of neutrophils in radical production during ischemia and reperfusion of the rat brain: Effect of neutrophil depletion on extracellular ascorbyl radical formation, *J. Cereb. Blood Flow Metab.* **15:**941–947.

McCord, J. M., 1985, Oxygen-derived free radicals in postischemic tissue injury, *N. Eng. J. Med.* **312:**159–163.

McNeill, H., Williams, C., Guan, J., Dragunow, M., Lawlor, P., Sirimanne, E., Nikolics, K., and Gluckman, P., 1994, Neuronal rescue with transforming growth factor-betal after hypoxic–ischemic brain injury, *Neuroreport* **5:**901–904.

McRae, A., Gilland, E., Bona, E., and Hagberg, H., 1995, Microglia activation after neonatal hypoxic-ischemia, *Dev. Brain Res.* **84:**245–252.

Meadows, M. E., Fisher, M., and Minematsu, K., 1994, Delayed treatment with a noncompetitive NMDA antagonist, CNS-1002, reduces infarct size in rats, *Cerebrovasc. Dis.* **4:**26–31.

Mishra, O. P., and Delivoria-Papadopoulos, M. D., 1988, Na+,K+-ATPase in developing fetal guinea pig brain and the effect of maternal hypoxia, *Neurochem. Res.* **13:**765–770.

Moison, R., Palinckx, J., Roest, M., Houdkamp, E., and Berger, H. M., 1993, Induction of lipid peroxidation of pulmonary surfactant by plasma of preterm babies, *Lancet* **341:**79–82.

Moorcraft, J., Bolas, N. M., Ives, N. K., Ouwerkerk, R., Smyth, J., Rajagopalan, B., Hope, P. L., and Radda, G. K., 1991, Global and depth resolved phosphorus magnetic resonance spectroscopy to predict outcome after birth asphyxia, *Arch. Dis. Child.* **66:**1119–1123.

Moos, T., 1995, Developmental profile of non-heme iron distribution in the rat brain during ontogenesis, *Dev. Brain Res.* **87:**203–213.

Mousa, S. A., Ritger, R. C., and Smith, R. D., 1992, Efficacy and safety of deferoxamine conjugated to hydroxyethyl starch, *J. Cardiovasc. Pharmacol.* **19:**425–429.

Murphy, T. H., Schnaar, R. L., and Coyle, J. T., 1990, Immature cortical neurons are uniquely sensitive to glutamate toxicity by inhibition of cysteine uptake, *FASEB J.* **4:**1624–1633.

Nathan, C., 1992, Nitric oxide as a secretory product of mammalian cells, *FASEB J.* **6:**3051–3064.

Nelson, C. W., Wei, E. P., Povlishock, J. T., Kontos, H. A., and Moskowitz, M. A., 1992, Oxygen radicals in cerebral ischemia, *Am. J. Physiol. Heart Circ. Physiol.* **263:**H1356–H1362.

Nishida, A., Misaki, Y., Kuruta, H., and Takashima, S., 1994, Developmental expression of copper, zinc-superoxide dismutase in human brain by chemiluminescence, *Brain Dev.* **16:**40–43.

Northington, F. J., Tobin, J. R., Koehler, R. C., and Traystman, R. J., 1995, In vivo production of nitric oxide correlates with NMDA-induced cerebral hyperemia in newborn sheep, *Am. J. Physiol. Heart Circ. Physiol.* **269:**H215–H221.

Nowicki, J. P., Duval, D., Poignet, H., and Scatton, B., 1991, Nitric oxide mediates neuronal death after focal cerebral ischemia in the mouse, *Eur. J. Pharmacol.* **204:**339–340.

Ohno, M., Aotani, H., and Shimada, M., 1995, Glial responses to hypoxic/ischemic encephalopathy in neonatal rat cerebrum, *Dev. Brain Res.* **84:**294–298.

Oka, A., Belliveau, M. J., Rosenberg, P. A., and Volpe, J. J., 1993, Vulnerability of oligodendroglia to glutamate: Pharmacology, mechanisms, and prevention, *J. Neurosci.* **13:**1441–1453.

Omene, J. A., Longe, A. C., Ihongbe, J. C., Glew, R. H., and Holzman, I. R., 1981, Decreased umbilical cord serum ceruloplasmin concentrations in infants with hyaline membrane disease, *J. Ped.* **99:**136–138.

Ostwald, K., Hagberg, H., Andiné, P., and Karlsson, J. O., 1993, Upregulation of calpain activity in neonatal rat brain after hypoxic-ischemia, *Brain Res.* **630:**289–294.

Palmer, C., 1996, Brain injury in the neonatal rat is reduced by neutrophil depletion induced before but not after a hypoxic ischemic insult, *Pediatr. Res.* **39:**378A.

Palmer, C., and Roberts, R. L., 1991, Reduction of perinatal brain damage with oxypurinol treatment after hypoxic–ischemic injury, *Pediatr. Res.* **29:**362A.

Palmer, C., Brucklacher, R. M., Christensen, M. A., and Vannucci, R. C., 1990, Carbohydrate and energy metabolism during the evolution of hypoxic–ischemic brain damage in the immature rat, *J. Cereb. Blood Flow Metab.* **10:**227–235.

Palmer, C., Pavlick, G., Karley, D., Roberts, R. L., and Connor, J. R., 1993a, The regional localization of iron in the cerebral cortex of the immature rat: relation to hypoxic–ischemic injury, *Pediatr. Res.* **33:** 374A.

Palmer, C., Towfighi, J., Roberts, R. L., and Heitjan, D. F., 1993b, Allopurinol administered after inducing hypoxia–ischemia reduces brain injury in 7-day-old rats, *Pediatr. Res.* **33:**405–411.

Palmer, C., Horrell, L., and Roberts, R. L., 1994a, Inhibition of nitric oxide synthase after cerebral hypoxia ischemia reduces brain swelling in neonatal rats: a dose response study, *Pediatr. Res.* **35:**385A.

Palmer, C., Roberts, R. L., and Bero, C., 1994b, Deferoxamine posttreatment reduces ischemic brain injury in neonatal rats, *Stroke* **25:**1039–1045.

Palmer, C., Roberts, R. L., and Young, P., 1996, The reduction of hypoxic–ischemic brain injury in the 7-day-old rat with PEG-SOD, *Pediatr. Res.* **39:**379A.

Panetta, J. A., and Clemens, J. A., 1994, Novel antioxidant therapy for cerebral ischemia-reperfusion injury, *Ann. NY Acad. Sci.* **723:**239–245.

Panetta, T., Marcheselli, V. L., Braquet, P., Spinnewyn, B., and Bazan, N. G., 1987, Effects of a platelet activating factor antagonist (BN 52021) on free fatty acids, diacylglycerols, polyphosphoinositides and blood flow in the gerbil brain: Inhibition of ischemia-reperfusion induced cerebral injury, *Biochem. Biophys. Res. Commun.* **149:**580–587.

Park, C. K., and Rudolphi, K. A., 1994, Antiischemic effects of propentofylline (HWA 285) against focal cerebral infarction in rats, *Neurosci. Lett.* **178:**235–238.

Parkinson, F. E., Rudolphi, K. A., and Fredholm, B. B., 1994, Propentofylline: a nucleoside transport inhibitor with neuroprotective effects in cerebral ischemia, *Gen. Pharmacol.* **25:**1053–1058.

Patt, A., Horesh, I. R., Berger, E. M., Harken, A. H., and Repine, J. E., 1990, Iron depletion or chelation reduces ischemia/reperfusion-induced edema in gerbil brains, *J. Pediatr. Surg.* **25:**224–228.

Phillis, J. W., 1989, Oxypurinol attenuates ischemia-induced hippocampus damage in the gerbil, *Brain Res. Bull.* **23:**467–470.

Phillis, J. W., and Sen, S., 1993, Oxypurinol attenuates hydroxyl radical production during ischemia/reperfusion injury of the rat cerebral cortex: An ESR study, *Brain Res.* **628:**309–312.

Phillis, J. W., Sen, S., and Cao, X., 1994, Amflutizole, a xanthine oxidase inhibitor, inhibits free radical generation in the ischemic/reperfused rat cerebral cortex, *Neurosci. Lett.* **169:**188–190.

Piantadosi, C. A., and Zhang, J., 1996, Mitochondrial generation of reactive oxygen species after brain ischemia in the rat, *Stroke* **27:**327–331.

Pourcyrous, M., Leffler, C. W., Bada, H. S., Korones, S. B., and Busija, D. W., 1993, Brain superoxide anion generation in asphyxiated piglets and the effect of indomethacin at therapeutic dose, *Ped. Res.* **34:**366–369.

Pulsinelli, W. A., Brierley, J. B., and Plum, F., 1982, Temporal profile of neuronal damage in a model of transient forebrain ischemia, *Ann. Neurol.* **11:**491–498.

Qi, Y., Jamindar, T. M., and Dawson, G., 1995, Hypoxia alters iron homeostasis and induces ferritin synthesis in oligodendrocytes, *J. Neurochem.* **64:**2458–2464.

Radi, R., Beckman, J. S., Bush, K. M., and Freeman, B. A., 1991, Peroxynitrite-induced membrane lipid peroxidation: The cytotoxic potential of superoxide and nitric oxide, *Arch. Biochem. Biophys.* **288:**481–487.

Ratan, R. R. Murphy, T. H., and Baraban, J. M., 1994, Oxidative stress induces apoptosis in embryonic cortical neurons, *J. Neurochem.* **62:**376–379.

Razdan, B., Marro, P. J., Tammela, O., Goel, R., Mishra, O. P., and Delivoria, P. M., 1993, Selective sensitivity of synaptosomal membrane function to cerebral cortical hypoxia in newborn piglets, *Brain Res.* **600:**308–314.

Reif, D. W., and Simmons, R. D., 1990, Nitric oxide mediates iron release from ferritin, *Archiv. Biochem. Biophys.* **283:**537–541.

Reilly, P. M., Schiller, H. J., and Bulkley, G. B., 1991, Pharmacologic approach to tissue injury mediated by free radicals and other reactive oxygen metabolites, *Am. J. Surg.* **161:**488–503.

Rice, J. E., Vannucci, R. C., and Brierley, J. B., 1981, The influence of immaturity on hypoxic–ischemic brain damage in the rat, *Ann. Neurol.* **9:**131–141.

Rordorf, G., Uemura, Y., and Bonventre, J. V., 1991, Characterization of phospholipase A2 (PLA2) activity in gerbil brain: Enhanced activities of cytosolic, mitochondrial and microsomal forms after ischemia and reperfusion, *J. Neurosci.* **11:**1829–1836.

Rosenberg, A. A., Murdaugh, E., and White, C. W., 1989, The role of oxygen free radicals in postasphyxia cerebral hypoperfusion in newborn lambs, *Pediatr. Res.* **26:**215–219.

Rosenthal, R. E., Chanderbhan, R., Marshall, G., and Fiskum, G., 1992a, Prevention of post-ischemic brain lipid conjugated diene production and neurological injury by hydroxyethyl starch-conjugated deferoxamine, *Free Rad. Biol. Med.* **12:**29–33.

Rosenthal, R. E., Williams, R., Bogaert, Y. E., Getson, P. R., and Fiskum, G., 1992b, Prevention of post-ischemic canine neurological injury through potentiation of brain energy metabolism by acetyl-L-carnitine, *Stroke* **23:**1312–1318.

Roskams, A., and Connor, J. R., 1994, Iron, transferrin, and ferritin in the rat brain during development and aging, *J. Neurochem.* **63:**709–716.

Roth, S. C., Edwards, A. D., Cady, E. B., Delpy, D. T., Wyatt, J. S., Azzopardi, D., Baudin, J., Townsend, J.,

Stewart, A. L., and Reynolds, E. O., 1992a, Relation between cerebral oxidative metabolism following birth asphyxia, and neurodevelopmental outcome and brain growth at one year [see comments], *Develop. Med. Child Neurol.* **34:**285–295.

Roth, S. C., Edwards, A. D., Cady, E. B., Delpy, D. T., Wyatt, J. S., Azzopardi, D., Baudin, J., Townsend, J., Stewart, A. L., and Reynolds, E. O. R., 1992b, Relation between cerebral oxidative metabolism following birth asphyxia and neurodevelopmental outcome and brain growth at one year, *Dev. Med. Child Neurol.* **34:**285–295.

Rothman, R., 1992, Cellular pool of transient ferric iron, chelatable by deferoxamine and distinct from ferritin, that is involved in oxidation cell injury, *Mol. Pharmacol.* **42:**703–710.

Rothman, S. M., and Olney, J. W., 1986, Glutamate and the pathophysiology of hypoxic–ischemic brain damage, *Ann. Neurol.* **19:**105–111.

Royall, J. A., Kooy, N. W., Ye, Y. Z., Kelly, D. R., and Beckman, J. S., 1994, Evidence of peroxynitrite in adult respiratory distress syndrome, *Pediatr. Res.* **35:**57A.

Sadrzadeh, S. M. H., and Eaton, J. W., 1988, Hemoglobin-mediated oxidant damage to the central nervous system requires endogenous ascorbate, *J. Clin. Invest.* **82:**1510–1515.

Sadrzadeh, S. M. H., and Eaton, J. W., 1992, Hemoglobin-induced oxidant damage to the central nervous system, in: *Free Radical Mechanisms of Tissue Injury* (M. T. Moslen and C. V. Smith, eds.), CRC Press, Boca Raton, Florida, pp. 23–32.

Sadrzadeh, S. M. H., Graf, E., Panter, S. S. Hallaway, P. E., and Eaton, J. W., 1984, Hemoglobin: A biological fenton reagent, *J. Biol. Chem.* **259:**14354.

Sadrzadeh, S. M. H., Anderson, D. K., Panter, S. S., Hallaway, P. E., and Eaton, J. W., 1987, Hemoglobin potentiates central nervous system damage, *J. Clin. Invest.* **79:**662–664.

Schiff, S. J., and Somjen, G. G., 1985, Hyperexcitability following moderate hypoxia in hippocampal tissue slices, *Brain Res.* **337:**337–340.

Schleien, C. L., Koehler, R. C., Shaffner, D. H., and Traystman, R. J., 1990, Blood–brain barrier integrity during cardiopulmonary resuscitation in dogs, *Stroke* **21:**1185–1191.

Schmid-Schonbein, G. W., and Lee, J., 1995, Leukocytes in capillary flow, *Int. J. Microcirc. Clin. Exp.* **15:**255–264.

Schraufstatter, I. U., Hyslop, P. A., Hinshaw, D. B., Spragg, R. G., Sklaar, L. A., and Cochrane, C. G., 1986, Hydrogen peroxide-induced injury of cells and its prevention by inhibitors of poly (ADP-ribose) polymerase, *Proc. Natl. Acad. Sci. USA* **83:**4908–4912.

Siesjo, B., and Bengtsson, F., 1989, Calcium fluxes, calcium antagonists, and calcium-related pathology in brain ischemia, hypoglycemia, and spreading depression: A unifying hypothesis, *J. Cereb. Blood Flow Metab.* **9:**127–140.

Siesjo, B. K., 1988, Mechanisms of Ischemic Brain Damage, *Crit. Care. Med.* **16:**954–963.

Siesjo, B. K., 1992, Pathophysiology and treatment of focal cerebral ischemia. Part II: Mechanisms of damage and treatment (review), *J. Neurosurg.* **77:**337–354.

Siesjo, B. K., Agardh, C.-D., and Bengtsson, F., 1989, Free radicals and brain damage, *Cerebrovasc. Brain Metab. Rev.* **1:**165–211.

Sirimanne, E. S., Blumberg, R. M., Bossano, D., Gunning, M., Edwards, A. D., Gluckman, P. D., and Williams, C. E., 1996, The effect of prolonged modification of cerebral temperature on outcome after hypoxic–ischemic brain injury in the infant rat, *Pediatr. Res.* **39:**591–597.

Smith, C. V., Hansen, T. N., Martin, N. E., McMicken, H. W., and Elliott, S. J., 1993, Oxidant stress responses in premature infants during exposure to hyperoxia, *Pediatr. Res.* **34:**360–365.

Smith, S. E., and Meldrum, B. S., 1996, Cerebroprotective effect of lamotrigine after focal ischemia in rats, *Stroke,* **26:**117–121.

Spinnewyn, B., Blavet, N., Clostre, F., Bazan, N., and Braquet, P., 1987, Involvement of platelet-activating factor (PAF) in cerebral post-ischemic phase in mongolian gerbils, *Prostaglandins* **34:**337–349.

Stein, D. T., and Vannucci, R. C., 1988, Calcium accumulation during the evolution of hypoxic–ischemic brain damage in the immature rat, *J. Cereb. Blood Flow Metab.* **8:**834–842.

Steinberg, G. K., Kunis, D., DeLaPaz, R., and Poljak, A., 1993, Neuroprotection following focal cerebral ischemia with the NMDA antagonist dextromethorphan, has a favourable dose response profile, *Neurol. Res.* **15:**174–180.

Subbarao, K. V., and Richardson, J. S., 1990, Iron-dependent peroxidation of rat brain: a regional study, *J. Neurosci. Res.* **26:**224–232.

Sullivan, J. L., 1992, *Iron Metabolism and Oxygen Radical Injury in Premature Infants,* CRC Press, Boca Raton, Florida.

Sweeney, M. I., Yager, J. Y., Walz, W., and Juurlink, B. H. J., 1995, Cellular mechanisms involved in brain ischemia, *Can. J. Physiol. Pharmacol.* **73:**1525–1535.

Takashima, S., Juruta, H., Mito,, T., Houdou, S., Konomi, H., Yao, R., and Onodera, K., 1990, Immunohistochemistry of superoxide dismutase-1 in developing human brain, *Brain Dev.* **12:**211–213.

Tan, S., Yokoyama, Y., Dickens, E., Cash, T. G., Freeman, B. A., and Parks, D. A., 1993, Xanthine oxidase activity in the circulation of rats following hemorrhagic shock, *Free Radic. Biol. Med.* **15:**407–414.

Tan, S., Gelman, S., Wheat, J. K., and Parks, D. A., 1995, Circulating xanthine oxidase in human ischemia reperfusion, *South. Med. J.* **88:**479–482.

Tan, W., Williams, C. E., Mallard, C. E., and Gluckman, P. D., 1994, Monosialoganglioside GM1 treatment after a hypoxic–ischemic episode reduces the vulnerability of the fetal sheep brain to subsequent injuries, *Am. J. Obstet. Gynecol.* **170:**663–670.

Thiringer, K., Hrbek, A., Karlsson, K., Rosen, K. G., and Kjellmer, I., 1987, Postasphyxial cerebral survival in newborn sheep after treatment with oxygen free radical scavengers and a calcium antagonist, *Ped. Res.* **22:**62–66.

Thoresen, M., Penrice, J., Lorek, A., Cady, E. B., Wylezinska, M., Kirkbride, B., Cooper, C. E., Brown, G. C., Edwards, A. D., Wyatt, J. S., and Reynolds, E. O. R., 1995, Mild hypothermia after severe transient hypoxia–ischemia ameliorates delayed cerebral energy failure in the newborn piglet, *Pediatr. Res.* **37:**667–670.

Tomic, D., Zobundzija, M., and Meáugorac, M., 1994, Postnatal development of nicotinamide adenine dinucleotide phosphate diaphorase (NADPH-d) positive neurons in rat prefrontal cortex, *Neurosci. Lett.* **170:**217–220.

Towfighi, J., Zec, N., Yager, J., Housman, C., and Vannucci, R. C., 1995, Temporal evolution of neuropathologic changes in an immature rat model of cerebral hypoxia: A light microscopic study, *Acta Neuropathol.(Berl.)* **90:**375–386.

Traystman, R. J., Kirsch, J. R., and Koehler, R. C., 1991, Oxygen radical mechanisms of brain injury following ischemia and reperfusion, *J. Appl. Physiol.* **71:**1185–1195.

Tsukahara, T., Yonekawa, Y., Tanaka, K., Ohara, O., Watanabe, S., Kimura, T., Nishijima, T., and Taniguchi, T., 1994, The role of brain-derived neurotrophic factor in transient forebrain ischemia in the rat brain, *Neurosurgery* **34:**323–331.

Turrens, J. F., and Boveris, A., 1980, Generation of superoxide anion by the NADH dehydrogenase of bovine heart mitochondria, *Biochem. J.* **191:**421–427.

Umemura, A., Mabe, H., and Nagai, H., 1992, A phospholipase C inhibitor ameliorates postischemic neuronal damage in rats, *Stroke* **23:**1163–1166.

Umemura, K., Wada, K., Uematsu, T., Mizuno, A., and Nakashima, M., 1994, Effect of 21-aminosteroid lipid peroxidation inhibitor, U74006F, in the rat middle cerebral artery occlusion model, *Eur. J. Pharmacol.* **251:**69–74.

Valentino, K., Newcomb, R., Gadbois, T., Singh, T., Bowersox, S., Bitner, S., Justice, A., Yamashiro, D., Hoffman, B. B., Ciaranello, R., Miljanich, G., and Ramachandran, J., 1993, A selective N-type calcium channel antagonist protects against neuronal loss after global cerebral ischemia, *Proc. Natl. Acad. Sci. USA* **90:**7894–7897.

van Bel, F., Dorrepaal, C. A., Benders, M. J. N. L., Houdkamp, E., van de Bor, M., and Berger, H. M., 1994, Neurologic abnormalities in the first 24 h following birth asphyxia are associated with increasing plasma levels of free iron and TBA-reactive species, *Pediatr. Res.* **35:**388A.

Vannucci, R. C., 1990, Experimental biology of cerebral hypoxia–ischemia: Relation to perinatal brain damage, *Pediatr. Res.* **27:**317–326.

Vannucci, R. C., 1993, Experimental models of perinatal hypoxic–ischemic brain damage, *APIMS Suppl.* **101:**89–95.

Vannucci, R. C., Christensen, M. A., and Yager, J. Y., 1993, Nature, time-course, and extent of cerebral edema in perinatal hypoxic–ischemic brain damage, *Pediatr. Neurol.* **9:**29–34.

Vannucci, R. C., Yager, J. Y., and Vannucci, S. J., 1994, Cerebral glucose and energy utilization during the evolution of hypoxic–ischemic brain damage in the immature rat, *J. Cereb. Blood Flow Metab.* **14:**279–288.

Volpe, J. J., 1987, Intracranial hemorrhage: Periventricular-intraventricular hemorrhage of the premature infant, in: *Neurology of the Newborn* (W. B. S. Staff, ed.), W. B. Saunders, Philadelphia, pp. 311–361.

Volterra, A., Trotti, D., Tromba, C., and Racagni, G., 1993, Additive inhibition of glutamate uptake by arachidonic acid and oxygen free radicals via two distinct mechanisms, *Neurosci. Abstr.* **19:**1350.

Walker, P. R., Weaver, V. M., Lach, B., LeBlanc, J., and Sikorska, M., 1994, Endonuclease activities associated with high molecular weight and internucleosomal DNA fragmentation in apoptosis, *Exp. Cell Res.* **213:**100–106.

Wallis, R. A., Panisson, K. L., Henry, D., and Wasterlain, C. G., 1993, Neuroprotection against nitric oxide injury with inhibitors of ADP-ribosylation, *Neuroreport* **5:**245–248.

Watanabe, T., Yuki, S., Egawa, M., and Nishi, H., 1994, Protective effects of MCI-186 on cerebral ischemia: Possible involvement of free radical scavenging and antioxidant actions, *J. Pharmacol. Exp. Ther.* **268:**1597–1604.

Watkins, M. T., Haudenschild, C. C., Al-Badawi, H., Velazquez, F. R., and Larson, D. M., 1995, Immediate responses of endothelial cells to hypoxia and reoxygenation: An in vitro model of cellular dysfunction, *Am. J. Physiol. Heart Circ. Physiol.* **268:**H749–H758.

Watkins, M. T., Al-Badawi, H., Cardenas, R., Dubois, E., and Larson, D. M., 1996, Endogenous reactive oxygen metabolites mediate sublethal endothelial cell dysfunction during reoxygenation, *J. Vasc. Surg.* **23:**95–103.

Wei, E. P., Ellison, M. D., Kontos, H. A., and Povlishock, J. T., 1986, O_2 radicals in arachidonate-induced increased blood–brain barrier permeability to proteins, *Am. J. Physiol.* **251:**H693–H699.

Weiss, G., Goossen, B., Doppler, W., Fuchs, D., Pantopoulos, K., Werner-Felmayer, G., Wachter, H., and Hentze, M. W., 1993, Translational regulation via iron-responsive elements by the nitric oxide/NO-synthase pathway, *EMBO J.* **12:**3651–3657.

Weiss, S. J., 1989, Tissue destruction by neutrophils, *N. Engl. J. Med.* **320:**365–376.

White, B. C., Daya, A., DeGracia, D. J., O'Neil, B. J., Skjaerlund, J. M., Trumble, S., Krause, G. S., and Rafols, J. A., 1993, Fluorescent histochemical localization of lipid peroxidation during brain reperfusion following cardiac arrest, *Acta Neuropathol.* **86:**1–9.

Whiteman, M., Tritschler, H., and Halliwell, B., 1996, Protection against peroxynitrite-dependent tyrosine nitration and a_1-antiproteinase inactivation by oxidized and reduced lipoic acid, *FEBS Lett.* **379:**74–76.

Wyatt, J. S., 1993, Near-infrared spectroscopy in asphyxial brain injury (review), *Clinics in Perinatology* **20:**369–378.

Yamakawa, T., Yamaguchi, S., Niimi, H., and Sugiyama, I., 1987, White blood cell plugging and blood flow maldistribution in the capillary network of cat cerebral cortex in acute hemorrhagic hypotension: An intravital microscopic study, *Circ. Shock.* **22:**323–332.

Yasuda, H., and Nakajima, A., 1993, Brain protection against ischemic injury by nizofenone, *Cerebrovasc. Brain Metab. Rev.* **5:**264–276.

Yoshida, T., Limmroth, V., Irikura, K., and Moskowitz, M. A., 1994, The NOS inhibitor, 7-nitroindazole, decreases focal infarct volume but not the response to topical acetylcholine in pial vessels, *J. Cereb. Blood Flow Metab.* **14:**924–929.

Yoshida, T., Tanaka, M., Sotomatsu, A., and Hirai, S., 1995, Activated microglia cause superoxide-mediated release of iron from ferritin, *Neurosci. Lett.* **190:**21–24.

Zaleska, M. M., and Floyd, R. A., 1995, Regional lipid peroxidation in rat brain in vitro: Possible role of endogenous iron, *Neurochem. Res.* **10:**397–410.

Zhang, J., and Piantadosi, C. A., 1994, Prolonged production of hydroxyl radical in rat hippocampus after brain ischemia-reperfusion is decreased by 21-aminosteroids, *Neurosci. Lett.* **177:**127–130.

Zhang, J., Dawson, V. L., Dawson, T. M., and Snyder, S. H., 1994, Nitric oxide activation of poly(ADP-ribose) synthetase in neurotoxicity, *Science* **263:**687–689.

Zhang, R. L., Chopp, M., Chen, H., and Garcia, J. H., 1994a, Temporal profile of ischemic tissue damage, neutrophil response, and vascular plugging following permanent and transient (2H) middle cerebral artery occlusion in the rat, *J. Neurol. Sci.* **125:**3–10.

Zhang, R. L., Chopp, M., Li, Y., Zalonga, C., Jiang, M., Jones, M., Miyasaka, M., and Ward, P., 1994b, Anti-ICAM-1 antibody reduces ischemic cell damage after transient middle cerebral artery occlusion in the rat, *Neurology* **44:**1747–1751.

Zhang, Z. G., Chopp, M., Tang, W. X., Jiang, N., and Zhang, R. L., 1995, Postischemic treatment (2–4 h) with anti-CD11b and anti-CD18 monoclonal antibodies are neuroprotective after transient (2 h) focal cerebral ischemia in the rat, *Brain Res.* **698:**79–85.

Zhao, Q., Pahlmark, K., Smith, M. L., and Siesjö, B. K., 1994, Delayed treatment with the spin trap a-phenyl-*N-tert*-butyl nitrone (PBN) reduces infarct size following transient middle cerebral artery occlusion in rats, *Acta Physiol. Scand.* **152:**349–350.

Chapter 13

The Role of Oxidative Processes and Metal Ions in Aging and Alzheimer's Disease

Leslie A. Shinobu and M. Flint Beal

13.1. AGE-RELATED CHANGES IN CELLULAR ENERGY METABOLISM

13.1.1. Introduction

Normal aging is associated with the slow progressive development of physiologic deficits accompanied by a subtle degree of cognitive involvement. A host of reviews summarizing the major biochemical changes associated with aging have been published over the last few years (Kehrer and Lund, 1994; Shigenaga and Ames, 1994; Ames *et al.*, 1993; Beal, 1993; Stadtman, 1992; Mooradian and Wong, 1991). Comprehensive lists of the panoply of age-related changes seen in man may be found in the *CRC Handbook of Biochemistry in Aging* (Florini, 1981) and the *CRC Handbook of Physiology in Aging* (Masoro, 1981).

Despite this impressive number of observations, the sequence of events that defines the aging process has not yet been determined. Traditionally, the relevant theories were roughly divided into those favoring some kind of genetically programmed process and the so-called "somatic" theories of aging. The latter emphasized the deleterious effects of random "hits" (free radical attack, changes in immunologic function, formation of molecular crosslinks, or mutations in the nuclear or mitochondrial genome), which accumulated over a lifetime and presaged the failure of critical organs. The historical aspects of these hypotheses have been reviewed by Harman (1993), Linnane *et al.* (1992), Richter (1992), Kirkwood (1991), Cerami (1985), Hayflick (1985), Walford (1974), and Goldstein (1971).

Leslie A. Shinobu and M. Flint Beal   Neurology Service, Massachusetts General Hospital, Boston, Massachusetts 02114.

Metals and Oxidative Damage in Neurological Disorders, edited by Connor. Plenum Press, New York, 1997.

In actual fact, different species clearly possess characteristic mean and maximum life-spans (Goldstein, 1971). Therefore, there must be some hereditary determinants of the aging process (Linanne *et al.,* 1992). On the other hand, even in the absence of obvious pathology, there can be wide variations in time to death amongst members of a given species, so other factors must be capable of modifying any underlying genetics. Over the last 40 years, the concepts of oxidative damage (Harman, 1956) and mitochondrial dysfunction (Shigenaga and Ames, 1994; Warner, 1994; Linanne *et al.,* 1992; Miquel, 1992) both culled from the somatic aging theories, have gained acceptance as prime examples of such modifying factors. The essential premises of their proposed role in the aging process are as follows: a) random mutations mediated by free radicals occur in the pool of mitochondrial DNA throughout life, b) cumulative damage to the mitochondrial genome eventually results in gene mutations and/or loss of various mitochondrial gene products, c) these changes manifest as a decline in mitochondrial respiratory function, d) the decreased bioenergetic capacity of the cell leads to decreased ATP-dependent synthesis of various macromolecules essential for cell maintenance and survival, and e) progressive loss of oxidative reserve is accompanied by a gradual decline in the performance of organs, and ultimately results in death (Yen *et al.,* 1994; Linanne *et al.,* 1992).

The major set of observations supporting the concept of diminishing oxidative reserve with advancing age centers around an empiric relationship between rate of oxygen metabolism and longevity (Shigenaga and Ames, 1994), i.e., there are strong correlations, which hold across species, between increased basal metabolic rate and more rapid aging, and between slower metabolism and a longer maximum life span potential (Freidlich and Butcher, 1994). Lowering ambient temperature, restricting physical activity, and severe caloric restriction without nutrient deficiency, all markedly prolong life—ostensibly by decreasing the metabolic rate (Ames *et al.,* 1993). Larger animals consume less oxygen per body mass and live longer than smaller ones (Schapira and Cooper, 1992). A rough correlation between protein half-life and body size has been attributed to increased turnover of proteins "marked" for degradation by oxidative processes in animals with high rates of oxygen consumption (Munro, 1969).

Metabolic rate is, by definition, a function of the total amount of oxygen consumed by the organism per unit time and is dependent on organ size and tissue-specific rates of oxygen consumption. The latter are dictated, to a large extent, by mitochondrial content (Shigenaga and Ames, 1994). Of note, mitochondria in different tissues can possess significantly different characteristics (Wallace, 1992) and the dynamic nature of their properties has been emphasized by Thorsness (1992).

13.1.2. The Mitochondria and Oxidative Phosphorylation

Mitochondria produce most of the energy (ATP) required by the cell via two interrelated sets of reactions: the Krebs or tricarboxylic acid cycle and oxidative phosphorylation. The Krebs cycle takes place in the mitochondrial matrix and generates reducing equivalents in the form of NADH and $FADH_2$. These compounds then undergo oxidative phosphorylation by the electron transport chain, whose constituent enzyme complexes are located in the inner mitochondrial membrane (Figure 13.1).

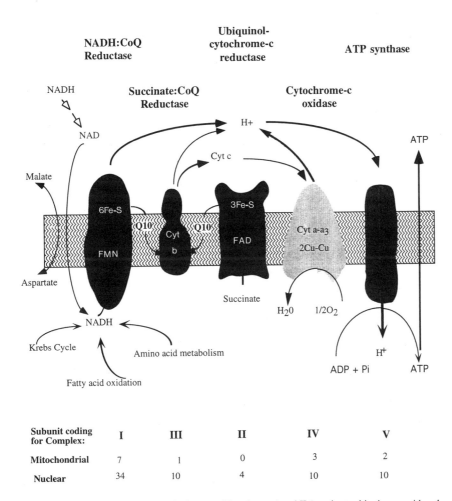

FIGURE 13.1. Complexes I (NADH:ubiquinone oxidoreductase) and II (succinate:ubiquinone oxidoreductase) collect electrons and transfer them to ubiquinone (coenzyme Q10). The electrons are sequentially transferred down an electrochemical gradient via complex III (ubiquinol:ferrocytochrome c oxidoreductase), cytochrome c, and complex IV (ferrocytochrome c:oxygen oxidoreductase or cytochrome c oxidase), finally reaching molecular oxygen, the terminal electron acceptor. Concomitant ejection of protons across the inner mitochondrial membrane results in the generation of an electrochemical proton gradient, which stores the potential energy used by complex V (ATP synthase) to condense ADP and inorganic phosphate into ATP. The ATP thus created can then be exchanged with ADP across the inner mitochondrial membrane by adenine nucleotide translocase and used in the cell cytoplasm Cyt C, cytochrome c; CoQ and Q10, Coenzyme Q10; FMN, flavin mononucleotide; FAD, flavin adenine dinucleotide, NADH, reduced nicotinamde adenine dinucleotide; NAD, nicotinamide adenine dinucleotide; ATP, Adenosine Triphosphate; ADP, Adenosine Diphosphate; Pi, inorganic phosphate; FeS and Cu, represent metal ion clusters. (Compliments of Dr. Anne B. Young, 1996.) (See Beal, 1995, for further details.)

Interestingly, all of the respiratory chain complexes, except complex II, are assemblies of polypeptide subunits encoded by both the nuclear and mitochondrial genomes (Johns, 1995). The mechanisms responsible for the coordinate transcription, translation, and transport of the nuclear DNA-encoded oxidative phosphorylation genes and their mitochondrial DNA-encoded counterparts, followed by their stoichiometric assembly in the mitochondrial membrane, are not yet understood (Johns, 1995). However, it is known that most of the mitochondrial DNA replication enzymes (i.e., those regulating the replication, transcription, and translation of mitochondrial genes) are encoded by the nucleus (Johns, 1995).

Within the mitochondria, NADH is formed by pyruvate dehydrogenase, the Krebs cycle enzymes (isocitrate, alpha-ketoglutarate, and the malate dehydrogenases), 3-hydroxyacyl-CoA dehydrogenase of the fatty acid oxidation pathway, beta-hydroxybutyrate-dehydrogenase (which participates in ketone body formation) and glutamate dehydrogenase (which plays a role in amino acid catabolism) (reviewed in Kehrer and Lund, 1994). NADH produced in the cytosol can also be transported into the mitochondria via the malate–aspartate shunt and used in the electron transport chain. Therefore, even though this chapter emphasizes the role of mitochondrial dysfunction in oxidative stress, it is important to keep in mind that defects at any of a number of levels (nuclear DNA, mitochondrial DNA, NADH production, NaK–ATPase, malate–aspartate shuttle, or changes to lipids or membrane-bound proteins that affect the mobility of ubiquinone or cytochrome c in the inner mitochondrial membrane) are potential contributors to altered oxidative metabolism and therefore, aging (Beal, 1995).

13.1.3.   Age-Related Changes in Mitochondrial DNA Structure

Under normal physiologic conditions, mitochondrial DNA (mtDNA) is continuously exposed to free radicals generated in the mitochondrial matrix as normal by-products of oxidative phosphorylation (Yen et al., 1994; Richter, 1992). MtDNA has been postulated to be particularly susceptible to free radical attack because of its paucity of non-coding sequences, lack of protective histones, proximity to the sites of free radical production in the mitochondria, and limited repair mechanisms (Richter et al., 1988). Compared to nuclear DNA, the steady-state level of oxidized bases in mtDNA seems to be at least 10-fold higher (Mecocci et al., 1994; Ames et al., 1993; Richter et al., 1988). Accordingly, in mammals, mtDNA has been shown to mutate much faster than nuclear DNA (Wallace, 1992).

The structural and functional properties of mtDNA change with advancing age. Deletions, point mutations, increases in dimeric catenated and circular forms of mtDNA, altered DNA-ethidium bromide binding properties, increased binding of proteins or peptides, and accumulation of base-pairs altered by oxidative damage, have all been described (Mecocci et al., 1993; Arnheim and Cortopassi, 1992; Randerath et al., 1992; Asano et al., 1991; Mooradian and Wong, 1991). Ozawa et al. (1990) proposed that accumulation of mtDNA mutations and their subsequent cytoplasmic segregation during the lifetime of an organism may be important for the aging process. Identification of DNA abnormalities "specific" to aging has been slow to follow, but at least 13 different mitochondrial DNA deletions have been reported in tissue from aged individuals without any of the known mitochondrial diseases (Wei, 1992). Most of these deletions seem to be clustered in the coding regions for cytochrome oxidase subunit III; the NADH dehydrogenase subunits 3, 4L, 4, 5, and 6; ATPase 6 and 8; and cytochrome *b* (Wei, 1992). One exception is the 3610 deletion that begins at base pair 1837 and includes the genes coding for 16S rRNA and NADH dehydrogenase subunits 1 and 2 (Wei, 1992).

Deletion 4977, originally identified in patients with Kearns–Sayre syndrome, is by far the most common and the best characterized of the currently accepted "age-related" mtDNA deletions (Corral-Debrinski *et al.*, 1992; Simonetti *et al.*, 1992; Soong *et al.*, 1992; Cortopassi and Arnheim, 1990). Evidence linking the presence of this deletion in the brain to increased oxidative stress consists of the following: a) Levels of dmtDNA 4977 seem to be negligible in neonatal tissue, appear by third decade of life, and increase exponentially in the seventh and eighth decades (Corral-Debrinski *et al.*, 1992; Cortopassi *et al.*, 1992; Soong *et al.*, 1992; Cortopassi and Arnheim, 1990), b) levels in different brain areas parallel known glucose utilization rates (Corral-Debrinski *et al.*, 1992), c) by far the highest ratios of dmtDNA 4977/total mtDNA are found in the caudate, putamen, and substantia nigra, where mtDNA is thought to be subject to free radicals generated by either monoamine oxidase during dopamine metabolism or significant levels of iron (Soong *et al.*, 1992), and d) the incidence of the 4977 deletion has been found to correlate with the presence of 8-hydroxy-2-deoxyguanosine, a well-accepted marker of oxidative damage to DNA (Hayakawa *et al.*, 1991). In other tissues, increased levels of the deletion are found under circumstances clearly associated with increased reactive oxygen species, e.g., in cardiac tissue of patients dying with ischemic heart disease (Corral-Debrinski *et al.*, 1992).

While discovery of these mutations has opened up a whole new concept in the neurodegenerative field, it is important to point out that acceptance of the relevance of these types of mutations in aging and diseases other than the so-called mitochondrial myopathies has been slow. This reluctance may be attributed to a number of factors. First, differences in PCR technique influence their detection rate, resulting in discrepancies between laboratories—shorter PCR cycle times have since been shown to result in the preferential amplification of the rarer deleted mitochondrial genomes over the longer and more common undeleted ones (Arnheim and Cortopassi, 1992). Second, some investigators have been perturbed by the fact that these mutations are found in only a limited number of tissues. In fact, it is not unreasonable that the largest numbers of mitochondrial deletions would accumulate in post-mitotic tissues (e.g. brain, heart, and muscle), as seems to be the case. Third, absolute levels of these deletions, even in aged tissues, are extremely low. Wei (1992) points out that since there are 2–10 copies of mtDNA per mitochondrion, and thousands of mitochondria per cell, a 0.1% incidence of a particular mtDNA deletion translates into only about 1 dmtDNA molecule per cell. However, measurements in brain homogenates may mask any striking focal regional accumulations. Fourth, even in the mitochondrial myopathies, where 20–80% of mtDNA may possess a specific 5 kb deletion, there is often no detectable effect on respiratory chain function (Holt *et al.*, 1988). Therefore, the deletions, in themselves, may not account for any observed decrease in mitochondrial oxidative capacity (Wei, 1992). Fifth, despite a definite age-related trend towards accumulation of deletions in some tissues, there is marked interindividual variation. For example, the 4977 bp deletion in liver mtDNA appeared in 62.5% of subjects between 31 and 40 years of age, but at age 50, 20% of people still did not show the deletion (Yen *et al.,* 1991). The 4977 bp deletion was not detected by Southern hybridization in the liver mtDNA of an 86–year-old subject (Wei, 1992).

In addition to mtDNA deletions, several mtDNA point mutations have recently been identified whose incidence seems to increase with age (Munscher *et al.,* 1993; Zhang *et al.,* 1993). Recent work suggests that these abnormalities are only the "tip of the iceberg" with regard to the range of mtDNA structural changes present in neuro-degenerative diseases. In fact, there may be a host of accompanying mtDNA structural alterations (large and small deletions; point mutations affecting tRNAs, rRNAs, and protein-encoding genes; and dynamic rearrangements) whose overall pattern turns out to correlate with the clinical, molecular, and histochemical phenotypes (Poulton and Holtz, 1995). The mitochondrial genome contains a great number of 4 bp or longer direct repeat sequences, all with the potential to mispair during mtDNA replication (Wei, 1992). Therefore, the list of age-associated mtDNA deletions should continue to grow (Wei, 1992). Recently, a nuclear gene defect leading to a respiratory chain complex deficiency (Bourgeron *et al.,* 1995), and an autosomal locus predisposing to deletions of mtDNA (Soumalainen *et al.,* 1995) have been described. The contribution of these types of genetic defects to the aging process has yet to be explored.

13.1.4.   Age-Related Changes in Mitochondrial Function

Many observations are consistent with an age-related decline in mitochondrial function—the data have been well-summarized by Shigenaga and Ames (1994). Importantly, respiratory chain function seems to decrease with age. The most consistent findings have been impaired complex I and/or IV activity (Boffoli *et al.,* 1994; Bowling *et al.,* 1993; Cooper *et al.,* 1992; Muller-Hocker, 1990; Trounce *et al.,* 1989). The fall in respiratory complex activity with age has been found to correlate, in certain tissues, with the general presence of mtDNA deletions (Linnane *et al.,* 1992; Sugiyama *et al.,* 1991; Yen *et al.,* 1991; Hayakawa *et al.,* 1991; Cortopassi and Arnheim, 1990). However, the sites of the age-related mtDNA deletions reported above predict abnormalities in NADH dehydrogenase, cytochrome III, or cytochrome *b* function. As suggested earlier, it may be that the full extent of mtDNA damage has just not been appreciated. The finding of age-related changes in complexes I and IV, far more than II, is consistent with mtDNA-mediated damage because, as indicated in Figure 13.1, complex II is encoded exclusively by nuclear DNA (Schapira and Cooper, 1992).

The major morphological alterations seen in aging mitochondria include enlargement, matrix vacuolization, shortened cristae, loss of dense granules, increased fragility, and changes in cardiolipin content (Shigenaga and Ames, 1994). Mitochondrial changes appear to account for essentially all the net loss of water that occurs with age in certain tissues (e.g., liver and heart), consistent with an observed age-associated increase in membrane rigidity (Shigenaga and Ames, 1994). The late-stage shrinkage of mitochondria can be mimicked by conditions of oxidative stress and may further impair mitochondrial function by limiting the lateral diffusion of proteins and impairing signal transduction (Shigenaga and Ames, 1994).

Recent studies in invertebrates have provided additional evidence for free radical involvement in aging. In *C. elegans,* the age-1 mutation is associated with increased longevity, increased resistance to oxidative stress, and increased activity of catalase and superoxide dismutase (Larsen, 1993). *Drosophila melanogaster* exposed to 100% oxygen develop progressive increases in mitochondrial protein carbonyl content and have much shorter lifespans than usual (Sohal and Dubey, 1994). Conversely, drosophila overexpressing both catalase and superoxide dismutase live longer and exhibit decreased protein oxidative damage and age-associated declines in physical performance (Orr and Sohal, 1994). Antioxidants and free radical spin traps have been shown to reverse the age-related accumulation of oxidized proteins and to restore enzyme activity in animal models of aging (Carney *et al.,* 1991). In lower species, long-term feeding of antioxidants (vitamins E, C, butylated hydroxytoluene, or mercaptoethylamine) has been shown to dramatically prolong lifespan (Shigenaga and Ames, 1994). In mammals, free radical spin traps were recently used to increase lifespan by as much as 50% (Edamatsu *et al.,* 1995).

The phenomenon of apoptosis or programmed cell death and its role in aging and neurodegeneration has been recently reviewed (Cotman and Anderson, 1995; Korsmeyer, 1995; Kroemer *et al.,* 1995; Thompson, 1995; Rothstein *et al.,* 1994). A detailed discussion is beyond the scope of this chapter, but suffice to say that it has become clear that at least two independent cytoplasmic systems can "trigger" apoptosis: one that requires the presence of mitochondria, and a second that involves the action of specific proteases (Kroemer *et al.,* 1995). In some paradigms, programmed cell death is defined as a process associated with decreasing mitochondrial energy potential that precedes loss of membrane integrity. The process can be inhibited by the anti-apoptotic proto-oncogene product Bcl-2 (Kroemer *et al.,* 1995). Bursts of reactive oxygen species production or inhibition of antioxidant pathways have been demonstrated to induce apoptosis in a number of systems. Quinone compounds that undergo redox cycles and cause superoxide radical formation, such as menadione or 2,3-dimethoxy-1,4-naphthoquinone, also provoke programmed cell death. In addition, mitochondrial function seems to be a determinant of whether cells undergo necrotic versus apoptotic cell death (Ankarcrona *et al.,* 1995; Bonfoco *et al.,* 1995).

In summary, substantial evidence supports the notion that impairments in oxidative phosphorylation result in increased free radical production and widespread oxidative damage to proteins, lipids, and nucleic acids. In addition, mitochondrial dysfunction and the level of various markers of oxidative damage in tissue clearly increase with age. A causal link between cumulative free radical-mediated damage to mtDNA and impaired mitochondrial respiratory function has been slower to follow, and the issue remains open for debate. In addition, it is possible that the so-called age-related mtDNA point mutations/deletions induce free radical-induced damage by mechanisms other than direct loss of catalytic activity in the subunits for which they code.

13.2. OXIDATIVE DAMAGE AND ALZHEIMER'S DISEASE

13.2.1. Introduction

The clinical entity referred to as Alzheimer's disease (AD) is currently being defined on a molecular basis with multiple genetic and other risk factors leading to a common phenotype with subtle variations. At present, the defining clinical features of this neurodegenerative disease include an insidious onset, impairment involving multiple areas of cognition and behavior, and progression leading to total debilitation and death (Katzman, 1993). The salient neuropathologic features include primary involvement (with atrophy) of neocortical association areas, the hippocampus and entorhinal cortex, amygdala, and nucleus basalis of Meynert, accompanied by more variable involvement of the medial nucleus of the thalamus, dorsal tegmentum, locus ceruleus, paramedian reticular areas and lateral hypothalamic nuclei (Katzman, 1989). By definition, involved regions must contain numerous neuritic plaques [focal structures consisting of an amyloid-β protein (Aβ) core surrounded by degenerating nerve terminals] and neurofibrillary tangles (neuronal cell bodies filled with paired helical filaments consisting mainly of tau protein) (Selkoe, 1994). These neuropathologic hallmarks typically appear first in the hippocampal CA1 field and subiculum, with increasing involvement of the association cortex, the basal nucleus, and other structures as the disease progresses (Katzman, 1989). The most striking neurochemical change associated with AD is a loss of choline acetyltransferase activity (Beal, 1995). Alterations in the levels of other neurotransmitters are less impressive and have been summarized by Terry and Davies (1980).

13.2.2. Risk Factors for Alzheimer's Disease

From a genetic point of view, defects in three different genes have thus far been associated with early-onset (< 60 years of age) autosomal dominant forms of familial AD with 100% penetrance. The ApoE ϵ4 phenotype seems to be a risk factor for increased Aβ deposition and sporadic AD, but is not sufficient in itself to cause the disease (Hyman and Tanzi, 1995; Roses, 1995). From an epidemiologic point of view, more than 150 factors have, at one time or another, been purported to be linked to the subsequent development of AD (Freidlich and Butcher, 1994). However, only age, family history, and past head trauma are consistently accepted as risk factors (Katzman, 1993). The relevant data have recently been reviewed by Katzman (1993) and Friedlich and Butcher (1994). A rapidly growing body of literature suggests that an underlying energy-based defect that renders specific neurons particularly prone to the sequelae of oxidative damage will turn out to be the common thread that links many of these predisposing conditions.

As described in the following sections, the results of descriptive studies already provide compelling support for a central role for oxidative damage in the pathophysiology of Alzheimer's disease.

13.2.3. Evidence for Impaired Energy Metabolism in Alzheimer's Disease

Data regarding cerebral hemodynamics, oxygen consumption, and brain function in dementia have been reviewed by Frackowiak *et al.* (1981) and Pettigrew *et al.* (1995). The majority of the studies suggest that there is a decline in both mean cerebral blood flow and mean cerebral oxygen utilization in Alzheimer's disease that correlates with the degree of dementia. In particular, positron emission tomographic studies (which provide information on glucose metabolism) and SPECT studies (which measure perfusion) have consistently shown decreases in the temporal and parietal cortices (Herholtz *et al.*, 1990; Duara *et al.*, 1986; Haxby *et al.*, 1986). There is at least one report that such changes precede significant cognitive impairment in patients examined early in their presentation (Haxby *et al.*, 1986). The parieto-temporal pattern tends to be replaced in the latest stages of the disease by profound depression in frontal regions, with relative sparing of occipital areas (Frackowiack *et al.*, 1981). The major criticism of these types of investigations has been that they do not allow differentiation between the primary disease process and effects secondary to vascular changes or neuronal cell loss.

Nuclear magnetic resonance (NMR) imaging techniques have been used to measure cerebral blood flow, lactate, n-acetylaspartate, myo-inositol, ATP and ATP metabolites, and soluble phospholipid membrane fragments in AD (Pettigrew *et al.*, 1995; Smith *et al.*, 1995; Miller *et al.*, 1993; Brown *et al.*, 1989). Differences in a number of the parameters used in these studies including the clinical definition of dementia, age of the patients, severity of the disease at the time of examination, and location and extent of the brain regions sampled, combined with technical factors involving data acquisition and processing, have all contributed to difficulties in interpreting and comparing the results of studies done by various groups. However, reductions in n-acetylaspartate and increases in myo-inositol have been reported (Pettigrew *et al.*, 1995; Miller *et al.*, 1993) and a recent study found reduced PCr/Pi ratio in the frontal cortex in patients with AD (Smith *et al.*, 1995). As technical difficulties are surmounted, it can be anticipated that NMR spectroscopic measurements of specific tissue metabolites and newer techniques such as functional NMR will provide exciting *in vivo* data on energy metabolism in AD.

13.2.4. Evidence of Mitochondrial Dysfunction in Alzheimer's Disease

The presence of mitochondrial defects that underlie changes in energy metabolism in AD is suggested by the following findings: a) abnormal mitochondria by electron microscopy in cortical biopsies showing increased matrix density and intercristal paracrystalline inclusions similar to those seen in patients with known mitochondrial diseases (Saraiva *et al.*, 1985), b) increased oxygen uptake in cerebral biopsies with unchanged levels of adenylate cyclase under conditions of submaximal metabolic activity, suggesting uncoupling of mitochondrial energy metabolism (Sims *et al.*, 1987), c) 70–100% reductions in postmortem brain tissue in the activity of the mitochondrial enzymes pyruvate dehydrogenase (Sorbi *et al.*, 1983) and 2-α-ketoglutarate dehydrogenase (Mastrogiacomo *et al.*, 1993; Butterworth and Besnard, 1990), d) reduced activity of the latter enzyme in fibroblasts from familial Alzheimer's disease patients with defined chromosome 14 defects (Sheu *et al.*, 1994), e) increased calcium uptake in mitochondria from Alzheimer fibroblasts following exposure to free radicals (Kumar *et al.*, 1994), and f) the recent discovery of mtDNA mutations with some specificity for Alzheimer's disease (Brown *et al.*, 1996; Davis *et al.*, 1997; Hutchin and Cortopassi, 1997; Shoffner *et al.*, 1993).

Three novel mtDNA point mutations have been reported in autopsy material from patients with either AD, Parkinson's disease (PD), or AD with PD (Brown *et al.*, 1996; Hutchin and Cortopassi, 1995; Shoffner *et al.*, 1993). The most common is a 4336G mutation that seems to be linked to the presence of a T → C polymorphism at position 16304 in the non-coding D-loop of the mitochondrial genome (Hutchin and Cortopassi, 1995). D-loops of these 4336G-bearing mtDNAs appeared to be evolutionarily more closely related to each other than to any other D-loop, with the mtDNA from the 6 patients examined clustering tightly into one node of the phylogenetic tree (Hutchin and Cortopassi, 1995). This raises the intriguing possibility of a heritable mitochondrial genomic contribution to "sporadic" neurodegenerative disease. In the populations examined thus far, 4336G is found at low frequencies in both young and old control groups, but at levels much lower than in the AD group (0.3–0.6% vs. 1.6–8.3%, depending on the cohort studied), and is therefore unlikely to be just a variant that is over-represented in older individuals (Hutchin and Cortopassi, 1995; Shoffner *et al.*, 1993). The significance of these particular mutations, however, still has to be clarified as they have not been found in all AD populations (Wragg *et al.*, 1995).

Complex IV defects were first reported by Parker *et al.* (1990) in AD platelets, and suggested the presence of a systemic energy-related defect. Despite some initial controversy (Van Zuylen *et al.*, 1992), this finding has now been confirmed and 20–50% reductions in the activity of this enzyme complex have been reported in affected cortical areas (Mutisya *et al.*, 1994; Parker *et al.*, 1994; Kish *et al.*, 1992), platelets from patients with Alzheimer disease (Davis *et al.*, 1997; Parker *et al.*, 1990, 1994), and cultured skin fibroblasts (Sheu *et al.*, 1994). In autopsy material, the concentration of cytochrome aa3 is normal, supporting the contention that the observed decrease in activity reflects a decline in catalytic ability, as opposed to merely being secondary to reduced neuronal activity and enzyme synthesis (Mutisya *et al.*, 1994). Alterations in the activity of other components of the electron transport chain have been less consistent.

An intriguing new piece of data and one that seems to hold considerable promise in providing a plausible link between mtDNA dysfunction, impaired energy production, and increased free radical production, is the finding of point mutations in mtDNA coding for complex IV (cytochrome oxidase, COX) of the mitochondrial electron transport chain in patients with AD. Davis *et al.* (1997) have described a distinctive pattern of six missense point mutations (three in Cox subunit I and three in subunit II coding regions), which appear with increased frequency in affected individuals. These mutations have, to date, been heteroplasmic in blood, with the relative proportion of mutated to wild-type genes ranging from 25–50%. Mutational burdens of at least 20% are found in over 60% of patients drawn from a rigorously, but clinically defined sporadic AD population, as opposed to 20% of several hundred age-matched asymptomatic individuals (Davis *et al.*, 1997). Of note, none of the familial Alzheimer disease patients examined so far have harboured this cluster of mtDNA mutations in their platelets, thus bolstering the theory that different "triggers" may lead to a final common pathway and a shared clinical phenotype.

Cytochrome oxidase function is tightly coupled to ATP production and has been used as a marker of neuronal activity (Hevner and Wong-Riley, 1993; Wong-Riley, 1989). The topographical distribution of this enzyme complex suggests that decreased cytochrome oxidase activity could contribute to the selective neuronal vulnerability seen in AD. Thus, preferential expression of subunit II mRNA in layer II and VI neurons of the entorhinal cortex, and preferential expression of cytochrome oxidase subunits I, II, and III in primate association cortex normally occurs (Chandrasekaran *et al.*, 1992). In Alzheimer's disease, there is a two-fold reduction in subunit I and III expression in the mid-temporal gyrus, with no accompanying change in mRNA levels for actin or lactate dehydrogenase (Chandrasekaran *et al.*, 1992, 1994; Hevner and Wong-Riley, 1993; Hyman *et al.*, 1990). Similarly, cytochrome oxidase subunit II mRNA is reduced throughout the hippocampal formation in AD, while expression of the nuclear encoded cytochrome oxidase subunit IV is unchanged (Simonian and Hyman, 1993). Cytochrome oxidase activity is also appropriately reduced in the perforant pathway terminal zone in AD, although this finding, in itself, could be a consequence of either neuronal loss or decreased neuronal activity (Simonian and Hyman, 1993). Interestingly, Alberts *et al.* (1992), reported that a gene homologous to cytochrome oxidase subunit II is overexpressed in AD, suggesting compensatory attempts at increased expression. In addition, as mentioned earlier, head trauma has been cited as a risk factor for AD—experimental concussive brain injury in rats is followed by a decrease in cortical cytochrome oxidase activity for up to 10 days after the insult (Hovda *et al.*, 1991). Finally, animals treated with inhibitors of cytochrome oxidase have shown cognitive deficits (Bennett *et al.*, 1992), suggesting that cytochrome oxidase deficiency could contribute to the behavioral manifestations of Alzheimer's disease (Beal, 1995).

13.2.5. Evidence for Oxidative Stress in Alzheimer's Disease

From a theoretical point of view, moderate decreases in cytochrome oxidase activity could contribute to the pathogenesis of AD by creating an underlying propensity for mitochondrial dysfunction and concomitant chronic, low levels of oxidative stress. Evidence of altered electron transport processes may not be present as long as normal homeostatic mechanisms can handle any requirements for either increased energy production or aberrant free radical generation.

Findings in autopsy tissue that could be interpreted as being consistent with impairment of mitochondrial function and/or increased oxidative stress in the AD brain include: a) a reduction in glutamine synthetase activity (Hensley *et al.*, 1995; Carney and Carney, 1994; Smith *et al.*, 1991, 1992), b) marked (80%) decreases in creatine kinase activity (Hensley *et al.*, 1995; Carney and Carney, 1994; Smith *et al.*, 1991,

1992), c) increased malondialdehyde levels (Hensley *et al.*, 1995; Lovell *et al.*, 1995; Balzacs and Leon, 1994; Hajimohammadreza and Brammer, 1990; Palmer and Burns, 1990; Subbarao *et al.*, 1990), d) increased levels of both nuclear and mitochondrial 8OH-dG, which are particularly striking in the parietal cortex (Mecocci *et al.*, 1994), e) a two-fold increase in DNA strand breaks in cerebral cortex, f) increased protein carbonyl content in the inferior parietal lobule (Hensley *et al.*, 1995), and g) changes in lipofuscin particle size in the frontal cortex (Dowson *et al.*, 1992). The significance of lipofuscin accumulation remains controversial, but the bulk of evidence seems to suggest that this process is free-radical mediated and occurs to a greater extent in AD than in normal aging.

Perhaps the most convincing evidence that the Alzheimer brain contains localized environments of elevated oxidative stress has been provided by immunohistochemical studies. Neuritic plaques and/or neurofibrillary tangles stain with antibodies to a wide variety of markers of oxidative damage (Table 13.1). Focal increases in copper-zinc superoxide dismutase (Pappolla *et al.*, 1992), manganese-superoxide dismutase (Furuta *et al.*, 1995); catalase (Pappolla *et al.*, 1992; Furuta *et al.*, 1995), ferritin and/or transferrin (Connor *et al.*, 1992a; Grundke-Iqbal *et al.*, 1990), and heme-oxygenase-1 (Schipper *et al.*, 1995; Yan *et al.*, 1994b) occur in these same areas. Several of these studies have demonstrated colocalization of two or more markers of oxidative damage. In addition, there is increased expression of receptors for advanced glycation end products in neurons and microglia near senile plaques and in the vasculature (Yan *et al.*, 1996), with clustering of activated microglia around neuritic, but not diffuse, amyloid plaques (Hensley *et al.*, 1995). A strong correlation between antioxidant enzyme activity and markers of lipid peroxidation in the medial temporal lobe (where histopathologic alterations are most severe), suggests that compensatory rises in antioxidant activity occur in response to increased free radical formation (Lovell *et al.*, 1995).

Data on the level of various endogenous intra- and extra-cellular antioxidants and enzymes with antioxidant activity in AD have been summarized in Table 13.2. Although there are many discrepancies, a trend suggests that circulating plasma antioxidants (Vitamin E, C, and A) are decreased, while tissue levels of most intracellular antioxidants and the antioxidant enzymes (glutathione peroxidase, glutathione reductase, catalase, CuZnSOD, MnSOD, and heme-oxygenase-1) are either unchanged or mildly increased.

Table 13.1

Plaques and/or Tangles as Foci of Increased
Oxidative Stress Immunohistochemical Evidence

Neurofilament-related protein carbonyls	Smith *et al.*, 1991
Tau modified by AGE products[a]	Schipper *et al.*, 1995; Smith *et al.*, 1994
3-nitrotyrosine	Beal *et al.*, unpublished
8-hydroxy-2′-deoxyguanosine	Beal *et al.*, unpublished
Pyrraline and pentosidine	Smith *et al.*, 1994b
Malondialdehyde	Yan *et al.*, 1994b; Beal *et al.*, unpublished

[a]AGE: advanced glycation end products

Table 13.2
Components of the Antioxidant System in Alzheimer's Disease

I. *Circulating Antioxidant Levels*

Vitamin E	Serum	Decreased	Zaman *et al.*, 1992
			Adams *et al.*, 1991
			Jeandel *et al.*, 1989
			Burns *et al.*, 1986
		Unchanged	Ahlskog *et al.*, 1995
Vitamin C	Serum	Decreased	Jeandel *et al.*, 1989
Vitamin A	Serum	Decreased	Jeandel *et al.*, 1989
Uric acid	Serum	Unchanged	Ahlskog *et al.*, 1995
Glutathione peroxidase	RBCs	Decreased	Jeandel *et al.*, 1989
		Unchanged	Sulkava *et al.*, 1986
		Increased	Anneren *et al.*, 1986
	CSF	Unchanged	Kaplan *et al.*, 1982
CuZnSOD	RBCs	Unchanged	Ceballos-Picot *et al.*, 1996
		Increased	Serra *et al.*, 1994
			de Lustig *et al.*, 1993

II. *The Antioxidant System in the Brain*

Vitamin E	Temporal cortex	Unchanged	Muller *et al.*, 1986
Glutathione	Many areas, frontal cortex	Unchanged	Balzacs and Leon, 1994;
			Subbarao *et al.*, 1990
			Perry *et al.*, 1987
	Hippocampus, midbrain	Increased	Adams *et al.*, 1991
Glutathione peroxidase	Many areas, hippocampus	Unchanged	Chen *et al.*, 1994
			Kish *et al.*, 1986
	Hippocampus	Increased	Lovell *et al.*, 1995
			Markesbery and Ehmann, 1994
Glutathione reductase	Hippocampus	Increased	Lovell *et al.*, 1995
			Markesbery and Ehmann, 1994
Glutathione transferase	Frontal cortex substantia inominata	Unchanged	Perry *et al.*, 1987
Catalase	Multiple areas	Decreased	Gsell *et al.*, 1995
		Unchanged	Chen *et al.*, 1993
	Hippocampus, and/or amygdala	Increased	Lovell *et al.*, 1995
			Markesbery and Ehmann, 1994
CuZnSOD	Frontal, hippocampus	Decreased	Chen *et al.*, 1994
			Richardson *et al.*, 1992
	Multiple areas	Unchanged	Gsell *et al.*, 1995
		Increased	Lovell *et al.*, 1995
			Balzacs and Leon, 1994
			Pappolla *et al.*, 1992
			Marklund *et al.*, 1985
	Fibroblasts	Increased	Zemlan *et al.*, 1989
MnSOD	Hippocampus	Unchanged	
		Increased	Marklund *et al.*, 1985
Hemeoxygenase-1	Temporal cortex, hippocampus	Increased	Schipper *et al.*, 1995
			Smith *et al.*, 1994b
			Yan *et al.*, 1994b

(*continued*)

Table 13.2

(*Continued*)

II. *The Antioxidant System in the Brain (cont.)*			
Metallothionein-II	Cortical areas	Unchanged	Erickson *et al.*, 1994
		Decreased	Palmiter *et al.*, 1992
			Tsuji *et al.*, 1992
			Uchida *et al.*, 1991
Ceruloplasmin	Temporal cortex	Decreased	Connor *et al.*, 1993

A wide variety of other enzyme activities known to be related to energy metabolism have also been screened in AD. Glucose-6-phosphate dehydrogenase (G6PD), whose main function is to supply reducing equivalents in the form of NADPH, is increased in AD cortex (Balzacs and Leon, 1994; Martins *et al.*, 1986). This has been postulated to be a secondary response to an increased demand for NADPH by the glutathione-dependent peroxide detoxifying system (Martins *et al.*, 1986) and is particularly interesting since the glycolytic enzymes and the rate of glucose utilization are markedly depressed in the AD brain (Balzacs and Leon, 1994). 6-Phosphogluconate dehydrogenase activity has also been reported to be increased (Martins *et al.*, 1986).

13.2.6. Peripheral Markers of Oxidative Damage in Alzheimer's Disease

To date, little effort has been directed at defining peripheral markers that reflect the ongoing neuropathologic damage associated with AD. Only fibroblasts from patients with AD have been examined in any detail, but a number of abnormalities consistent with oxidative stress have been found. These include decreased adhesiveness (Ueda *et al.*, 1989), approximately 30% increases in CuZnSOD activity (Zemlan *et al.*, 1989; Martins *et al.*, 1986), mitochondria showing basal impairments in calcium transport processes and increased sensitivity to oxygen-containing free radicals (Kumar *et al.*, 1994); reduced ability to repair DNA strand breaks (Kumar *et al.*, 1994), and immunoreactivity to monoclonal antibodies raised against paired helical filaments.

The only so-called markers of oxidative damage that have been examined, thus far, in patient's with AD are those that reflect lipid peroxidation. Variable results have been obtained with regard to plasma malondialdehyde levels as reflected by thiobarbituric acid reactive substances (Ahlskog *et al.*, 1995; Kalman *et al.*, 1994; Andorn *et al.*, 1990; Jeandel *et al.*, 1989). Single, unconfirmed reports of increases in lipid-conjugated dienes and trienes and decreases in n-3 and n-6 essential fatty acids in serum have been published (Corrigan *et al.*, 1993; Jeandel *et al.*, 1989). Most recently, free and bound levels of 4-hydroxy-2-nonenal, another toxic aldehydic by-product of lipid peroxidation, have been shown to be elevated in CSF taken at autopsy from AD patients (Lovell *et al.*, 1995). This finding has not yet been confirmed *in vivo*, but 4-hydroxy-2-nonenal may have some pathologic significance in AD as it has been shown to be cytotoxic to microglial cultures and to cause covalent modification and cross-linking of tau into abnormally high-molecular-weight protein species with immunoreactivity for ubiquitin reminiscent of the key components of neurofibrillary tangles (Montine *et al.,* 1996).

Other reliable markers of oxidative damage (Appendix) have not been rigorously examined in patients with Alzheimer's disease. In fact, until recently, there has been limited enthusiasm for studies emphasizing the detection of such compounds in the plasma, urine, or CSF from patients with AD. Nonetheless, there is good precedent for finding elevated levels of these markers in other disease processes, whose pathophysiology is known to be tied to the presence of reactive oxygen species. The compelling immunohistochemical data showing increased oxidative damage in the AD brain, the detection of elevated levels of markers of oxidative damage in cortical areas, and the recent suggestion of the presence of a systemic mitochondrial COX IV defect in patients with AD, all make peripheral evidence of ongoing oxidative stress an important missing link in the oxidative damage theory of AD. While any observed changes may not turn out to be specific enough for diagnostic purposes, it is anticipated that characterization of the level of such compounds in body fluids could be helpful in providing end points and measuring efficacy in clinical trials designed to slow disease progression.

13.2.7. Specific Roles for Reactive Oxygen Species in the Pathogenesis of Alzheimer's Disease

A number of reviews on neuritic plaque and neurofibrillary tangle formation have been published recently (LaFerla *et al.,* 1995; Cotman and Anderson, 1995; Poirier, 1994; Freidlich and Butcher, 1994; Kalaria *et al.,* 1992; Sisodia and Price, 1993; Trojanowski *et al.,* 1993). Therefore, only evidence consistent with a central role of altered oxidative processes in the pathogenesis of these hallmarks of AD will be presented here.

13.2.7.1. Senile Plaque Formation and β-Amyloid Toxicity

Cotman and Anderson (1995), emphasize that the plaques seen in AD are dynamic entities (as opposed to merely being the end result of a neurodegenerative process), whose composition changes over the natural history of the disease. They are composed primarily of β-amyloid protein (Aβ), accompanied by a host of other compounds (e.g., apo E, a fragment of synuclein, acetylcholinesterase, immunoglobulins, heparin sulfate proteoglycans, variable levels of different metal ions) (Brookes and St. Clair, 1994). Mutations of the APP (amyloid precursor protein) gene, which codes for the large parent polypeptide from which Aβ is derived, are determinants for AD. On the other hand, increased Aβ production is not, in itself, sufficient to cause the formation of neuritic plaques, but may confer some degree of risk. For example, there seems to be a gene dosage effect, as evidenced by the fact that individuals with trisomy 21 invariably and prematurely develop the plaques and tangles associated with AD (Rumble *et al.,* 1989), and have a high ($\approx 50\%$) incidence of developing dementia, but do not invariably become demented (Katzman, 1993). In addition, increased Aβ deposition can be a sequelae of head trauma.

The body normally produces isoforms of APP generated by alternate splicing of mRNAs. The most common amyloidogenic polypeptides have been designated APP770, APP751, APP714, and APP695. The major components of plaques, Aβ1-40 and Aβ1-42 are, in turn, produced by proteolytic cleavage of APP. The cellular function of APP is unknown, but work with different cell lines has shown that the secreted or membrane-associated forms of APP can regulate cell growth and neurite length and participate in cell–cell and cell–matrix adhesion (Multhaup *et al.,* 1996).

Cleavage of APP at the cell surface usually occurs at a site within the Aβ residue and is thought to be mediated by an enzyme (not yet isolated), referred to as α-secretase. This process releases soluble fragments with negligible potential for forming amyloid deposits. However, APP can also theoretically be cleaved by alternative enzymes, designated β- and γ-secretases, producing peptides that contain the entire Aβ sequence. The most common fragment produced is Aβ1-40. This peptide is found in almost all cells, is a normal constituent of CSF and plasma, and forms amyloid fibrils, *in vitro,* only rarely (Motter *et al.,* 1995). However, if cleavage occurs at a position only two amino acids away, the product (Aβ1-42) possesses a striking ability to form fibrillar amyloid structures. Even small amounts of Aβ1-42 have been shown to effectively seed the deposition of large amounts of Aβ1-40 (Maggio *et al.,* 1995). Whether the predominent Aβ formed is 1-40 or 1-42 appears to be a major determinant for early onset AD (Younkin, 1995).

Two major lines of evidence suggest that impaired energy metabolism and oxidative stress may contribute to the development of senile plaques. First, ATP depletion can alter APP695 processing in transfected COS cells (Gabuzda *et al.,* 1994). Inhibition of the mitochondrial electron transport chain by sodium azide or the mitochondrial uncoupler carbonyl cyanide m-chlorophenylhydrazone results in a large increase in the basal production of an 11-kDa COOH-terminal derivative containing the full length β-amyloid sequence (Gabuzda *et al.,* 1994). This processing appears to occur during transport of the precursor protein through the endoplasmic reticulum or proximal golgi apparatus on its way to the cell surface (Gabuzda *et al.,* 1994). The second link between oxidative stress and amyloid deposition comes from the demonstration that Aβ neurotoxicity seems to be mediated, at least in part, by free radicals. The introduction of Aβ fibrils into culture media causes growing neurons to produce hydrogen peroxide (Behl *et al.,* 1994b). Generation of hydrogen peroxide is blocked by the addition of flavin-type NADPH oxidase inhibitors (Schubert *et al.,* 1995). In addition, the Aβ toxicity seen in some cell cultures can be eliminated by the introduction of catalase (which destroys any peroxide formed) or free-radical scavengers such as vitamin E (Schubert *et al.,* 1995).

Aβ, and portions of Aβ, are also capable of directly producing free radicals as detected by EPR spin trapping techniques, fluorescent dyes, or salicylate trapping (Harris *et al.,* 1995; Hensley *et al.,* 1994). In hippocampal neurons in culture, no EPR radical signal was detected and toxicity was minimal when cultures were treated with freshly dissolved Aβ (Harris *et al.,* 1995). However, if Aβ1-40 was allowed to sit with cells in solution, peptide aggregation was accompanied by a transient burst of free radical production that maximized after about 6 hrs and then faded (Harris *et al.,* 1995). Parallel increases in cell protein carbonyl content (preventable by the antioxidant propyl gallate), and concurrent decreases in the activity of glutamine synthetase and creatine kinase were observed (Harris *et al.,* 1995). Schubert *et al.* (1995) demonstrated that a number of proteins associated with the human amyloidoses (amylin, calcitonin, and atrial natriuretic peptide) are all toxic in culture systems via a free radical pathway indistinguishable from that triggered by Aβ. These investigators suggested that the amphiphilic character of these proteins, rather than beta structure per se, was the key mediator of their toxicity. The latter can be abolished by the addition of the NADPH oxidase inhibitor diphenylene iodonium, the antioxidant vitamin E, or the free radical spin trap N-t-butyl-α-phenylnitrone (Schubert *et al.,* 1995).

Two other oxidative processes may play a role in Aβ toxicity. First, the presence of amino acid residues (histidine, tyrosine, methionine) that can undergo oxidation-dependent protein cross-linking reactions seems to be important for Aβ aggregation *in vitro* (Dyrks *et al.,* 1992, 1993). Soluble A4CT (an APP fragment) precipitates following the addition of hemoglobin/hydrogen peroxide, a process inhibited by the addition of free histidine, tyrosine, vitamin C, or the vitamin E derivative, trolox (Dyrks *et al.,* 1993). Second, Aβ25-35 is a potent initiator of lipoperoxidation (Butterfield *et al.,* 1994) and cell surface-bound Aβ induces oxidant stress as indicated by the generation of thiobarbituric acid reactive substances (Yan *et al.,* 1996).

Other proposed roles for Aβ in the pathogenesis of AD are linked to the process of apoptosis. Cotman and Anderson (1995) and Kroemer *et al.* (1995) have reviewed the literature supporting a causal link between oxidative damage and apoptosis in aging and neurodegeneration. Although it has been difficult to find direct evidence that cell death is mediated by apoptosis in neurodegenerative diseases, there is increasing precedent for this assumption (see Roy *et al.,* 1995; Rothstein *et al.,* 1994).

In general, apoptotic cell death seems to be associated with long-term, milder, cumulative, or synergistic insults. With regard to AD, chronic low-level triggers could include factors such as loss of neurotrophic support, Aβ accumulation, or oxidative damage (Cotman and Anderson, 1995). High ambient oxygen levels, glutathione depletion, and exposure of cells to hydrogen peroxide—all phenomenon associated with free radical production—have been shown to trigger apoptotic cell death *in vitro*. In cell culture systems, low concentrations of Aβ induce apoptotic death, while high concentrations induce necrosis (Watt *et al.,* 1994; Behl *et al.,* 1994a; Loo *et al.,* 1993). This is a phenomenon shared with a number of other neurotoxins (Lennon *et al.,* 1991; Kunimoto, 1994).

Jun expression can be induced in hippocampal neuronal cell cultures treated with Aβ. Rapid induction of jun, followed by increased membrane permeability (as reflected by loss of the ability to exclude trypan blue dye) has been demonstrated following exposure of cells to Aβ (Cotman and Anderson, 1995). In addition, jun and fos-related immunoreactivity has been reported to be increased in AD tissue compared to controls with interesting colocalization in a small percentage of NFTs and plaque-associated microglia (Anderson *et al.,* 1994).

A "knock-out" of the anti-apoptotic gene product Bcl-2 has been observed in AD cybrid cells with well-defined COX IV defects (Davis *et al.,* 1996). These cells, derived from a parental SH-SY5Y line, contain the host nuclear DNA, but have had their endogenous mtDNA replaced by mtDNA from either patients with AD or age-matched controls (Miller *et al.,* 1995). Phorbol ester-induced differentiation into AD cybrid cells with neuronal-like features resulted in a progressive decrease in COX activity and a large increase (up to 75%) in the production of reactive oxygen species (Miller *et al.,* 1995; Glasco, 1995; Lakis *et al.,* 1995). Since inhibition of cytochrome oxidase with sodium azide leads to a marked increase in the production of amyloidogenic fragments in cultured cells (Gabuzda *et al.,* 1994) a link between the COX IV defect and APP processing in the hybrid cells is currently being sought.

Finally, transgenic mouse models of AD are just beginning to be developed. Overexpression of APP or Aβ using a neurofilament light chain (NF-L) promoter in transgenic mice led to early death with extensive neuronal degeneration accompanied by morphological and biochemical evidence of apoptotic cell death (LaFerla *et al.,* 1995). However, these mice showed no evidence of extracellular amyloid, neurofibrillary tangles or senile plaques. In contrast, overexpression of APP containing the 717 mutation in transgenic mice results in amyloid deposits closely resembling the senile plaques of AD, accompanied by synaptic loss, gliosis, and neuritic dystrophy (Games *et al.,* 1995). These *in vivo* models should be useful in helping to sort out the relevance of the various *in vitro* findings described above.

13.2.7.2. Reactive Oxygen Species and tau

The degree of neurofibrillary tangle formation, seen histopathologically, correlates strongly with clinical measures of dementia, duration, and severity (Gomez-Isla, 1996). Like plaques, neurofibrillary tangles have multiple constituents but are composed of one major aberrant protein—in this case, what is thought to be an abnormally phosphorylated form of tau protein, designated A68.

A number of links between impaired oxidative stress and altered tau metabolism have recently been proposed. For example, phosphorylation of tau protein following the activation of specific kinases by decreasing cellular ATP levels, followed by neurofibrillary tangle formation (Roder *et al.,* 1993). Alternatively, Troncoso *et al.* (1993) demonstrated that iron-catalyzed oxidation of tau *in vitro* promotes the dimerization and polymerization of tau monomers, allowing the formation of insoluble bundles composed of parallel arrays of 3–5 nm filaments. Oxidation of histidine residues and tryptophan deamination were suggested to be key determinants of this process.

The capacity for self-polymerization of synthetic taulike proteins, or tau itself, also seems to depend on the availability of certain cysteine residues for oxidation (Guttman *et al.,* 1995; Schweers *et al.,* 1995). Tau–tau interactions can be decreased by blocking cysteine residues or by exposing tau to reducing (nitrogen and dithiothreitol) agents (Guttman *et al.,* 1995). Thus, tau self association decreases when the redox state of tau is increased. These authors suggest that under conditions where oxidative stress is increased, tau may form dimers through disulfide linkages, which potentiate their intermolecular interactions and encourage paired helical filament aggregation and subsequently neurofibrillary formation (Guttman *et al.,* 1995).

In summary, markers of oxidative damage are elevated in the same areas of the brain pathologically affected in AD. In particular, the plaque and/or tangle regions clearly seem to be focal regions of high oxidative stress. The presence of COX IV systemic defects with significant specificity for AD, coupled with a growing appreciation of the ability of radical-based reactions to influence critical steps in APP process-ing, Aβ handling, and even tau formation, suggests that oxidative dysfunction is important in the pathogenesis of this neurodegenerative disease.

13.3. THE ROLE OF METAL IONS IN AGING
AND ALZHEIMER'S DISEASE

The ability of metal ions to facilitate the production of reactive oxygen species and trigger lipid peroxidation, DNA damage, depletion of sulfhydryl groups, modifica-tion of amino acid residues, or perturbations of calcium homeostasis has been empha-sized in earlier chapters. However, despite the obvious potential for metal-related toxicity, there is still no direct evidence that allows the assignment of a causal role to a specific metal ion in either the aging process or the development of clinical AD.

Technical difficulties inherent in the analytic methods used to define normal endogenous levels, biochemical forms, and specific locations of metal ions within cells have been reviewed (Iyengar *et al.,* 1993; Lovell *et al.,* 1993; Markesbery and Ehmann, 1993; Xu *et al.,* 1992; Chan and Gerson, 1990) and emphasized in other chapters in this book. This information is considered to be critical in establishing causation because the inherently site-specific nature of most metal ion-facilitated reactions requires proof of their compartmentalization with a given target molecule. At the same time, it is impor-tant to remember that not all oxidative damage is mediated by metals. In addition, the metabolism of various metal ions is intertwined; zinc deficiency can cause iron accu-mulation in several tissues, aluminum binds to the iron carrier ferritin, homeostatic mechanisms seem to maintain large pools of various metalloenzymes with relatively strict stoichiometric ratios of more than one metal ion, and a number of divalent or trivalent metals are able, at least *in vitro,* to substitute for one another as cofactors of certain enzymes. Therefore, studies that examine the distribution of only one metal ion at a single point in the natural history of a disease could run the risk of erroneous bias and inaccurate conclusions.

As mentioned in section 13.1, sorting out the biochemical pathways that are key determinants of the aging process has been a difficult task. The role of metal ions, other than in facilitating oxidative reactions and contributing in a general way to a slow decline of either metabolic function or metabolic reserve with age, remains highly speculative and will not be discussed further. On the other hand, for Alzheimer's disease, specific functions for metal ions are now beginning to be postulated based on the results of *in vitro* experiments and their presumed importance has been bolstered by the recent identification of multiple metal-binding sites on APP.

Of note, several laboratories have attempted comprehensive surveys of metal ion concentrations in the brains of patients with AD compared to age-matched controls. The following global increases have been reported: confirmed—bromide, mercury, and silicon (Nikaido *et al.,* 1972; Candy *et al.,* 1986; Wenstrup *et al.,* 1990; Ehmann *et al.,* 1986), and unconfirmed—chloride and lead (Ehmann *et al.,* 1986; Hess and Straub, 1974). It is difficult, at present, to assign major significance to any of these findings. In addition, hypotheses centered around metal ions such as mercury, selenium, silicon, and trimethyl tin, have also been proposed in the literature (Marksbery and Ehmann,

1994; Chazot and Brouselle, 1993; Wenstrup *et al.*, 1990; Prohaska, 1987), but in our opinion are not likely to be key determinants of the processes involved in the development of AD. The major current theories proposing a pathogenic role for specific metal ions in AD are summarized next.

13.3.1. Aluminum

The historical aspects of the proposed connection between aluminum and Alzheimer's disease have been reviewed by Markesbery and Ehmann (1994), Xu *et al.* (1992), Chan and Gerson (1990), Hewitt *et al.* (1990), and Perl and Pendlebury (1986).

Aluminum is normally present in brain tissues at a 1–2 ppm level (Good *et al.*, 1992). Aluminum levels in the cerebrospinal fluid, serum, or hair from individuals with Alzheimer's disease are normal (Delaney, 1979; Hershey *et al.*, 1983; Shore and Wyatt, 1983). No physiological role has yet been found for this metal ion, nor can it, by itself, catalyze oxyradical generation (Good *et al.*, 1992). Early reports of increased aluminum in bulk samples from Alzheimer autopsy tissue (Crapper *et al.*, 1973) have been largely superceded by findings of highly localized accumulations of aluminum (up to 80 ppm) within hippocampal neurofibrillary tangles (Good *et al.*, 1992; Xu *et al.*, 1992) and/or senile plaques (Candy *et al.*, 1986; Tokutake *et al.*, 1995). These results have been disputed by Dedman *et al.* (1992), Landsberg *et al.* (1992) and Chafi *et al.* (1991), but the bulk of evidence seems to suggest that focal elevations of aluminum do exist. Tokutake *et al.* (1995), recently suggested that the accumulation of aluminum as aluminosilicate complexes in senile plaques and neurofibrillary tangles depends on their lipofuscin content, a variable not previously examined in any detail.

Regardless of the technical difficulties in documenting levels of aluminum in postmortem tissue or biological fluids, advocates of the aluminum hypothesis cite a number of *in vitro* observations as being supportive of some role for this metal ion in the pathogenesis of AD. Of these, the most plausible mechanism of aluminum neurotoxicity is via enhancement of iron-dependent free-radical-mediated damage. The presence of aluminum clearly exacerbates iron-catalyzed lipoperoxidation, purportedly via indirect physical effects that facilitate cross-linking of phospholipids (Hensley *et al.*, 1995). Second, aluminum has been shown to be a particularly potent inducer of protein aggregation and precipitation. This phenomenon has been demonstrated with Aβ, tau, tubulin, and a variety of endogenous and synthetic neurofilaments (Abdel-Ghanye *et al.*, 1993; Mantyh *et al.*, 1993; Bush *et al.*, 1993; Hollosi *et al.*, 1992). Mixtures of copper, zinc and aluminum, or zinc, aluminum, and iron, seem to be particularly potent "metal ion cocktails" for inducing precipitation of Aβ (Multhaup *et al.*, 1996). Third, dephosphorylation of tau protein in synaptosomal cytosolic fractions can be inhibited by aluminum and the formation of neurofibrillary tangles in cultured rat neurons has been reported following exposure to this metal ion. Fourth, acidic endocytic compartments containing enzymes capable of processing APP to amyloidogenic fragments are also sites of cellular Al and Fe uptake (Connor *et al.*, 1992b; Frederickson *et al.*, 1984). Finally, aluminum seems to be capable of mediating a variety of functions that could turn out to have some relevance in neurodegeneration. These include expression of an immunologic epitope associated with AD on human neuroblastoma cells, cross-linking of DNA strands, alteration of the activity of a number of enzyme systems, including those involved in neurotransmitter function, activation of G proteins, modulation of adenylate cyclase activity, inhibition of calcium influx via Aβ channels in synthetic lipid bilayers, and competition with calcium for calmodulin (Perl and Pendlebury, 1986).

In summary, aluminum shows some ability to modify tau/neurofilament interactions, Aβ toxicity, and iron-induced oxidative processes *in vitro.* Although the presence of this polyvalent cation in certain regions of the brain may allow it to alter some processes involved in the pathogenesis of AD, the question of whether aluminum accumulation is merely a secondary effect of neurodegeneration remains open to debate, and further work is needed to clarify any major role for this metal ion.

13.3.2. Copper

APP contains multiple cysteine and carboxylic acid residues capable of binding metal ions in a non-specific manner. However, a high affinity zinc (II) binding site within residues 181–200, and a copper (II)-binding site located between residues 135–155 have also been recently defined (Bush *et al.,* 1994a). Copper is an important component of various redox enzymes (reviewed by Stohs and Bagchi, 1995). The physiologic importance of this metal is reflected by the fact that both copper deficiency and excess are associated with distinct clinical syndromes (i.e. Wilson's disease and Menkes syndrome). The candidate genes for these diseases encode membrane proteins that share homology with bacterial membrane Cu(II) and Cd(II) transporters and with CuZnSOD. Transmembrane APP binds Cu(II), therefore, Multhaup *et al.* (1996) have proposed that APP is involved in copper homeostasis. Accumulation of APP in neurites could then lead to disruption of copper compartmentalization and toxicity (Multhaup *et al.,* 1996).

Both Zn(II) and Cu(II) binding seem to influence APP conformation, stability, and homophilic interactions (Multhaup *et al.,* 1996). Multhaup *et al.* (1996) have hypothesized that APP functions as a transcytotic receptor in neurons, facilitating the transport of Zn(II) and Cu(II) in the brain. In addition, this same group has recently shown, using matrix-assisted laser desorption mass spectrometry, that Cu(II) specifically induces the oxidation of the single cysteine residue contained within APP135–155. Of interest, there is one report of loss of more than one-third of the normal ceruloplasmin levels in the superior temporal gyrus in AD (Connor *et al.,* 1993). This protein is not only a major copper transporter, but possesses antioxidant properties, i.e., by oxidizing iron, the redox cycles necessary for the formation of either hydroxyl radical, ferryl ion, or the Fe(II): Fe(III) complex are disrupted and cellular injury is minimized (Krsek-Staples and Webster, 1993). Ceruloplasmin also functions as a weak scavenger of superoxide radicals (Connor *et al.,* 1993). Therefore, a plausible sequelae of abnormal copper metabolism would be a secondary increase in Fenton-type chemistry.

13.3.3. Iron

Alterations in iron content, transferrin, transferrin receptors, and ferritin have all been reported in AD (Loeffler *et al.,* 1995; Connor *et al.,* 1992b). The data, to date, suggest that an abnormality of iron metabolism accompanies the neuropathologic changes seen in this disease (Connor et al., 1992b). Elevated iron levels in one or more brain regions affected neuropathologically in AD have been detected using bulk brain samples by several groups (Dedman et al., 1992; Thompson, 1995). Focal increases in iron content have been reported to be associated with neurofibrillary tangles (Good et al., 1992) and senile plaques (Connor et al., 1992a; Nikaido et al., 1972).

Two patterns of iron immunoreactivity seem to be associated with plaques: a diffuse, homogenous surround, and distinct cellular staining (Connor et al., 1992b; Goodman, 1953). Dramatic increases (fivefold) in the ratio of the heavy chain isoform of ferritin to the light chain isoform are seen in AD (Connor and Menzies, 1995). In addition, a robust ferritin immunoreaction is found in close proximity to senile plaques and many blood vessels in the AD brain (Connor et al., 1992a). Interestingly, ferritin is known to be located primarily in microglial cells (Connor et al., 1992a; Grundke-Iqbal et al., 1990; Kaneko et al., 1989), and activated microglia are seen clustering around neuritic, but not diffuse, plaques. One study has noted that iron-laden processes from microglia seem to be extending into the core of some plaques (Connor et al., 1992a). Wisnieswski et al. (1989) found amyloid fibrils within the distended cisternae of the rough endoplasmic reticulum of microglia. Two possible roles for the ferritin/microglia system in amyloid formation have been proposed: (1) participating in the formation of amyloid via free radical generation, or (2) assisting in the removal and processing of amyloid (Grundke-Iqbal et al., 1990). The latter observation may explain the highly selective formation of vascular amyloid around parenchymal and meningeal blood vessels (Dyrks et al., 1992). In contrast to plaques, neurofibrillary tangles seem to contain little, if any, ferritin. Grundke-Iqbal et al. (1990), found that only tangles in the extracellular space stained ferritin positive and no ferritin was detected by Western blots in paired helical filaments isolated from AD brain.

Abnormalities of other iron carriers have also been noted. For example, transferrin immunoreactivity is also altered in AD (Connor et al., 1992a). These authors found a diffuse, homogenous reaction product surrounding the core of senile plaques, and an unexpected transferrin-positive astrocyte-associated staining in cortical areas other than the hippocampus. Loss of cortical transferrin receptors in AD has been reported by Kalaria et al. (1992). Elevated levels of another iron transporter, lactoferrin, have also been suggested by immunohistochemical studies.

Most recently, the presence of iron has been shown to modulate alpha-secretase cleavage of the amyloid precursor protein (Bodovitz et al., 1995). An iron-responsive element is present in mRNA encoding for APP (Zubenko et al., 1992). Fe(II) has been reported to induce Aβ 1-40 aggregation (Mantyh et al., 1993) by an oxidative mechanism that is abolished by the presence of antioxidants like ascorbate—unlike the Zn(II)-induced Aβ aggregation described next, which appears to be mediated primarily by changes in tertiary structure (Dyrks et al., 1992).

In summary, the potential for iron-induced lipid peroxidation in the Alzheimer brain has been emphasized by Subbarao et al. (1990) and the many empiric observations described in section 13.2. Since both ceruloplasmin and ferritin levels tend to be decreased in the brains of patients with AD, despite relatively unchanged total levels of iron, there may be some risk of the sequelae of increased levels of unbound iron (Connor et al., 1993).

13.3.4. Zinc

Zinc is actively taken up by the brain (Wensink *et al.*, 1988) and stored in synaptic vesicles in nerve terminals throughout the telencephalon (Ibata and Otsuka, 1969; Perez-Clausell and Danscher, 1985; Friedman and Price, 1984). Zinc levels in normal brain appear to be highest in the hippocampus (Hock *et al.*, 1975; Frederickson *et al.*, 1984). No statistically significant changes in zinc concentration have been reported with aging (Markesbery and Ehmann, 1994).

A number of laboratories have reported decreased levels of zinc in the temporal lobe and/or hippocampus in Alzheimer's brain (Emard *et al.*, 1995; Corrigan *et al.*, 1993; Constantinidis, 1990; Wenstrup *et al.*, 1990) although this finding is controversial (Markesbery and Ehmann, 1994). Other zinc-related changes reported in AD include an 80% elevation in cerebrospinal fluid (Hershey *et al.*, 1983), increased hepatic zinc with reduced zinc bound to metallothionein (Lui *et al.*, 1990, and decreased levels of astrocytic growth-inhibitory factor, a metallothionein-like protein that chelates zinc (Uchida *et al.*, 1991; Palmiter *et al.*, 1992; Tsuji *et al.*, 1992; Pountney *et al.*, 1994). However, there is also some dispute about these findings (Fitzgerald, 1995; Erickson *et al.*, 1994). Clinical zinc deficiency is common in Down's syndrome where, as mentioned earlier, premature AD pathology is invariable (Franceschi *et al.*, 1990). Vener *et al.* (1993) reported that Zn(II)- and Mg(II)-stimulated tyrosine kinase activities, distinct from activity of p60c-src, are decreased in AD hippocampal preparations. Despite the discrepancies, there is sufficient data to suggest that there may be an abnormality of zinc uptake in AD that results in high extracellular and low intracellular concentrations in the brain.

A pathophysiologic role for zinc in Alzheimer's disease was initially suggested by Burnet (1981) who hypothesized that many of the neurologic symptoms associated with dementia are related to a zinc deficiency and secondary impairment of the metalloenzymes associated with DNA and RNA metabolism. Chazot and Broussolle (1993) reviewed the literature and suggested several interpretations of some of the zinc data in the literature. First, intracerebroventricular administration of a cholinelike neurotoxin in rats produces nonspecific neuronal destruction and reduces the tissue levels of zinc in the hippocampus (Szerdahelyi and Kasa, 1984). Therefore, the decrease in zinc levels in the hippocampus could be a secondary phenomenon. Second, Westbrook and Mayer (1987) demonstrated that zinc selectively blocks the action of N-methyl-D-aspartate (NMDA) on cortical neurons, so endogenous zinc may normally serve a protective function, preventing the baseline level of excitatory synaptic activity from becoming neurotoxic. If this is the case, a deficit leading to reduced zinc release at excitatory synapses might then produce gradual NMDA receptor-mediated neuronal death. Zinc is known to affect immune processes, so the third hypothesis is that focal zinc deficiencies in the brain lead to local immune deficiencies and increased susceptibility to amyloid formation. Fourth, zinc deficiency can cause a secondary iron accumulation in several tissues. Finally, zinc may be involved in several components of the oxidant defense system: 1) as an essential component of CuZnSOD, 2) through its association with metallothionein, which is rich in thiolate groups with zinc binding affinity, and 3) through competition with copper and iron for membrane bindings sites and an ability to reduce the potential for hydroxyl radical formation via redox recycling (Oteiza *et al.*, 1995; Bray and Bettger, 1990).

Two observations have recently generated considerable interest. The first of these is the identification of the zinc binding site (residues 181–188) on APP and the finding that it is highly conserved among members of the APP superfamily of proteins (Bush *et al.*, 1993). This area appears to be strategically positioned between two functionally important regions: the alternately spliced Kunitz protease inhibitor sequence and the heparin-binding domain (Bush *et al.*, 1993). Therefore, the presence of zinc could keep enzymes from breaking down amyloid by sterically inhibiting access to normal cleavage sites (Bush *et al.*, 1994b).

The second observation is that Aβ solubility is dramatically lowered by the combined presence of micromolar concentrations of Zn(II) and a mildly acidotic pH (Constantinidis, 1990; Mantyh *et al.*, 1993; Bush *et al.*, 1993, 1994a). Zinc binding results in the accelerated aggregation of synthetic βA4 peptides and the formation of insoluble precipitates with the tinctorial properties of amyloid plaque (Bush *et al.*, 1994a). These findings suggest that zinc not only regulates normal APP function, but may also play a crucial role in amyloid deposition in the brain (Bush *et al.*, 1995). In support of this hypothesis, extracellular zinc concentrations are known to fluctuate up to 200–300 μMs during synaptic transmission (Frederickson *et al.*, 1984). The plausibility of this hypothesis lies in the physiologic possibility of zinc encountering soluble Aβ in the brain.

Focal differences in zinc levels may be linked with abnormal mitochondrial energy metabolism. For example, failing energy reserves in the presynaptic terminal may inhibit normal zinc reuptake mechanisms. Transient elevations in extracellular zinc levels, in combination with a mild local acidosis, would then inhibit Aβ catabolism and induce Aβ precipitation (Bush *et al.*, 1994b). Thus, a link between the impaired oxidative phosphorylation, zinc uptake in cells, and Aβ secretion is currently being studied.

13.4. SUMMARY

Peripheral Markers of Oxidative Damage in Alzheimer's Disease and Antioxidant-Based Therapeutic Strategies for Alzheimer's Disease

In conclusion, there is now a substantial body of literature indicating that free-radical-mediated processes contribute to both cytoskeletal pathology and cell death in AD. A simplified scheme showing the factors thought to be most important in this model is shown in Figure 13.2. The diagram suggests a number of treatment strategies. These include:

1) modifying the metabolism of APP /or Aβ1-42
2) decreasing free radical production
3) administering spin traps or free radical scavengers
4) replacing electron transport chain cofactors
5) inhibiting neuroinflammatory processes

Only a handful of pilot studies designed to test these hypotheses have been carried out so far. Initial encouraging results have been reported using three different ap-

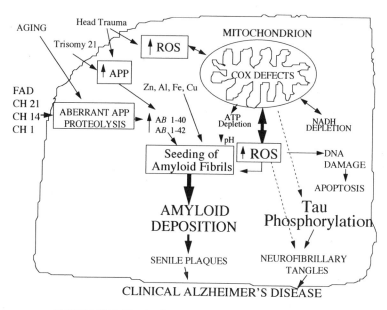

FIGURE 13.2. The development of clinical Alzheimer's disease.

proaches. Treatment with deprenyl (Eldepryl), an inhibitor of the enzyme monoamine oxidase-B, has been reported to slow the rate of cognitive decline (Finali *et al.,* 1991; Agnoli *et al.,* 1992; Knoll, 1992). The use of desferrioxamine (Desferal), an iron chelator that *in vitro* is an excellent inhibitor of iron-catalyzed generation of reactive oxygen species, also may slow the clinical progression of AD (Crapper-McLachlan *et al.,* 1991). However, desferrioxamine does not cross the blood–brain barrier, and these authors did not determine the amount or species of metal ion chelated. Nonetheless, the results were sufficiently intriguing that the study is being repeated with some modifications (Ross, 1995). Finally, patients with a clinical diagnosis of AD have been treated with a combination of coenzyme Q10 (60 mg/day), vitamin B6 (180 mg/day) and iron citrate (150 mg/day) (Imagawa *et al.,* 1992). Two of these patients had documented clinical improvement over at least six months, accompanied by increased blood flow by PET and improved electroencephalograms (EEGs) (Imagawa *et al.,* 1992). Thus, there is now considerable optimism that a therapy, based on modifying oxidative processes in the brain, will eventually prove to be useful as either the mainstay or an important adjuvant in the treatment of AD.

13.5.   REFERENCES

Abdel-Ghany, M., El-Sabae, A. K., and Shalloway, D., 1993, Aluminum-induced nonenzymic phospho-incorporation into human tau and other proteins, *J. Biol. Chem.* **268**:11976–11981.
Adams, J. D., Jr., Kaidman, L. K., Odunze, I. N., Shen, H. C., and Miller, C. A., 1991, Alzheimer's and

Parkinson's disease. Brain levels of glutathione, glutathione disulfide, and vitamin E, *Mol. Chem. Neuropathol.* **14**:213–226.

Agnoli, A., Fabbrini, G., Fioravanti, M., and Marucci, N., 1992, CBF and cognitive evaluation of Alzheimer type patients before and after IMAO-B treatment: A pilot study, *Eur. Neuropsychopharmacol.* **2**:31–35.

Ahlskog, J. E., Uitti, R. J., Low, P. A., Tyce, G. M., Nickander, K. K., Petersen, R. C., and Kokmen, E., 1995, No evidence for systemic oxidant stress in Parkinson's or Alzheimer's disease, *Mov. Dis.* **10**:566–573.

Alberts, M. J., Ioannu, P., Deucher, R., Gilbert, J., Lee, J., Middleton, L., Roses, A. D., 1992, Isolation of a cytochrome oxidase gene overexpressed secretase cleavage of amyloid precursor protein, *J. Neurochem.* **64**:307–315.

Ames, B. N., Shigenaga, M. K., and Hagen, T. M., 1993, Oxidants, antioxidants, and the degenerative diseases of aging, *Proc. Natl. Acad. Sci. USA* **90**:7915–7922.

Anderson, A. J., Cummings, B. J., and Cotman, C. W., 1994, Increased immunoreactivity for jun- and fos-related proteins in Alzheimer's disease: Association with pathology, *Exp. Neurol.* **125**:286–295.

Andorn, A. C., Britton, R. S., and Bacon, B. R., 1990, Evidence that lipid peroxidation and total iron are increased in Alzheimer's brain, *Neurobiol. Aging* **1**:316.

Ankarcrona, M., Dypbukt, J. M., Bonfoco, E., Zhivotovsky, B., Orrenius, S., Lipton, S. A., and Nicotera, P., 1995, Glutamate-induced neuronal death: A succession of necrosis or apoptosis depending on mitochondrial function, *Neuron* **15**:961–973.

Anneren, G., Gardner, A., and Lundin, T., 1986, Increased glutathione peroxidase activity in erythrocytes in patients with Alzheimer's disease/senile dementia of Alzheimer's type, *Acta. Neurol. Scand.* **73**:586–589.

Arispe, N., Rojas, E., and Pollard, H. B., 1993, Alzheimer disease amyloid beta protein forms calcium channels in bilayer membranes: Blockade by tomethamine and aluminum, *Proc. Natl. Acad. Sci. USA* **90**:567–571.

Arnheim, N., and Cortopassi, G., 1992, Deleterious mitochondrial DNA mutations accumulate in aging human tissues, *Mutat. Res.* **275**:157–167.

Asano, K., Amagase, S., Matsura, E. T., and Yamagishi, H., 1991, Changes in the rat liver mitochondrial DNA upon aging, *Mech. Ageing Dev.* **60**:275–284.

Balzacs, L., and Leon, M., 1994, Evidence of an oxidative challenge in the Alzheimer's brain, *Neurochem. Res.* **19**:1131–1137.

Beal, M. F., 1993, Neurochemical aspects of aging in primates, *Neurobiol. Aging* **14**:707–709.

Beal, M. F., 1994, Energy, oxidative damage, and Alzheimer's disease: Clues to the underlying puzzle, *Neurobiol. Aging* **15**:171–174.

Beal, M. F., 1995, Aging, energy, and oxidative stress in neurodegenerative diseases, *Ann. Neurol.* **38**:357–366.

Behl, C., Davis, J. B., Klier, F. G., and Schubert, D., 1994a, Amyloid beta peptide induces necrosis rather than apoptosis, *Brain Res.* **645**:253–264.

Behl, C., Davis, J. B., Lesley, R., and Schubert, D., 1994b, Hydrogen peroxide mediates amyloid beta protein toxicity, *Cell* **77**:817–827.

Bennett, M. C., Diamond, D. M., Stryker, S. L., Parks, J. K., and Parker, W. D. Jr., 1992, Cytocyhrome oxidase inhibition: A novel animal model of Alzheimer's disease, *J. Geriatr. Psych. Neurol.* **5**:93–101.

Bittles, A. H., 1992, Evidence for and against the causal involvement of mitochondrial DNA mutation in mammalian ageing, *Mutat. Res.* **275**:217–225.

Bodovitz, S., Dalduto, M. T., Frail, D. E., and Klein, W. L., 1995, Iron levels modulate alpha-secretase cleavage of amyloid precursor protein, *J. Neurochem.* **64**:307–315.

Boffoli, D., Scacco, S. C., Vergari, R., Solaruno, G., Santancroce, G., Papa, S., 1994, Decline with age of the respiratory chain activity in human skeletal muscle, *Biochim. Biophys. Acta* **1226**:73–82.

Bonfoco, E., Drainic, D., Ankarcrona, M., Nicotera, P., and Lipton, S. A., 1995, Apoptosis and necrosis: Two distinct events induced, respectively, by mild and intense insults with N-methyl-D-aspartate or nitric oxide/superoxide in cortical cell cultures, *Proc. Natl. Acad. Sci. USA* **92**:7162–7166.

Bourgeron, T., Rustin, P., Chretien, D., Birch-machin, M., Bourgeois, M., Viegas-Pequignot, E., Munnich, A., and Rotig, A., 1995, Mutation of a nuclear succinate dehydrogenase gene results in mitochondrial respiratory chain deficiency, *Nature Genet.* **11**:144–149.

Bowling, A. C., and Beal, M. F., 1994, Aging, energy and Alzheimer's disease, in: *Amyloid Protein*

Precursor in Development, Aging and Alzheimer's Disease (C. L. Masters, K. Beyreuther, M. Trillet, and Y. Christen, eds.), Springer-Verlag, Berlin, pp. 216–227.

Bowling, A. C., and Beal, M. F., 1995, Bioenergetic and oxidative stress in neurodegenerative diseases, *Life Sci.* **56:**1151–1171.

Bowling, A. C., Mutisya, E., Walker, L. C., Price, D. L., Cork, L. C., and Beal, M. F., 1993, Age-dependent impairment of mitochondrial function in primate brain, *J. Neurochem.* **60:**1964–1967.

Bray, T. M., and Bettger, W. J., 1990, The physiological role of zinc as an antioxidant, *Free Rad. Biol. Med.* **8:**281–291.

Brookes, A. J., and St. Clair, D., 1994, Synuclein proteins and Alzheimer's disease, *Trends Neurosci.* **17:**404–405.

Brown, G. G., Levine, S. R., Gorell, J. M., Pettegrew, J. W., Gdowski, J. W., Bueri, J. A., Helpern, J. A., and Welch, K. M., 1989, In vivo 31P NMR profiles of Alzheimer's disease and multiple subcortical infarct dementia, *Neurology* **39:**1423–1427.

Brown, M. D., Shoffner, J. M., Kim, Y. L., Jun, A. S., Graham, B. H., Cabell, M. F., Buley, D. S., and Wallace, D. C., 1996, Mitochondrial DNA sequence analysis of four Alzheimer's and Parkinson's disease patients, *Am. J. Med. Genet.* **61:**283–289.

Busciglio, J., and Yankner, B. A., 1995, Apoptosis and increased generation of reactive oxygen species in Down's syndrome neurons in vitro, *Nature* **378:**776–779.

Burnet, F. M., 1981, A possible role of zinc in the pathology of dementia, *Lancet* **i:**186–188.

Bush, A. I., Multhaup, G., Moir, R. D., Williamson, T. G., Small, D. H., Rumble, B., Pollwein, P., Beyreuther, K., and Masters, C. L., 1993, A novel zinc(II) binding site modulates the function of the βA4 amyloid protein precursor of Alzheimer's disease, *J. Biol. Chem.* **268:**16109–16112.

Bush, A. I., Pettingell, W. H., Multhaup, G., Paradis, M., Vonsattel, J.-P., Gusella, J. F., Beyreuther, K., Masters, C. L., and Tanzi, R. E., 1994a, Rapid induction of Alzheimer Aβ formation by zinc, *Science* **265:**1464–1467.

Bush, A. I., Pettingell, W. H., Paradis, M. D., and Tanzi, R. E., 1994b, Modulation of Aβ adhesiveness and secretase site cleavage by zinc, *J. Biol. Chem.* **269:**12152–12158.

Bush, A. I., Moir, R. D., Rosenkranz, K. M., and Ranzi, R. E., 1995, Zinc and Alzheimer's disease, *Science* **268:**1921–1922.

Butterfield, D. A., Hensley, K., Harris, M., Mattson, M., and Carney, J., 1994, β-amyloid peptide free radical fragments initiate synaptosomal lipoperoxidation in a sequence-specific fashion: Implications to Alzheimer's disease, *Biochem. Biophys. Res. Comm.* **200:**710–715.

Butterworth, R. F., and Besnard, A. M., 1990, Thiamine-dependent enzyme changes in temporal cortex of patients with Alzheimer's disease, *Metab. Brain Dis.* **5:**179–184.

Candy, J. M., Oakley, A. E., Klinowski, J., Carpenter, R. A., Perry, R. H., Atack, J. R., Perry, E. K., Blessed, G., Fairbairn, A., and Edwardson, J. A., 1986, Aluminosilicates and senile plaque formation in Alzheimer's disease, *Lancet* **i:**354–357.

Carney, J. M., Starke-Reed, P. E., Oliver, C. N., Landum, R. W., Cheng, J. M., Wu, J. F., and Floyed, R. A., 1991, Reversal of age-related increase in brain-protein oxidation, decrease in enzyme activity, and loss in temporal and spatial memory by chronic administration of the spin-trapping compound N-tert-butyl-α-phenylnitrone, *Proc. Natl. Acad. Sci. USA* **88:**3833–3636.

Carney, J. M., and Carney, A. M., 1994, Role of protein oxidation in aging and in age-associated neurodegenerative diseases, *Life Sci.* **55:**2097–2103.

Ceballos-Picot, I., Merad-Boudia, M., Nicole, A., Thevenin, M., Hellier, G., Legrain, S., and Berr, C., 1996, Peripheral antioxidant enzyme activities and selenium in elderly subjects and in dementia of Alzheimer's type–place of the extracellular glutathione peroxidase, *Free Radic. Biol. Med.* **20:**579–587.

Cerammi, A., 1985, Hypothesis: Glucose as a mediator of aging, *J. Am. Geri. Soc.* **33:**626–634.

Chafi, A. H., Haw, J.-J., Rancurel, G., Berry, J.-P., and Galle, C., 1991, Absence of aluminum in Alzheimer's disease brain tissue: Electron probe and ion microprobe studies, *Neuroscience Lett.* **123:**61–64.

Chan, S., and Gerson, B., 1990, Technical aspects of quantification of aluminum, *Clin. Lab. Med.* **10:**423–433.

Chandrasekaran, K., Stoll, J., Rapoport, S. I., and Brady, D. R., 1992, Localization of cytochrome oxidase (COX) activity and COX mRNA in the hippocampus and entorhinal cortex of the monkey brain: correlation with specific neuronal pathway, *Brain Res.* **579:**333–336.

Chandrasekaran, K., Giordano, R., Brady, D. R., Stoll, J., Martin, L. J., and Rapoport, S. I., 1994, Impairment in mitochondrial cytochrome oxidase gene expression in Alzheimer disease, *Mol. Br. Res.* **24:**336–340.

Chazot, G., and Broussolle, E., 1993, Alterations in trace elements during brain aging and in Alzheimer's dementia, *Prog. Clin. Biol. Res.* **380:**269–281.

Chen, L., Richardson, J. S., Caldwell, J. E., and Ang, L. C., 1994, Regional brain activity of free radical defense enzymes in autopsy samples from patients with Alzheimer's disease and from nondemented controls, *Intern. J. Neuroscience* **75:**83–90.

Connor, J. R., Menzies, S. L., St. Marin, S. M., and Mufson, E. J., 1992a, A histochemical study of iron, transferrin, and ferritin in Alzheimer's diseased brains, *J. Neurosci. Res.* **31:**75–83.

Connor, J. R., Snyder, B. S., Beard, J. L., Fine, R. E., and Mufson, E. J., 1992b, The regional distribution of iron in aging and Alzheimer's disease, *J. Neurosci. Res.* **31:**327.

Connor, J. R., Tucker, P., Johnson, M., and Snyder, B., 1993, Ceruloplasmin levels in the human superior temporal gyrus in aging and Alzheimer's disease, *Neuroscience Lett.* **159:**88–90.

Connor, J. R., and Menzies, S. L., 1995, Cellular management of iron in the brain, *J. Neurol. Sci.* **134S:**33–44.

Constaninidis, J., and Tissot, R., 1981, Role of glutamate and zinc in the hippocampal lesions of Picks disease, in: *Glutamate as a Neurotransmitter* (G. Dichiara and L. Gessa, eds.), Raven Press, New York. p. 413.

Constantinidis, J., 1990, Alzheimer's disease and the zinc theory, *Encephale* **16:**231–2399.

Cooper, J. M., Mann, V. M., and Schapira, A. H. V., 1992, Analyses of mitochondrial respiratory chain function and mitochondrial DNA deletion in human skeletal muscle: Effect of ageing, *J. Neurol. Sci.* **113:**91–98.

Corral-Debrinski, M., Stepien, G., Shoffner, J. M., Lott, M. T., Kanter, K., and Wallace, D., 1991, Hypoxemia is associated with mitochondrial DNA damage and gene induction: Implications for cardiac disease, *JAMA* **266:**1812–1816.

Corral-Debrinski, M., Horton, R., Lott, M. T., Shoffner, J. M., Beal, M. F., and Wallace, D. C., 1992, Mitochondrial DNA deletions in human brain: Regional variability and increase with advanced age, *Nature Genet.* **2:**324–329.

Corrigan, F. M., Van Rhijn, A., and Horrobin, D. F., 1993, Essential fatty acids in Alzheimer's disease, *Ann. New York Acad. Sci.* **640:**250–252.

Cortopassi, G. A., and Arnheim, N., 1990, Detection of a specific mitochondrial DNA deletion in tissues of older humans, *Nucleic Acids Res.* **18:**6927–6933.

Cortopassi, G. A., Shibata, D., Soong, N. W., and Arnheim, N. A., 1992, A pattern of accumulation of a somatic deletion in mitochondrial DNA of various tissues in ageing in human tissue, *Proc. Natl. Acad. Sci. USA* **89:**7370–7374.

Cotman, C. W., and Anderson, A. J., 1995, A potential role for apoptosis in neurodegeneration and Alzheimer's disease, *Mol. Neurobiol.* **10:**19–45.

Cotton, P., 1994, Constellation of risks and processes seen in the search for Alzheimer's clues, *JAMA* **271:**89–91.

Crapper, D. R., Kirshnan, S. S., and Dalton, A. J., 1973, Aluminum distribution in Alzheimer's disease and experimental neurofibrillary degeneration, *Science* **180:**511–513.

Crapper-McLachlan, D. R., Dalton, A. J., Kruck, T. P. A., Bell, M. Y., Smith, W. L., Kalow, W., and Andrews, D. R., 1991, Intramuscular desferrioxamine in patients with Alzheimer's disease, *Lancet* **337:**1304–1308.

Davis, R. E., Miller, S., Herrnstadt, C., Ghosh, S. S., Fahy, E., Shinobu, L. A., Galasko, D., Thal, L. J., Beal, M. F., Howell, N., and Parker, W. D., 1997, Mutations in mitochondrial cytochrome *c* oxidase genes segregate with late-onset Alzheimer disease, *Proc. Natl. Acad. Sci. USA* **94:**4526–4531.

Dedman, D. J., Trefry, M., Candy, J. M., Taylor, G. A. A., Morris, C. M., Bloxham, C. A., Perry, R. H., Edwardson, J. A., and Harrison, P. M., 1992, Iron and aluminum in relation to brain ferritin in normal individuals and Alzheimer's disease and chronic renal-dialysis patients, *Biochem. J.* **287:**509–514.

Delaney, J. R., 1979, Spinal fluid aluminum levels in patients with Alzheimer's disease, *Ann. Neurol.* **5:**580–581.

de Lustig, E., Serra, J. A., Kohan, S., Canziani, G. A., Famulari, A. L., and Dominguez, R. O., 1993, Copper-zinc superoxide dismutase activity in red blood cells and serum in demented patients and in aging, *J. Neurol. Sci.* **115:**18–25.

De Stefano, N., Mathews, P. M., Ford, B., Genge, A., Karpati, G., and Arnold, D. L., 1995, Short-term dicholoroacetate treatment improves indices of cerebral metabolism in patients with mitochondrial disorders, *Neurology* **45:**1193–1198.

Dexter, D. T., Sian, J., Jenner, P., and Marsden, C. D., 1993, Implications of alterations in trace element levels in brain in Parkinson's disease and other neurological disorders affecting the basal ganglia, *Adv. Neurol.* **60:**273–281.

Dowson, J. H., Mountjoy, C. Q., Cairns, M. R., and Wilton-Cox, H., 1992, Changes in intraneuronal lipopigment in Alzheimer's disease, *Neurobiol. Aging* **13:**493–500.

Duara, R., Grady, C., Haxby, J., Sundaram, M., Cutler, N. R., Heston, L., Moore, A., Schlageter, N., Larson, S., and Rapoport, S. I., 1986, Positron emission tomography in Alzheimer's disease, *Neurology* **36:**879–887.

Dyrks, T., Dyrks, E., Hartmann, T., Masters, C., and Beyreuther, K., 1992, Amyloidogeneicity of βA4 and βA4-bearing amyloid protein precursor fragments by metal-catalyzed oxidation, *J. Biol. Chem.* **267:**18210–18217.

Dyrks, T., Dyrks, E., Masters, C. L., and Beyreuther, K., 1993, Amyloidogeneicity of rodent and human βA4 sequences, *FEBS* **324:**231–236.

Eckert, A., Hartmann, H., Forstl, H., and Muller, W. E., 1994, Alterations in intracellular calcium regulation during aging and Alzheimer's disease in nonneuronal cells, *Life Sci.* **55:**2019–2029.

Edamatsu, R., Mori, A., and Packer, L., 1995, The spin-trap N-tert-α-phenyl-butylnitrone prolongs the life span of the senscence accelerated mouse, *Biochem. Biophys. Res. Comm.* **211:**847–849.

Ehmann, W. D., Markesbery, W. R., Alauddin, M., Hossain, T. I. M., and Brubaker, E. H., 1986, Brain trace elements in Alzheimer's disease, *Neurotoxicology* **7:**197–206.

Emard, J.-F., Thouez, J.-P., and Gauvreau, D., 1995, Neurodegenerative diseases and risk factors: A literature review, *Soc. Sci. Med.* **40:**847–858.

Erickson, J. C., Sewall, A. K., Jensen, L. T., Winge, D. R., and Palmiter, R. D., 1994, Enhanced neurotrophic activity in Alzheimer's disease cortex is not associated with down-regulation of metallothionein-III (GIF), *Brain Res.* **649:**297–304.

Finali, G., Piccirilli, M., Oliani, C., and Piccinin, G. L., 1991, L-deprenyl therapy improves verbal memory in amnesic Alzheimer patients, *Clin. Neuropharmacol.* **14:**523–36.

Fitzgerald, D. J., 1995, Zinc and Alzheimer's disease, *Science* **268:**1920.

Fleming, J. E., Miquel, J., and Bensh, K. G., 1985, Age-dependent changes in mitochondria, *Basic Life Sciences,* **35:**143–156.

Florini, J. R., ed., 1981, *CRC Handbook in Biochemistry of Aging,* CRC Press, Boca Raton, Florida.

Frackowiak, R. S. J., Pozzilli, C., Legg, N. G., Du Boulay, G. Y. H., Marshal, J., Lenzi, G. L., and Jones, T., 1981, Regional cerebral oxygen supply and utilization in dementia: A clinical and physiological study with oxygen-15 and positron tomography, *Brain* **104:**753–778.

Franceschi, M., Comola, M., Piattoni, R., Gualandri, W., and Canal, N., 1990, Prevalence of dementia in adult patients with trisomy 21, *Am. J. Med. Genet.* **7:**306–308.

Frederickson, C. J., Howell, G. A., and Kasarskis, E. J., 1984, *The Neurobiology of Zinc, Volumes 11A and 11B,* Alan R. Liss, New York.

Friedlich, A. L., and Butcher, L. L., 1994, Involvement of free oxygen radicals in β-amyloidosis: An hypothesis, *Neurobiol. Aging* **15:**443–455.

Friedman, B., and Price, J. L., 1984, Fiber systems in the olfactory bulb and cortex: A study in adult and developing rats, using the Timms method with the light and electron microscope, *J. Comp. Neurol.* **223:**88–109.

Fukuyama, H., Ogawa, M., Yamaguchi, H., Yamaguchi, S., Kimura, J., Yonekawa, Y., and Konishi, J. H., 1995, Altered cerebral energy metabolism in Alzheimer's disease: a PET study, *J. Nucl. Med.* **35:**1–6.

Furuta, A., Price, D. L., Pardo, C. A., Troncoso, J. C., Xu, Z.-S., Taniguchi, N., and Martin, L. J., 1995, Localization of superoxide dismutases in Alzheimer's disease and Down's syndrome neocortex and hippocampus, *Am. J. Pathol.* **146:**357–367.

Gabuzda, D., Busciglio, J., Chen, L. B., Matsudaira, P., and Yankner, B. A., 1994, Inhibition of energy metabolism alters the processing of amyloid precursor protein and induces a potentially amyloidogenic derivative, *J. Biol. Chem.* **6:**13623–13628.

Gadaleta, M. N., Rainaldi, G., Lezza, A. M., Milella, F., Fracasso, F., and Cantatore, P., 1992, Mitochondrial DNA copy number and mitochondrial DNA deletion in adult and senescent rats, *Mutat. Res.* **275:**181–193.

Games, D., Adams, D., Alessandrini, R., Barbour, R., Berthelette, P., Blackwell, C., Carr, T., Clemens, J., Donaldson, T., Gillespie, F., Guido, R., Hagoplan, S., Johnson-Wood, D., Khan, K., Lee, M., Leibowitz, P., Leiberburg, I., Little, S., Masliah, E., McConiogue, L., Montoya-Zavala, M., Mucke, L., Paganini, L., Schenk, K., Seubert, P., Snyder, B., Soriano, F., Tan, H., Vitale, J., Wadsworth, S., Wolozin, B., and Zhao, J., 1995, Alzheimer-type neuropathology in transgenic mice overexpressing V717F β-amyloid precursor protein, *Nature* **373:**523–527.

Gerlach, M., Ben-Schachter, D., Riederer, P., and Youdim, M. B. H., 1994, Altered brain metabolism of iron as a cause of neurodegenerative diseases?, *J. Neurochem.* **63:**793–807.

Glasco, S., Miller, S. W., Thal, L. J., and Davis, R. E., 1995, Alzheimer's disease cybrids manifest a cytochrome oxidase defect, *Soc. Neuroscience* **21:**979.

Goldstein, S., 1971, The biology of aging, *New Engl. J. Med.* **285:**1120–1129.

Gomez-Isla, T., West, H. L., Rebeck, G. W., Harr, S. D., Growdon, J. H., Locasio, J. J., Perls, T. T., Lipsitz, L. A., and Hyman, B. T., 1996, Clinical and pathological correlates of apoplipoprotien E ε4 in Alzheimer's disease, *Ann. Neurol.* **39:**62–70.

Good, P. F., Perl, D. P., Bierer, L. M., and Schmeidler, J., 1992, Selective accumulation of aluminum and iron in the neurofibrillary tangles of Alzheimer's disease: a laser microprobe (LAMMA) study, *Ann. Neurol.* **31:**286–292.

Goodman, L., 1953, Alzheimer's disease: A clinicopathologic analysis of twenty-three cases with a theory on pathogenesis, *J. Nerv. Ment. Dis.* **118:**97–130.

Grundke-Iqbal, I., Fleming, J., Tung, Y.-C., Lassmann, H., Iqbal, K., and Joshi, J. G., 1990, Ferritin is a component of the neuritic (senile) plaque in Alzheimer dementia, *Acta Neuropathol.* **81:**105–110.

Gsell, W., Conrad, R., Hickethier, M., Sofic, E., Frolich, L., Wichart, I., Jellinger, K., Moll, G., Ransmayr, Beckmann, H., and Riederer, P., 1995, Decreased catalase activity but unchanged superoxide dismutase activity in brains of patients with dementia of Alzheimer type, *J. Neurochem.* **64:**1216–1223.

Guttman, R. P., Erickson, A. C., and Johnson, G. V. W., 1995, τ Self-association: stabilization with a chemical cross-linker and modulation by phosphorylation and oxidation state, *J. Neurochem.* **64:**1209–1215.

Hajimohammadreza, I., and Brammer, M., 1990, Brain membrane fluidity and lipid peroxidation in Alzheimer's disease, *Neuroscience Lett.* **112:**333–337.

Harman, D., 1956, Role of free radical and radiation chemistry, *J. Gerontol.* **11:**298–300.

Harman, D., 1993, Free radical theory of aging: A hypothesis on pathogenesis of senile dementia of the Alzheimer's type, *Age* **16:**23–30.

Harris, M. E., Hensley, K., Butterfield, D. A., Leedle, R. A., and Carney, J. M., 1995, Direct evidence of oxidative injury produced by the Alzheimer's β-amyloid peptide (1-40) in cultured hippocampal neurons, *Exp. Neurol.* **131:**193–202.

Hartmann, M., Eckert, A., and Muller, W. E., 1994, Disturbances of the neuronal calcium homeostasis in the aging nervous system, *Life Sci.* **55:**2011–2018.

Haxby, J. V., Grady, C. L., and Duara, R., 1986, Neocortical metabolic abnormalities precede nonmemory cognitive deficits in early Alzheimer-type dementia, *Arch. Neurol.* **43:**882–885.

Hayakawa, M., Torii, K., Sugiyama, S., Tanaka, M., and Ozawa, T., 1991, Age-associated accumulation of 8-hydroxydeoxyguanosine in mitochondrial DNA of human diaphragm, *Biochem. Biophys. Res. Comm.* **179:**1023–1029.

Hayflick, L., 1985, Theories of biological aging, *Exp. Geront.* **20:**145–159.

Hensley, K., Carney, J. M., Mattson, M. P., Aksenova, M., Harris, M., Wu, J. F., Floyd, R. A., and Butterfield, D. A., 1994, A model for β-amyloid aggregation and neurotoxicity based on free radical generation by the peptide: relevance to Alzheimer disease, *Proc. Natl. Acad. Sci. USA* **91:**3270–3274.

Hensley, K., Hall, N., Subramanian, R., Cole, P., Harris, M., Aksenov, M., Aksenova, M., Gabbita, S. P., Wu, J. F., Carney, J. M., Lovell, M., Markesbery, W. R., and Butterfield, D. A., 1995, Brain regional correspondence between Alzheimer's disease histopathology and biomarkers of protein oxidation, *J. Neurochem.* **65:**2146–2156.

Herholtz, K., Heindel, W., Rackl, A., Neubauer, I., Steinbrich, W., Peitrzyk, U., Erasmi-Korber, H., and

Heiss, W. D., 1990, Regional cerebral blood flow in patients with leuko-ariosis and atherosclerotic carotid artery disease, *Arch. Neurol.* **47**:392–296.

Hershey, C. O., Hershey, L. A., Varnes, A., Vibhakar, S. D., Lavin, P., and Strain, W. H., 1983, Cerebrospinal fluid trace element content in dementia: clinical, radiologic, and pathologic correlations, *Neurology* **33**:1350–1353.

Hess, K., and Straub, P. W., 1974, Chronic lead poisoning, *Schweizerische Rundschau fur Medizin Praxis* **63**:177–183.

Hevner, R. F., and Wong-Riley, M. T. T., 1993, Entorhinal cortex of the human, monkey, and rat: Metabolic map as revealed by cytochrome oxidase, *J. Comp. Neurol.* **326**:451–469.

Hewitt, C. D., Savory, J., and Wills, M. R., 1990, Aspects of aluminum toxicity, *Clin. Lab. Med.* **10**:403–422.

Hock, A., Demmel, U., Schicka, H., Kasperek, K., and Feinendegen, L. E., 1975, Trace element concentration in human brain: Activation analysis of cobalt, iron, rubidium, selenium, zinc, chromium, silver, cesium, antimony and scandium, *Brain* **98**:44–64.

Hockenbery, D. M., Oltvai, Z. N., Yin, X. M., Milliman, C. L., and Korsmeyer, S. J., 1993, Bcl-2 functions in an antioxidant pathway to prevent apoptosis, *Cell* **75**:241–251.

Hollosi, M., Urge, L., Perczel, A., Kajtar, J., Teplan, I., Otvos, L. Jr., and Fasman, G. D., 1992, Metal ion-induced conformational changes of phosphorylated fragments of human neurofilament (NF-M) protein, *J. Mol. Biol.* **223**:673–682.

Holt, I. J., Harding, A. E., and Morgan-Hughes, J. A., 1988, Deletions of mitochondrial DNA in patients with mitochondrial myopathies, *Nature* **331**:717–719.

Hovda, D. A., Yoshino, A., Kawamata, T., Katayama, Y., and Becker, D. P., 1991, Diffuse prolonged depression of cerebral oxidative metabolism following concussive brain injury in the rat: a cytochrome oxidase histochemistry study, *Brain Res.* **567**:1–10.

Hoyer, S., Oestereich, K., and Wagner, O., 1988, Glucose metabolism as the site of the primary abnormality in early-onset dementia of Alzheimer type, *J. Neurol.* **235**:143–148.

Hruszkewycz, A. M., 1992, Lipid peroxidation and mtDNA degeneration: A hypothesis, *Mutat. Res.* **275**:243–248.

Hutchin, T., and Cortopassi, G., 1995, A mitochondrial DNA clone is associated with increased risk for Alzheimer disease, *Proc. Natl. Acad. Sci. USA* **92**:6892–6895.

Hyman, B. T., and Tanzi, R., 1995, Molecular epidemiology of Alzheimer's disease, *New Eng. J. Med.* **333**:1283–1284.

Hyman, B. T., Van Hoesen, G. W., and Damasiao, A. R., 1990, Memory-related neuronal systems in Alzheimer's disease: an anatomic study, *Neurology* **40**:1721–1730.

Ibata, Y., and Otsuka, N., 1969, Electron microscopic demonstration of zinc in the hippocampal formation using Timms' sulfide silver technique, *J. Histochem. Cytochem.* **17**:171–175.

Imagawa, M., Naruse, S., Tsuji, S., Fujioka, A., and Yamaguchi, H., 1992, Coenzyme Q10, iron, and vitamin B6 in genetically-confirmed Alzheimer's disease, *Lancet* **340**:671.

Ishitani, R., Sunaga, K., Hirano, A., Saunders, P., Katsube, N., and Chuang, D.-M., 1996, Evidence that glyceraldehyde-3-phosphate dehydrogenase is involved in age-induced apoptosis in mature cerebellar neurons in culture, *J. Neurochem.* **66**:929–935.

Iyengar, V., Kumpulainen, J., Okamoto, K., Morita, M., Hirai, S., and Nomoto, S., 1993, Recent trends in analytical approaches for trace element determinations in biomedical investigations, *Prog. Clin. Biol. Res.* **380**:329–354.

Jeandel, C., Nicolas, M. B., Dubois, F., Nabet-Belleville, F., Penin, R., and Cuny, G., 1989, Lipid peroxidation and free radical scavengers in Alzheimer's disease, *Gerontology* **35**:275–282.

Johns, D. R., 1995, Seminars in medicine of the Beth Israel Hospital, Boston, mitochondrial DNA and disease, *New Engl. J. Med.* **333**:638–644.

Kaiser, J., 1994, Alzheimer's: Could there be a zinc link?, *Science* **265**:1365.

Kalaria, R. N., Sromek, S. M., Grahovac, I., and Harik, S. I., 1992, Transferrin receptors of rat and human brain and cerebral microvessels and their status in Alzheimer's disease, *Brain Res.* **585**:87–93.

Kalman, J., Dey, I., Ilona, S. V., Markovics, B., Brown, D., Janka, Z., Farkas, T., and Joos, F., 1994, Platelet membrane fluidity and plasma malondialdehyde levels in Alzheimer's demented patients with and without family history of dementia, *Soc. Biol. Psychiatry* **35**:190–194.

Kaplan, E., Bigelow, D., Vatassery, G., and Ansari, K., 1982, Glutathione peroxidase in human cerebrospinal fluid, *Brain Res.* **252**:391–393.

Kaneko, Y., Kitamoto, R., Tateisha, J., and Yamaguchi, K., 1989, Ferritin immunohistochemistry as a marker for microglia, *Acta Neuropathol.* **79**:129–136.

Katzman, R., 1989, The dementias, in: *Merritt's Textbook of Neurology* (L. P., Rowland, ed.), 8th edition, Lea and Febiger, Philadelphia, pp. 637–644.

Katzman, R., 1993, Clinical and epidemiological aspects of Alzheimer's disease, *Clin. Neuroscience* **1**:165–170.

Kehrer, J. P., and Lund, L. G., 1994, Cellular reducing equivalents and oxidative stress, *Free Rad. Biol. Med.* **17**:65–75.

Kirkwood, T. B., 1991, Genetic basis of limited cell proliferation, *Mutat. Res.* **256**:323–328, 1991.

Kish, S. J., Morito, C. L. H., and Hornykiewicz, O., 1986, Brain glutathione peroxidase in neurodegenerative disorders, *Neurochem. Path.* **4**:23–38.

Kish, S. J., Bergeron, C., Rajput, A., Dozic, S., Mastrogiacomo, F., Chang, L. J., Wilson, J. M., DiStefano, L. M., and Nobrega, J. N., 1992, Brain cytochrome oxidase in Alzheimer's disease, *J. Neurochem.* **59**:776–779.

Knoll, J., 1992, (−)Deprenyl-medication: a strategy to modulate the age-related decline of the striatal dopaminergic system, *J. Am. Ger. Soc.* **40**:839–847.

Konig, G., Masters, C. L., and Beyreuther, K., 1990, Retinoic acid induced differentiated neuroblastoma cells show increased expression of the βA4 amyloid gene of Alzheimer's disease and an altered splicing pattern, *FEBS* **269**:305–310.

Korsmeyer, S. J., 1995, Regulators of cell death, *Trends in Genetics* **11**:101–105.

Kroemer, G., Petit, P., Zamzami, N., Vayssiere, J.-J., and Mignotte, B., 1995, The biochemistry of programmed cell death, *FASEB J.* **9**:1277–1287.

Krsek-Staples, J. A., and Webster, R. O., 1993, Ceruloplasmin inhibits carbonyl formation in endogenous cell proteins, *Free Rad. Biol. Med.* **14**:115–125.

Ku, H. H., Brunk, U. T., and Sohal, R. S., 1993, Relationship between mitochondrial superoxide and hydrogen peroxide production and longevity of mammalian species, *Free Rad. Biol. Med.* **15**:621–627.

Kumar, U., Dunlop, D. M., and Richardson, J. S., 1994, Mitochondria from Alzheimer's fibroblasts show decreased uptake of calcium and increased sensitivity to free radicals, *Life Sciences* **24**:1855–1860.

Kunimoto, M., 1994, Methylmercury induces apoptosis of rat cerebellar neurons in primary culture, *Biochem. Biophys. Res. Comm.* **204**:310–317.

LaFerla, F. M., Tinkle, B. T., Bieberich, C. J., Haudenschild, C. C., and Jay, G., 1995, The Alzheimer's Aβ peptide induces neurodegeneration and apoptotic cell death in transgenic mice, *Nature Genet.* **9**:21–30.

Lakis, J., Galasco, S., Miller, S. W., Thal, L. J., and Davis, R. E., 1995, Production of reactive oxygen species correlates with decreased cytochrome oxidase activity in Alzheimer's disease cybrids, *Soc. Neuroscience* **21**:979.

Landsberg, J., McDonald, B., Grime, G., and Watt, F., 1992, Microanalysis of senile plaques using nuclear microscopy, *J. Geriatric Psych. Neurol.* **6**:97–104.

Lamb, B. T., 1995, Making models for Alzheimer's disease, *Nature Genet.* **9**:4–6.

Larsen, P. L., 1993, Aging and resistance to oxidative damage in Caenorhabditis elegans, *Proc. Natl. Acad. Sci. USA* **90**:8905–8909.

Lee, H.-C., Pang, C.-Y., Hsu, H.-S., and Wei, Y. H., 1994, Differential accumulations of 4,977 bp deletions in mitochondrial DNA of various tissues in human ageing, *Biochem. Biophys. Acta* **1226**:37–43.

Lennon, S. V., Martin, S. J., and Cotter, T. G., 1991, Dose-dependent induction of apoptosis in human tumour cell lines by widely diverging stimuli, *Cell Prolif.* **24**:203–214.

Li, J. J., Surini, M., Carsicas, S., Kawashima, E., and Bouras, C., 1995, Age-dependent accumulation of advanced glycosylation end products in human neurons, *Neurobiol. Aging* **16**:69–76.

Liu, Y., Hernandez, A. M., Shibata, D., and Cortopassi, G. A., 1994, BCL2 translocation frequency rises with age in humans, *Proc. Natl. Acad. Sci. USA* **91**:89810–8914.

Linnane, A. W., Zhang, C., Baumer, A., and Nagley, P., 1992, Mitochondrial DNA mutation and the ageing process: bioenergy and pharmacological intervention, *Mutat. Res.* **275**:195–208.

Lippa, C. R., Smith, R. W., Smith, J. M., Swearer, D. A., Drachman, B., Ghetti, L., Nee, D., Pulaski-salo, D., Dickson, D., Robitaille, Y., Bergeron, C., Crain, B., Benson, M. D., Farlow, M., Hyman, B. T., St. George-Hyslop, P., Roses, A. D., and Pollen, D. A., 1996, Familial and sporadic Alzheimer's disease: Neuropathology cannot exclude a final common pathway, *Neurology* **46**:406–412.

Loeffler, D. A., Connor, J. R., Juneaue, P. L., Snyder, B. S., Kanaley, L., DeMaggio, A. J., Nguyen, H., Brickman, C. M., and LeWitt, P. A., 1995, Transferrin and iron in normal, Alzheimer's disease, and Parkinson's disease brain regions, *J. Neurochem.* **65**:710–716.

Loo, D. R., Copani, A. G., Pike, C. J., Whittemore, E. R., Walencewicz, A. J., and Cotman, C. W., 1993, Apoptosis is induced by beta-amyloid in cultured central nervous system neurons, *Proc. Natl. Acad. Sci. USA* **90**:7951–7955.

Lovell, M. A., Ehmann, W. E., and Markesbery, W. R., 1993, Laser microprobe analysis of brain aluminum in Alzheimer's disease, *Ann. Neurol.* **33**:36–42.

Lovell, M. A., Ehmann, W. D., Butler, S. M., and Markesbery, W. R., 1995, Elevated thiobarbituric acid-reactive substances and antioxidant enzyme activity in the brain in Alzheimer's disease, *Neurology* **45**:1594–1601.

Lowe, S. L., Francis, P. T., Procter, A. W., Palmer, A. M., Davison, A. N., and Bowen, D. M., 1988, Gamma-aminobutyric acid concentration in brain tissue at two stages of Alzheimer's disease, *Brain* **111**:785–799.

Lui, E., Fisman, M., Wong, C., and Diaz, F., 1990, Metals and the liver in Alzheimer's disease. an investigation of hepatic zinc, copper, cadmium, and metallothionein, *J. Am. Ger. Soc.* **38**:633–639.

Maggio, J. E., Esler, W. P., Stimson, E. R., Jennings, J. M., Ghilari, J. R., and Mantyh, P. W., 1995, Zinc and Alzheimer's disease, *Science* **268**:1920–1921.

Mantyh, P. W., Ghilardi, J. R., Rogers, S., DeMaster, E., Allen, C. J., Stimson, E. R., and Maggio, J. E., 1993, Aluminum, iron, and zinc ions promote aggregation of physiological concentrations of β-amyloid peptide, *J. Neurochem.* **61**:1171–1174.

Markesbery, W. R., and Ehmann, W. D., 1993, Aluminum and Alzheimer's disease, *Clin. Neuroscience* **1**:212–218.

Markesbery, W. R., and Ehmann, W. D., 1994, Brain trace elements in Alzheimer disease, in: *Alzheimer Disease* (R. D. Terry, R. Katzman, and K. L. Bick, eds.), Raven Press, New York, pp. 353–367.

Marklund, S. L., Adolfsson, R., Gottfries, C. G., And Winglad, B., 1985, Superoxide dismutase isoenzymes in normal brains and in brains from patients with dementia of Alzheimer type, *J. Neurol. Sci.* **67**:319–325.

Martins, R. N., Harper, C. G., Stokes, G. B., and Masters, C. L., 1986, Increased cerebral glucose-6-phosphate dehydrogenase activity in Alzheimer's disease may reflect oxidative stress, *J. Neurochem.* **46**:1042–1045.

Masoro, E. J., ed., 1981, *CRC Handbook of Physiology in Aging*, CRC Press, Boca Raton, Florida.

Mastrogiacomo, F., Bergeron, C., and Kish, S. J., 1993, Brain α-ketoglutarate dehydrogenase complex activity in Alzheimer's disease in Alzheimer's brain, *Mol. Cell Neurosci.* **3**:461–470.

Mattson, M. P., Cheng, B., Davis, D., Bryant, K., Leiberburg, I., and Rydd, R. E., 1992, β-amyloid peptides destabilize calcium homeostasis and render human cortical neurons vulnerable to excitotoxicity, *J. Neurosci.* **12**:376–389.

McLachlan, D. R. C., Bergeron, C., Smith, J. E., Boome, D., and Rifat, S. L., 1996, Risk for neuropathologically confirmed Alzheimer's disease and residual aluminum in municipal drinking water employing weighted residential histories, *Neurology* **46**:401–405.

Mecocci, P., MacGarvey, U., Kaufman, A. E., Koontz, D., Shoffner, J. M., Wallace, D. C., and Beal, M. F., 1993, Oxidative damage to mitochondrial DNA shows marked age-dependent increases in human brain, *Ann. Neurol.* **34**:609–616.

Mecocci, P., MacGarvey, U., and Beal, M. F., 1994, Oxidative damage to mitochondrial DNA is increased in Alzheimer's disease, *Ann. Neurol.* **36**:747–751.

Miller, B. L., Moats, R. A., Shonk, T., Ernst, T., Woolley, S., and Ross, B. D., 1993, Alzheimer disease: Depiction of increased cerebral myo-inositol with proton MR spectroscopy, *Radiology* **187**:433–437.

Miller, S. W., Herrnstadt, C., Parker, W. D., Jr., and Davis, R. E., 1995, Creation of mitochondrial DNA deficient neuroblastoma cell lines: Rescue of aerobic phenotype by human mitochondrial transfer, *Soc. Neuroscience* **21**:21.

Miquel, J., 1992, An update on the mitochondrial-DNA mutation hypothesis of cell aging, *Mutat. Res.* **275**:209–216.

Montine, T. J., Amarnath, V., Martin, M. E., Strittmatter, W. J., and Graham, D. G., 1996a, E-4-hydroxy-2-nonenal is cytotoxic and cross-links cytoskeletal proteins in P19 neuroglial cultures, *Am. J. Pathol.* **148**:89–93.

Montine, T. J., Huang, D. Y., Valentine, W. M., Amamath, V., Saunders, A., Weisgraber, K. H., Graham, D. G., and Strittmatter, W. J., 1996b, Crosslinking of apolipoprotein E by products of lipid peroxidation, *J. Neuropathol Exp. Neurol.* **55**:202–210.

Mooradian, A. D., and Wong, N., 1991, Molecular biology of aging Part II: a synopsis of current research, *J. Am. Ger. Soc.* **39**:717–723.

Motter, R., Vigo-Pelfrey, C., Kholodenko, D., Barbour, R., Johnson-Wood, K., Galasko, D., Chang, L., Miller, B., Clark, C., Green, R., Olson, D., Southwick, P., Wolfert, R., Munroe, B., Lieberburg, I., Seubert, P., and Schenk, D., 1995, Reduction of β-amyloid peptide 42 in the cerebrospinal fluid of patients with Alzheimer's disease, *Ann. Neurol.* **38**:643–648.

Muller, D. P. R., Metcalf, R., and Baren, D. M., 1986, Vitamin E in brains of patients with Alzheimer's disease and Down's syndrome, *Lancet* **i**:1093–1094.

Muller-Hocker, J., 1990, Cytochrome c oxidase deficient fibres in the limb muscle and diaphragm of man without muscular disease: an age-related alteration, *J. Neurol. Sci.* **100**:14–21.

Multhaup, G., Bush, A., Pollwein, P., and Masters, C. L., 1994, Interaction between the zinc(II) and the heparin binding site of the Alzheimer's disease βA4 amyloid precursor protein (APP), *FEBS Letters* **355**:151–154.

Multhaup, G., Schlicksupp, A., Hesse, L., Beher, D., Ruppert, T., Masters, C. L., and Beyreuther, K., 1996, The amyloid precursor protein of Alzheimer's disease in the reduction of copper (II) to copper (I), *Science* **271**:1406–1409.

Munro, H. N., 1969, Evolution of protein metabolism in mammals, in: *Mammalian Protein Metabolism, Volume 3* (H. N. Munro and J. B. Allison, eds.), Academic Press, New York, pp. 133–182.

Munscher, C., Muller-Hocker, J., and Kadenback, B., 1993, Human aging is associated with various point mutations in tRNA genes of mitochondrial DNA, *Biol. Chem. Hoppe-Seyler* **374**:1099–2003.

Murray, A. M., W. F., Marshall, J. R., Hurtig, H. I., Gottleib, G. L., and Joyce, J. N., 1995, Damage to dopamine systems differs between Parkinson's disease and Alzheimer's disease with Parkinsonism, *Ann. Neurol.* **37**:300–312.

Mutisya, E. M., Bowling, A. C., and Beal, M. F., 1994, Cortical cytochrome oxidase activity is reduced in Alzheimer's disease, *J. Neurochem.* **63**:2179–2184.

Nikaido, T., Austin, J., Trueb, L., and Rinehart, T. R., 1972, Studies in ageing of the brain. II. Microchemical analyses of the nervous system in Alzheimer patients, *Arch. Neurol.* **27**:549–554.

Orr, W. C., and Sohal, R. S., 1994, Extension of life-span by overexpression of superoxide dismutase and catalase in Drosophila melanogaster, *Science* **263**:1128–1130.

Oteiza, P. I., Oline, K. L., Fraga, C. G., and Keen, C. L., 1995, Zinc deficiency causes oxidative damage to proteins, lipids and DNA in rat testes, *J. Nutr.* **125**:823–829.

Ozawa, T., Tanaka, M., Sugiyama, S., Hattori, K., Ito, T., Ohno, K., Takahashi, A., Sato, W., Takada, G., Mayumi, B., Yamamoto, K., Adachi, K., Koga, Y., and Toshima, H., 1990, Multiple mitochondrial DNA deletions exist in cardiomyocytes of patients with hypertrophic or dilated cardiomyopathy, *Biochem. Biophys. Res. Commun.* **170**:830–836.

Palmer, A. M., and Burns, M., 1990, Selective increase in lipid perioxidation in the inferior temporal cortex in Alzheimer's disease, *Brain Res.* **645**:338–342.

Palmiter, R. D., Findley, S. D., Whitmore, T. E., and Durnam, D. M., MT-III, a brain-specific member of the metallothionein gene family, *Proc. Natl. Acad. Sci. USA* **89**:6333–6337.

Pappolla, M. A., Omar, R. A., Kim, K. S., and Robakis, N. K., 1992, Immunohistochemical evidence of antioxidant stress in Alzheimer's disease, *Am. J. Pathol.* **140**:621–628.

Parker, W. D. Jr., Filley, C. M., and Parks, J. K., 1990, Cytochrome oxidase deficiency in Alzheimer's disease, *Neurology* **40**:1302–1303.

Parker, W. D., Parks, J., and Filley, C. M., 1994, Electron transport chain defects in Alzheimer's disease brain, *Neurology* **44**:1090–1096.

Partridge, R. S., Monroe, S. M., Parks, J. K., Johnson, K., Parker, W. D. Jr., Eaton, G. R., and Eaton, S. S., 1994, Spin trapping of azidyl and hydroxyl radicals in azide-inhibited rat brain submitochondrial particles, *Arch. Biochem. Biophys.* **310**:210–217.

Perez-Clausell, J., and Danscher, G., 1985, Intravesicular localization of zinc in rat telencephalic boutons: a histochemical study, *Brain Res.* **337**:91–98.

Perl, D. P., and Pendlebury, W. W., 1986, Aluminum neurotoxicity: potential role in the pathogenesis of neurofibrillary tangle formation, *Can. J. Neurol. Sci.* **13**:441–445.

Perry, G., and Smith, M. A., 1993, Senile plaques and neurofibrillary tangles: What role do they play in Alzheimer's disease? *Clin. Neurosci.* **1**:199–203.

Perry, T. L., Yong, V. W., Bergeron, C., Hansen, S., and Jones, K., 1987, Amino acids, glutathione, and glutathione transferase activity in the brains of patients with Alzheimer's disease, *Ann. Neurol.* **21**:331–336.

Pettigrew, J. W., Klunk, W. E., Panchalingam, K., Kanfer, J. N., and Mclure, R. J., 1995, Clinical and neurochemical effects of acetyl-L-carnitine in Alzheimer's disease, *Neurobiol. Aging* **16**:1–4.

Pike, C. J., and Cotman, C. W., 1993, Cultured GABA-immunoreactive neurons are resistant to toxicity induced by β-amyloid, *Neuroscience* **56**:269–274.

Polvikoski, T., Sulvaka, R., Haltia, M., Kainulainen, K., Vuorio, A., Verkkoniemi, Niinisto, L., Halonen, P., and Kontula, K., 1995, Apolipoprotein E, dementia, and cortical deposition of β-amyloid protein, *New Engl. J. Med.* **333**:1242–1247.

Poirier, J., 1994, Apolipoprotein E in animal models of CNS injury and in Alzheimer's disease, *Trends Neurosci.* **17**:525–530.

Poulton, J., and Holtz, I. J., 1995, Mitochondrial DNA: Does more lead to less?, *Nature Genet.* **8**:313–315.

Pountney, D. L., Fundel, S. M., Faller, P., Birchler, N. E., Hunziker, P., and Vasak, M., 1994, Isolation, primary structures and metal binding properties of neuronal growth inhibitory factor (GIF) from bovine and equine brain, *FEBS Lett.* **345**:193–197.

Prohaska, J. R., 1987, Functions of trace elements in brain metabolism, *Physiol. Rev.* **67**:858–901.

Pullen, R. G. L., Candy, J. M., Morris, C. M., Taylor, G., Keith, A. B., and Edwardson, J. A., 1990, Gallium-67 as a potential marker for aluminum transport in rat brain: implications for Alzheimer's disease, *J. Neurochem.* **55**:251–259.

Randerath, K., Putnam, K. L., Osterburg, H. H., Johnson, S. A., Morgan, D. G., and Finch, C. E., 1992, Age-dependent increases of DNA adducts (I-compounds) in human and rat brain DNA, *Mutat. Res.* **295**:11–18.

Reichman, H., Florke, S., Hebenstreit, G., Schrubar, H., and Riederer, P., 1993, Analyses of energy metabolism and mitochondrial genome in post-mortem brain from patients with Alzheimer's disease, *J. Neurol.* **240**:377–380.

Richardson, J. S., Subbarao, K. V., and Ang, L. C., 1992, On the possible role of iron-induced free radical peroxidation in neural degeneration in Alzheimer's disease, *Ann. New York Acad. Sci.* **648**:326–327.

Richter, C., 1992, Reactive oxygen and DNA damage in mitochondria, *Mutat. Res.* **275**:249–255.

Richter, C., Park, J. W., and Ames, B. N., 1988, Normal oxidative damage to mitochondrial and nuclear DNA is extensive, *Proc. Natl. Acad. Sci. USA* **85**:6465–6467.

Roder, H. M., Eden, P. A., and Ingram, V. M., 1993, Brain protein kinase PK40erk converts Tau into a PHF-like form as found in Alzheimer's disease, *Biochem. Biophys. Res. Comm.* **193**:639–647.

Roses, A., 1995, Apolipoprotein E genotyping in the differential diagnosis, not prediction of Alzheimer's disease, *Ann. Neurol.* **38**:6–14.

Ross, M., 1995, Many questions but no clear answers on link between aluminum, Alzheimer's disease, *Can. Med. Assoc. J.* **150**:68–69.

Rothstein, J. D., Bristol, L. A., Hosler, B., Brown, R. H. Jr., and Kuncl, R. W., 1994, Chronic inhibition of superoxide dismutase produces apoptotic death of spinal neurons, *Proc. Natl. Acad. Sci. USA* **91**:4155–4159.

Roy, N., Mahadevan, M. S., McLean, M., Shutler, G., Yaraghi, A., Farahani, R., Baird, S., Besner-Johnston, A., Lefebvre, C., Kang, X., Salih, M., Aubry, H., Tamai, K., Guan, X., Ioannou, P., Crawford, T. O., de Jong, P. J., Surh, L., Ikeda, J.-E., Korneluk, R. G., and MacKenzie, A., 1995, The gene for neuronal apoptosis inhibitory protein is partially deleted in individuals with spinal muscular atrophy, *Cell* **80**:167–178.

Rumble, B., Retallack, R., Hilbich, D., Simms, G., Multhaup, G., Marins, R., Hockey, A., Montogomery, P., Beyreuther, K., and Masters, C. L., 1989, Amyloid β4 protein and its precursor in Down's syndrome and Alzheimer's disease, *New Engl. J. Med.* **320**:1446–1452.

Saraiva, A. A., Borges, M. M., Madeira, M. D., Tavares, M. A., and Paula-Barbosa, M. M., 1985, Mitochondrial abnormalities in cortical dendrites from patients with Alzheimer's disease, *J. Submicrosc. Cytol.* **17**:459–464.

Sato, M., and Bremner, I., 1993, Oxygen free radicals and metallothionein, *Free Rad. Biol. Med.* **14**:325–337.

Schapira, A. H. V., and Cooper, J. M., 1992, Mitochondrial function in neurodegeneration and ageing, *Mutat. Res.* **275**:133–143.

Schipper, H. M., Cisse, S., and Stopa, E. G., 1995, Expression of heme oxygenase-1 in the sensecent and Alzheimer-diseased brain, *Ann. Neurol.* **37**:758–768.

Schubert, D., Behl, C., Lesley, R., Brack, A., Dargusch, R., Sagara, Y., and Kimura, H., 1995, Amyloid peptides are toxic via a common oxidative mechanism, *Proc. Natl. Acad. Sci. USA* **92**:1989–1993.

Schwartz, B. L., Hashtroudi, S., Herting, R. L., Schwartz, P., and Deutsch, S. I., 1996, d-Cycloserine enhances implicit memory in Alzheimer patients, *Neurology* **46**:420–424.

Schweers, O., Mandelkow, E.-M., Biernat, J., and Mandelkow, E., 1995, Oxidation of cysteine-322 in the repeat domain of microtubule-associated protein τ controls the in vitro assembly of paired helical filaments, *Proc. Natl. Acad. Sci. USA* **92**:8463–8467.

Selkoe, D. J., 1994, Cell biology of the amyloid beta-protein precursor and the mechanism of Alzheimer's disease, *Ann. Rev. Cell Biol.* **10**:373–403.

Serra, J. A., Famulari, A. L., Kohan, S., Marschoff, E. R., Dominguez, R. O., and de Lustig, E. S., 1994, Copper-zinc superoxide dismutase activity in red blood cells in probable Alzheimer's patients and their first-degree relatives, *J. Neurol. Sci.* **122**:179–188.

Sewell, A. K., Jensen, L. T., Erickson, J. C., Palmiter, R. D., and Winge, D. R., 1995, Bioactivity of metallothionein-3 correlates with its novel beta domain sequence rather than metal binding properties, *Biochem.* **34**:4740–4747.

Sheu, K. F., Cooper, A. J., Koike, K., Koike, M., Lindsay, J. G., and Blass, J. P., 1994, Abnormality of the α-ketoglutarate dehydrogenase complex in fibroblasts from familial Alzheimer's disease, *Ann. Neurol.* **35**:312–318.

Sheu, S.-S., and Jou, M.-J., 1994, Mitochondrial free Ca2+ concentration in living cells, *J. Bioenerget. Biomem.* **26**:487–493.

Shigenaga, M. K., and Ames, B., 1994, Oxidants and mitochondrial decay in aging, in: *Natural Antioxidants in Human Health and Disease* (B. Frei, ed), Academic Press, New York, pp. 63–106.

Shigenaga, M. K., Parks, J.-W., Cundy, K. C., Gimeno, C. J., and Ames, B. N., 1990, In vivo oxidative DNA damage: Measurement of 8-hydroxy-2′-deoxyguanosine in DNA and urine by High-performance liquid chromatography with electrochemical detection, *Methods in Enzymology* **186**:521–530.

Shoffner, J. M., Brown, M. D., Torroni, A., Lott, M. T., Cabell, P. Mirra, S. S., Beal, M. F., Yang, C.-C., Gearing, M., Salvo, R., Watts, R. L., Juncos, J. L., Hanson, L. A., Crain, B. J., Fayad, M., and Wallace, D. C., 1993, Mitochondrial DNA mutations associated with Alzheimer's and Parkinson's disease, *Genomics* **17**:171–184.

Shore, D., and Wyatt, R. J., 1983, Aluminum and Alzheimer's disease, *J. Nerv. Mental Dis.* **171**:553–558.

Simonetti, S., Chen, X., DiMauro, S., and Schon, E. A., 1992, Accumulation of deletions in human mitochondrial DNA during normal aging: analysis by quantitative PCR, *Biochem. Biophys. Acta* **1180**:113–122.

Simonian, N. A., and Hyman, B. T., 1993, Functional alterations in Alzheimer's disease: Diminution of cytochrome oxidase in hippocampal formation, *J. Neuropathol. Exp. Neurol.* **52**:580–585.

Sims, N. R., Finegan, M. M., Blass, J. P., Bowe, D. M., and Neray, D., 1987, Mitochondrial function in brain tissue in primary degenerative dementia, *Brain Res.* **436**:30–38.

Sisodia, S. S., and Price, D. L., 1993, Amyloidogenesis in Alzheimer's disease, *Cllin. Neuroscience* **1**:176–183.

Smith, C. D., Carney, J. M., Starke-Reed, P. E., Oliver, C. N., Stadtman, E. R., Floyd, R. A., and Markesbery, W. R., 1991, Excess brain protein oxidation and enzyme dysfunction in normal aging and in Alzheimer disease, *Proc. Natl. Acad. Sci. USA* **88**:10540–10543.

Smith, C. D., Carney, J. M., Tatsumo, T., Stadtman, E. R., Floyd, R. A., and Markesbery, W. R., 1992, Protein oxidation in aging brain, *Ann. New York Acad. Sci.* **663**:110–119.

Smith, C. D., Pettigrew, L. C., Avison, M. L., Kirsch, J. E., Tinkhtman, A. J., Schmitt, F. A., Wemerling, D. P., Wekstein, D. R., and Markesbery, W. R., 1995, Frontal lobe phosphorus metabolism and neuropsychological function in aging and in Alzheimer's disease, *Ann. Neurol.* **38**:194–201.

Smith, M. A., Kutty, K., Richey, P. L., Yan, S.-D., Stern, D., Chader, G. J., Wiggert, B., Petersen, R. B., and Perry, G., 1994a, Heme oxygenase-1 is associated with the neurofibrillary pathology of Alzheimer's disease, *Am. J. Pathol.* **145**:42–47.

Smith, M. A., Taneda, S., Richey, P. L., Miyata, S., Yan, S.-D., Stern, D., Sayre, L. M., Monnier, V. M., and Perry, G., 1994b, Advanced maillard reaction end products are associated with Alzheimer disease pathology, *Proc. Natl. Acad. Sci. USA* **91**:5710–5714.

Smith, M. A., Richey, P. L., Kutty, R. K., Wiggert, B., and Perry, G., 1995a, Ultrastructural localization of heme oxygenase-1 to the neurofibrillary pathology of Alzheimer disease, *Mol. Chem. Neuro.* **24**:227–230.

Smith, M. A., Rudnicka-Nawrot, M., Richey, P. L., Praprotnik, D., Mulvihill, P., Miller, C. A., Sayre, L. M., and Perry, G., 1995b, Carbonyl-related postranslational modification of neurofilament protein in the neurofibrillary pathology of Alzheimer disease, *J. Neurochem.* **64**:1–7.

Smith, M. A., Sayre, L. M., Monnier, V. M., and Perry, G., 1995c, Radical ageing in Alzheimer's disease, *Trends Neurosci.* **18**:1–7.

Sohal, R., 1993, Aging, cytochrome oxidase activity, and hydrogen peroxide release by mitochondria, *Free Rad. Biol. Med.* **14**:583–588.

Sohal, R. S., and Sohal, B. H., 1991, Hydrogen peroxide release by mitochondria increases during aging, *Mech. Ageing Dev.* **57**:187–202.

Sohal, R. S., and Brunk, U. T., 1992, Mitochondrial production of pro-oxidants and cellular senescence, *Mutat. Res.* **275**:295–304.

Sohal, R. S., and Dubey, A., 1994, Mitochondrial oxidative damage, hydrogen peroxide release, and aging, *Free Rad. Biol. Med.* **16**:621–626.

Soumalainen, A., Kaukonen, J., Amati, P., Timonen, R., Haltia, M., Weissenback, J., Zeriani, M., Somer, H., and Pettonen, L., 1995, An autosomal locus predisposing to deletions of mitochondrial DNA, *Nature Genet.* **9**:146–151.

Soong, N. W., Hinton, D. R., Cortopassi, G., and Arnheim, N., 1992, Mosaicism for a specific somatic mitochondrial DNA mutation in adult human brain, *Nature Genet.* **2**:318–323.

Sorbi, S., Bird, E. D., and Blass, J. P., 1983, Decreased pyruvate dehydrogenase complex activity in Huntington and Alzheimer brain, *Ann. Neurol.* **13**:72–78.

Sparks, D. L., Huaichen, L., Scheff, S. W., Coyne, C. M., and Hunsaker, J. C., 1993, Temporal sequence of plaque formation in the cerebral cortex of non-demented individuals, *J. Neuropath. Exper. Neurol.* **52**:135–142.

Stadtman, E. R., 1992, Protein oxidation and aging, *Science* **257**:1220–1224.

Stadtman, E. R., 1995, Role of oxidized amino acids in protein breakdown and stability, *Methods in Enzymology* **258**:379–393.

Stohs, S. J., and Bagchi, D., 1995, Oxidative mechanisms in the toxicity of metal ions, *Free Rad. Biol. Med.* **18**:321–336.

Strittmatter, W. J., Weisgraber, K. H., Huang, D. Y., Dong, L. M., Salvesen, G. S., Pericak-Vance, M., Schemechel, D., Saunders, A. M., Godgaber, D., and Roses, A. D., 1993, Binding of human apolipoprotien E to synthetic amyloid b peptide: isoform-specific effects and implications for late-onset Alzheimer disease, *Proc. Natl. Acad. Sci. USA* **91**:8098–8102.

Subbarao, D. V., Richardson, J. S., and Ang, L. C., 1990, Autopsy samples of Alzheimer's cortex show increased peroxidation in vitro, *J. Neurochem.* **55**:342–345.

Sugiyama, S., Hattori, K., Hayakawa, M., and Ozawa, T., 1991, Quantitative analysis of age-associated accumulation of mitochondrial DNA with deletion in human hearts, *Biochem. Biophys. Res. Comm.* **180**:894–899.

Sulkova, R., Nordberg, U.-R., Erkinjuntti, T., and Westermarck, T., 1986, Erythrocyte glutathione peroxidase and superoxide dismutase in Alzheimer's disease and other dementias, *Acta. Neurol. Scand.* **73**:487–489.

Szerdahelyi, P., and Kasa, P., 1984, Histochemistry of zinc and copper, *Int. Rev. of Cytology* **89**:1–33.

Tappel, A. L., 1973, Lipid peroxidation damage to cell components, *Fed. Proc.* **32**:1870–1874.

Terry, R. D., and Davies, P., 1980, Dementia of the Alzheimer type, *Ann. Rev. Neuroscience* **3**:77–95.

Thompson, C. B., 1995, Apoptosis in the pathogenesis and treatment of disease, *Science* **267**:1456–1462.

Thorsness, P. E., 1992, Structural dynamics of the mitochondrial compartment, *Mutat. Res.* **275**:237–241.

Tokutake, S., Nagase, H., Morisaki, S., and Oyanagi, S., 1995, Aluminium detected in senile plaques and neurofibrillary tangles is contained in lipofusin granules with silicon, probably as aluminosilicate, *Neuroscience Lett.* **185**:99–102.

Trojanowski, J. Q., Schmidt, M. L., Shin, R.-W., Bramblett, G. T., Goedert, M., and Lee, V. M.-Y., 1993, PHFτ(A68): From pathological marker to potential mediator of neuronal dysfunction and degeneration in Alzheimer's disease, *Clin. Neuroscience* **1**:184–191.

Troncoso, J. C., Costello, A., Watson, A. L. Jr., and Johnson, G. V. W., 1993, In vitro polymerization of oxidized tau into filament, *Brain Res.* **613**:313–316.

Trounce, I., Byrne, E., and Marzuki, S., 1989, Decline in skeletal muscle mitochondrial respiratory chain function: Possible factor in ageing, *Lancet* **I**:637–739.

Tsuji, S., Kobayashi, H., Uchida, Y., Ihara, Y., and Miyatake, T., 1992, Molecular cloning of human growth inhibitory factor cDNA and its down-regulation in Alzheimer's disease, *EMBO J.* **11**:4843–4850.

Uchida, Y., Takio, K., Titani, K., Ihara, Y., and Tomonaga, M., 1991, The growth inhibitory factor that is deficient in the Alzheimer's disease brain is a 68 amino acid metallothionein-like protein, *Neuron* **7**:337–347.

Ueda, K., Cole, G., Sundsmo, M., Katzman, R., and Saitoh, T., 1989, Decreased adhesiveness of Alzheimer's disease fibroblasts: Is amyloid beta-protein precursor involved?, *Ann. Neurol.* **25**:246–251.

Van Zuylen, A. J., Bosman, G. J., Ruitenbeek, W., Van Kalmthout, P. J., and DeGrip, W. J., 1992, No evidence for reduced thrombocyte oxidase activity in Alzheimer's disease, *Neurology* **42**:1246–1247.

Vener, A. V., Aksenova, M., and Burbaeva, G. S., 1993, Drastic reduction of the zinc- and magnesium-stimulated protein tyrosine kinase activities in Alzheimer's disease hippocampus, *FEBS Lett.* **328**:6–8.

Vitek, M. P., Ghattacharya, K., Glendening, J. M., Stopa, E., Vlassara, H., Bucala, R., Maogue, K., and Cerami, A., 1994, Advanced glycation end products contribute to amyloidosis in Alzheimer disease, *Proc. Natl. Acad. Sci. USA* **91**:4766–4770.

Volicer, L., and Crino, P. B., 1990, Involvement of free radicals in dementia of the Alzheimer type: a hypothesis, *Neurobiol. Aging* **11**:567–571.

Walford, R. L., 1974, Immunologic theory of aging: current status, *Fed. Proc.* **33**:2020–2027.

Wallace, D. C., 1992, Mitochondrial genetics: A paradigm for aging and degenerative diseases?, *Science* **256**:628–632.

Wallace, D. C., Ye, J. H., Neckelmann, S. N., Singh, G., Webster, K. A., and Greenberg, B. D., 1987, Sequence analysis of cDNAs for the human and bovine ATP synthase beta subunit: Mitochondrial DNA genes sustain seventeen times more mutations, *Curr. Genet.* **12**:81–90.

Wallace, D. C., Shoffner, J. M., Trounce, I., Brown, M. D., Ballinger, S. W., Corral-Debrinksi, M., Horton, T., Jun, A. S., and Lott, M. T., 1995, Mitochondrial DNA mutations in human degenerative diseases and aging, *Biochim. Biophys. Acta.* **1271**:141–151.

Warner, H. R., 1994, Superoxide dismutases, aging and degenerative disease, *Free Rad. Biol. Med.* **17**:249–256.

Watt, J., Pike, C. J., Walencewicz, A. J., and Cotman, C. W., 1994, Ultrastructural analysis of beta-amyloid-induced apoptosis in cultured hippocampal neurons, *Brain Res.* **661**:147–156.

Wei, Y.-H., 1992, Mitochondrial DNA alterations as ageing-associated molecular events, *Mutat. Res.* **275**:145–155.

Wensink, J., Molnaar, A. J., Woroniecka, U. D., and Van den Hamer, D. J., 1988, Zinc uptake into synaptosomes, *J. Neurochem.* **50**:782–789.

Wenstrup, D., Ehmann, W. D., and Markesbery, W. R., 1990, Trace element imbalances in isolated subcellular fractions of Alzheimer's disease brains, *Brain Res.* **533**:125–131.

Westbrook, G. L. and Mayer, M. L., 1987, Micromolar concentrations of Zn^{2+} antagonize NMDA and GABA responses of hippocampal neurons, *Nature* **328**:640–643.

Wisniewski, H. M., 1994, Aluminum, tau protein, and Alzheimer's disease, *Lancet* **344:**203–205.

Wisniewski, H. M., Wegiel, J., Wang, K. C., Kujawa, M., and Lack, B., 1989, Ultrastructural studies of the cells forming amyloid fibers in classical plaques, *Can. J. Neurol. Sci.* **16:**535–542.

Wong-Riley, M. T. T., 1989, Cytochrome oxidase: an endogeneous metabolic marker for neuronal activity, *Trends Neurosci.* **12:**94–101.

Wragg, M. A., Talbot, C. J., Morris, J. C., Lendon, C. L., and Goate, A. M., 1995, No association found between Alzheimer's disease and a mitochondrial tRNA glutamine gene variant, *Neuroscience Lett.* **201:**107–110.

Xu, N., Majidi, V., Markesbery, W. R., and Ehmann, W. D., 1992, Brain aluminum in Alzheimer's disease using an improved GFAAS method, *Neurotoxicol.* **13:**735–744.

Yan, S.-D., Chen, X., Schmidt, A.-M., Brett, J., Godman, G., Zou, Y.-S., Scott, C. W., Caputo, C., Frappier, T., Smith, M. A., Perry, G., Yen, S.-H., Stern, D., 1994a, Glycated tau protein in Alzheimer disease: a mechanism for induction of oxidant stress, *Proc. Natl. Acad. Sci. USA* **91:**7787–7791.

Yan, S.-D., Schmidt, A. M., Anderson, G. M., Shang, J., Brett, J., Zou, Y. S., Pinsky, D., and Stern, D., 1994b, Enhanced cellular oxidant stress by the interaction of advanced glycation end products with their receptor binding proteins, *J. Biol. Chem.* **269:**9889–9897.

Yan, S.-D., Chen, X., Fu, J., Chen, M., Godman, G., Gern, D., and Schmidt, A.-M., 1996, RAGE: a receptor upregulated in Alzheimer's disease on neurons, microglia, and cerebrovascular endothelium that binds amyloid-β-peptide and mediates induction of oxidant stress, *Neurology* **46A:**A276.

Yen, T.-C., Chen, Y.-S., and King, K.-L., 1989, Liver mitochondrial respiratory functions decline with age, *Biochem. Biophys. Res. Comm.* **165:**994–1003.

Yen, T.-C., Su, J. H., King, K. L., and Wei, Y. H., 1991, Ageing-associated 5kb deletion in human liver mitochondrial DNA, *Biochem. Biophys. Res. Commun.* **178:**124–131.

Yen, T.-C., King, K. L., Lee, H. C., Yeh, S. H., and Wei, Y. H., 1994, Age-dependent increase of mitochondrial DNA deletions together with lipid peroxides and superoxide dismutase in human liver mitochondria, *Free Rad. Biol. Med.* **16:**207–214.

Youdim, M. B., and Lavie, L., 1994, Selective MAO-A and B inhibitors, radical scavengers and nitric oxide synthase inhibitors in Parkinson's disease, *Life. Sci.* **55:**2077–2082.

Younkin, S. G., 1995, Evidence that Aβ42 is the real culprit in Alzheimer's disease, *Ann. Neurol.* **37:**287–288.

Zaman, Z., Roche, S., Fielden, P., Frost, P. G., Niriella, D. C., and Cayley, A. C., 1992, Plasma concentrations of vitamins A and E and carotenoids in Alzheimer's disease, *Age Ageing* **21:**91–94.

Zemlan, F. P., Thienhaus, O. J., and Bosmann, H. B., 1989, Superoxide dismutase activity in Alzheimer's disease: Possible mechanism for paired helical filament formation, *Brain Res.* **476:**160–162.

Zhang, C., Baumer, A., Maxwell, R. J., Linnane, A. W., and Nagley, P., 1992, Multiple mitochondrial deletions in an elderly human individual, *FEBS Letters* **297:**34–38.

Zhang, C., Linnane, A. W., and Nagley, P., 1993, Basic FGF, NGF, and IGFs protect hippocampal and cortical neurons against iron-induced degeneration, *J. Cereb. Blood Flow Metab.* **13:**378–388.

Zubenko, G. S., Farr, J., Stiffler, J. S., Hughes, H. B., and Kaplan, B. B., 1992, Clinically-silent mutation in the putative iron-responsive element in exon 17 of the beta-amyloid precursor protein gene, *J. Neuropathol. Exper. Neurol.* **51:**459–463.

Chapter 14

Oxidative Stress with Emphasis on the Role of LAMMA in Parkinson's Disease

Paul F. Good, Daniel P. Perl, and C. Warren Olanow

14.1. INTRODUCTION

Until recently, the causes and underlying pathogenetic mechanisms responsible for cell degeneration in the neurodegenerative disorders have remained obscure. Neuro-pathological analysis have revealed the presence of cytoskeletal markers such as the Lewy body and neurofibrillary tangle, which are associated with the neuronal degeneration, but the cellular events underlying these changes were unknown. More recently, it has been proposed that oxidative stress may be a unifying factor in the pathogenesis of the major neurodegenerative diseases (Olanow, 1993). This hypothesis proposes that an increase in free radicals due to excess formation or diminished antioxidant defenses is responsible for damage to critical biological molecules such as cytoskeletal proteins, membrane lipids, and DNA. This chapter will review the evidence that oxidative stress contributes to cell damage in Parkinson's disease (PD), with particular emphasis on the role of laser microprobe mass analysis (LAMMA) and the therapeutic implications of these observations.

PD has provided key insights into the possibility that oxidative stress may contribute to neurodegeneration. PD is a progressive, age-related neurodegenerative disorder characterized clinically by resting tremor, cog-wheel rigidity, and bradykinesia. Pathologically, there is a selective loss of melanized neurons in the substantia nigra pars compacta (SNc), intracellular inclusions known as Lewy bodies, and a reduction in striatal dopamine. SNc neurons contain the pigment neuromelanin, which is a product

Paul F. Good and Daniel P. Perl   Department of Pathology and Fishberg Research Center for Neurobiology, Mount Sinai School of Medicine, New York, New York 10029.   **C. Warren Olanow**   Department of Neurology, Mount Sinai School of Medicine, New York, New York 10029.

Metals and Oxidative Damage in Neurological Disorders, edited by Connor. Plenum Press, New York, 1997.

of the auto-oxidation of the neurotransmitter dopamine employed by SNc neurons in their nigrostriatal projection. The oxidative stress hypothesis in PD was conceived based on the potential of the metabolism of dopamine to yield highly reactive and cytotoxic free radicals and other oxidizing species, as shown in following equations.

1) Auto-oxidation of dopamine:

$$DA + O_2 \rightarrow SQ^{\cdot} + {}^{\cdot}O_2^- + H^+$$

$$DA + {}^{\cdot}O_2^- + 2H^+ \rightarrow SQ^{\cdot} + H_2O_2$$

The auto-oxidation of dopamine produces semiquinones (SQ) and superoxide (${}^{\cdot}O_2^-$) radicals, as well as hydrogen peroxide (H_2O_2), which is readily converted to hydroxyl radical (${}^{\cdot}OH$) in the presence of ferrous iron (Fe^{+2}) according to the Fenton reaction shown in equation 3.

2) Enzymatic oxidation of dopamine:

$$DA + O_2 + H_2O \xrightarrow{\text{MAO}} 3,4 \text{ DHPA} + NH_3 + H_2O_2$$

The enzymatic oxidation of dopamine by monoamine oxidase yielding $H_2O_2 \cdot H_2O_2$ can react with Fe^{+2} to generate the highly toxic ${}^{\cdot}OH$. Dopamine is a substrate for both the A and B isoforms of the MAO enzyme. MAO-B has been considered to be the more important, but recent information suggests that MAO-A is largely responsible for the enzymatic oxidation of dopamine in the SNc.

3) The Fenton Reaction

$$H_2O_2 + Fe^{+2} \rightarrow {}^{\cdot}OH + OH^- + Fe^{+3}$$

Generation of the cytotoxic hydroxyl radical by a reaction in which H_2O_2 is reduced to ${}^{\cdot}OH$ by an electron provided by the conversion of iron from its ferrous (Fe^{+2}) to its ferric (Fe^{+3}) state.

14.2. OXIDATIVE STRESS

Free radicals contain an unpaired electron in one or more of their orbitals (Halliwell and Gutteridge, 1985). They have the potential to react with and extract an electron from neighboring molecules so as to complete the electron requirement of their own orbitals. In the process they can damage other molecules through this oxidative reaction as well as initiate chain reactions. These reactions can be identified by the formation of oxidized products such as carbonyl proteins, lipid hydroperoxides, and hydroxy-DNA adducts, which form as a consequence of hydroxyl radical interaction with, and damage to, proteins, lipids, and DNA molecules, respectively.

Oxidative reactions are catalyzed by transition metals such as iron, copper, and manganese, which exist in multiple valence states and have the capacity to provide or accept an electron to promote redox reactions and the formation of free radicals (Halliwell and Gutteridge, 1988). Iron is the most important of these in biological systems and its transport is tightly regulated in order to avoid toxic injury (Aisen, 1992). The potential for iron to promote the formation of the cytotoxic hydroxyl radical was illustrated in equation 3.

There are normally a series of protective mechanism that defend against the production of free radicals (equation 4). Glutathione (GSH) and a series of associated enzymes, (glutathione peroxidase, glutathione reductase, and others), are the major brain systems responsible for preventing the formation of free radicals from the metabolism of dopamine. Glutathione peroxidase catalyzes the reduction of hydrogen peroxide (H_2O_2) to water by oxidizing GSH to GSSG. This prevents H_2O_2 accumulation and the potential of its reaction with iron leading to hydroxyl radical formation as shown in equation 3. Superoxide dismutase (SOD) catalyzes the detoxification of the superoxide anion and prevents direct tissue damage due to this radical species. In addition, SOD prevents superoxide radical from reacting with nitric oxide leading to the formation of peroxynitrite (equation 5). These reactions can be identified by the presence of nitrotyrosines, which result from an interaction between peroxynitrite and tyrosine-containing molecules. SOD has been implicated in the development of neurodegeneration by the demonstration of a series of point mutations in the gene that encodes for SOD-1 in patients with familial amyotrophic lateral sclerosis (FALS) (Rosen et al., 1993). Transgenic mice that express the mutant SOD gene develop a neurodegenerative syndrome with an ALS phenotype (Gurney et al., 1994). However, it is not clear that the mutation leads to decreased SOD activity in this condition as recent evidence points to the mutation causing a gain of function or toxic effect (Wiedau-Pazos et al., 1996).

4) Antioxidant systems

$$\text{i) } 2GSH + H_2O_2 \rightarrow GSSG + 2H_2O$$

$$\text{ii) } \cdot O_2^- + \cdot O_2^- \rightarrow H_2O_2 + O_2$$

These reactions demonstrate the role of GSH in detoxifying H_2O_2 and of SOD in dismutating $\cdot O_2^-$ to H_2O_2. In this way, these antioxidants clear potentially damaging oxidant species from the brain.

5) Formation of peroxynitrite

$$\cdot O_2^- + NO\cdot \rightarrow ONOO^-$$

Peroxynitrite is formed by the interaction of superoxide radical and nitric oxide radical. Peroxynitrite can by protonated to form peroxynitrous acid, which can decompose to form the hydroxyl radical. In addition, peroxynitrite can react with tyrosine molecules to generate nitrotyrosine. This can lead to degenerative changes in cytoskeletal molecules as well as the potential for impairing the phosphorylation of tyrosine-containing trophic factor receptors and loss of a critical trophic effect.

There are a number of naturally occurring free radical scavengers, including ascorbate and α-tocopherol, that can prevent tissue damage by detoxifying free radicals and/or terminating free radical chain reactions such as lipid peroxidation. Finally, there are a group of iron-binding proteins such as transferrin, and the 450000 kD multimeric protein ferritin that sequester iron and prevent it from participating in oxidative reactions. Transferrin receptor and ferritin mRNAs normally respond to an increase in cellular iron by shifting their translation-competent state, resulting in increased levels of ferritin protein, which can bind iron and prevent iron-related tissue damage and decreased transferrin receptor thereby limiting the transfer of iron across membranes.

14.3. EVIDENCE OF OXIDATIVE STRESS AND OXIDATIVE DAMAGE IN PARKINSON'S DISEASE

Parkinsonism related to the overproduction of free radicals in PD has been described in association with the use of a meperidinelike street drug (Langston *et al.*, 1983). The compound was determined to be 1-methyl-4-phenyl-1,2,3,6-tetrahydropyridine (MPTP) (Chiba *et al.*, 1984). MPTP is a protoxin that exerts its neurotoxic effects via its oxidation to the toxic metabolite 1-methyl-4-phenyl-pyridinium (MPP$^+$) within brain astrocytes. MPP$^+$ is taken up into dopaminergic neurons by the dopamine transporter system and selectively damages mitochondrial complex I of dopaminergic neurons. A selective mitochondrial complex I defect has been described in PD (Schapira *et al.*, 1990). Agents that protect against MPP$^+$ formation have been shown to protect against the development of MPTP-parkinsonism in animal models (Cohen *et al.*, 1985). However, MPTP-like molecules have not been identified in association with PD.

Cell degeneration in PD could also result from oxidant stress due to dysregulation of dopamine metabolism with consequent free radical formation. This might occur as a result of i) an increase in dopamine turnover with increased H_2O_2 formation, ii) a decrease in GSH levels, which normally clears H_2O_2 from the brain and prevents it from reacting with iron, or iii) an increase in the availability of reactive iron, which can provide electrons to catalyze oxidation reactions and promote the conversion of H_2O_2 to ˙OH. Indeed, postmortem studies provide evidence suggesting that each of these factors is operative in PD, that the SNc is in a state of oxidative stress, and that numerous critical molecules have undergone oxidative damage (reviewed in Jenner *et al.*, 1992; Olanow *et al.*, 1992; Jenner and Olanow, 1996).

The hypothesis that free-radical-induced damage to key biological molecules may be critical to the cell death that occurs in PD has led to an extensive search for evidence of alterations in markers of oxidative stress and oxidative damage in the SNc of these patients. A series of imaging, biochemical, and LAMMA (see below) studies has demonstrated that bulk iron levels are elevated in the SNc of PD patients (Dexter *et al.*, 1989a; Sofic *et al.*, 1991; Olanow *et al.*, 1992) and that this elevation occurs within neuromelanin granules of SNc neurons (Good *et al.*, 1992). If the excess iron is in the ferrous state and in a reactive form, it might then catalyze the conversion of hydrogen peroxide derived from dopamine metabolism to hydroxyl radicals. Levels of ferritin, the major system of protection against reactive iron, have been examined to provide indirect information as to the likely reactivity of iron in the SNc. Unfortunately, results have been conflicting, possibly because one study used antibodies directed against hepatic ferritin and the other against splenic ferritin (Riederer *et al.*, 1989; Dexter *et al.*, 1991). More recently, using specific antibodies to L and H chain brain ferritin, evidence has been provided that neuronal ferritin does not show the anticipated rise to an increase in brain iron, supporting the notion that at least some of the increased iron in the SNc in PD may be in a reactive form (Connor *et al.*, 1995). In addition, we have demonstrated that infusion of iron into the SNc of rodents can induce a parkinsonian syndrome with histologic biochemical, and behavioral features of PD (Sengstock *et al.*, 1992, 1993). These changes are dose dependent and progressive following single iron infusions (Sengstock *et al.*, 1994). These findings support the concept that iron accumulation in PD may promote oxidant stress and lead to neuronal degeneration (Olanow and Youdim, 1996). However, we have also demonstrated that lesions placed at sites remote from the nigrostriatal tract can lead to a secondary increase in SNc iron (Oestreicher *et al.*, 1994). Thus, it is possible that the iron accumulation found in PD may arise secondary to an alternate etiologic process and may not be directly responsible for initiating the neurodegenerative process. Nonetheless, if the iron is in a reactive form, it may still contribute to neuronal degeneration.

The finding that a selective mitochondrial complex I toxin can lead to parkinsonism has led to a detailed analysis of mitochondria in PD. Both Mizuno and Schapira and their coworkers have demonstrated that a mitochondrial complex I defect is also present in the SNc in PD (Schapira *et al.*, 1990; Mizuno *et al.*, 1989). This defect has also been demonstrated in mitochondria from skeletal muscle, platelets, and fibroblasts of PD patients (Parker *et al.*, 1989; DiMauro *et al.*, 1993; Mytilineou *et al.*, 1994) raising the possibility that these patients possess a generalized mitochondrial defect, possibly due to a susceptibility to complex I damage. The source of such a defect in complex I in PD patients is currently unknown but could be secondary to MPTP-like toxins, oxidative stress, increased cytosolic calcium, or a primary mitochondrial DNA mutation. However, it has been noted that oxidative damage to mitochondria results in more widespread damage with involvement of complexes I and IV (Hartley *et al.*, 1993). Defects in mitochondria could lead to diminished ATP production and a rise in intracellular calcium due to decreased calcium extrusion and sequestration. This is a particular concern in view of recent studies demonstrating parallel reduction in α-ketoglutarate dehydrogenase, which would enhance the functional consequences of a complex I defect (Mizuno *et al.*, 1994). Additionally, recent studies suggest that mitochondrial defects are a primary factor in apoptotic cell death, which has been implicated in PD (Anglade *et al.*, 1995).

There is also evidence of decreased GSH in the SNc in PD. A number of studies have reported a decrease in reduced glutathione (GSH) levels in PD (Riederer *et al.*, 1989), but unequivocal results have not emerged until recently. Early investigations by Perry *et al.* (1982) reported a dramatic reduction in GSH levels to virtually zero, however these studies were considered flawed for a number of technical reasons. More recently, carefully controlled studies have demonstrated a decrease in GSH levels in the SNc of PD patients with no concomitant increase in GSSG (Sian *et al.*, 1994a). The cause of this decrease is unknown but two likely possibilities stand out. One is based on the recent finding of an increase in the levels of γ-glutamyl transpeptidase (γ-GTP), an enzyme responsible for clearing GSSG from cells (Sian *et al.,* 1994b). The oxidation product of GSH, GSSG is itself associated with cellular toxicity through the formation of mixed protein thiols. Increased levels of GSSG secondary to oxidative stress could provoke a rise in levels of γ-GTP, resulting in less GSSG available for reduction by glutathione reductase and accordingly lower levels of GSH. Alternatively, mitochondrial complex I defects could lead to a reduction in the energy-dependent synthesis of GSH by the synthetic enzyme, γ-glutamyl cystenyl synthase. Analyses of SOD, ascorbate, and alpha tocopherol have not demonstrated any alterations in these molecules (reviewed in Jenner and Olanow, 1996).

Defects in systems with the potential to lead to oxidative stress do not establish that oxidative damage to critical molecules capable of causing neurodegeneration has occurred. However, investigations of postmortem specimens of PD brains demonstrate that a number of molecules have undergone oxidative damage in the SNc. Increased lipid peroxidation has been demonstrated based on the findings of increased levels of the intermediates malonyldialdehyde and lipid hydroperoxides (Dexter *et al.,* 1989b, 1994a). The finding of increased levels of 8-OH deoxyguanosine, a marker of the hydroxylation of deoxyguanosine by hydroxyl radicals, provides evidence of oxidative damage to DNA (Sanchez-Ramos *et al.,* 1994). More recently, new evidence has been put forth for oxidized proteins and nitrated tyrosine groups in the SNc of PD patients, providing further support for a role of oxidant stress in the pathogenesis of cell degeneration in PD (reviewed in Jenner and Olanow, in press).

More recently, a novel mechanism for oxidative damage mediated by peroxynitrite ($ONOO^-$) has been implicated in the pathogenesis of PD. As shown in equation 4, $ONOO^-$ is formed at a near diffusion-limited rate by the reaction of superoxide radical with nitric oxide ($^.NO$). Peroxynitrite is capable of preferentially nitrating tyrosine residues in reactions that are catalyzed by either superoxide anion or iron. In addition, peroxynitrite is protonated to peroxynitrous acid (HONOO), which can spontaneously decompose into the cytotoxic hydroxyl radical and the nitrogen dioxide radical ($NO_2^.$), which is also capable of protein nitration. Conditions such as excitotoxicity that result in a rise in neuronal intracytoplasmic calcium can result in activation of calcium-dependent protease and nitric oxide synthase (NOS) enzymes leading to overproduction of superoxide and nitric oxide radicals. Under these circumstances neighboring structures could be susceptible to damage by protein nitration. Point mutations in superoxide dismutase, as have been described in familial amyotrophic lateral sclerosis (FALS) are associated with selective degeneration of motor neurons (Rosen *et al.,* 1993). More recently, electron spin resonance studies have demonstrated that these SOD mutations promote the formation of free radicals and apoptosis of cultured neurons (Wiedau-Pazos *et al.,* 1996). These reactions are promoted by the transition metal copper and inhibited by copper chelation. We have recently developed an antibody that recognizes nitrotyrosine and demonstrates markedly increased nitrotyrosne staining in neurofibrillary tangles in AD, as well as in neurofibrillary tangles found in substantia nigra neurons of patients with Guamanian Parkinson dementia (Good *et al.,* 1996). These findings provide direct evidence of $^.NO$-mediated free radical toxicity in these neurodegenerative disorders.

14.4. MICROPROBE STUDIES OF TRACE ELEMENTS IN PARKINSON'S DISEASE

Biochemical measures of trace metals rely on bulk analytic techniques that do not provide information on their subcellular localization and may miss regional metal accumulations, particularly in the early stages of disease. Attempts to determine the subcellular distribution of iron and other trace metals in PD using conventional techniques are relatively insensitive and imprecise. Initial studies using Perls stain suggested that in PD, iron is localized to areas devoid of neuromelanin (Sofic *et al.,* 1991). Further studies using X-ray microprobe provided conflicting results, suggesting that iron both does, and does not, accumulate within neuromelanin granules (Hirsch *et al.,* 1991, Jellinger *et al.,* 1992). Further, there are substantial limitations of this technique as well. X-ray microprobe analysis generally employs scanning electron microscopy (SEM) for imaging the specimen and identification of targets for analysis. This technique employs relatively thick brain sections, and the surface images provided by SEM do not provide sufficient cellular detail to enable definitive identification of specific cell type, subcellular component, or pathological inclusion. X-ray spectrometry also has a relatively high minimum detection level for most elements of interest [approximately 200–300 parts per million (ppm) for iron] and thus may not detect subtle increases of trace elements in pathologic conditions. Additionally, because of the surface irregularities in sectioned biologic tissues and variability in the depth of penetration of the electron beam, this technique does not provide consistent quantitative measures of the amount of element present in a tissue sample. Accordingly, one cannot reliably compare the concentration of a particular element in different brain specimens.

14.4.1. The Laser Microprobe

A determination of the precise cellular and subcellular distribution of iron and other metals is of critical importance in evaluating their potential role in the pathogenesis of oxidative stress in conditions such as PD. In our laboratory, we have employed the laser microprobe mass analyzer, or LAMMA, technique to analyze trace elements in biological tissues. The LAMMA provides precise histological detail because it employs semithin plastic-embedded sections, elemental detection limits on the order of 1–5 ppm, and the ability with properly prepared standardized samples to quantitate the concentration of a given element of interest within the experimental tissue sample. The LAMMA incorporates a high resolution (100 ×, N.A. 1.4) quartz objective that permits precise histologic identification of target sites measuring as little as 1 micron in diameter. After determination of a specific target site, elements are separated by a pulsed laser and subsequently detected by a time-of-flight mass spectrometer. This produces a mass spectrum with simultaneous identification of all elements throughout the periodic table. Additionally, the intensity of the spectral signal is proportional to the concentration of the element present within the target site. Thus, relative concentrations can be determined by comparisons of peak intensity (Verbueken *et al.,* 1988, Good *et al.,* 1991). Alternatively, by using comparisons to assayed biological standards, quantitative estimates of the intracellular concentration of individual elements can be determined (Good *et al.,* 1994). In contrast to bulk analytic methods, the LAMMA collects elemental data from precisely defined subcellular targets, thereby providing the means to identify pathologic accumulations of elements of interest and associate such accumulations with specific cellular sites.

LAMMA samples are prepared as semi-thin (typically 0.75 μm. thick), plastic-embedded sections that are stained with toluidine blue and mounted on a standard 3 mm electron microscopy grid. The grid is mounted on an X–Y stage traverse and placed in a high vacuum behind a quartz coverglass. The LAMMA is equipped with a high power pulsed Nd:YAG laser that can be focused through the 100X objective of the light microscope to a target site on the specimen measuring approximately 1 μm in diameter. The high energy laser is then directed to the target area of interest by a collinear continuous low-energy He:Ne laser, which is visible as a red dot on the section. A brief (15 nsec) pulse of the high-energy laser vaporizes and ionizes a small area of the tissue specimen, creating a minute perforation in the section. The ions produced by the laser perforation are attracted by a series of charged ion lenses into a 1.8 meter long time-of-flight mass spectrometry column where the ions are separated and detected according to their atomic mass number.

The tissues that are probed by LAMMA are examined as stained histologic sections with high-resolution light microscopy. In contrast to bulk analytic methods, the LAMMA collects elemental data by targeting and sampling specific cellular features. Accordingly, cells of interest are identified and targeted for analysis based on their light microscopic appearance. The size of the probe site (laser perforation) is sufficiently small that specific intracellular sites can be targeted and analyzed selectively. The site of microprobe analysis can be immediately confirmed by observing the location of the laser perforation responsible for the production of each individual spectrum.

14.4.2. Laser Microprobe Studies in Parkinson's Disease

Based on the findings of increased SNc iron in PD using MRI and bulk analytic methods, we examined dopaminergic neurons in the SNc in PD patients and controls to determine the cellular and subcellular distribution of iron and other trace elements (Good *et al.,* 1992). We evaluated brain specimens from three PD patients with a classic clinical presentation and a history of a good response to levodopa. In all three cases, neuropathologic evaluation demonstrated prominent loss of pigmented neurons in the SNc accompanied by Lewy bodies in remaining nigral neurons. Cases employed in this study were free of dementia and showed no Alzheimer's-disease-related changes on neuropathologic examination. Three age-matched controls were chosen for comparison based on normal extrapyramidal and cognitive functions during life and an absence of AD or PD changes on neuropathologic examination. Following fixation in 10% buffered formalin, tissue blocks measuring $1 \times 2 \times 2$ mm were dissected from the SNc, osmicated in 2% aqueous OsO_4, and embedded in Spurr's epoxy resin. Semithin sections (0.75 μm thick) of each block were cut, stained with toluidine blue, and mounted for LAMMA study.

Intact SNc neurons containing neuromelanin granules were targeted for LAMMA analysis. Laser probe sites were directed to neuromelanin granules, non-melanized portions of the cytoplasm, and the neuropil adjacent to each selected neuron. Multiple laser shots were directed to each subcellular target and a total of 10 neurons were

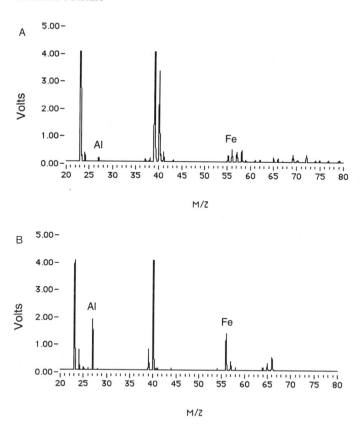

FIGURE 14.1. Representative partial mass spectra derived from a single probe site to neuromelanin of SNc neurons of A) autopsy control and B) Parkinson's disease. Mass state (m/z = mass to charge ratio) is indicated and signal strength is in volts. In the PD case note the elevated peaks at mass 27 (aluminum) and 56 (major isotope of iron).

examined in each case. The control cases were prepared and probed in a similar manner using identical instrumental parameters. Representative spectra are presented in Figure 14.1. The spectra for each target area were averaged and a multivariant ANOVA was performed to compare the intensity of iron and aluminum in the various target areas of PD patients and controls. The mean integrated peak intensities for iron within probe sites directed to neuromelanin and adjacent neuropil for the PD patients and the controls are presented in Figure 14.2. Each data point represents the mean value of the total number of LAMMA acquisitions obtained for each target site. In specimens from control patients, peaks related to the presence of iron were more pronounced within intraneuronal neuromelanin granules than in non-melanized neuronal cytoplasm or in the adjacent neuropil ($p < 0.05$). In patients with PD, a similar pattern was detected. However, the iron-related signal in neuromelanin granules was significantly increased in comparison to controls ($p < 0.0001$). There were no significant differences in iron-related peaks in the non-melanized neuronal cytoplasm or the adjacent neuropil be tween PD patients and controls.

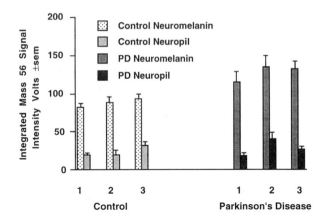

FIGURE 14.2. Mean integrated mass 56 signal intensity ($\pm$ sem) of probe sites to neuromelanin and adjacent neuropil of SNc neurons of autopsy controls and PD patients. In patients with PD, the iron-related signal in the neuromelanin granules was significantly increased in comparison to controls ($p < 0.01$ ANOVA and Student's t test).

Interestingly, in the PD cases, aluminum(Al)-related peaks were also dramatically increased in the neuromelanin granules in comparison to controls ($p < 0.005$). Aluminum is not a transition metal and does not participate in redox reactions because of its constant Al^{3+} valence state. However, aluminum salts have been shown to increase iron-catalyzed lipid peroxidation by eightfold (Gutteridge *et al.,* 1985). The mechanism responsible for this effect is not known, but it has been proposed that aluminum may displace iron from its binding site and/or rearrange membrane lipids so as to facilitate peroxidative attack by iron. Regardless of the mechanism, the finding of a marked increase in iron and aluminum within dopamine neurons that preferentially degenerate in PD, further supports the notion that oxidative stress contributes to the pathogenesis of cell death in this disorder.

To assess the possibility that iron can contribute to the neurodegenerative process in PD, we stereotactically infused iron citrate into the SNc of rodents. We demonstrated that iron induces a dose-related alteration in the number of residual nigral neurons, the striatal dopamine concentration, and the behavioral response to apomorphine (Sengstock *et al.,* 1992, 1993). Further, we and others have shown that SNc infusion of low concentrations of iron induces a progressive neuronal degeneration (Sengstock *et al.,* 1994; Riederer *et al.,* in press) and in this way more closely mirrors PD than does the parkinsonism induced by other neurotoxins such as 6-OHDA or MPTP. However, it has also been shown that lesions of the nigro-striatal tract induced by 6-OHDA or MPTP can result in a secondary accumulation of iron in the SNc (Oestreicher *et al.,* 1994; Temlett *et al.,* 1994). It is thus not possible to state with confidence whether the increased SNc iron found in PD is primary or secondary to an alternate etiologic process. Nonetheless, excess iron, regardless of its origin, if in a reactive form could contribute to the neurodegeneration that characterizes PD.

The possibility that GSH alterations might be the initiating step in the development of oxidative stress in PD must also be considered. As described above, GSH is one of the brain's principal defenses against oxidant stress and is deficient in the SNc of PD patients. The cause of this deviation is not known, but, it is noteworthy that GSH deficiency has been detected in patients with incidental Lewy bodies who are thought to have preclinical PD (Dexter *et al.*, 1994). As no alterations in mitochondria or iron concentration have been detected in this population of patients, it raises the possibility that a reduction in GSH may be the first alteration that occurs in PD. In this regard it is interesting that animals rendered GSH deficient by buthionine sulfoximine (BSO) do not develop parkinsonism but are more sensitive to dopaminergic toxins such as 6 OHDA and MPP$^+$ (Pilebad *et al.*, 1989; Wüllner *et al.*, 1996). It is therefore interesting to speculate that GSH may be the earliest alteration in PD patients, giving rise to symptomatology because it renders dopamine neurons vulnerable to later exposure to neurotoxins and oxidative damage.

14.5. PARKINSONISM ASSOCIATED WITH ACCUMULATION OF TRACE METALS

Oxidant stress and the pathologic accumulation of heavy metals have been implicated in a variety of other neurodegenerative disorders in which parkinsonism may be a major clinical manifestation. There has been considerable interest in the role of manganese (Mn) and oxidant stress in the pathogenesis of a parkinsonian disorder. Mn is a transition metal that is capable of promoting redox reactions and inducing tissue damage (Donaldson *et al.*, 1981). Specifically, a parkinsonian disorder resembling PD has been described in association with Mn neurointoxication. The syndrome is characterized by gait disturbance with particular difficulty walking backwards, rigidity, bradykinesia, and dystonia. It has primarily been described in workers who have had massive occupational exposure to Mn (Huang *et al.*, 1993). In addition, Mn intoxication has been seen in association with chronic liver disease, presumably due to the failure of the diseased liver to satisfactorily clear the normal dietary load of Mn (Hauser *et al.*, 1994).

Mn intoxication can be associated with elevated levels of Mn in the blood and urine as well as abnormal signal hyperintensity in the globus pallidus (GP) and substantia nigra pars reticularis (SNr) on high field strength, T1-weighted, MRI (Newland *et al.*, 1989). Positron emission tomography (PET) studies show normal striatal fluorodopa (FD) uptake indicating relative preservation of the integrity of the nigro-striatal tract (Wolters *et al.*, 1989). Pathologically, Mn intoxication is associated with damage that is primarily localized to the GP and relatively spares the SNc. It can be differentiated from PD by the clinical picture (early involvement of gait with dystonia with relative absence of tremor and asymetry), lack of response to levodopa, MRI pattern, and normal striatal FD uptake on PET scan (Calne *et al.*, 1994).

Brain uptake of Mn occurs by way of transferrin–transferrin receptor-mediated endocytosis (Aschner and Gannon, 1994) in a manner similar to that which occurs with iron (Roberts *et al.*, 1992). Recently, we have produced a primate model of manganese intoxication with parkinsonian features by employing intravenous injections of manganese chloride (Olanow *et al.*, 1996b). Animals developing symptoms of manganese intoxication showed bradykinesia, rigidity, and facial grimacing suggestive of dystonia. The clinical symptoms were not responsive to levodopa. On neuropathological examination, pathological changes were noted in the GP and substantia nigra pars reticulata (SNr). Microscopically, there was severe perivascular and parenchymal mineralization that stained positively for iron. Trace elemental analysis by LAMMA, as described above, showed massive iron and aluminum accumulation in the mineral deposits, but, interestingly, Mn levels were not increased above background levels. While further studies to follow the time course of these mineralizations is necessary to obtain a complete picture of their appearance, this raises the possibility that Mn levels are transiently increased in the brain, result in damage to the blood–brain barrier, and the subsequent accumulation of iron and aluminum from exogenous sources. Together, these transition metals could induce damage as a result of oxidative stress. In addition, studies of the direct effect of Mn on striatal function (Brouillet *et al.*, 1993) demonstrate a drop in ATP production and an increase in lactate formation consistent with a mitochondrial lesion and a consequent impairment of energy metabolism. Interestingly, the effects of Mn-induced neurotoxicity in this model system can be blocked by removal of the cortical input to the striatum or by blockade of N-methyl-D-aspartate (NMDA) receptors implicating an excitotoxic mechanism, possibly secondary to a primary mitochondrial defect.

These findings suggest that the parkinsonism associated with Mn intoxication might be the result of oxidative stress consequent to any or all of: i) the redox-promoting properties of Mn, ii) Mn-induced mitochondrial damage, and iii) the pathologic accumulation of iron and aluminum possibly due to a breakdown in the blood–brain barrier. The selective involvement of the GP and SNr found pathologically in Mn intoxication most likely reflects the preferential accumulation of Mn and other heavy metals in these brain regions (Hallgren and Sourander, 1958) rather than any specific vulnerability of these structures to oxidant stress or Mn.

Hallervorden–Spatz syndrome is a disorder of iron metabolism characterized by the young onset of an akinetic-rigid syndrome and a variety of other neurological problems that include dystonia, chorea, retinitis pigmentosa, retardation, and seizures. Pathology is primarily localized to the GP and SNr and consists of massive iron accumulation with pigmentary degeneration and axonal spheroids (Olanow, 1994). This pattern of pathologic changes strongly suggests that iron-induced oxidative stress is implicated in the pathogenesis of cell degeneration, although the specific etiology of this condition is still not known. Wilson's disease is known to be due to a genetically determined defect in copper metabolism (Bull *et al.*, 1993; Tanzi *et al.*, 1993; Yamaguchi *et al.*, 1993) that results in a reduction in levels of the copper binding protein ceruloplasmin and a consequent rise in brain copper. Copper, like iron, is a transition metal that has the capacity to induce oxidative stress. Untreated, copper accumulation in Wilson's disease can lead to degeneration in the basal ganglia and a parkinsonian syndrome, particularly in young individuals. Multiple system atrophy (MSA) is associated with a parkinsonian syndrome and widespread degeneration that can involve the SNc, striatum, and brain stem nuclei. Clinically, these syndromes differ from PD by the early onset of speech and balance dysfunction, lack of resting tremor, and the relatively poor response to levodopa therapy (Olanow, 1992a). Pathologically,

Lewy bodies are only rarely encountered but relatively specific glial inclusions have been described (Wenning *et al.,* 1994). Both MRI and pathological studies have demonstrated a large increase in iron accumulation in degenerated regions and particularly the striatum (Olanow, 1992a; Dexter *et al.,* 1991). The significance of iron accumulation in these conditions is not known and it is likely that its is a secondary effect and not directly related to the cause of the disease.

14.6. THERAPEUTIC IMPLICATIONS

From a clinical perspective, the free-radical-oxidative stress hypothesis in PD provides a scientific rationale for testing trials of antioxidants in the hope that they might be "neuroprotective" and modify the natural progression of the disease. The recently completed DATATOP study (Parkinson Study Group, 1993), demonstrated that the free radical scavenger vitamin E did not delay the development of disability in patients with early PD, although, it is not certain than brain levels of vitamin E were significantly increased. In contrast, the relatively selective MAO-B inhibitor deprenyl significantly delayed the development of clinical disability in this same population of untreated PD patients. However, subsequent analyses demonstrated that deprenyl was associated with symptomatic benefits that confounded delineation of any neuroprotective effect the drug could have masked, rather than inhibited neuronal degeneration (Olanow and Calne, 1991). The recently completed Sindepar study (Olanow *et al.,* 1995) was designed to try and clarify this dilemma. One hundred and one patients with untreated PD were randomized to treatment with deprenyl or placebo. After approximately 12 months, deprenyl was discontinued and a final untreated evaluation was performed 14 months after initial baseline. A comparison of change in Parkinson status between the untreated baseline and final visits demonstrated that deprenyl-treated patients had significantly less deterioration than those receiving placebo. This study provides strong support for the notion that deprenyl acts through other than a symptomatic mechanism and is consistent with the notion that it is neuroprotective.

There are a number of mechanisms that might explain the deprenyl-related benefits seen in PD. As an MAO-B inhibitor, deprenyl might reduce oxidant stress by blocking H_2O_2 formation generated from the MAO-B oxidation of dopamine as shown in equation 3. However, in the doses employed, deprenyl does not prevent the formation of peroxides generated by MAO-A or auto-oxidation of dopamine. Both of these mechanisms might play an important role in dopamine metabolism, raising the possibility that more generic inhibition of dopamine metabolism might have a more potent neuroprotective effect, and more profoundly influence the natural progression of PD.

There is also evidence to suggest that the therapeutic action of deprenyl may result from mechanisms other than MAO-B inhibition. Tatton and coworkers demonstrated that delayed administration of deprenyl still protects nigral neurons from MPTP even though it is likely that such a delay prevents inhibition of MPTP to MPP^+ (Tatton *et al.,* 1991). In addition, deprenyl has been shown to protect cultured dopamine neurons from age-related degeneration, direct application of MPP^+, trophic factor withdrawal, and excitotoxicity (reviewed in Tatton *et al.,* 1996). Deprenyl benefits are not confined to dopaminergic nerve cells. Neuronal rescue has also been seen following optic crush injury and axotomy of facial motorneurons. Interestingly, these benefits can be seen at concentrations of deprenyl too low to inhibit MAO-B. It has been postulated that deprenyl acts by inducing transcriptional events with consequent upregulation of molecules that protect against oxidant stress and apoptotic cell death (Ansari *et al.*, 1993; Tatton *et al.*, 1994). These include SOD, catalase, bcl-2, and bcl-xl. The significance of such findings is that axotomized neurons undergoing degeneration as a result of the loss of trophic factors do so by the mechanism of apoptosis. This hypothesis is supported by recent work from our laboratory demonstrating that deprenyl protects against excitotoxicity through central mechanisms that are independent of its capacity to block MAO-B (Mytilineou *et al.*, 1996). Apoptosis, often thought of as "cellular suicide" contrasts with necrosis or "cellular homicide," is postulated as a possible mechanism in neurodegenerative diseases, and is thought to be mediated by free radical mechanisms (Hockenbery *et al.*, 1993). While a clear link between PD and apoptosis has not been established, evidence demonstrating this possibility has been forthcoming (Hartley *et al.*, 1994; Ziv *et al.*, 1994; Anglade *et al.*, 1995) and indicates that deprenyl and other strategies designed to counter apoptotic events is worth pursuing. There are numerous other antioxidant approaches that also might prove useful in PD including free radical scavengers, glutathione-like agents, iron chelators, and trophic factors that also might prove to be neuroprotective in PD and will likely be tested in the near future (reviewed in Olanow and Schapira, in press).

14.7.   REFERENCES

Anglade, P., Michel, P., Marquez, J., Mouatt-Prient, A., Ruberg, M., Hirsch, E. C., and Agid, Y., 1995, Apoptotic degeneration of nigral dopaminergic neurons in Parkinson's disease, *Proc. Soc. Neurosci.* **21**:489–493.

Aisen, P., 1992, Entry of iron into cells: A new role for the transferrin receptor in modulating iron release from transferrin, *Ann. Neurol.* **32**:62–68.

Ansari, K. S., Yu, P. H., Kruck, T. P., and Tatton, W. G., 1993, Rescue of axotomized immature rat facial motorneurons by R(-)-deprenyl: Stereospecificity and independence from monoamine oxidase inhibition, *J. Neurosci.* **13**:4042–53.

Aschner, M., and Gannon, M., 1994, Manganese (Mn) transport across the rat blood–brain barrier; saturable and transferrin dependent transport mechanisms, *Brain Res. Bull.* **33**:345–349.

Brouillet, E. P., Shinobu, L., McGarvey, U., Hochberg, F., and Beal, M. F., 1993, Manganese injection into the rat striatum produces excitotoxic lesions by impairing energy metabolism, *Exp. Neurol.* **120**:89–94.

Bull, P. C., Thomas, G. R., Rommens, J. M., Forbes, J. R., and Cox, D. W., 1993, The Wilson disease gene is a putative copper transporting P-type ATPase similar to the Menkes gene, *Nat. Genet.* **5**:327–337.

Calne, D. B., Chu, N. S., Hung, C. C., Lu, C. S., and Olanow, C. W., 1994, Manganism and idiopathic parkinsonism; similarities and differences, *Neurology* **44**:1583–1586.

Chandra, S. V., and Shukla, G. S., 1981, Concentrations of striatal catecholamines in rats given manganese chloride through drinking water, *J. Neurochem.* **32**:683–687.

Chiba, K., Trevor, A., and Castagnoli, N. Jr., 1984, Metabolism of the neurotoxic tertiary amine, MPTP, by brain monoamine oxidase, *Biochem. Biophys. Res. Comm.* **120**:457–478.

Cohen, G., Pasik, P., Cohen, B., Leist, L., Mytilingou, C., and Yahr, M. D., 1985, Pargyline and deprenyl prevent the neurotoxicity of 1-methyl-4-phenyl-1,2,3,6-tetrahydropyridine (MPTP) in monkeys, *Eur. J. Pharmacol.* **106**:209–210.

Connor, J. R., Snyder, B. S., Arosio, P., Loeffler, D. A., and LeWitt, P., 1995, A quantitative analysis of isoferritins in select regions of aged, parkinsonian, and Alzheimer's diseased brains, *J. Neurochem.* **65:**717–724.

Dexter, D. T., Wells, F. R., Lees, A. J., Agid, F., Agid, Y., Jenner, P., Marsden, C. D., 1989a, Increased nigral iron content and alterations in other metal ions occurring in brain in Parkinson's disease, *J. Neurochem.* **52:**1830–1836.

Dexter, D. T., Carter, C. J., Wells, F. R., Javoy-Agid, F., Agid, Y., Lees, A., Jenner, P., and Marsden, C. D., 1989b, Basal lipid peroxidation in substantia nigra is increased in Parkinson's disease, *J. Neurochem.* **52:**381–389.

Dexter, D. T., Carayon, A., Javoy-Agid, F., Agid, Y., Wells, F. R., Daniel, S. E., Lees, A. J., Jenner, P., and Marsden, C. D., 1991, Alterations in the levels of iron, ferritin and other trace metals in Parkinson's disease and other neurodegenerative diseases affecting the basal ganglia, *Brain* **114:**1953–1975.

Dexter, D. T., Holley, A. E., Flitter, W. D., Slater, T. F., Wells, F. R., Daniel, S. E., Lees, A. J., Jenner, P., and Marsden, C. D., 1994a, Increased levels of lipid hydroperoxides in the parkinsonian substantia nigra: An HPLC and ESR study, *Movement Disorders* **9:**92–97.

Dexter, D. T., Sian, J., Rose, S., Hindmarsh, J. G., Mann, V. M., Cooper, J. M., Wells, F. R., Daniel, S. E., Lees, A. J., Schapira, A. H. V., Jenner, P., and Marsden, C. D., 1994b, Indices of oxidative stress and mitochondrial function in individuals with incidental Lewy body disease, *Ann. Neurol.* **35:**38–44.

DiMauro, S., 1993, Mitochondrial involvement in Parkinson's disease: the controversy continues, *Neurology* **43:**2170–2172.

Donaldson, J., Labella, F. S., and Gesser, D., 1981, Enhanced autooxidation of dopamine as a possible basis of manganese neurotoxicity, *Neurotoxicology* **2:**53–64.

Good, P. F., Katz, R. N., and Perl, D. P., 1991, Calculation of intraneuronal concentration of aluminum in neurofibrillary tangle (NFT) bearing and NFT free hippocampal neurons of Alzheimer's disease using laser microprobe mass analysis (LAMMA), *J. Neuropathol. Exp. Neurol.* **50:**300.

Good, P. F., Olanow, C. W., and Perl, D. P., 1992, Neuromelanin-containing neurons of the substantia nigra accumulate iron and aluminum in Parkinson's disease: A LAMMA study, *Brain Res.* **593:**343–346.

Good, P. F., and Perl, D. P., 1994, A quantitative comparison of aluminum concentration in neurofibrillary tangles of Alzheimer's disease and parkinsonian dementia complex of Guam by laser microprobe mass analysis, *Neurobiol. Aging* **15:**S28.

Good, P. F., Werner, P., Hsu, A., Olanow, C. W., and Perl, D. P., 1996, Evidence for neuronal oxidative damage in Alzheimer's disease, *Am. J. Path.* **149:**21–28.

Gurney, M. E., Pu, H., Chiu, A. Y., Dal Canto, M. C., Polchow, C. Y., Alexander, D. D., Caliendo, J., Hentati, A., Kwon, Y. W., Deng, H.-X., Chen, W., Zhai, P., Sufit, R. L., and Siddique, T., 1994, Motor neuron degeneration in mice that express a human Cu, Zn superoxide dismutase mutation, *Science* **264:**1772–1775.

Gutteridge, J. M. C., Quinlan, G. J., Clark, I., and Halliwell, B., 1985, Aluminum salts accelerate peroxidation of membrane lipids stimulated by iron salts, *Biochem. Biophys. Acta* **835:**441–447.

Hallgren, B., and Sourander, P., 1958, The effect of age on the non-haemin iron in the human brain, *J. Neurochem.* **3:**41–51.

Halliwell, B., and Gutteridge, J., 1985, Oxygen radicals and the nervous system, *Trends Neurol. Sci.* **8:**22–29.

Halliwell, B., and Gutteridge, J. M. C., 1988, Iron as a biological prooxidant, *ISI Atlas Sci. Biochem.* **1:**48–52.

Hallgren, B., and Sourander, P., 1958, The effect of age on the non-haemin iron in the human brain, *J. Neurochem.* **3:**41–51.

Hartley, A., Cooper, J. M., and Schapira, A. H. V., 1993, Iron-induced oxidative stress and mitochondrial dysfunction: Relevance to Parkinson's disease, *Brain Res.* **627:**349–353.

Hartley, A., Stone, J. M., Heron, C., Cooper, J. M., and Schapira, A. H., 1994, Complex I inhibitors induce dose-dependent apoptosis in PC12 cells: Relevance to Parkinson's disease, *J. Neurochem.* **63:**1987–1990.

Hauser, R. A., Zesiewicz, T. A., Rosemurgy, A. S., Martinez, C., and Olanow, C. W., 1994, Manganese intoxication and chronic liver failure, *Ann. Neurol.* **36:**871–875.

Hirsch, E. C., Brandel, J. P., Galle, P., Javoy-Agid, F., and Agid, Y., 1991, Iron and aluminum increase in the substantia nigra of patients with Parkinson's disease: An x-ray microanalysis, *J. Neurochem.* **56:**446–51.

Hockenbery, D. M., Oltvai, Z. N., Yin, X. M., Milliman, C. L., and Korsmeyer, S. J., 1993, Bcl-2 functions in an antioxidant pathway to prevent apoptosis, *Cell* **75**:241–51.

Huang, C. C., Lu, C. S., Chu, N. S., Hochberg, F., Lilenfeld, D., Olanow, C. W., and Calne, D. B., 1993, Progression after chronic manganese exposure, *Neurology* **43**:1479–1483.

Jellinger, K., Kienzl, E., Rumpelmair, G., Riederer, P., Stachelberger, H., Ben-Shachar, D., and Youdim, M. B., 1992, Iron-melanin complex in substantia nigra of parkinsonian brains: An x-ray microanalysis, *J. Neurochem.* **59**:1168–71.

Jenner, P., Schapira, A. H. V., and Marsden, C. D., 1992, New insights into the cause of Parkinson's disease, *Neurology* **42**:2241–2250.

Jenner, P., and Olanow, C. W., 1996, Pathological evidence for oxidative stress in Parkinson's disease and related degenerative disorders, in: *Neurodegeneration and Neuroprotection in Parkinson's Disease* (C. W. Olanow, P. Jenner, and M. H. B. Youdim, eds.), Academic Press, London, pp. 24–45.

Jenner, P., and Olanow, C. W., 1996, Oxidative stress and the pathogenesis of Parkinson's disease, *Neurology* **47**:S161–S170.

Langston, J. W., Ballard, P. A., Tetrud, J. W., and Irwin, I., 1983, Chronic parkinsonism in humans due to a product of meperidine analog synthesis, *Science* **219**:979–980.

Mizuno, Y., Ohta, S., Tanaka, M., Takamiya, S., Suzuki, K., Sato, T., Oya, H., Ozawa, T., and Kagawa, Y., 1989, Deficiencies in complex I subunits of the respiratory chain in Parkinson's disease, *Biochem. Biophys. Res. Commun.* **163**:1450–1455.

Mizuno, Y., Matuda, S., Yoshino, H., Mori, H., Hattori, N., and Ikebe, S.-I., 1994, An immunohistochemical study on α-ketoglutarate dehydrogenase complex in Parkinson's disease, *Ann. Neurol.* **35**:204–210.

Munro, H., 1993, The ferritin genes: Their response to iron status, *Nutr. Rev.* **51**:65–73.

Mytilineou, C., Werner, P., Molinari, S., DiRocco, A., Cohen, G., and Yahr, M. D., 1994, Impaired oxidative decarboxylation of pyruvate in fibroblasts from patients with Parkinson's disease, *J. Neural Transm.* **8**:223–228.

Mytilineou, C., Radcliffe, P., Leonardi, E. K., Werner, P., and Olanow, C. W., L-deprenyl protects mesencephalic dopamine neurons from glutamate receptor-mediated toxicity, *J. Neurochem.* (in press).

Newland, M. C., Ceckler, T. L., Kordower, J. H., and Weiss, B., 1989, Visualizing manganese in the primate basal ganglia with magnetic resonance imaging, *Exp. Neurol.* **106**:251–258.

Oestreicher, E., Sengstock, G. J., Riederer, P., Olanow, C. W., Dunn, A. J., and Arendash, G., 1994, Degeneration of nigrostriatal dopaminergic neurons increases iron within the substantia nigra: A histochemical and neurochemical study, *Brain Res.* **660**:8–18.

Olanow, C. W., 1992, Magnetic resonance imaging in parkinsonism, *Neurol. Clin.* **10**:405–20.

Olanow, C. W., 1993, A radical hypothesis for neurodegeneration, *Trends Neurosci.* **16**:439–444.

Olanow, C. W., 1994, Hallervorden Spatz syndrome, in: *Neurodegenerative Diseases* (D. B. Calne, ed.), W. B. Saunders, Philadelphia, pp. 807–823.

Olanow, C. W., and Calne, D., 1991, Does selegiline monotherapy in Parkinson's disease act by symptomatic or protective mechanisms?, *Neurology* **42**:13–26.

Olanow, C. W., and Schapira, A. V. M., Neuroprotection and Parkinson's disease, in: *Neuroprotection: Fundamental and Clinical Aspects* (P. R. Bar and F. Beal, eds.), Marcel Decker, New York (in press).

Olanow, C. W., Cohen, G., Perl, D. P., and Marsden, C. D., 1992, Role of iron and oxidant stress in the normal and parkinsonian brain, *Ann. Neurol.* **32**:1–145.

Olanow, C. W., Hauser, R. A., Gauger, L., Malapira, T., Koller, W., Hubble, J., Bushenbark, K., Lilenfeld, D., and Esterlitz, J., 1995, The effect of deprenyl and levodopa on the progression of signs and symptoms in Parkinson's disease, *Ann. Neurol.* **38**:771–777.

Olanow, C. W., and Youdim, M. H. B., 1996a, Iron and neurodegeneration: Prospects for neuroprotection, in: *Neurodegeneration and Neuroprotection in Parkinson's Disease* (C. W. Olanow, P. Jenner, and M. H. B. Youdim, eds.), Academic Press, London, pp. 55–67.

Olanow, C. W., Good, P. F., Shinotoh, H., Hewitt, K. A., Vingerhoets, F., Snow, B. J., Beal, M. F., Calne, D. B., and Perl, D. P., 1996b, Manganese intoxication in the rhesus monkey: A clinical, imaging, pathologic, and biochemical study, *Neurology* **46**:492–498.

Parker, W. D., Boyson, S. J., and Parks, J. K., 1989, Abnormalities of the electron transport chain in idiopathic Parkinson's disease, *Ann. Neurol.* **26**:719–723.

Parkinson Study Group, 1993, Effects of tocopherol and deprenyl on the progression of disability in early Parkinson's disease, *New Eng. J. Med* **328:**176–183.

Perry, T. L., Godin, D. V., and Hansen, S., 1982, Parkinson's disease: A disorder due to nigral glutathione deficiency?, *Neurosci. Lett.* **33:**305–310.

Pileblad, E., Magnusson, T., and Fornstedt, B., 1989, Reduction of brain glutathione by L-buthionine sulfoximine potentiates the dopamine-depleting action of 6-hydroxydopamine in rat striatum, *J. Neurochem.* **52:**978–980.

Riederer, P., and Wasserman, A., Progressive lesioning of nigrostriatal dopamine neurons by intranigral iron injections, *J. Neural Trans.* (in press).

Riederer, P., Sofic, E., Rausch, W.-D., Schmidt, B., Reynolds, G. P., Jellinger, K., and Youdim, M. B. H., 1989, Transition metals, ferritin, glutathione, and ascorbic acid in parkinsonian brains, *J. Neurochem.* **52:**515–520.

Roberts, R., Sandra, A., Siek, G. C., Lucas, J. J., and Fine, R. E. 1992, Studies of the mechanism of iron transport across the blood-brain barrier, *Ann. Neurol.* **32:**S43–S50.

Rosen, D. R., Siddique, T., Patterson, D., Figlewicz, D. A., Sapp, P., Hentai, A., Donaldson, D., Goto, J., O'Regan, J. P., Deng, H.-X., Rahmani, Z., Krizus, A., McKenna-Yosek, D., Cayabyab, A., Gaston, S. M., Berger, B., Tanzi, R. E., Halperin, J. J., Herzfeldt, B., Van den Bergh, R., Hung, W.-Y., Bird, T., Deng, G., Mulder, D. W., Smyth, C., Laihg, N. G., Soriano, E., Pericak-Vance, M. A., Haines, J., Rouleau, G. A., Gusella, J. S., Horvitz, H. R., and Brown, R. H., 1993, Mutations in Cu/Zn superoxide dismutase gene are associated with familial amyotrophic lateral sclerosis, *Nature* **362:**59–62.

Sanchez-Ramos, J. R., Overvik, E., and Ames, B. N., 1994, A marker of oxyradical-mediated DNA damage (8-hydroxy-2′-deoxyguanosine) is increased in nigro-striatum of Parkinson's disease brain, *Neurodegeneration* **3:**197–204.

Schapira, A. H. V., Cooper, J. M., Dexter, D., Clark, J. B., Jenner, P., and Marsden, C. D., 1990, Mitochondrial complex I deficiency in Parkinson's disease, *J. Neurochem.* **54:**823–827.

Sengstock, G., Olanow, C. W., Dunn, A. J., and Arendash, G. W., 1992, Iron induces degeneration of nigrostriatal neurons, *Brain Res. Bull.* **28:**645–649.

Sengstock, G. J., Olanow, C. W., Menzies, R. A., Dunn, A. J., and Arendash, G., 1993, Infusion of iron into the rat substantia nigra: Nigral pathology and dose-dependent loss of striatal dopaminergic markers, *J. Neurosci. Res.* **35:**67–82.

Sengstock, G., Olanow, C. W., Dunn, A. J., Barone, S., and Arendash, G., 1994, Progressive changes in striatal dopaminergic markers, nigral volume, and rotational behavior following iron infusion into the rat substantia nigra, *Exp. Neurol.* **130:**82–94.

Sian, J., Dexter, D. T., Lees, A. J., Daniel, S., Agid, Y., Javoy-Agis, F., Jenner P., and Marsden, C. D., 1994a, Alterations in glutathione levels in Parkinson's disease and other neurodegenerative disorders affecting basal ganglia, *Ann. Neurol.* **36:**348–355.

Sian, J., Dexter, D. T., Lees, A. J., Daniel, S., Jenner, P., and Marsden, C. D., 1994b, Glutathione-related enzymes in brain in Parkinson's disease, *Ann. Neurol.* **36:**356–361.

Sofic, E., Paulus, W., Jellinger, K., Riederer, P., and Youdim, M. B., 1991, Selective increase of iron in substantia nigra zona compacta of parkinsonian brains, *J. Neurochem.* **56:**978–82.

Tanzi, R. E., Petrukhin, K., Chernov, I., Pelleguer, J. L., Wasco, W., Ross, B., Romano, D. M., Parano, E., Pavone, L., Brzustowicz, L. M., Devoto, M., Peppercorn, J., Bush, A. I., Sternlieb, I., Pirasto, M., Grusella, J. F., Evgrafov, O., Penchaszadeh, G. K., Homig, B., Edelman, I. S., Soares, M. B., Scheinberg, I. H., and Gilliom, C., 1993, The Wilson gene is a copper transporting ATPase with homology to the Menkes disease gene, *Nat. Genet.* **5:**344–350.

Tatton, W. G., and Greenwood, C. E., 1991, Rescue of dying neurons: A new action for deprenyl in MPTP parkinsonism, *J. Neurosci. Res.* **30:**666–677.

Tatton, W. G., Ju, W. Y. L., Holland, D. P., Tai, C., and Kwan, M., 1994, (-)-Deprenyl reduces PC12 cell apoptosis by inducing new protein synthesis, *J. Neurochem.* **63:**1572–1575.

Tatton, W. G., Ju, W. Y. H., Wadia, J., and Tatton, N. A., 1996, Reduction of neuronal apoptosis by small molecules: promise for new approaches to neurological therapy, in: *Neurodegeneration and Neuroprotection in Parkinson's Disease* (C. W. Olanow, P. Jenner, and M. H. B. Youim, eds.), Academic Press, London, pp. 202–220.

Temlett, J. A., Landsberg, J. P., Watt, F., and Grime, G. W., 1994, Increased iron in the substantia nigra compacta of the MPTP-lesioned hemiparkinsonian African Green monkey: Evidence from proton microprobe elemental microanalysis, *J. Neurochem.* **62:**134–146.

Wenning, G. K., Quinn, N., Magalhaes, M., Mathias, C., and Daniel, S. E., 1994, "Minimal Change" multiple system atrophy, *Mov. Disord.* **9:**161–166.

Wolters, E. C. H., Huang, C. C., Clark, C., Peppard, R. F., Okada, J., Chu, N.-S., Adam, M. J., Ruth, T. J., Li, D., and Calne, D. B., 1989, Positron emission tomography in manganese intoxication, *Ann. Neurol.* **26:**647–651.

Wüllner, U., Löschmann, P.-A., Schulz, J. B., Schmid, A., Dringen, R., Eblen, F., Turski, L., and Klockgether, T., 1996, Glutathione depletion potentiates MPTP and MPP$^+$ toxicity in nigral dopaminergic neurones, *Neuroreport* **7:**921–923.

Verbueken, A. H., Bruynseels, F. J., Van Grieken, R., and Adams, F., 1988, Laser microprobe mass spectrometry, in: *Inorganic Mass Spectrometry* (F. Adams, R. Gibels and R. Van Grieken eds.), New York, John Wiley, pp. 173–256.

Wiedau-Pazos, M., Goto, J. J., Rabizadeh, S., *et al.*, 1996, Altered reactivity of superoxide dismutase in familial amyotrophic lateral sclerosis, *Science* **271:**515–518.

Yamaguchi, Y., Heiny, M. E., and Gitlin, J. D., 1993, Isolation and characterization of a human liver cDNA as a candidate gene for Wilson disease, *Biochem. Biophys. Res. Commun.* **197:**271–277.

Ziv, I., Melamed, E., Nardi, N., Luria, D., Achiron, A., Offen, D., and Barzilai, A., 1994, Dopamine induces apoptosis-like cell death in cultured chick sympathetic neurons—a possible novel pathogenetic mechanism in Parkinson's disease, *Neurosci. Lett.* **170:**136–40.

Chapter 15

Perspectives on the Mechanisms of Familial Amyotrophic Lateral Sclerosis Caused by Mutations in Superoxide Dismutase 1

David R. Borchelt, Philip C. Wong, Mark W. Becher,
Lucie I. Bruijn, Don W. Cleveland, Neal G. Copeland,
Valeria C. Culotta, Nancy A. Jenkins, Michael K. Lee,
Carlos A. Pardo, Donald L. Price, Sangram S. Sisodia,
and Zhou-Shang Xu

15.1. INTRODUCTION

The mechanisms that lead to the selective degeneration of motor neurons in familial and sporadic amyotrophic lateral sclerosis (FALS and ALS) are not clearly understood. A subset of FALS cases are caused by mutations in Cu/Zn superoxide dismutase 1

David R. Borchelt, Philip C. Wong, Mark W. Becher, Michael K. Lee, and Carlos A. Pardo   Department of Pathology and the Neuropathology Laboratory, The Johns Hopkins University School of Medicine and School of Hygiene and Public Health, Baltimore, Maryland 21205.   **Lucie I. Bruijn, Don W. Cleveland, and Zhou-Shang Xu**   Department of Biological Chemistry, The Johns Hopkins University School of Medicine and School of Hygiene and Public Health, Baltimore, Maryland 21205; *current address of LIB and DWC:* Departments of Neurology and Neuroscience, the Ludwig Institute, University of California at San Diego, La Jolla, California 92093; *current address of Z-SX:* The Worcester Foundation for Experimental Biology, Shrewsbury, Massachusetts 01545.   **Neal G. Copeland and Nancy A. Jenkins**   Mammalian Genetics Laboratory, ABL-Basic Research Program, NCI-Frederick Cancer Center Research and Development, Frederick, Maryland 21702.   **Valeria C. Culotta**   Departments of Biochemistry and Environmental Health Sciences, The Johns Hopkins University School of Medicine and School of Hygiene and Public Health, Baltimore, Maryland 21205.   **Donald L. Price**   Departments of Pathology, Neurology, and Neuroscience, and the Neuropathology Laboratory, The Johns Hopkins University School of Medicine and School of Hygiene and Public Health, Baltimore, Maryland 21205.   **Sangram S. Sisodia**   Departments of Pathology and Neuroscience and the Neuropathology Laboratory, The Johns Hopkins University School of Medicine and School of Hygiene and Public Health, Baltimore, Maryland 21205.

Metals and Oxidative Damage in Neurological Disorders, edited by Connor. Plenum Press, New York, 1997.

(SOD1) (Deng *et al.*, 1993; Rosen *et al.*, 1993), but the mechanism by which mutant SOD1 injure motor neurons is unclear. We have suggested that abnormalities in Cu^{++} metabolism may play a role in disease (Wong *et al.*, 1995). This speculation derives from a number of observations: Cu^{++} can catalyze the formation of toxic radical species (Olanow, 1993; Brown, 1995); SOD1 is abundant in nervous tissues (Pardo *et al.*, 1995; Tsuda *et al.*, 1994); and the metal binding domains of the enzyme generally lack mutations (Brown, 1995; Wong and Borchelt, 1995). Moreover, several investigations have demonstrated that FALS-linked mutations do not necessarily compromise free radical scavenging activity (Borchelt *et al.*, 1994; Fujii *et al.*, 1995; Rabizadeh *et al.*, 1995), and transgenic mice expressing mutant SOD1 develop motor neuron disease (MND) despite elevated or unchanged levels of superoxide scavenging activity (Gurney *et al.*, 1994; Ripps *et al.*, 1995; Wong *et al.*, 1995). Although these data suggest that abnormal Cu^{++} metabolism could play a role in the disease, it is not clear whether mutations in SOD1 alter the levels of free Cu^{++} by diminishing the functions of an abundant Cu^{++}-binding protein or whether mutations alter SOD1 structure to allow enzyme-bound Cu^{++} to catalyze deleterious reactions, such as protein nitration (Beckman *et al.*, 1993) or peroxidation (Stadtman, 1990; Yim *et al.*, 1990; Stadtman and Oliver, 1991; Yim *et al.*, 1993; Wiedau-Pazos *et al.*, 1996).

15.2. CLINICAL AND NEUROPATHOLOGICAL PHENOTYPES OF FALS

MND, including sporadic ALS and FALS, results from the selective dysfunction and death of subsets of motor neurons in the spinal cord and brainstem (Figure 15.1) (Harding, 1993; Smith, 1992; Williams and Windebank, 1993; Rowland, 1994a; Haverkamp *et al.*, 1995). Large motor neurons in these regions degenerate, leading to weakness, paralysis, and spasticity (Brownell *et al.*, 1970; Delisle and Carpenter, 1984; Tandan and Bradley, 1985a; Banker, 1986; Hirano, 1991; Kuncl *et al.*, 1992; Oppenheimer and Esiri, 1992; Price *et al.*, 1992a, 1992b; Harding, 1993; Williams and Windebank, 1993). The classical neuropathological features of ALS include intracytoplasmic abnormalities of neurofilaments (NF), axonal spheroids, Bunina bodies, ubiquitin-positive inclusions, and fragmentation of the Golgi apparatus (Figures 15.2

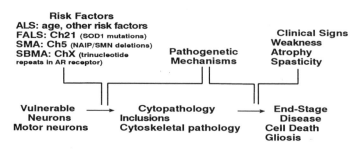

FIGURE 15.1. Motor neuron disease.

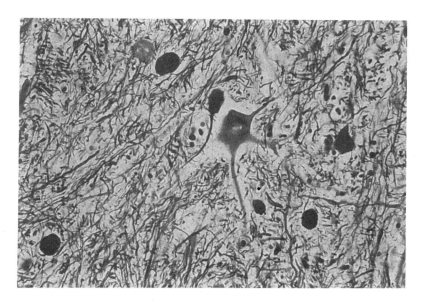

FIGURE 15.2. Case of ALS. Argentophilic axonal spheroids, silver stain.

and 15.3) (Carpenter, 1968; Hirano *et al.,* 1969, 1984; Hirano, 1973, 1984, 1991; Schmidt *et al.,* 1987; Chou, 1992; Gonatas *et al.,* 1992; Hirano and Kato, 1992; Matsumoto *et al.,* 1992; Leigh, 1994; Lee and Cleveland, 1994). Similar abnormalities, including SOD1-immunoreactive inclusions, occur in some cases of SOD1-linked FALS (Figure 15.4). In end-stage disease, there is a significant loss of large myelinated

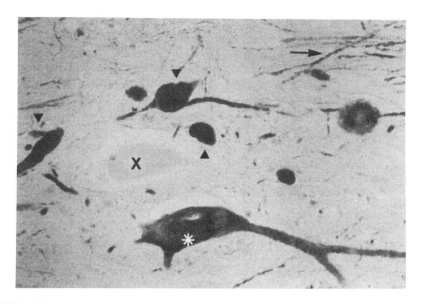

FIGURE 15.3. Case of ALS. Phosphorylated NF in the motor neuron cell body (*) and axonal spheroids (arrowheads). Immunocytochemistry for phosphorylated NF (*). Note normal cell body (X) and axons (arrow).

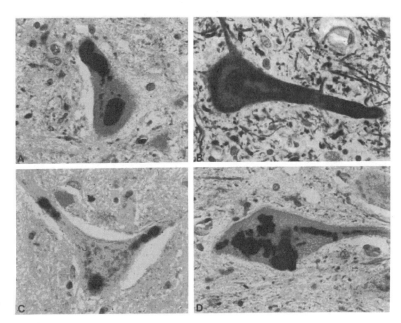

FIGURE 15.4. Inclusions showing immunoreactivity for ubiquitin (A), NF (B), and SOD1 (C,D). (Case of A4V FALS supplied by Dr. T. Siddique.)

fibers in corticospinal tracts and ventral roots as well as evidence of Wallerian degeneration and atrophy of axonal myelinated fibers (Delisle and Carpenter, 1984; Tandan and Bradley, 1985a; Banker, 1986; Manetto *et al.*, 1988; Munoz *et al.*, 1988; Sasaki *et al.*, 1988; Harding, 1993; Lowe *et al.*, 1993; Oppenheimer and Esiri, 1992; Williams and Windebank, 1993). Many end plates have small or absent axonal terminals, and skeletal muscles show scattered grouped atrophy (denervation) as well as fiber type grouping (reinnervation sprouting of surviving motor neurons).

In cases of ALS and FALS, abnormalities may extend beyond the large motor neurons of the brainstem and spinal cord to include nonmotor systems, such as the dorsal root ganglia and sensory nerves. Other populations of neurons that control eye movements and sphincters (i.e., Onuf's nucleus), as well as cells in Clarke's column and the intermediolateral cell column, are usually spared in both disorders. With heroic life support efforts, long-term survivors of ALS show more widespread abnormalities involving nonmotor systems (Kato *et al.*, 1993; Lowe, 1994). These data indicate that the insults that cause ALS and FALS are not restricted to motor neurons, but rather that motor neurons are intrinsically more vulnerable to the degenerative processes occurring in these diseases.

15.3. ETIOLOGICAL FACTORS/MECHANISMS IN ALS AND FALS

A variety of etiologies and risk factors have been proposed to explain why motor neurons dysfunction and die (Table 15.1) (Tandan and Bradley, 1985a, 1985b; Young,

Table 15.1
Postulated Mechanisms Associated with Motor Neuron Dysfunction/Death

Abnormalities of the cytoskeleton
Excitotoxicity mediated by glutamate receptors
Loss of superoxide scavenging activity (SOD1 mutations)
Oxidative injury/free radical damage
Nitration of proteins, particularly NF proteins
Calcium-related cell damage
Mitochondria injury
Metal toxicity
Autoimmune mechanisms
Loss of neurotrophic influences
Genetic abnormalities in proteins coded for by survival motor neuron (SMN) or neuronal apoptosis
  inhibitory protein (NAIP) genes

1990; Kurtzke, 1991; Harding, 1993; Alexianu *et al.*, 1994; Leigh, 1994; Rowland, 1994b; Beal, 1995; Beal *et al.*, 1995; Brady, 1995; Engelhardt *et al.*, 1995; Rothstein and Kuncl, 1995; Smith and Appel, 1995). A major advance in understanding the etiologies of FALS has come from molecular genetic investigations. Approximately 10% of adult-onset cases of ALS are familial, with autosomal dominant inheritance and an age-dependent penetrance (Siddique *et al.*, 1989, 1991). Recent studies have demonstrated that a subset set of FALS is caused by mutations in the metalloenzyme SOD1 (Deng *et al.*, 1993; Rosen *et al.*, 1993). More than 29 different missense mutations in SOD1 have been discovered to date (Table 15.2; Figure 15.5).

The importance of SOD1 in free radical scavenging and its neuroprotective activities has fueled the hypothesis that SOD1-linked FALS results from free radical damage caused by diminished superoxide scavenging activity (Deng *et al.*, 1993). For example, in several neuronal cell culture paradigms, diminished SOD1 activity is associated with increased vulnerability, whereas increased activity is associated with greater viability (Rothstein *et al.*, 1994a, 1994b; Troy and Shelanski, 1994; Greenlund *et al.*, 1995; Jordan *et al.*, 1995; Rabizadeh *et al.*, 1995). Moreover, initial studies reported that levels of SOD1 activity in a variety of tissues (e.g., red blood cells, lymphocytes, and brain) were reduced in individuals with SOD1-linked FALS (Bowling *et al.*, 1993; Deng *et al.*, 1993; Robberecht *et al.*, 1994). Thus, the disease was initially suggested to result from a reduction in SOD1 activity, a loss of free radical scavenging, and the burden of enhanced oxidative injury.

Despite the association between SOD1 and oxidative metabolism and the evidence that oxidative damage plays a role in neurodegenerative diseases (Dexter *et al.*, 1986; Hajimohammadreza and Brammer, 1990; Subbarao *et al.*, 1990; McIntosh *et al.*, 1991; Mecocci *et al.*, 1993; Yan *et al.*, 1994), there are several problems associated with the idea that reduced scavenging activity underlies the degeneration of motor neurons in SOD1-linked FALS. First, as described below, SOD1 mutations linked to FALS do not eliminate SOD1 activity (Borchelt *et al.*, 1994; Fujii *et al.*, 1995). Second, SOD1-linked FALS is a dominantly inherited disease, a condition not usually associated with loss of function. Finally, studies from three groups (Gurney *et al.*, 1994; Ripps *et al.*, 1995; Wong *et al.*, 1995) have demonstrated that mutant SOD1, when

Table 15.2
SOD1 Mutations Linked to FALS

Mutation	Exon	Reference
A4V,T	1	Deng *et al.*, 1993; Nakano *et al.*, 1994
V7E		Hirano *et al.*, 1994
G37R	2	Rosen *et al.*, 1993
L38V		
G41S,D		Rosen *et al.*, 1993; Rainero *et al.*, 1994
H43R		Rosen *et al.*, 1993
H46R		Aoki *et al.*, 1994
H48Q		Enayat *et al.*, 1995
L84V	4	Aoki *et al.*, 1995
G85R		Rosen *et al.*, 1993
G93C,A,R		Rosen *et al.*, 1993; Elshafey *et al.*, 1994
E100G		Rosen *et al.*, 1993
I104F		Ikeda *et al.*, 1995
L106V		Rosen *et al.*, 1993; Kawamata *et al.*, 1994
I112T		Enayat *et al.*, 1995
I113T		Rosen *et al.*, 1993
R115G		Kostrzewa *et al.*, 1994
v118...v119		Sapp *et al.*, 1995
D125H		Enayat *et al.*, 1995
L126....[a]	4,5	Pramatarova *et al.*, 1994
L144F,S	5	Deng *et al.*, 1993; Sapp *et al.*, 1995
A145T		Sapp *et al.*, 1995
V148G		Deng *et al.*, 1993
I149T		Pramatarova *et al.*, 1995

[a]This mutation is a 2-bp deletion that introduces a premature stop at codon 130

expressed in transgenic mice, leads to MND, despite unchanged or increased SOD1 activity levels relative to nontransgenic controls. Mice expressing high levels of wild-type (wt) Hu (human) SOD1 do not develop disease. These data are consistent with the hypothesis that the disease is caused by a dominant gain of an adverse property by mutant enzyme subunits (Wong *et al.*, 1995).

15.4. FUNCTIONS OF Cu/Zn SOD1

Cu/Zn SOD1 is encoded by an ~15-kD gene distributed in five exons on chromosome 21 (Rosen *et al.*, 1993). The superoxide dismutases (Table 15.3), a family of metalloenzymes, act as free radical scavengers by catalyzing the formation of H_2O_2 through the dismutation of $\cdot O_2$ (Fridovich, 1986; Imlay and Linn, 1988; Halliwell, 1992; Stadtman, 1992). SOD1 appears to be essential for aerobic life in many lower organisms; ablation of the SOD1 gene in yeast, *Drosophila,* and nematodes leads to increased sensitivity to $\cdot O_2$, and diminished growth and survival (Carlioz and Touati, 1986; Farr *et al.*, 1986; Bermingham-McDonogh *et al.*, 1988; Phillips *et al.*, 1989;

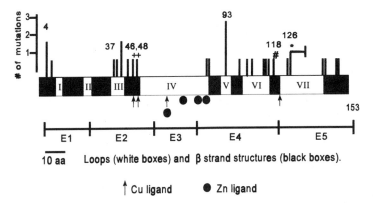

FIGURE 15.5. Distribution of FALS mutations in SOD1. Loops IV and VII fold to form the active site of the enzyme. Mutations in histidine ligands of Cu^{++} (+). Splicing mutations results in the in-frame insertion of three residues (#). Frameshift mutation alters the reading frame to terminate four residues from residue 126 (*). Numbers at the top of this figure mark the positions of mutations of particular interest because of frequency (4), novelty (118,126), introduction into transgenic mice (37,85,93), or importance in Cu^{++} binding.

Chary *et al.*, 1994; Orr and Sohal, 1994; Reveillaud *et al.*, 1994). However, it is noteworthy that a variety of metal binding proteins and/or free metals, such as Mn^{++} and Cu^{++}, can replace the activities of SOD1 in yeast (Liu and Cizewski-Culotta, 1994; Lapinskas *et al.*, 1995). Moreover, in yeast, SOD1 plays an important role in Cu^{++} homeostasis (Cizewski-Culotta *et al.*, 1995) indicating that superoxide scavenging activity is not the only function performed by the protein. Similarly, in mammalian cultures, Cu/Zn SOD1 is among the first molecules to bind radiolabeled Cu^{++} upon introduction to culture medium (Harris and Percival, 1989; Freedman and Peisach, 1989; Percival and Harris, 1989). Finally, the high abundance of SOD1 in tissues (0.1– 1.0% of total protein) (Pardo *et al.*, 1995; Tsuda *et al.*, 1994), more abundant than many typical enzymes, is consistent with the notion that SOD1 may be involved in Cu^{++} homeostasis. By virtue of the bound Cu^{++} ion, SOD1 can catalyze the dismutation of $\cdot O_2$, but whether catalysis of this reaction is the sole or even principal function of the molecule in all cells is unclear because $\cdot O_2$ can be metabolized by other cellular

Table 15.3
Members of the Superoxide Dismutase Family[a]

SOD1, a soluble homodimeric Cu,Zn enzyme, is distributed in the cytoplasm and
    nucleus
SOD2, a homotetrameric Mn enzyme in the mitochondrial matrix, is encoded by a
    gene on chromosome 6
SOD3, a homotetrameric, glycosylated Cu,Zn enzyme, is encoded by a gene on
    chromosome 4 and localized to the extracellular space

[a]From Hendrickson *et al.*, 1990.

proteins (Table 15.3). Thus, it is possible that one of the major functions of SOD1 is to regulate intercellular levels of free Cu^{++}.

15.5. PROPERTIES OF FALS MUTANT SOD1

Several studies have established that FALS mutations do not uniformly reduce enzymatic activities or diminish polypeptide stability (Borchelt *et al.*, 1994; Fujii *et al.*, 1995; Rabizadeh *et al.*, 1995). For example, native polyacrylamide gel assays of G37R mutant SOD1 expressed in COS-1 cells indicated that this mutant possesses a specific activity $\geq$ wt enzyme (Table 15.4) (Borchelt *et al.*, 1994). SOD1 with the G93C and I113T mutations showed similarly high levels of superoxide scavenging activity (Borchelt *et al.*, 1994). Although the G85R mutant appeared to be inactive in gel assays (Borchelt *et al.*, 1994), G85R HuSOD1 produced in insect cells showed full specific activity when assayed in solution (Table 15.4) (Fujii *et al.*, 1995). Two mutations of particular relevance to the hypothesis that deleterious Cu^{++} chemistry plays a role in the disease (see below) are the H46R and H48O mutations (Aoki *et al.*, 1994; Enayat *et al.*, 1995) that eliminate histidine ligands for the Cu^{++} cofactor (Parge *et al.*, 1992). However, one investigation of the H46R mutant demonstrated that, although this mutation reduces the affinity of the enzyme for Cu^{++}, enzymatic activity is not eliminated (Carri *et al.*, 1994).

Although the intrinsic activities of the mutants do not appear to be uniformly affected, a drastic reduction in the half-life of the enzyme could also diminish overall levels of SOD1 activity. In mammalian cells, the half-lives of mutant polypeptides

Table 15.4
Specific Activities of FALS Mutant SOD1 Assessed
via Two Different Methods

Mutant	Host	Specific activity[a]	Reference
A4V	COS-1 cells[b]	0.5	Borchelt *et al.*, 1994
G37R		1.0–1.5	
G41D		0.5	
G85R		0	
G93C		0.75	
I113T		0.75	
G41D	Sf9 insect cells[c]	0.5	Fujii *et al.*, 1995
H43R		0.65	
G85R		1.0	

[a]Specific activity of mutant homodimers relative to wild-type homodimers.
[b]Primate COS-1 cells were transiently transfected with pEF.BOS expression plasmids encoding mutant HuSOD1. Specific activities were estimated by determining enzyme activity in gel assays and quantifying HuSOD1 polypeptide levels by immunoblot.
[c]Insect Sf9 cells were infected with recombinant *Baculovirus* encoding mutant HuSOD1 (culture medium was supplemented with excess $CuCl_2$). Specific activities were determined on purified protein in solution assays.

show modest (25%) to significant (75%) reductions (Borchelt *et al.*, 1994). The half-life of wt HuSOD1 has been estimated to be ~30 hrs when transiently expressed in COS-1 cells (Borchelt *et al.*, 1994) and ≥48 hrs when constitutively expressed in mouse neuroblastoma N2a cells (Borchelt *et al.*, 1995). Thus, although some or all mutations may compromise polypeptide stability in motor neurons, the destabilizing effects of the mutations are superimposed on a relatively long-lived molecule.

To assess further whether FALS mutant SOD1 can provide normal function, mutant enzymes have been assayed for the complementation of growth defects in yeast in which the endogenous SOD1 gene has been disrupted (*sod1Δ* yeast) (Rabizadeh *et al.*, 1995). These yeast fail to grow in normal atmosphere in medium lacking lysine and methionine (Bilinski *et al.*, 1985; Chang and Kosman, 1990). SOD1 with FALS mutations G93C, L38V, G93A, and A4V retained the ability to rescue aerobic growth, obviated the requirements for lysine and methionine, and restored resistance to paraquat, a potent generator of ˙O_2 (Rabizadeh *et al.*, 1995). These data are consistent with the idea that mutant enzymes retain significant levels of normal function, including SOD1 activity.

Finally, it has been argued that mutant SOD1 may, upon the formation of heterodimeric enzymes, exert dominant negative effects on wt subunits (Deng *et al.*, 1993; Orrell *et al.*, 1995; Phillips *et al.*, 1995). Investigations measuring reductions of SOD1 levels to 30% of normal in erythrocytes from individuals with SOD1-linked FALS and reductions in SOD1 activities in *Drosophila* that harbor experimental SOD1 mutations have been interpreted to support this idea (Deng *et al.*, 1993; Orrell *et al.*, 1995; Phillips *et al.*, 1995). However, the coexpression of mutant subunits (G41D and G85R) with wt subunits in mammalian cells has demonstrated that neither the level of activity nor the stability of wt subunits was altered by the presence of excess mutant subunits (Borchelt *et al.*, 1995). Moreover, wt mammalian SOD1 subunits appear to exchange and reassort relatively rapidly at physiological temperatures and osmolarity (Borchelt *et al.*, 1995). Thus, to date, all available data suggest that many mutant SOD1 subunits retain significant levels of normal function, that SOD1 subunits with FALS-linked mutations do not transduce adverse structural changes to wt subunits, and that each subunit of the heterodimer is metabolized independently.

15.6. TRANSGENIC MODELS OF SOD1-LINKED FALS

Studies from three groups (Gurney *et al.*, 1994; Ripps *et al.*, 1995; Wong *et al.*, 1995) have demonstrated that mutant SOD1, when expressed in transgenic mice, leads to MND, whereas mice expressing high levels of wt HuSOD1 do not develop disease (Wong *et al.*, 1995). In mice expressing mutant enzymes, levels of SOD1 activity were either unchanged or elevated relative to nontransgenic controls (Table 15.5). For example, in the spinal cords of mice expressing high levels of the G37R variant, levels of the transgene-encoded SOD1 polypeptide in spinal cord were 5–12x levels of endogenous SOD1, leading to comparable increases in levels of SOD1 activity (Wong *et al.*, 1995).

Mice expressing high levels of G37R and G93A HuSOD1 develop a number of clinical signs, including reduced spontaneous movements, hind limb weakness, and

Table 15.5
Transgenic Mice with Wt or Mutant SOD1

| Mutant | Line | Strain | SOD1 Activity[a] | | SOD1[b] in spinal cord | Onset of MND (months) | Reference |
			Spinal cord	Brain			
HuWT	76	B6/C3[c]	10	no	10	none	Wong *et al.*, 1995
	30		8	data	7		
HuG37R	42		14		12	3.5–4	
	9		9		6	5–6	
	106		7		5	5.5–7.5	
	29		7		5	6–8	
HuG93A	G1	B6/SJL[d]	no	4	5	4–6	Gurney *et al.*, 1994
MoG85R[e]	M1	FVB	data	1	nd	3–4	Ripps et al., 1995

[a]SOD1 activity level relative to nontransgenic control.
[b]HuSOD1 polypeptide levels relative to endogenous MoSOD1.
[c]The original founder was an F2 hybrid of C57BL/6J × C3H/HEJ, with subsequent generations produced by mating of transgenic males to C57BL/6J × C3H/HEJ F1 females.
[d]The original founder was an F2 hybrid of C57BL/6J and SJL mice.
[e]The original founder and subsequent generations were FVB mice. The duration of disease in FVB mice was considerably more abbreviated than other transgenics listed, three days vs. 3–6 weeks, respectively. Whether this difference is caused by the strain of mouse, mutation tested, or species of SOD1 used remains to be determined.

muscle wasting. Electromyography disclose patterns consistent with MND. Eventually, the forelimbs become weak, and the hind limbs are completely paralyzed (Figure 15.6). In G37R mice expressing high levels of the transgene product, pathology is evident at two months of age and evolves rapidly. The most conspicuous abnormality initially is the presence of vacuoles in dendrites and axons of large motor neurons, with later involvement of the cell bodies. These vacuoles are usually associated with damaged mitochondria. Degenerating motor neurons also possess ubiquitin immunoreactivity, as occurs in the human disease, as well as altered patterns of phosphorylated NF. Motor nerve roots show characteristics of Wallerian degeneration.

In another transgenic model, produced by introducing a murine SOD1 genomic DNA fragment with a FALS mutation (G85R), MND similar to that of mice expressing mutant HuSOD1 was observed (Ripps *et al.*, 1995). In this model, no significant changes were noted in levels of total mouse SOD1 polypeptide or activity relative to controls. Clinically, mice develop generalized weakness at 3–4 months, progressing to spastic paralysis involving hind limbs and associated muscle wasting. Large spinal motor neurons were lost, and remaining neurons exhibited "pyknosis and karyorrhexis" associated with dystrophic neurites, large and small perikarya, and swollen fragmented processes (Ripps *et al.*, 1995).

The results of the foregoing investigations have established that multiple lines of transgenic mice expressing different FALS-linked mutant SOD1 develop some of the major pathological hallmarks of ALS, despite normal or elevated free radical scavenging activity. In our experience, multiple lines of mice, accumulating high levels of wt SOD1 in indistinguishable tissue distributions as mutant mice, do not develop clinical

FIGURE 15.6. G37R SOD1 mouse with MND.

signs or pathology, at least by 16 months of age (Wong *et al.,* 1995). Thus, in mice, MND does not appear to involve an imbalance of SOD1 activities as had been proposed (McCord, 1994). Moreover, there is no correlation between SOD1 activity and the age of onset or duration of disease in SOD1-linked individuals with FALS (Cleveland *et al.,* 1995). These findings, coupled with data indicating FALS mutations do not uniformly diminish SOD1 activity or stability, indicate that the disease in mice is caused by a dominant gain of a toxic property by mutant enzyme subunits (Wong *et al.,* 1995).

15.7. NATURE OF THE TOXIC PROPERTY

The central question in the pathogenesis of SOD1-linked FALS is how so many different mutations cause SOD1 to acquire one "adverse" property that leads to FALS (Wong *et al.,* 1995). The basic mechanism of disease in FALS mice is unlikely to involve a global abnormality in Cu^{++} metabolism, leading to increased levels of free Cu^{++} because, in models of free Cu toxicity, many other organs are affected in addition to the nervous system (Howell *et al.,* 1974; Gopinath and Howell, 1975). In the FALS transgenic mice, mutant Cu/Zn SOD1 is present in all tissues, but motor neurons are selectively damaged. One hypothesis is that mutant subunits fail to shield the Cu^{++} cofactor of the enzyme appropriately (Wong *et al.,* 1995), allowing inappropriate substrates to contact the catalytic Cu^{++}, thereby resulting in aberrant enzyme-catalyzed toxic products (Table 15.6). Consistent with the view that bound Cu^{++} plays a role in the pathogenicity of the mutant enzyme, among the 29 reported mutations linked to disease, none preclude the synthesis of a polypeptide with the Cu^{++}-binding binding domain (loop IV), and this domain is relatively free of mutations (Figure 15.5). Transition metals such as Cu^{++} can catalyze the formation of ·OH from H_2O_2 via the Fenton reaction (Olanow, 1993; Brown, 1995). Indeed, wt SOD1 can catalyze the formation of ·OH from peroxide (Yim *et al.,* 1990, 1993; Stadtman, 1990; Stadtman and Oliver, 1991). Recent investigations indicate that FALS-mutant SOD1 act as peroxidases, increasing the rate of hydroxyl radical formation, presumably by diminishing the binding/shielding of Cu^{++} (Wiedau-Pazos *et al.,* 1996). Hydroxyl species could damage a variety of cellular targets, including SOD1, mitochondrial

Table 15.6
SOD1, Cu/Zn, Motor Neuron, and Cell Damage

SOD1, a metal-binding/buffering protein, is abundant in neural tissue
Motor neurons have a high content of SOD1
Different mutations variably alter SOD1 activity and stability
Mutant polypeptide, which may not buffer/shield metals normally, can
    generate toxic radicals, nitrate tyrosines on proteins (i.e., SOD1, NF),
    promote lipid peroxidation, etc.

membranes (via lipid peroxidation), and glutamate transporters (via oxidation) (Figure 15.7). For example, both $\cdot O_2$ and $\cdot OH$ are known to damage glutamate transporters (Volterra *et al.,* 1992, 1994; Pogun *et al.,* 1994), and reductions in glutamate transporters, as observed in patients with ALS (Rothstein *et al.,* 1992, 1995), would lead to increases in the concentration of extracellular glutamate, activation of glutamate receptors, and increased Ca^{++} influx, enhancing the potential for excitotoxic injury (Table 15.7) (Coyle and Puttfarcken, 1993; Rothstein *et al.,* 1993; Rothstein and Kuncl, 1995).

In addition to catalyzing the formation of hydroxyl radicals, mutant SOD1 may interact with $\cdot OONO$ to catalyze the nitration of tyrosine residues (Beckman *et al.,* 1993) (Figure 15.7). The appearance of nitrotyrosine immunoreactivity in the spinal cords of individuals with sporadic ALS provides support for a possible role of this chemical reaction in the disease (Abe *et al.,* 1995). However, whether either of these potentially deleterious reactions is the principal mechanism by which mutant SOD1 causes disease remains to be determined. Considerable evidence indicates that significant fractions of cellular SOD1 lack bound Cu^{++} (Uauy *et al.,* 1985; Dameron and Harris, 1987; DiSilvestro, 1989; Levieux *et al.,* 1991; Steinkuhler *et al.,* 1991, 1994; Harris, 1992; Percival *et al.,* 1993). Thus, the bioavailability of Cu^{++} in tissues damaged by mutant SOD1 is an important variable to consider in defining the role of Cu^{++} in the disease.

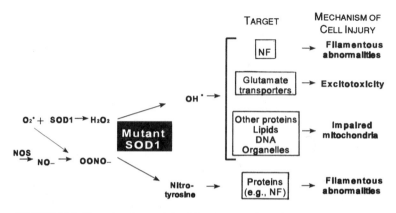

FIGURE 15.7. Normal and potentially deleterious reactions of wt and mutant SOD1.

Table 15.7
Postulated Events in Vulnerable Neurons

Primary event
    Adverse property of mutant SOD1
Secondary events
    Oxidative damage (free radicals)
    Damage to mitochondrial proteins/lipids
    Reduced energy supply
    Enhanced vulnerability to weak excitotoxicity
Amplifying cascade
    Ca^{++} influx
    Oxidative/excitotoxic injuries
    Alterations in cytoskeleton (i.e., nitration of NF) and impairment in axonal transport
Cell death/dissolution

15.8. FACTORS GOVERNING THE SEVERITY OF DISEASE

A number of studies have attempted to link the clinical severity of SOD1-linked FALS, as measured by the duration of clinical abnormalities, to the level of superoxide scavenging activity in circulating erythrocytes of individuals with mutations. However, to date, no clear correlation has emerged (Deng *et al.*, 1993; Cleveland *et al.*, 1995; Orrell *et al.*, 1995). This finding may, in part, be caused by influences of other genes on the clinical phenotype. However, it is anticipated that, in inbred transgenic mouse models of FALS, it will be possible to discern the intrinsic pathogenicity of FALS mutant SOD1 apart from the influences of other genes. Within the context of the hypothesis that deleterious Cu^{++} chemistry underlies disease, three intrinsic properties of the enzyme may influence pathogenicity: polypeptide half-life, Cu^{++} affinity, and protein structure surrounding the Cu^{++}-binding site. Enzymes with very short half-lives, very poor affinity for Cu^{++}, and a closed structure around the Cu^{++}-binding site would be expected to be less toxic. If these hypotheses prove true, then therapies that target the destruction of SOD1 or, in some manner, bind to exposed Cu^{++} within the active site, may slow the progression or delay the onset of SOD1-linked FALS.

15.9. CONCLUSIONS

The discovery of mutations in SOD1 in individuals with FALS has provided an unparalleled opportunity to explore the mechanisms by which specific populations of motor neurons in the cortex, brainstem, and spinal cord are damaged in FALS and ALS. Because all mutants examined to date retain some level of Cu^{++} affinity, as indicated by superoxide scavenging activity, we have proposed that damage to motor neurons may be mediated by deleterious reactions involving enzyme-bound Cu^{++}. Biochemical studies, such as those recently conducted by Wiedau-Pazos *et al.*, (1996) indicate that mutant SOD1 may indeed be more prone to produce toxic reactive substances such as ˙OH. However, whether such reactions play significant roles in the disease is presently unclear. The availability of transgenic models of SOD1-linked FALS provides a means to assess the extent of cellular damage by such reactive molecules and, perhaps through the use of specific therapeutic agents, to determine the role of reactive species in the pathogenesis of the disease.

ACKNOWLEDGMENTS. This work was supported by grants from the U.S. Public Health Service (NIH NS 20471, AG 05146, AG 07914, NS 27036[DWC]) as well as The Robert L. and Clara G. Patterson Trust, The Metropolitan Life Foundation, the American Health Assistance Foundation, the ALS Association, the Develbiss Fund, and funds from the Claster family. Drs. Price, Borchelt, and Wong are the recipients of a Leadership and Excellence in Alzheimer's Disease (LEAD) award (NIH AG 07914). Dr. Price is the recipient of a Javits Neuroscience Investigator Award (NIH NS 10580). Drs. Borchelt, Wong, Lee, and Cleveland are supported by grants from the ALS Association. Drs. Copeland and Jenkins are supported by the National Cancer Institute, DHHS, under contract with the ABL-Basic Research Program.

15.10. REFERENCES

Abe, K., Pan, L. H., Watanabe, M., Kato, T., and Itoyama, Y., 1995, Induction of nitrotyrosine-like immunoreactivity in the lower motor neuron of amyotrophic lateral sclerosis, *Neurosci. Lett.* **199**:152–154.

Alexianu, M. E., Ho, B.-K., Mohamed, A. H., La Bella, V., Smith, R. G., and Appel, S. H., 1994, The role of calcium-binding proteins in selective motoneuron vulnerability in amyotrophic lateral sclerosis, *Ann. Neurol.* **36**:846–858.

Aoki, M., Ogasawara, M., Matsubara, Y., Narisawa, K., Nakamura, S., Itoyama, Y., and Abe, K., 1994, Familial amyotrophic lateral sclerosis ALS in Japan associated with H46R mutation in Cu/Zn superoxide dismutase gene: A possible new subtype of familial ALS, *J. Neurol. Sci.* **126**:77–83.

Banker, B. Q., 1986, The pathology of motor neuron disorders, in *Myology* (A. G. Engel and B. Q. Banker, eds.), McGraw Hill, New York, p. 2031.

Beal, M. F., 1995, Aging, energy, and oxidative stress in neurodegenerative diseases. *Ann. Neurol.* **38**:357–366.

Beal, M. F., Hyman, B. T., and Koroshetz, W., 1995, Do defects in mitochondrial energy metabolism underlie the pathology of neurodegenerative diseases?, *Trends Neurosci.* **16**:125–131.

Beckman, J. S., Carson, M., Smith, C. D., and Koppenol, W. H., 1993, ALS, SOD and peroxynitrite, *Nature* **364**:584.

Bermingham-McDonogh, O., Gralla, E. B., and Valentine, J. S., 1988, The copper, zinc-superoxide dismutase gene of *Saccharomyces cerevisiae:* Cloning, sequencing, and biological activity, *Proc. Natl. Acad. Sci. USA* **85**:4789–4793.

Bilinski, T., Krawiec, Z., Liczmanski, A., and Litwinska, J., 1985, Is hydroxyl radical generated by the Fenton reaction in vivo?, *Biochem. Biophys. Res. Commun.* **130**:533–539.

Borchelt, D. R., Lee, M. K., Slunt, H. H., Guarnieri, M., Xu, Z.-S., Wong, P. C., Brown, R. H., Jr., Price, D. L., Sisodia, S. S., and Cleveland, D. W., 1994, Superoxide dismutase 1 with mutations linked to familial amyotrophic lateral sclerosis possesses significant activity, *Proc. Natl. Acad. Sci. USA* **91**:8292–8296.

Borchelt, D. R., Guarnieri, M., Wong, P. C., Lee, M. K., Slunt, H. S., Xu Z., Sisodia, S. S., Price, D. L., and Cleveland, D. W., 1995, Superoxide dismutase 1 subunits with mutations linked to familial amyotrophic lateral sclerosis do not affect wild-type subunit function, *J. Biol. Chem.* **270**:3234–3238.

Bowling, A. C., Schulz, J. B., Brown, R. H., Jr., and Beal, M. F., 1993, Superoxide dismutase activity, oxidative damage, and mitochondrial energy metabolism in familial and sporadic amyotrophic lateral sclerosis, *J. Neurochem.* **61**:2322–2325.

Brady, S., 1995, Interfering with the runners, *Nature* **375:**12–13.

Brown, R. H., Jr., 1995, Amyotrophic lateral sclerosis: recent insights from genetics and transgenic mice, *Cell* **80:**687–692.

Brownell, B., Oppenheimer, D. R., and Hughes, J. T., 1970, The central nervous system in motor neurone disease, *J. Neurol. Neurosurg. Psych.* **33:**338–357.

Carlioz, A., and Touati, D., 1986, Isolation of superoxide dismutase mutants in *Escherichia coli:* Is superoxide dismutase necessary for aerobic life?, *EMBO J.* **5:**623–630.

Carpenter, S., 1968, Proximal axonal enlargement in motor neuron disease, *Neurology* **18:**841–851.

Carri, M. T., Battistoni, A., Polizio, F., Desideri, A., and Rotilio, G., 1994, Impaired copper binding by the H46R mutant of human Cu,Zn superoxide dismutase, involved in amyotrophic lateral sclerosis, *FEBS Lett.* **356:**314–316.

Chang, E. C., and Kosman, D. J., 1990, O_2-dependent methionine auxotrophy in Cu,Zn superoxide dismutase-deficient mutants of *Saccharomyces cerevisiae, J. Bacteriol.* **172:**1840–1845.

Chary, P., Dillon, D., Schroeder, A. L., and Natvig, D. O., 1994, Superoxide dismutase *sod-1* null mutants of *Neurospora crassna:* Oxidative stress sensitivity, spontaneous mutation rate and response to mutagens, *Genetics* **137:**723–730.

Chou, S. M., 1992, Pathology-light microscopy of amyotrophic lateral sclerosis, in: *Handbook of Amyotrophic Lateral Sclerosis* (R. A. Smith, ed.), Marcel Dekker, New York, p. 133.

Cizewski-Culotta, V., Joh, H.-D., Lin, S.-J., Hudak-Slekar, K., and Strem, J., 1995, A physiological role for saccharomyces cerevisice copper/zinc superoide dismutase in copper buffering, *J. Biol. Chem.* **270:**29991–29997.

Cleveland, D. W., Laing, N., Hurse, P. V., and Brown, R. H., Jr., 1995, Toxic mutants in Charcot's sclerosis, *Nature* **378:**342.

Coyle, J. T., and Puttfarcken, P., 1993, Oxidative stress, glutamate, and neurodegenerative disorders, *Science* **262:**689–695.

Dameron, C. T., and Harris, E. D., 1987, Regulation of aortic CuZn-superoxide dismutase with copper. Caeruloplasmin and albumin re-activate and transfer copper to the enzyme in culture, *Biochem. J.* **248:**669–675.

Delisle, M. B., and Carpenter, S., 1984, Neurofibrillary axonal swellings and amyotrophic lateral sclerosis, *J. Neurol. Sci.* **63,** 241–250.

Deng, H.-X., Hentati, A., Tainer, J. A., Iqbal, Z., Cayabyab, A., Hung, W.-Y., Getzoff, E. D., Hu, P., Herzfeldt, B., Roos, R. P., Warner, C., Deng, G., Soriano, E., Smyth, C., Parge, H. E., Ahmed, A., Roses, A. D., Hallewell, R. A., Pericak-Vance, M. A., and Siddique, T., 1993, Amyotrophic lateral sclerosis and structural defects in Cu,Zn superoxide dismutase, *Science* **261:**1047–1051.

Dexter, D., Carter, C., Agid, F., Agid, Y., Lees, A. J., Jenner, P., and Marsden, C. D., 1986, Lipid peroxidation as cause of nigral cell death in Parkinson's disease, *Lancet* **2:**639–640.

DiSilvestro, R. A., 1989, Copper activation of superoxide dismutase in rat erythrocytes, *Arch. Biochem. Biophys.* **274:**298–303.

Elshafey, A., Lanyon, W. G., and Connor, J. M., 1994, Identification of a new missense point mutation in exon 4 of the Cu/Zn superoxide dismutase SOD-1 gene in a family with amyotrophic lateral sclerosis, *Hum. Mol. Genetics* **3:**363–364.

Enayat, Z. E., Orrell, R. W., Claus, A., Ludolph, A., Bachus, R., Brockmüller, J., Ray-Chaudhuri, K., Radunovic, A., Shaw, C., Wilkinson, J., King, A., Swash, M., Leigh, P. N., de Belleroche, J., and Powell, J., 1995, Two novel mutations in the gene for copper zinc superoxide dismutase in UK families with amyotrophic lateral sclerosis, *Hum. Mol. Genet.* **4:**1239–1240.

Engelhardt, J. I., Siklós, L., Komuves, L., Smith, R. G., and Appel, S. H., 1995, Antibodies to calcium channels from ALS patients passively transferred to mice selectively increase intracellular calcium and induce ultrastructural changes in motoneurons, *Synapse* **20:**185–199.

Farr, S. B., D'Ari, R., and Touati, D., 1986, Oxygen-dependent mutagenesis in *Escherichia coli* lacking superoxide dismutase, *Proc. Natl. Acad. Sci. USA* **83:**8268–8272.

Freedman, J. H., and Peisach, J., 1989, Intracellular copper transport in cultured hepatoma cells, *Biochem. Biophys. Res. Commun.* **164:**134–140.

Fridovich, I., 1986, Superoxide dismutases, *Adv. Enzymol. Relat. Areas Mol. Biol.* **58:**61–97.

Fujii, J., Myint, T., Seo, H. G., Kayanoki, Y., Ikeda, Y., and Taniguchi, N., 1995, Characterization of wild-type and amyotrophic lateral sclerosis-related mutant Cu,Zn superoxide dismutases overproduced in baculovirus-infected insect cells, *J. Neurochem.* **64**:1456–1461.

Gonatas, N. K., Stieber, A., Mourelatos, Z., Chen, Y., Gonatas, J. O., Appel, S. H., Hays, A. P., Hickey, W. F., and Hauw, J.-J., 1992, Fragmentation of the Golgi apparatus of motor neurons in amyotrophic lateral sclerosis, *Am. J. Pathol.* **140**:731–737.

Gopinath, C., and Howell, J. M., 1975, Experimental chronic copper toxicity in sheep. Changes that follow the cessation of dosing at the onset of haemolysis, *Res. Vet. Sci.* **19**:35–43.

Greenlund, L. J. S., Deckwerth, T. L., and Johnson, E. M., Jr., 1995, Superoxide dismutase delays neuronal apoptosis: A role for reactive oxygen species in programmed neuronal death, *Neuron* **14**:303–315.

Gurney, M. E., 1994, Mutant mice, Cu,Zn superoxide dismutase, and motor neuron degeneration, *Science* **266**:1587.

Gurney, M. E., Pu, H., Chiu, A. Y., Dal Canto, M. C., Polchow, C. Y., Alexander, D. D., Caliendo, J., Hentati, A., Kwon, Y. W., Deng, H.-X., Chen, W., Zhai, P., Sufit, R. L., and Siddique, T., 1994, Motor neuron degeneration in mice that express a human Cu,Zn superoxide dismutase mutation, *Science* **264**:1772–1775.

Hajimohammadreza, I., and Brammer, M., 1990, Brain membrane fluidity and lipid peroxidation in Alzheimer's disease, *Neurosci. Lett.* **112**:333–337.

Halliwell, B., 1992, Oxygen radicals as key mediators in neurological disease: Fact or fiction, *Ann. Neurol.* **32**:S10–S15.

Harding, A. E., 1993, Inherited neuronal atrophy and degeneration predominantly of lower motor neurons, in: *Peripheral Neuropathy* (P. J. Dyck, P. K., Thomas, J. W. Griffin, P. A. Low, and J. F. Poduslo, eds.), W. B. Saunders, Philadelphia, pp. 1051–1064.

Harris, E. D., 1992, Copper as a cofactor and regulator of copper, zinc superoxide dismutase, *J. Nutr.* **122**:636–640.

Harris, E. D., and Percival, S. S., 1989, Copper transport: Insights into a ceruloplasmin-based delivery system, *Adv. Exp. Med. Biol.* **258**:95–102.

Haverkamp, L. J., Appel, V., and Appel, S. H., 1995, Natural history of amyotrophic lateral sclerosis in a database population. Validation of a scoring system and a model for survival prediction, *Brain* **118**:707–719.

Hendrickson, D. J., Fisher, J. H., Jones, C., and Ho, Y.-S., 1990, Regional localization of human extracellular superoxide dismutase gene to 4pter-q21, *Genomics* **8**:736–738.

Hirano, A., Malamud, N., Kurland, L. T., and Zimmerman, H. M., 1969, A review of the pathologic findings in amyotrophic lateral sclerosis, in: *Motor Neuron Diseases. Research on Amyotrophic Lateral Sclerosis and Other Disorders* (F. H. Norris and L. T. Kurland, eds.), Grune & Stratton, New York, pp. 51–60.

Hirano, A., 1973, Progress in the pathology of motor neuron diseases, in: *Progress in Neuropathology* (H. M. Zimmerman, ed.), Grune & Stratton, New York, pp. 181–215.

Hirano, A., 1984, Neuropathology of ALS, in: *Amyotrophic Lateral Sclerosis in Asia and Oceania* (K. M. Chen and Y. Yase, eds.), Shyan-Fu Chou, Taiwan, pp. 23–30.

Hirano, A., 1991, Cytopathology of amyotrophic lateral sclerosis, in: *Amyotrophic Lateral Sclerosis and Other Motor Neuron Diseases* (L. P. Rowland, ed.), Raven Press, New York, pp. 91–101.

Hirano, A., and Kato, S., 1992, Fine structural study of sporadic and familial amyotrophic lateral sclerosis, in: *Handbook of Amyotrophic Lateral Sclerosis* (R. A. Smith, ed.), Marcel Dekker, New York, pp. 183–192.

Hirano, A., Donnenfeld, H., Sasaki, S., and Nakano, I., 1984, Fine structural observations of neurofilamentous changes in amyotrophic lateral sclerosis, *J. Neuropathol. Exp. Neurol.* **43**:461–470.

Hirano, M., Fujii, J., Nagai, Y., Sonobe, M., Okamoto, K., Araki, H., Taniguchi, N., and Ueno, S., 1994, A new variant Cu/Zn superoxide dismutase $Val^7 \rightarrow Glu$ deduced from lymphocyte mRNA sequences from Japanese patients with familial amyotrophic lateral sclerosis, *Biochem. Biophys. Res. Commun.* **204**:572–577.

Howell, J. M., Blakemore, W. F., Gopinath, C., Hall, G. A., and Parker, J. H., 1974, Chronic copper poisoning and changes in the central nervous system of sheep, *Acta Neuropathol. Berl.* **29**:9–24.

Ikeda, M., Abe, K., Aoki, M., Sahara, M., Watanabe, M., Shoji, M., St. George-Hyslop, P. H., Hirai, S., and

Itoyama, Y., 1995, Variable clinical symptoms in familial amyotrophic lateral sclerosis with a novel point mutation in the Cu/Zn superoxide dismutase gene, *Neurology* **45**:2038–2042.

Imlay, J. A., and Linn, S., 1988, DNA damage and oxygen radical toxicity, *Science* **240**:1302–1309.

Jordan, J., Ghadge, G. D., Prehn, J. H., Toth, P. T., Roos, R. P., and Miller, R. J., 1995, Expression of human copper/zinc-superoxide dismutase inhibits the death of rat sympathetic neurons caused by withdrawal of nerve growth factor, *Mol. Pharmacol.* **47**:1095–1100.

Kato, S., Oda, M., and Hayashi, H., 1993, Neuropathology in amyotrophic lateral sclerosis patients on respirators: uniformity and diversity in 13 cases, *Neuropathology* **13**:229–236.

Kawamata, J., Hasegawa, H., Shimohama, S., Kimura, J., Tanaka, S., and Ueda, K., 1994, Leu[106] → Val CTC → GTC mutation of superoxide dismutase-1 gene in patient with familial amyotrophic lateral sclerosis in Japan, *Lancet* **343**:1501.

Kostrzewa, M., Burck-Lehmann, U., and Müller, U., 1994, Autosomal dominant amyotrophic lateral sclerosis: A novel mutation in the Cu/Zn superoxide dismutase-1 gene, *Hum. Mol. Genet.* **3**:2261–2262.

Kuncl, R. W., Crawford, T. O., Rothstein, J. D., and Drachman, D. B., 1992, Motor neuron diseases, in: *Diseases of the Nervous System* (A. K. Asbury, G. M. McKhann and W. I. McDonald, eds.), W. B. Saunders, Philadelphia, pp. 1179–1208.

Kurtzke, J. F., 1991, Risk factors in amyotrophic lateral sclerosis, in: *Amyotrophic Lateral Sclerosis and Other Motor Neuron Diseases* (L. P. Rowland, ed.), Raven Press, New York, pp. 245–270.

Lapinskas, P. J., Cunningham, K. W., Liu, X. F., Fink, G. R., and Cizewski-Culotta, V., 1995, Mutations in PMR1 suppress oxidative damage in yeast cells lacking superoxide dismutase, *Mol. Cell. Biol.* **15**:1382–1388.

Lee, M. K., and Cleveland, D. W., 1994, Neurofilament function and dysfunction: Involvement in axonal growth and neuronal disease, *Curr. Opin. Cell Biol.* **6**:34–40.

Leigh, P. N., 1994, Pathogenic mechanisms in amyotrophic lateral sclerosis and other motor neuron disorders, in: *Neurodegenerative Diseases* (D. B. Calne, ed.), W. B. Saunders, Philadelphia, pp. 473–488.

Levieux, A., Levieux, D., and Lab, C., 1991, Immunoquantitation of rat erythrocyte superoxide dismutase: Its use in copper deficiency, *Free Radic. Biol. Med.* **11**:589–595.

Liu, X. F., and Cizewski-Culotta, V., 1994, The requirement for yeast superoxide dismutase is bypassed through mutations in *BSD2*, a novel metal homeostasis gene, *Mol. Cell. Biol.* **14**:7037–7045.

Lowe, J., 1994, New pathological findings in amyotrophic lateral sclerosis, *J. Neurol. Sci.* **124**:38–51.

Lowe, J., Mayer, R. J., and Landon, M., 1993, Ubiquitin in neurodegenerative diseases, *Brain Pathol.* **3**:55–65.

Manetto, V., Sternberger, N. H., Perry, G., Sternberger, L. A., and Gambetti, P., 1988, Phosphorylation of neurofilaments is altered in amyotrophic lateral sclerosis, *J. Neuropathol. Exp. Neurol.* **47**:642–653.

Matsumoto, S., Kusaka, H., Murakami, N., Hashizume, Y., Okazaki, H., and Hirano, A., 1992, Basophilic inclusions in sporadic juvenile amyotrophic lateral sclerosis: An immunocytochemical and ultrastructural study, *Acta Neuropathol.* **83**:579–583.

McCord, J. M., 1994, Mutant mice, Cu,Zn superoxide dismutase, and motor neuron degeneration, *Science* **266**:1586–1587.

McIntosh, L. J., Trush, M. A., and Troncoso, J. C., 1991, Oxygen-free radical mediated processes in Alzheimer's disease, *Soc. Neurosci. Abstr.* **17**:1071 (abstract).

Mecocci, P., MacGarvey, U., Kaufman, A. E., Koontz, D., Shoffner, J. M., Wallace, D. C., and Beal, M. F., 1993, Oxidative damage to mitochondrial DNA shows marked age-dependent increases in human brain, *Ann. Neurol.* **34**:609–616.

Munoz, D. G., Greene, C., Perl, D. P., and Selkoe, D. J., 1988, Accumulation of phosphorylated neurofilaments in anterior horn motoneurons of amyotrophic lateral sclerosis patients, *J. Neuropathol. Exp. Neurol.* **47**:9–18.

Nakano, R., Sato, S., Inuzuka, T., Sakimura, K., Mishina, M., Takahashi, H., Ikuta, F., Honma, Y., Fujii, J., Taniguchi, N., and Tsuji, S., 1994, A novel mutation in Cu/Zn superoxide dismutase gene in Japanese familial amyotrophic lateral sclerosis, *Biochem. Biophys. Res. Commun.* **200**:695–703.

Olanow, C. W., 1993, A radical hypothesis for neurodegeneration, *Trends Neurosci.* **16**:439–444.

Oppenheimer, D. R., and Esiri, M. M., 1992, Diseases of the basal ganglia, cerebellum and motor neurons, in:

Greenfield's Neuropathology (J. H. Adams and L. W. Duchen, eds.), Oxford University Press, New York, p. 988.

Orr, W. C., and Sohal, R. S., 1994, Extension of life-span by overexpression of superoxide dismutase and catalase in *Drosophila melanogaster, Science* **263:**1128–1130.

Orrell, R., de Belleroche, J., Marklund, S., Bowe, F., and Hallewell, R., 1995, A novel SOD mutant and ALS, *Nature* **374:**504–505.

Pardo, C. A., Xu, Z., Borchelt, D. R., Price, D. L., Sisodia, S. S., and Cleveland, D. W., 1995, Superoxide dismutase is an abundant component in cell bodies, dendrites, and axons of motor neurons and in a subset of other neurons, *Proc. Natl. Acad. Sci. USA* **92:**954–958.

Parge, H. E., Hallewell, R. A., and Tainer, J. A., 1992, Atomic structures of wild- type and thermostable mutant recombinant human Cu,Zn superoxide dismutase, *Proc. Natl. Acad. Sci. USA* **89:**6109–6113.

Percival, S. S., and Harris, E. D., 1989, Ascorbate enhances copper transport from ceruloplasmin into human K562 cells, *J. Nutr.* **119:**779–784.

Percival, S. S., Bae, B., and Patrice, M., 1993, Copper is required to maintain Cu/Zn-superoxide dismutase activity during HL-60 cell differentiation, *Proc. Soc. Exp. Biol. Med.* **203:**78–83.

Phillips, J. P., Campbell, S. D., Michaud, D., Charbonneau, M., and Hilliker, A. J., 1989, Null mutation of copper/zinc superoxide dismutase in *Drosophila* confers hypersensitivity to paraquat and reduced longevity, *Proc. Natl. Acad. Sci. USA* **86:**2761–2765.

Phillips, J. P., Tainer, J. A., Getzoff, E. D., Boulianne, G. L., Kirby, K., and Hilliker, A. J., 1995, Subunit-destabilizing mutations in *Drosophila* copper/zinc superoxide dismutase: Neuropathology and a model of dimer dysequilibrium, *Proc. Natl. Acad. Sci. USA* **92:**8574–8578.

Pogun, S., Dawson, V., and Kuhar, M., 1994, Nitric oxide inhibits ^{3}H-glutamate transport in synaptosomes, *Synapse* **18:**21–26.

Pramatarova, A., Goto, J., Nanba, E., Nakashima, K., Takahashi, K., Takagi, A., Kanazawa, I., Figlewicz, D. A., and Rouleau, G. A., 1994, A two basepair deletion in the SOD 1 gene causes familial amyotrophic lateral sclerosis, *Hum. Mol. Genet.* **3:**2061–2062.

Price, D. L., Borchelt, D. R., Walker, L. C., and Sisodia, S. S., 1992a, Toxicity of synthetic Aβ peptides and modeling of Alzheimer's disease, *Neurobiol. Aging* **13:**623–625.

Price, D. L., Martin, L. J., Clatterbuck, R. E., Koliatsos, V. E., Sisodia, S. S., Walker, L. C., and Cork, L. C., 1992b, Neuronal degeneration in human diseases and animal models, *J. Neurobiol.* **23:**1277–1294.

Rabizadeh, S., Butler-Gralla, E., Borchelt, D. R., Gwinn, R., Selverstone-Valentine, J., Sisodia, S., Wong, P., Lee, M., Hahn, H., and Bredesen, D. E., 1995, Mutations associated with amyotrophic lateral sclerosis convert superoxide dismutase from an antiapoptotic gene to a proapoptotic gene: Studies in yeast and neural cells, *Proc. Natl. Acad. Sci. USA* **92:**3024–3028.

Rainero, I., Pinessi, L., Tsuda, T., Vignocchi, M. G., Vaula, G., Calvi, L., Cerrato, P., Rossi, B., Bergamini, L., McLachlan, D. R. C., and St. George-Hyslop, P. H., 1994, SOD1 missense mutation in an Italian family with ALS, *Neurology* **44:**347–349.

Reveillaud, I., Phillips, J., Duyf, B., Hilliker, A., Kongpachith, A., and Fleming, J. E., 1994, Phenotypic rescue by a bovine transgene in a Cu/Zn superoxide dismutase-null mutant of *Drosophila melanogaster, Mol. Cell. Biol.* **14:**1302–1307.

Ripps, M. E., Huntley, G. W., Hof, P. R., Morrison, J. H., and Gordon, J. W., 1995, Transgenic mice expressing an altered murine superoxide dismutase gene provide an animal model of amyotrophic lateral sclerosis, *Proc. Natl. Acad. Sci. USA* **92:**689–693.

Robberecht, W., Sapp, P., Viaene, M. K., Rosen, D., McKenna-Yasek, D., Haines, J., Horvitz, R., Theys, P., and Brown, R., Jr., 1994, Cu/Zn superoxide dismutase activity in familial and sporadic amyotrophic lateral sclerosis, *J. Neurochem.* **62:**384–387.

Rosen, D. R., Siddique, T., Patterson, D., Figlewicz, D. A., Sapp, P., Hentati, A., Donaldson, D., Goto, J., O'Regan, J. P., Deng, H.-X., Rahmani, Z., Krizus, A., McKenna-Yasek, D., Cayabyab, A., Gaston, S. M., Berger, R., Tanzi, R. E., Halperin, J. J., Herzfeldt, B., Van den Bergh, R., Hung, W.-Y., Bird, T., Deng, G., Mulder, D. W., Smyth, C., Laing, N. G., Soriano, E., Pericak-Vance, M. A., Haines, J., Rouleau, G. A., Gusella, J. S., Horvitz, H. R., and Brown, R. H., Jr., 1993, Mutations in Cu/Zn superoxide dismutase gene are associated with familial amyotrophic lateral sclerosis, *Nature* **362:**59–62.

Rothstein, J. D., and Kuncl, R. W., 1995, Neuroprotective strategies in a model of chronic glutamate-mediated motor neuron toxicity, *J. Neurochem.* **65:**643–651.

Rothstein, J. D., Martin, L. J., and Kuncl, R. W., 1992, Decreased glutamate transport by the brain and spinal cord in amyotrophic lateral sclerosis, *N. Engl. J. Med.* **326:**1464–1468.

Rothstein, J. D., Jin, L., Dykes-Hoberg, M., and Kuncl, R. W., 1993, Chronic inhibition of glutamate uptake produces a model of slow neurotoxicity, *Proc. Natl. Acad. Sci. USA* **90:**6591–6595.

Rothstein, J. D., Bristol, L. A., Hosler, B., Brown, R. H., Jr., and Kuncl, R. W., 1994a, Chronic inhibition of superoxide dismutase produces apoptotic death of spinal neurons, *Proc. Natl. Acad. Sci. USA* **91:**4155–4159.

Rothstein, J. D., Martin, L., Dykes-Hoberg, M., Jin, L., Levey, A., and Kuncl, R. W., 1994b, Glutamate transporter subtypes: Role in excitotoxicity and amyotrophic lateral sclerosis, *Ann. Neurol.* **36:**282 (abstract).

Rothstein, J. D., Van Kammen, M., Levey, A. I., Martin, L. J., and Kuncl, R. W., 1995, Selective loss of glial glutamate transporter GLT-1 in amyotrophic lateral sclerosis, *Ann. Neurol.* **38:**73–84.

Rowland, L. P., 1994a, Natural history and clinical features of amyotrophic lateral sclerosis and related motor neuron diseases, in: *Neurodegenerative Diseases* (D. B. Calne, ed.), W. B. Saunders, Philadelphia, pp. 507–521.

Rowland, L. P., 1994b, Amyotrophic lateral sclerosis: theories and therapies, *Ann. Neurol.* **35:**129–130.

Sapp, P. C., Rosen, D. R., Hosler, B. A., Esteban, J., Mckennayasek, D., Oregan, J. P., Horvitz, H. R., and Brown, R. H., 1995, Identification of three novel mutations in the gene for Cu/Zn superoxide dismutase in patients with familial amyotrophic lateral sclerosis, *Neuromusc. Disord.* **5:**353–357.

Sasaki, S., Kamei, H., Yamane, K., and Maruyama, S., 1988, Swelling of neuronal processes in motor neuron disease, *Neurology* **38:**1114–1118.

Schmidt, M. L., Carden, M. J., Lee, V. M.-Y., and Trojanowski, J. Q., 1987, Phosphate dependent and independent neurofilament epitopes in the axonal swellings of patients with motor neuron disease and controls, *Lab. Invest.* **56:**282–294.

Siddique, T., Pericak-Vance, M. A., Brooks, B. R., Roos, R. P., Hung, W.-Y., Antel, J. P., Munsat, T. L., Phillips, K., Warner, K., Speer, M., Bias, W. B., Siddique, N. A., and Roses, A. D., 1989, Linkage analysis in familial amyotrophic lateral sclerosis, *Neurology* **39:**919–925.

Siddique, T., Figlewicz, D. A., Pericak-Vance, M. A., Haines, J. L., Rouleau, G., Jeffers, A. J., Sapp, P., Hung, W.-Y., Bebout, J., McKenna-Yasek, D., Deng, G., Horvitz, H. B., Gusella, J. F., Brown, R. H., Jr., Roses, A. D., Roos, R. P., Williams, D. B., Mulder, D. W., Watkins, P. C., Noore, R., Nicholson, G., Reed, R., Brooks, B. R., Festoff, B., Antel, J. P., Tandan, R., Munsat, T. L., Laing, N. G., Halperin, J. J., Norris, F. H., Van den Bergh, R., Swerts, L., Tanzi, R. E., Jubelt, B., Mathews, K. D., and Bosch, E. P., 1991, Linkage of a gene causing familial amyotrophic lateral sclerosis to chromosome 21 and evidence of genetic-locus heterogeneity, *N. Engl. J. Med.* **324:**1381–1384.

Smith, R. A., 1992, *Handbook of Amyotrophic Lateral Sclerosis*, Marcel Dekker, New York.

Smith, R. G., and Appel, S. H., 1995, Molecular approaches to amyotrophic lateral sclerosis, *Annu. Rev. Med.* **46:**133–145.

Stadtman, E. R., 1990, Metal ion-catalyzed oxidation of proteins: biochemical mechanism and biological consequences, *Free Radic. Biol. Med.* **9:**315–325.

Stadtman, E. R., 1992, Protein oxidation and aging, *Science* **257:**1220–1224.

Stadtman, E. R., and Oliver, C. N., 1991, Metal-catalyzed oxidation of proteins, *J. Biol. Chem.* **266:**2005–2008.

Steinkuhler, C., Sapora, O., Carri, M. T., Nagel, W., Marcocci, L., Ciriolo, M. R., Weser, U., and Rotilio, G., 1991, Increase of Cu,Zn-superoxide dismutase activity during differentiation of human K562 cells involves activation by copper of a constantly expressed copper-deficient protein, *J. Biol. Chem.* **266:**24580–24587.

Steinkuhler, C., Carri, M. T., Micheli, G., Knoepfel, L., Weser, U., and Rotilio, G., 1994, Copper-dependent metabolism of Cu,Zn-superoxide dismutase in human K562 cells. Lack of specific transcriptional activation and accumulation of a partially inactivated enzyme, *Biochem. J.* **302:**687–694.

Subbarao, K. V., Richardson, J. S., and Ang, L. C., 1990, Autopsy samples of Alzheimer's cortex show increased peroxidation in vitro. *J. Neurochem.* **55:**342–345.

Tandan, R., and Bradley, W. G., 1985a, Amyotrophic lateral sclerosis: Part 1. Clinical features, pathology, and ethical issues in management, *Ann. Neurol.* **18:**271–280.

Tandan, R., and Bradley, W. G., 1985b, Amyotrophic lateral sclerosis: Part 2. Etiopathogenesis. *Ann. Neurol.* **18:**419–431.

Troy, C. M., and Shelanski, M. L., 1994, Down-regulation of copper/zinc superoxide dismutase causes apoptotic death in PC12 neuronal cells, *Proc. Natl. Acad. Sci. USA* **91:**6384–6387.

Tsuda, T., Munthasser, S., Fraser, P. E., Percy, M. E., Rainero, I., Vaula, G., Pinessi, L., Bergamini, L., Vignocchi, G., McLachlan, D. R. C., Tatton, W. G., and St. George-Hyslop, P., 1994, Analysis of the functional effects of a mutation in *SOD1* associated with familial amyotrophic lateral sclerosis, *Neuron* **13:**727–736.

Uauy, R., Castillo-Duran, C., Fisberg, M., Fernandez, N., and Valenzuela, A., 1985, Red cell superoxide dismutase activity as an index of human copper nutrition, *J. Nutr.* **115:**1650–1655.

Volterra, A., Trotti, D., Cassutti, P., Tromba, C., Salvaggio, A., Melcangi, R. C., and Racagni, G., 1992, High sensitivity of glutamate uptake to extracellular free arachidonic acid levels in rat cortical synaptosomes and astrocytes, *J. Neurochem.* **59:**600–606.

Volterra, A., Trotti, D., Tromba, C., Floridi, S., and Racagni, G., 1994, Glutamate uptake inhibition by oxygen free radicals in rat cortical astrocytes, *J. Neurosci.* **14:**2924–2932.

Wiedau-Pazos, M., Goto, J. J., Rabizadeh, S., Gralla, E. B., Roe, J. A., Lee, M. K., Valentine, J. S., and Bredesen, D. E., 1996, Altered reactivity of superoxide dismutase in familial amyotrophic lateral sclerosis, *Science* **271:**515–518.

Williams, D. B., and Windebank, A. J., 1993, Motor neuron disease, in: *Peripheral Neuropathy* (P. J. Dyck, P. K. Thomas, J. W. Griffin, P. A. Low, and J. F. Poduslo, eds.), W. B. Saunders, Philadelphia, pp. 1028–1050.

Wong, P. C., and Borchelt, D. R., 1995, Motor neuron disease caused by mutations in superoxide dismutase 1. *Curr. Opin. Neurol.* **8:**294–301.

Wong, P. C., Pardo, C. A., Borchelt, D. R., Lee, M. K., Copeland, N. G., Jenkins, N. A., Sisodia, S. S., Cleveland, D. W., and Price, D. L., 1995, An adverse property of a familial ALS-linked SOD1 mutation causes motor neuron disease characterized by vacuolar degeneration of mitochondria, *Neuron* **14:**1105–1116.

Yan, S.-D., Chen, X., Schmidt, A.-M., Brett, J., Godman, G., Zou, Y.-S., Scott, C. W., Caputo, C., Frappier, T., Smith, M. A., Perry, G., Yen, S.-H., and Stern, D., 1994, Glycated tau protein in Alzheimer disease: A mechanism for induction of oxidant stress, *Proc. Natl. Acad. Sci. USA* **91:**7787–7791.

Yim, M. B., Chock, P. B., and Stadtman, E. R., 1990, Copper,zinc superoxide dismutase catalyzes hydroxyl radical production from hydrogen peroxide, *Proc. Natl. Acad. Sci. USA* **87:**5006–5010.

Yim, M. B., Chock, P. B., and Stadtman, E. R., 1993, Enzyme function of copper,zinc superoxide dismutase as a free radical generator, *J. Biol. Chem.* **268:**4099–4105.

Young, A. B., 1990, What's the excitement about excitatory amino acids in amyotrophic lateral sclerosis?, *Ann. Neurol.* **28:**9–11.

Chapter 16

Tardive Dyskinesia and Oxidative Stress

Jean Lud Cadet

16.1. INTRODUCTION

The use of antipsychotic drugs is associated with a number of neurologic complications. Some of these complications occur after acute administration, while others occur after chronic use of the drugs (Table 16.1). The treatment of the acute side-effects often involves decreasing the amount of neuroleptics. When this is not possible because of the patients' clinical status, the addition of anticholinergic drugs can provide significant relief to the patients. The treatment of the long-term complications of neuroleptics is more complicated. Thus, the purpose of this chapter is to touch briefly on the role of some therapeutic approaches to tardive dyskinesia and to elaborate on the idea that oxidative stress plays a role in the manifestations of this neurologic syndrome.

16.2. DESCRIPTION

Tardive dyskinesia (TD) is characterized by various abnormal movements that develop after the chronic use of antipsychotic drugs (Dynes, 1970; Jeste and Wyatt, 1981, 1982; Kane *et al.,* 1984; Sigwald *et al.,* 1959). Initially described as "dyskinesie facio-buccio-lingui-masticatrice" (Sigwald *et al.,* 1959), the concept now includes nonrepetitive choreic and choreoathetoid movements of the fingers ("piano playing"), hands, legs, and trunk. The abnormal facial movements can include opening of the mouth, protrusion of the tongue, lip smacking, and grimacing. The differential diagnosis of tardive dyskinesia includes Huntington's chorea, drug-induced chorea, and levodopa-induced dyskinesia.

Jean Lud Cadet   Molecular Neuropsychiatry Section, NIH/NIDA, Intramural Research Program, Baltimore, Maryland 21224.

Metals and Oxidative Damage in Neurological Disorders, edited by Connor. Plenum Press, New York, 1997.

Table 16.1
Drug-Induced Movement Abnormalities

Acute neuroleptic-induced movement disorders
    Acute dystonic reaction
    Parkinsonism
    Akathisia
    Neuroleptic malignant syndrome
Chronic neuroleptic-induced movement disorders
    Tardive dyskinesia (stereotypies)
    Tardive dystonia
    Tardive akathisia
    Tardive tics
    Tardive myoclonus
    Tardive tremor

16.3. NEUROPATHOLOGICAL FINDINGS ASSOCIATED WITH NEUROLEPTIC USE

Neuropathological abnormalities have been reported secondary to long-term administration of neuroleptics (Aksel, 1956; Benes *et al.,* 1983, 1985; Dom, 1967; Gerlach, 1975; Gross and Kaltenbach, 1969; Hunter, 1968; Koizuma and Shiraishi, 1970, 1973a, 1973b; Mackiewicz and Gershon, 1964; Nielsen and Lyon, 1978; Pakkenberg and Fog, 1974; Pakkenberg *et al.,* 1973; Roisin *et al.,* 1959). Specifically, chronic use of chlorpromazine results in the degeneration of the cytoplasm and the nuclei, as well as accumulation of chromatin around the nuclear membrane in striatal and thalamic cells. This drug also causes moderate chromatolytic changes, increased satellitosis, and neuronophagia in the cortex, the lenticular nuclei, the thalamus, and the hypothalamus in monkeys (Roisin *et al.,* 1959). Other brain regions such as the pons, the medulla, and the cerebellum are also affected. Chlorpromazine and haloperidol alter the synaptic areas of the basal ganglia (Koizumi and Shiraishi, 1973a, 1973b), while haloperidol can cause deposition of fibrillary material in the postsynaptic dendrites and vacuolization of axon terminals (Koizumi and Shiraishi, 1973a). Benes *et al.,* (1985) have reported a significant shift in the distribution of axon terminals in the brains of rats chronically treated with haloperidol. These pathological findings are summarized in Table 16.2.

Table 16.2
Some Neuropathological Effects
of Neuroleptics

Chromatolysis	Neuronal loss
Satellitosis	Atrophy
Neuronophagia	Gliosis
Necrosis	

Postmortem tissues of patients who had been treated with neuroleptics show chromatolysis, satellitosis, and neuronophagia in the basal ganglia, the cerebellum, and the cerebral cortex. Patients who had suffered from persistent dyskinesia showed more gliosis in the midbrain and in the brainstem and more cell degeneration in the substantia nigra (Christensen *et al.,* 1970). Patients with persistent dyskinesia showed swelling of large neurons and satellitosis, as well as proliferation of astroglia in the caudate nucleus (Jellinger, 1977). The putamen and globus pallidus also showed evidence of chromatolysis and cytoplasmic ballooning. Pathologic abnormalities were reported in 57% of patients who suffered from dyskinesia prior to death, but in only 7% of patients without movement disorders and 4% of age-matched psychotics who had no history of long-term neuroleptics. Although the exact manner by which long-term use of antipsychotic drugs might cause pathologic changes in the brain are still uncertain, these pathological abnormalities might play a role in the manifestation of these changes. I will later develop the thesis that reactive toxic species are responsible for these changes.

16.4. MODELS OF TARDIVE DYSKINESIA

The first developed hypothesis of TD is that of postsynaptic dopamine receptor supersensitivity. This idea is supported by a number of reports. For example, chronic treatment with a neuroleptic does result in increases in striatal dopamine receptors (Jeste and Wyatt, 1981; Waddington and Gamble, 1980a, 1980b, 1981, 1982). In addition, neuroleptic-treated animals develop behavioral supersensitivity to dopamine agonists such as apomorphine and amphetamine (Waddington and Gamble, 1980b, 1981, 1982).

On the basis of this idea, therapeutic approaches that aim at decreasing DA tone in the basal ganglia have been tried. These include the use of reserpine and of tetrabenazine (Fahn, 1978; Table 16.3). Nevertheless, the dopamine receptor hypothesis lacks explanatory power for a number of reasons. For example, DA binding sites increase within a few days of drug administration, whereas TD develops from several months to years after initiation of drug treatment. Moreover, the dopamine receptor hypothesis does not explain the persistence of TD because, after withdrawal of neuroleptics, the number of dopamine receptors returns to normal in animals that have been treated with these agents. Some of these issues have been discussed at length by others (Fibiger and Lloyd, 1984; Jeste and Wyatt, 1981; Waddington and Gamble, 1980a, 1980b, 1981, 1982).

Table 16.3
Agents Used in the Treatment
of Tardive Dyskinesia

Reserpine	GABA mimetics
Tetrabenazine	Vitamin E

The possibility that the GABAergic system may also play a role in the development of tardive dyskinesia has also been discussed (Tamminga *et al.,* 1980; Fibiger and Lloyd, 1984). Chronic neuroleptic administration results in a significant reduction of GAD, the enzyme that catalyzes the synthesis of GABA (Gunne and Haggstrom, 1983). Only animals that showed the reduction in GAD had dyskinetic movements (Gunne *et al.,* 1984). The role of the GABAergic system is supported by the finding that some GABA agonists may have beneficial effects on TD (Tamminga *et al.,* 1980). It is possible that some of the biochemical and neuropathological abnormalities reported with the use of neuroleptics might be secondary to oxidative stress, as described below.

16.5. FREE-RADICAL HYPOTHESIS OF TARDIVE DYSKINESIA

Neuroleptics can cause the production of free radicals via nonenzymatic and enzymatic breakdown of catecholamines. They can also be metabolized through reactive metabolites. These active species could cause abnormalities in GABAergic and cholinergic systems. From another perspective, because neuroleptics can cause accumulation of manganese in the brain (Bird *et al.,* 1967; Weiner *et al.,* 1980), reactions with H_2O_2 produced during increased catecholamine metabolism could lead to production of the very toxic hydroxyl radicals. An increased concentration of iron has been reported in the brain of a patient who suffered from TD (Campbell *et al.,* 1985).

Neuroleptic-induced cell death might not be the most important causal factor in the development of TD. Destabilization of cell membrane function might be more important in view of the chaotic nature of the clinical manifestation of the syndrome. For example, Cohen and Zubenko have demonstrated that *in vitro* exposure of normal human platelets to psychotropic drugs caused changes in the structural order of the cell membranes (Zubenko and Cohen, 1984, 1985). Striatal cells of rats treated with neuroleptics were reported to show abnormal physicochemical properties (Cohen and Zubenko, 1985). Patients who suffered from TD after chronic neuroleptic treatment showed similar abnormalities, whereas patients without TD did not (Zubenko and Cohen, 1986). Since not all patients who are treated with neuroleptics develop TD, it is conceivable that there might be identifiable characteristics between susceptible and nonsusceptible individuals. For example, subjects who develop TD might have low levels of antioxidants in their brains so that they are unable to protect their brains against the oxidative stress caused by neuroleptics.

Because free radicals are thought to be involved in the development and course of idiopathic Parkinson's disease (Cadet, 1988, 1993; Cohen, 1984; see also Table 16.4 and Chapter 14), it is possible that a similar mechanism may be involved in neuroleptic-induced parkinsonism. Such an argument implies that patients who showed this side effect early during their treatment with neuroleptics may be more susceptible to develop TD. Some studies have reported such an occurrence (Chouinard *et al.,* 1986; Crane, 1971, 1972; Richardson *et al.,* 1986). For example, Crane (1972) reported that 12.5% of 180 patients had both neuroleptic-induced parkinsonism and TD. Richardson and

Table 16.4
Free Radical Mechanisms
and Neuropsychiatric Illnesses

Parkinsonism	Huntington's disease
Schizophrenia	Manganese neurotoxicity
Tardive dyskinesia	Stroke
Wilson's disease	Amyotrophic lateral sclerosis

Craig (1982) also reported 26.7% of 86 patients who suffered from neuroleptic-induced abnormalities had both parkinsonism and tardive dyskinesia. Furthermore, the presence of severe or worsening Parkinsonism is associated with the development of TD (Chouinard et al., 1986).

Another potential risk factor for the development of TD is poor prognosis schizophrenic patients who show poor response to neuroleptic treatment. One confounding factor with this group of patients relates to the observation of spontaneous involuntary movements in chronic schizophrenic patients who have not been treated with neuroleptics (Altrocchi, 1972; Barnes et al., 1983; Brandon et al., 1971; Crow et al., 1984; Owens et al., 1982). The free-radical hypothesis of tardive dyskinesia may help to explain the occurrence of these movement abnormalities in severe chronic schizophrenics. In essence, in some schizophrenic patients there may exist, in the course of the illness, a state of excess dopamine in the basal ganglia that causes, in time, the formation of toxic by-products of catecholamines. Thus, related pathogenetic mechanisms may be responsible for the development of idiopathic movements in schizophrenia (spontaneous dyskinesia), the progression of the schizophrenic illness, and the appearance of tardive dyskinesia in some patients.

The free-radical theory of TD predicts that older patients and patients with cognitive deficits or abnormal CT scans should be at greater risk to develop TD. A number of studies support this notion (Bourgeois et al., 1980; Jeste and Wyatt, 1981; Kane et al., 1984; Lieberman et al., 1984; Myslobodky et al., 1985; Struve and Wilner, 1983). The association between persistent dyskinesia and brain damage has also been discussed (Edwards, 1980). More recently, Richardson and colleagues (Richardson et al., 1986) reported that there was a positive correlation between increasing age and the development of TD in a mentally retarded population.

Waddington and Gamble (1980b, 1981, 1982) have also reported that rats receiving long-term treatment with neuroleptics had a higher incidence of abnormal orofacial movement than controls. These abnormal movements were comparable to those observed in older animals.

Schizophrenic patients with TD were older and were more cognitively impaired than schizophrenic patients without TD (Waddington and Gamble, 1980a). It has been suggested that some patients may develop a condition termed "tardive dysmentia" during chronic treatment with neuroleptics (Myslobodsky, 1985; Wilson et al., 1983).

In summary, when taken together, the reviewed data suggest that chronic use of neuroleptics may be accelerating a mechanism that is common to the development of aging and TD. This process may be the formation of toxic active species during chronic use of neuroleptics and might be associated with secondary neuronal loss. Some animal studies have documented the fact that neuroleptics can cause changes in the oxidative status of the rat brain (Cadet and Perumal, 1990). Furthermore some of the neuroleptic-induced biochemical changes were prevented by treatment with vitamin E (Cadet, 1993; Gattaz et al., 1993).

16.6. CLINICAL IMPLICATIONS OF THE FREE-RADICAL HYPOTHESIS

The present argument suggests that treatment with antioxidants such as vitamin E (Chapter 10) and selenium may be beneficial in alleviating or preventing the development of TD. The first reported double-blind study revealed that vitamin E can indeed be beneficial in the treatment of TD (Cadet and Lohr, 1989). These findings have been replicated (Adler, 1993; Bell, 1965; Egan et al., 1992; Elkashef et al., 1990; Peet et al., 1993; Spivak et al., 1992; but see Shrigui et al., 1992, for a negative study). It will be essential to document if vitamin E may actually delay or prevent the appearance of TD.

Postmortem and cerebrospinal fluid studies of patients with TD may help to characterize the systems that are more affected by chronic use of neuroleptics (Lohr et al., 1990). Specifically, the status of glutathione, superoxide dismutase, and catalase need to be evaluated (Table 16.5). It will be important to relate these to the premorbid status of the patients.

16.7. CONCLUSION

In conclusion, free radicals appear to play an important role in the pathobiology of a number of neurological disorders. Some drugs used in the treatment of these diseases might cause side effects by producing free radicals and oxidative stress. It is important to point out that in addition to their action on DA receptors, neuroleptics can affect calmodulin, which is important to the action of nitric oxide (NO) synthase (Bradt and

---

Table 16.5
Possible Research Approaches

Measures of free radical production
Measures of scavenging systems (vitamin E, vitamin C, glutathione, etc.)
Molecular controls of enzymes after chronic neuroleptic treatment
Effects of neuroleptic on molecular machinery
Treatment with scavenging agents

Snyder, 1990). Decreased levels of NO might lead to increased action of the superoxide radical, which is known to react with NO to form peroxynitrite (Huie and Padmaja, 1992). It has recently been suggested that NO may act both as a toxic agent in its own right (Dawson *et al.*, 1991; see also Chapter 11) or as a scavenger of the superoxide radical (Oury *et al.*, 1992). Thus, the role of NO in TD needs to be evaluated. A multidisciplinary approach to these disorders might help to dissect the cellular and molecular mechanisms involved in the initial appearance and the progression of some of these neuropsychiatric disorders (Melhorn and Cole, 1985). Finally, the possibility that chronic treatment with neuroleptics might cause apoptotic changes in the brains of rodents needs to be evaluated.

16.8. REFERENCES

Adler, L. A., Preselow, E., Rotrosen, J., Duncan, E., Lee, M., Rosenthal, M., and Angrist, B., 1993, Vitamin E treatment of tardive dyskinesia, *Am. J. Psych.* **150:**1405–1407.

Aksel, J. S., 1956, Etude clinique et experimentale del'hibernotherapie, *Encephale* **45:**566.

Alpert, M., Friedhoff, A. J., and Diamond, F., 1983, Use of dopamine receptor number as treatment for tardive dyskinesia, in: *Advances in Neurology: Experimental Therapeutics of Movement Disorders.* Vol. 37 (S. Fahn, D. B. Calne, and I. Shoulson, eds.) Raven Press, New York, pp. 253–258.

Altrocchi, P. H., 1972, Spontaneous oral-facial dyskinesia, *Arch. Neurol.* **26:**506–512.

Barnes, T. R. E., Rossor, M., and Trauer, T., 1983, A comparison of purposeless movements in psychiatric patients treated with antipsychotic drugs, and normal individuals, *J. Neurol. Neurosurg. Psych.* **46:**540–546.

Bell, D. S., 1965, Comparison of amphetamine psychosis and schizophrenia, *Br. J. Psych.* **111:**701–707.

Benes, F. M., Paskevich, P. A., and Domesick, V. B., 1983, Haloperidol-induced plasticity of axon terminals in rat substantia nigra, *Science* **221:**969–971.

Benes, F. M., Paskevich, P. A., Davidson, J., and Domesick, V. B., 1985, The effects of haloperidol on synaptic patterns in the rat striatum, *Brain Res.* **329:**265–274.

Bird, E. D., Collins, G. H., Dodson, M. H., and Grant, L. G., 1967, The effect of phenothiazine on the manganese concentration in the basal ganglia of subhuman primates, in: *Progress in Neurogenetics* (A. Barbeau and J. R. Burnette, eds.), Excerpta Medica, Montreal, pp. 600–605.

Bourgeois, M., Bouilh, P., Tignol, J., and Yesavage, J., 1980, Spontaneous dyskinesia vs. neuroleptic-induced dyskinesia in 270 elderly subjects, *J. Nerv. Ment. Dis.* **168:**177–178.

Brandon, S., McClellan, H. A., and Protheroe, C., 1971, A study of facial dyskinesia in a mental population, *Br. J. Psych.* **118:**171–184.

Bredt, D. S., and Snyder, S. H., 1990, Isolation of nitric oxide synthetase, a calmodulin-requiring enzyme, *Proc. Natl. Acad. Sci. USA* **87:**682–685.

Cadet, J. L., 1988, A unifying hypothesis of movement and madness; involvement of free radical in disorders of the isodenderitic core, *Med. Hypoth.* **27:**5963.

Cadet, J. L., 1993, Movement disorders: therapeutic role of vitamin E, *Toxicol. Ind. Health* **9:**337–347.

Cadet, J. L., and Lohr, I. B., 1989, Possible involvement of free radicals in neurolepticinduced movement disorders: Evidence from treatment of tardive dyskinesia with vitamin E, *Ann. N.Y. Acad. Sci.* **570:**176–185.

Cadet, J. L., and Perumal, A. S., 1990, Chronic treatment with prolixin causes oxidative stress in the rat brain, *Biol. Psych.* **28:**738–740.

Campbell, W. G., Raskind, M. A., Gordon, T., and Shaw, C. M., 1985, Iron pigment in the brain of a man with tardive dyskinesia, *Am. J. Psych.* **142:**364–365.

Chouinard, G., Annable, L., Mercier, P., and Ross-Chouinard, A., 1986, A five-year followup study of tardive dyskinesia, *Psychopharm. Bull.* **22:**259–263.

Christensen, E., Moller, J. D., and Faurbye, A., 1970, Neuropathological investigation of 28 brains from patients with dyskinesia, *Acta Psychiatr. Scand.* **46:**14–23.

Clow, A., Theordorou, A., Jenner, P., and Marsden, C. D., 1980, Cerebral dopamine function in rats following withdrawal from one year of continuous neuroleptic administration, *Eur. J. Pharmacol.* **63:**135–144.

Cohen, B. M., and Zubenko, G. S., 1985, In vivo effects of psychotropic agents on the physical pertest of cell membranes in the rat brain, *Psychopharmacology* **86:**365–368.

Cohen, G., 1984, Oxyradical toxicity in catecholamine neurons, *Neurotoxicology* **5:**7782.

Crane, G. E., 1971, Persistence of neurological symptoms due to neuroleptic drugs, *Am. J. Psych.* **127:**1407–1410.

Crane, G. E., 1972, Pseudoparkinsonism and tardive dyskinesia, *Arch. Neurol.* **27:**426–430.

Crow, T. J., Bloom, S. R., Cross, A. J., Ferrier, I. N., Johnstone, E. C., Woen, F., Owens, D. G. C., and Roberts, G. W., 1984, Abnormal involuntary movements schizophrenia: Neuro-chemical correlates and relation to the disease process, in: *Catecholamines: Neuropharmacology and Central Nervous System I. Therapeutic Aspects* (E. Usdin, A. Carlsson, A. Dahlstrom, and J. Engel, eds.), Alan R. Liss, New York, pp. 61–67.

Dawson, V. L., Dawson, T. D., London, E. D., Bredt, D. S., and Snyder, S. H., 1991, Nitric oxide mediates glutamate neurotoxicity in primary cortical cultures, *Proc. Natl. Acad. Sci. USA* **88:**6368–6371.

Dom, S., 1967, Local glial reaction in the CNS of albino-rats in response to the administration of a neuroleptic drug (butyrophenone), *Acta Neurol. Belg.* **67:**755–762.

Dynes, J. B., 1970, Oral dyskinesia—occurrence and treatment, *Dis. Nerv. Syst.* **31:**854–950.

Edwards, H., 1980, The significance of brain damage in persistent oral dyskinesia, *Br. J. Psych.* **116:**271–275.

Egan, M-F., Hyde, T. M., Albers, G. W., Elkashef, A., Alexander, R., Reeve, A., Blum, A., Saenz, R. E., and Wyatt, R. J., 1992, Treatment of tardive dyskinesia with vitamin E, *Am. J. Psych.* **149:**773–777.

Elkashef, A. M., Ruskin, P. E., Bacher, N., and Barrett, D., 1990, Vitamin E in the treatment of tardive dyskinesia, *Am. J. Psych.* **147:**505–506.

Fahn, S., 1978, Treatment of tardive dyskinesia with combined reserpine and alpha-methyltyrosine, *Trans. Am. Neurol. Assoc.* **103:**100–103.

Fibiger, H. C., and Lloyd, K. G., 1984, Neurobiological substrates of tardive dyskinesia: the GABA hypothesis, *Trends Neurosci.* **7:**462–464.

Gattaz, W. F., Emrich, A., and Behrens, S., 1993, Vitamin E attenuates the development of haloperidol-induced dopaminergic hypersensitivity in rats: Possible implications for tardive dyskinesia, *J. Neural Transm. (GenSect)* **92:**197–201.

Gerlach, J., 1975, Long-term effect of perphenazine on the substantia nigra in rats, *Psychopharmacologia* (Berlin) **45:**51–54.

Gross, H., and Kaltenbach, E., 1969, Neuropathological findings in persistent hyperkinesia after neuroleptic long-term therapy, in: *The Present Status of Psychotropic Drugs* (A. Cerletti and F. J. Bove, eds.), Excerpta Medica, Amsterdam, pp. 474–480.

Gunne, L. M., and Haggstrom, J. E., 1983, Reduction of nigral glutamic acid decarboxylase in rats with neuroleptic-induced oral dyskinesia, *Psychopharmacology* **81:**191–194.

Gunne, L. M., Haggstrom, J. E., and Sjoquist, G., 1984, Association with persistent neuroleptic-induced dyskinesia of regional changes in brain GABA synthesis, *Nature* **309:**347–349.

Huie, R. E., and Padmaja, S., 1992, The reaction of NO with superoxide, *Free Radical Res. Commun.* **18:**195–199.

Hunter, R., Blackwood, E., Smith, M. C., and Cumings, J. N., 1968, Neuropathological findings in three cases of persistent dyskinesia following phenothiazines, *J. Neurol. Sci.* **7:**263–273.

Jellinger, K., 1977, Neuropathological findings after neuroleptic long-term therapy, in: *Neurotoxicology* (L. Roizin, H. Shiraki, and N. Greevic, eds.), Raven Press, New York, pp. 25–42.

Jeste, D. V., and Wyatt, R. J., 1981, Dogma disputed: Is tardive dyskinesia due to postsynaptic dopamine receptor supersensitivity?, *J. Clin. Psych.* **42:**455–457.

Jeste, D. V., and Wyatt, R. J., 1982, *Understanding and Treating Tardive Dyskinesia,* Guildford Press, New York.

Kane, J., Woerner, M., Weinhold, P., Wegner, J., Kinon, B., and Borenstein, M., 1984, Incidence of tardive dyskinesia: five-year data from a perspective study, *Psychopharm. Bull.* **20:**387–389.

King, R., Barchas, J. D., and Huberman, B. A., 1984, Chaotic behavior in dopamine neurodynamics, *Proc. Natl. Acad. Sci. USA* **81:**1244–1247.

Koizumi, J., and Shiraishi, H., 1970, Glycogen accumulation in astrocytes of the striatum and palladium of the rabbit following administration of psychotropic drugs, *J. Electron Microsc.* **19:**182–187.

Koizumi, J., and Shiraishi, H., 1973a, Synaptic changes in the rabbit palladium following long-term haloperidol administration, *Folia Psychiatr. Neurol. Jap.* **27:**51–57.

Koizumi, J., and Shiraishi, H., 1973b, Synaptic alteration in the hypothalamus of the rabbit following long-term chlorpromazine administration, *Folia Psychiatr. Neurol. Jap.* **27:**59–67.

Lieberman, J., Kane, J. M., Woerner, M., Weinhold, P., Basavaraju, N., Kurucz, J., and Bergmann, K., 1984, Prevalence of tardive dyskinesia in elderly samples, *Psychopharm. Bull.* **20:**382–386.

Lohr, J. B., Kuczenski, R., Bracha, H. S., Moir, M., and Jeste, D. V., 1990, Increased indices of free radical activity in the cerebrospinal fluid of patients with tardive dyskinesia, *Biol. Psychiatry* **28:**535–539.

Mackiewicz, J., and Gershon, S., 1964, An experimental study of the neuropathological and toxicological effects of chlorpromazine and reserpine, *J. Neuropsych.* **5:**159–169.

Melhorn, R. J., and Cole, G., 1985, The free radical theory of aging: A critical review, *Adv. Free Rad. Biol. Med.* **1:**165–223.

Myslobodsky, M. S., 1986, Anosognosia in tardive dyskinesia: "tardive dysmentia" or "tardive dementia"?, *Schizophr. Bull.* **12:**1–6.

Myslobodky, M. S., Tomer, R., Holden, T., Kempler, S., and Sigal, M., 1985, Cognitive impairment in patients with tardive dyskinesia, *J. Nerv. Ment. Dis.* **173:**156–160.

Nielsen, E. B., and Lyon, M., 1978, Evidence for cell loss in corpus striatum after long-term treatment with a neuroleptic drug (flupenthixol in rats), *Psychopharmacology* **59:**85–87.

Oury, T. D., Ho, Y.-S., Piantadosi, C. A., and Crapo, J. D., 1992, Extra-cellular superoxide dismutase, nitric oxide, and central nervous system of O_2 toxicity, *Proc. Natl. Acad. Sci. USA* **89:**9715–9719.

Owens, D. G. C., Johnstone, E. C., and Frith, C. D., 1982, Spontaneous involuntary disorders of movement in neuroleptic treated and untreated chronic schizophrenics: prevalence, severity and distribution, *Arch. Gen. Psych.* **39:**452–461.

Pakkenberg, H., and Fog, R., 1974, Short-term effect of perphenazine enanthate on the rat brain, *Psychopharmacologia* (Berlin) **40:**165–169.

Pakkenberg, H., Fog, R., and Nilakantan, B., 1973, The long-term effect of perphenazine enanthate on the rat brain. Some metabolic and anatomical observations, *Psychopharmacologia* (Berlin) **29:**329–336.

Peet, M., Laugharne, J., Rangarajan, N., and Reynolds, G. P., 1993, Tardive dyskinesia, lipid peroxidation, and sustained amelioration with vitamin E treatment, *Int. Clin. Psychopharmacol.* **8:**151–153.

Richardson, M. A., and Craig, T. J., 1982, The coexistence of parkinsonism-like symptoms and tardive dyskinesia, *Am. J. Psych.* **139:**341–343.

Richardson, M. A., Haughland, G., Pass, R., and Craig, T. J., 1986, The prevalence of tardive dyskinesia in a mentally retarded population, *Psychopharmacol. Bull.* **22:**243–249.

Roisin, L., True, C., and Knight, M., 1959, Structural effects of tranquilizers, *Res. Publ. Assoc. Res. Nerv. Ment. Dis.* **37:**285–324.

Shriqui, C. L., Bradwejn, J., Annable, L., and Jones, B. D., 1992, Vitamin E in the treatment of tardive dyskinesia: A double-blind placebo-controlled study, *Am. J. Psych.* **149:**391–393.

Sigwald, J., Bouttier, D., Raymondeaud, C., and Piot, C., 1959, Quatre cas de dyskinesie facio-buccio-lingui-masticatrice a revolution prolongee secondaire a un traitement par les neuroleptiques, *Rec. Neurol.* **100:**751–755.

Spivak, B., Schwartz, B., Radwan, M., and Weizman, A., 1992, a-Tocopherol treatment of tardive dyskinesia, *J. Nerv. Ment. Dis.* **180:**400–401.

Struve, F. A., and Wilner, W. E., 1983, Cognitive dysfunction and tardive dyskinesia, *Br. J. Psych.* **143:**597–600.

Tamminga, C. A., Crayton, J. W., and Chase, T. N., 1980, Improvement in tardive dyskinesia after muscimol therapy, *Arch. Gen. Psych.* **37:**1376–1379.

Waddington, J. L., and Gamble, S. J., 1980a, Neuroleptic treatment for a substantial proportion of adult life: Behavioral sequelae of 9 months haloperidol administration, *Eur. J. Pharmacol.* **67:**363–369.

Waddington, J. L., and Gamble, S. J., 1980b, Spontaneous activity and apomorphine stereotypy during and after withdrawal from 3½ months continuous administration of haloperidol, *Psychopharmacology* **71:**75–77.

Waddington, J. L., and Gamble, S. J., 1980, Emergence of apomorphine-induced "vacuous chewing" during 6 months continuous treatment with fluphenazine decanoate, *Eur. J. Pharmacol.* **68:**387–388.

Waddington, J. L., Gamble, S. J., and Blourne, R. C., 1981, Sequelae of 6 months continuous administration of *cis* (Z)- and *trans*(E)-flupenthixol in the rat, *Eur. J. Pharmacol.* **69:**511–513.

Weiner, W. J., Nausieda, P. A., and Klawans, H. L., 1980, Regional brain manganese in an animal model of tardive dyskinesia, in: *Tardive Dyskinesia: Research and SP* (W. E. Fann, R. C. Smith, J. M., Davis, and E. F. Domino, eds.), Medical and Scientific Books, New York, pp. 159–163.

Wilson, I. C., Garbutt, J. C., Lanier, C. F., Moylan, J., Nelsoln, W., and Prange, Jr., A. J., 1983, Is there a tardive dysmentia?, *Schizopkr. Bull.* **9:**187–192.

Zubenko, G. S., and Cohen, B. M., 1984, In vitro effects of psychotropic agents on the microviscosisty of platelet membranes, *Psychopharmacology* **84:**289–292.

Zubenko, G. S., and Cohen, B. M., 1985, Effects of phenothiazine treatment on the physical properties of platelet membranes from psychiatric patients, *Biol. Psych.* **20:**384–396.

Zubenko, G. S., and Cohen, B. M., 1986, A cell membrane correlate of tardive dyskinesia in patients treated with phenothiazines, *Psychopharmacology* **88:**230–236.

Chapter 17

Antioxidant Therapeutic Strategies in CNS Disorders

Edward D. Hall

17.1. INTRODUCTION

A major role of oxygen-radical-induced cellular injury in the pathophysiology of acute central nervous system (CNS) injuries and in the pathogenesis of the chronic neuro-degenerative disorders has been increasingly recognized. While proteins, nucleic acids, and carbohydrates are all susceptible to oxygen radical damage, perhaps the most avid targets of oxygen-radical-induced injury are cell membrane lipids, including choles-terol and, in particular, polyunsaturated fatty acids. The process of lipid damage by oxygen radicals is known as lipid peroxidation (LP). Central nervous tissue provides an especially avid environment for the occurrence of LP reactions. One reason for this is the high content of iron found in many brain regions, which varies in parallel with the regional sensitivity to *ex vivo* LP (Zaleska and Floyd, 1985). Iron participates in both the initiation and propagation of LP (see section 17.2). Additionally, brain and spinal cord membrane phospholipids contain a higher proportion of polyunsaturated fatty acids, such as linoleic acid (18:2) and arachidonic acid (20:4), that are sensitive to LP (White, 1973) in comparison to other tissues.

Indeed, iron-catalyzed, oxygen-radical-induced LP has been strongly implicated in the pathophysiology of acute CNS traumatic and ischemic injuries (Braughler and Hall, 1989; Hall and Braughler, 1989, 1993; Siesjo *et al.*, 1989; Traystman *et al.*, 1991) and subarachnoid hemorrhage (SAH) (Sano *et al.*, 1980; Asano *et al.*, 1991). Much of the evidence in support of this statement is derived from studies demonstrating the early occurrence of LP in injured CNS tissue together with the neuroprotective efficacy

Edward D. Hall   CNS Diseases Research, Pharmacia and Upjohn, Inc., Kalamazoo, Michigan 49001.

Metals and Oxidative Damage in Neurological Disorders, edited by Connor. Plenum Press, New York, 1997.

of antioxidant pharmacological compounds. In addition, LP has been increasingly recognized as a significant player in Parkinson's disease (Cohen, 1986; Spina and Cohen, 1989; Jenner *et al.*, 1992; Youdim *et al.*, 1993), Alzheimer's disease (Subbarao *et al.*, 1990; Smith *et al.*, 1994), and motor neuron diseases like amyotrophic lateral sclerosis (Rosen *et al.*, 1993; Gurney *et al.*, in press). Based on the biochemical and pharmacological evidence for the involvement of LP in acute and chronic CNS disorders, there has been growing interest in the potential treatment of these conditions with antioxidant compounds, i.e., agents that either scavenge oxygen radicals or that inhibit LP.

17.2. CHEMISTRY OF LIPID PEROXIDATION

The chemistry and sources of oxygen free radical production and the process of LP have been reviewed in detail in prior publications (Braughler and Hall, 1989; Halliwell and Gutteridge, 1991; Hall and Braughler, 1993). Thus, these issues will only be presented here in sufficient detail to provide the reader with a basis for appreciating the mechanisms by which oxygen radical damage can be inhibited.

Initiation of LP occurs when a radical species with significant oxidizing capability, such as the hydroxyl radical ($^{\cdot}OH$), removes an allylic hydrogen from an unsaturated fatty acid (LH), resulting in a radical chain reaction:

$$LH + {}^{\cdot}OH \rightarrow L^{\cdot} \text{ (alkyl radical, conjugated diene)} + H_2O$$

$$L^{\cdot} + O_2 \rightarrow LOO^{\cdot} \text{ (peroxyl radical)}$$

$$LOO^{\cdot} + LH \rightarrow LOOH \text{ (lipid hydroperoxide)} + L^{\cdot}$$

Once LP begins, iron may participate in driving the process as lipid hydroperoxides (LOOH) formed through initiation are decomposed by reactions with either Fe^{+2}, Fe^{+3}, or their chelates:

$$LOOH + Fe^{+2} \rightarrow LO^{\cdot} \text{ (alkoxyl)} + OH + Fe^{+3}$$

or

$$LOOH + Fe^{+3} \rightarrow LOO^{\cdot} + Fe^{+2}$$

Both of the reactions of LOOH with iron have acidic pH optima, and are thus more likely to occur in the context of ischemic acidosis. Under normal circumstances, low molecular weight forms of redox-active iron in the brain are maintained, as in other tissues, at extremely low levels. Extracellularly, the iron transport protein, transferrin, tightly binds iron in the Fe^{+3} form. Intracellularly, Fe^{+3} is sequestered by the iron storage protein, ferritin. While both ferritin and transferrin have very high affinity for iron at neutral pH and effectively maintain iron in a non-catalytic state, both proteins readily give up their iron at pH values of 6.0 or less. Therefore, within the ischemic CNS environment where pH is typically lowered, conditions are favorable for the potential release of iron from storage proteins. Additionally, the bound iron can be released from storage proteins by reduction by superoxide radicals (Halliwell and

Gutteridge, 1991). Whether by either or both mechanisms, the inappropriate release of iron so that it can catalyze LP has been suggested as being critically important in promoting peroxidative injury in acute CNS insults (Braughler and Hall, 1989; Hall and Braughler, 1993) and Parkinson's disease (Youdim *et al.*, 1993).

A second source of catalytically active iron is hemoglobin, which may enter the brain secondary to post-traumatic, post-ischemic or post-SAH blood–brain barrier compromise, or by intracerebral hemorrhage. While hemoglobin itself has been reported to stimulate oxygen radical reactions, it is more likely that iron, released from hemoglobin, is responsible for hemoglobin-mediated lipid peroxidation. Iron is released from hemoglobin by either H_2O_2 or LOOHs. As with transferrin and ferritin, this release is further enhanced as the pH falls to 6.5 or below. Either alkoxyl (LO˙) or peroxyl (LOO˙) radicals arising from LOOH decomposition by iron can initiate so-called lipid hydroperoxide-dependent LP, resulting in chain branching reactions.

$$LOO˙ + LH \rightarrow LOOH + L˙$$
or
$$LO˙ + LH \rightarrow LOH + L˙$$

Thus, it may be unnecessary to distinguish between true initiation by a primordial radical (˙OH) and LOOH-dependent LP since sufficient LOOH and iron may pre-exist in normal membranes to allow for the latter to occur.

17.3. POTENTIAL MECHANISMS OF LIPID PEROXIDATION INHIBITION

There are a variety of mechanisms by which oxygen radical damage can be inhibited. From a chemical perspective, the designation of a compound as an antioxidant implies that it prevents the loss of an electron from another compound. Indeed, chemical antioxidants, like butylated hydroxytoluene (BHT), prevent oxidation by donating an electron to a compound that has just been oxidized, thus reducing it back to its non-oxidized state. However, from a pharmacological or biological perspective, it is reasonable to more broadly apply the term antioxidant to any compound that prevents oxidation, whether by chemical (i.e., electron donation) mechanisms, by removal of the oxidizing species, by inhibition of the formation of the oxidizing radical, or perhaps by physicochemical effects on membrane phospholipid dynamics such that the propagation of LP reactions is inhibited. Accordingly, the category of pharmacological antioxidants may include any compound that inhibits oxidant damage to lipids or other molecular species, whether it acts by a direct chemical mechanism or by an indirect non-chemical mechanism.

17.3.1. Inhibition of Oxygen Radical Formation

Based on this broader pharmacological definition, the first potential antioxidant neuroprotective mechanism to be considered involves the inhibition of specific sources of oxygen radical production. In that regard, compounds that act as inhibitors of enzymatic mechanisms that generate free radicals may be classed as antioxidants. For example, an important source of oxygen radicals in the context of acute CNS traumatic and ischemic injuries (Hall and Braughler, 1993) is prostaglandin and leukotriene synthesis by prostaglandin synthase and 5-lipoxygenase, respectively. Each of these enzymes produces the superoxide radical as a by-product of their enzymatic function (Kukreja et al., 1986). Thus, the numerous available inhibitors of these enzymes will have a secondary antioxidant action. However, it is seldom possible to clearly delineate the extent that their antioxidant mechanism is important against the background of the inhibited formation of potentially deleterious eicosanoids (e.g., $PGF_{2\alpha}$, TXA_2, LTB_4, LTC_4). Nevertheless, the various prostaglandin synthase inhibitors (e.g., ibuprofen, indomethacin) and 5-lipoxygenase inhibitors (e.g., BW755C, AA-861), while inhibiting eicosanoid formation, are at the same time attenuating oxygen radical formation.

Similarly, another enzyme whose pharmacological inhibition leads to an inhibition of superoxide production is xanthine oxidase that is localized primarily in vascular endothelial cells. Indeed, this enzyme, which is inhibited by agents such as allopurinol or its active metabolite oxypurinol, has been extensively invoked as a major source of oxidative stress in the context of ischemia/reperfusion injury to brain and other tissues (McCord, 1985; Phillis, 1989; Lin and Phillis, 1992).

17.3.2. Enzymatic Radical Scavenging

A second mechanism for interrupting radical damage involves the scavenging or interception of inorganic oxygen radicals after they are formed. The endogenous enzyme superoxide dismutase (SOD) specifically converts two moles of superoxide ($+2$ H^+) to a mole of hydrogen peroxide, which is then decomposed to water by the enzyme catalase (Halliwell and Gutteridge, 1991). Superoxide dismutase or its longer-acting polyethylene glycol-conjugated form (PEG-SOD) has been the subject of several investigations in models of cerebral ischemia and SAH where vascular damage plays an important part in the overall pathophysiology (Uyama et al., 1990, 1992; Imaizumi et al., 1990; Araki et al., 1992; MacDonald et al., 1992). However, the application of antioxidant enzymes to the chronic neurodegenerative disorders is impractical due to lack of oral bioavailability and/or their inability to penetrate the blood–brain barrier.

17.3.3. Chemical Radical Scavenging

A third antioxidant mechanism has to do with the chemical scavenging (i.e., adduct formation) of oxygen radical species. The "spin-trapping" compound N-tert-butyl-α-phenyl nitrone (or PBN), which can form adducts with hydroxyl radicals, falls into this category. PBN has been shown to have protective efficacy in models of focal (Phillis and Cao, 1994) and global (Phillis and Clough-Helfman, 1990) cerebral ischemia. Moreover, the compound has been shown to reduce the level of oxidized proteins in the aged gerbil brain in parallel with an enhancement in the action of the enzyme glutamine synthetase and an improvement in short-term memory function (Smith et al., 1991). Another compound said to be a hydroxyl radical scavenger is dimethylthiourea (DMTU), which has also been shown to be protective in a rat model of global cerebral ischemia when administered in massive doses (Pahlmark et al., 1993).

A number of compounds have, in fact, been referred to as ˙OH scavengers. However, it should be understood that ˙OH radicals are highly reactive and capable of reacting with almost any chemical they come in contact with. Consequently, any target molecule can be viewed as a ˙OH scavenger. Nevertheless, there are select compounds that very effectively attenuate post-ischemic ˙OH concentrations in brain, including the non-glucocorticoid 21-aminosteroids ("lazaroids") tirilazad mesylate (U-74006F) (Althaus *et al.*, 1993), and U-74389G (Zhang and Piantadosi, 1994), which may qualify them as bona fide ˙OH scavengers in the *in vivo* setting. The apparent fact that some compounds seem to be particularly effective hydroxyl scavengers in the *in vivo* context may depend on their ability to localize at the sites at which hydroxyl radicals are being generated. For instance, tirilazad is largely localized in the vascular endothelium (Audus *et al.*, 1991), an important site of early radical generation, after traumatic or ischemic insults (Hall *et al.*, 1993; Grammas *et al.*, 1993). Another ˙OH scavenger that has been examined in the context of cerebral ischemia models is 1,2-bis(nicotinamide)-propane (AVS, nicaraven) (Asano *et al.*, 1984).

17.3.4. Peroxynitrite Scavenging

A fourth scavenging mechanism that possesses potential neuroprotective efficacy has to do with the concept of pharmacologically scavenging peroxynitrite (Beckman, 1991). It is known that endothelial cells, neutrophils, macrophages, and microglia can produce both superoxide and nitric oxide (NO˙), the latter from nitric oxide synthetase. The two species can combine to form peroxynitrite anion (OONO-). The peroxynitrite will then undergo protonation (pKa = 6.8), thus becoming peroxynitrous acid (ONOOH). However, ONOOH is an unstable acid that can readily decompose to give hydroxyl radical and nitrogen dioxide (˙NO_2):

$$O_2^- + NO˙ \rightarrow ONOO^- + H^+ \rightarrow ONOOH \rightarrow ˙NO_2 + ˙OH$$

Via this mechanism, another source of highly reactive radicals may be operative within brain tissue. A particularly attractive aspect of this scenario is that peroxynitrite has a relatively long half-life and, thus, is potentially more diffusible compared to either superoxide or ˙OH. Therefore, it may offer a mechanism by which free radical damage may occur at a site remote from the actual location of initial oxygen radical formation. However, the relative stability of peroxynitrite also suggests the therapeutic possibility of peroxynitrite scavenging compounds. In that regard, penicillamine and cysteamine are reasonably potent peroxynitrite scavengers (Althaus *et al.*, 1994).

17.3.5. Peroxyl Radical Scavenging

Once LP is initiated by an oxidizing radical, one of the most efficient ways to prevent its propagation is by scavenging LOO˙. The prototypical peroxyl scavenger is alpha tocopherol (vitamin E), which donates an electron to the LOO˙ species, converting it to a LOOH that can then be decomposed by the enzyme glutathione peroxidase. In the process of vitamin E's antioxidant action, it is converted to a non-reactive tocopherol radical, which can potentially be reconverted to alpha tocopherol by other intracellular reducing agents, such as ascorbic acid and, perhaps, glutathione (McCay, 1985). A number of pharmacological antioxidants that function as peroxyl radical scavengers more effectively than vitamin E have been developed. These include the 21-aminosteroids tirilazad, U-74389G, and U-74500A (Braughler and Pregenzer, 1989; Hall et al., 1994), the 2-methylaminochroman U-78517F (Hall et al., 1991a), the thiazolidinone LY178002 (Clemens et al., 1991), dihydrolipoic acid (Prehn et al., 1992) and, very recently, the pyrrolopyrimidines, such as U-101033E (Hall et al., 1995; Andrus et al., 1995; Oostveen et al., 1995a). All of these have been shown to be neuroprotective in cerebral ischemia models. Consistent with the definitions provided above, compounds that scavenge peroxyl radicals qualify as true chemical antioxidants.

17.3.6.   Iron Chelation

As noted earlier, iron plays an important role in the propagation of LP reactions by catalyzing the decomposition of LOOH to either a LOO˙ or an equally reactive LO˙. Therefore, an indirect approach to stopping lipid peroxidative reactions involves scavenging or "chelating" iron with compounds like desferal (also known as desferrioxamine or deferoxamine). Desferal has been reported to have beneficial effects in some models of acute cerebral ischemia or SAH (Kumar et al., 1988; Vollmer et al., 1991).

Recently, efforts have been made to enhance the plasma retention and duration of action of desferal by bonding it to dextran. This dextran–desferal combination compound has shown better efficacy than desferal in an acute head injury model (Rosenthal et al., 1992; Panter et al., 1992).

17.3.7.   Membrane Stabilization

Lastly, it is possible to limit the propagation of lipid peroxidation in a membrane with compounds that decrease membrane phospholipid fluidity. For a peroxyl radical to effectively react with a neighboring polyunsaturated fatty acid, it must have a degree of mobility within the membrane. Compounds that have been shown to effectively decrease membrane fluidity, such as the 21-aminosteroids tirilazad or U-74500A (Audus et al., 1991; Hall et al., 1994), can have a profound physicochemical antioxidant effect by that mechanism. In addition, the glucocorticoid steroid methylprednisolone has been documented to inhibit post-traumatic lipid peroxidation, most likely via a membrane stabilizing mechanism (Hall, 1992).

Figure 17.1 provides the structures of various natural (e.g., vitamin E) or synthetic antioxidants that have been explored in models of acute and/or chronic neurodegenerative disorders.

17.4.   TIRILAZAD: AN EXAMPLE OF A MULTI-MECHANISTIC ANTIOXIDANT NEUROPROTECTIVE

Several years ago, a novel group of 21-aminosteroids (nicknamed "lazaroids") were discovered that are potent inhibitors of post-traumatic or post-ischemic LP in nervous tissue, and possess both neuroprotective and vasoprotective activity in various

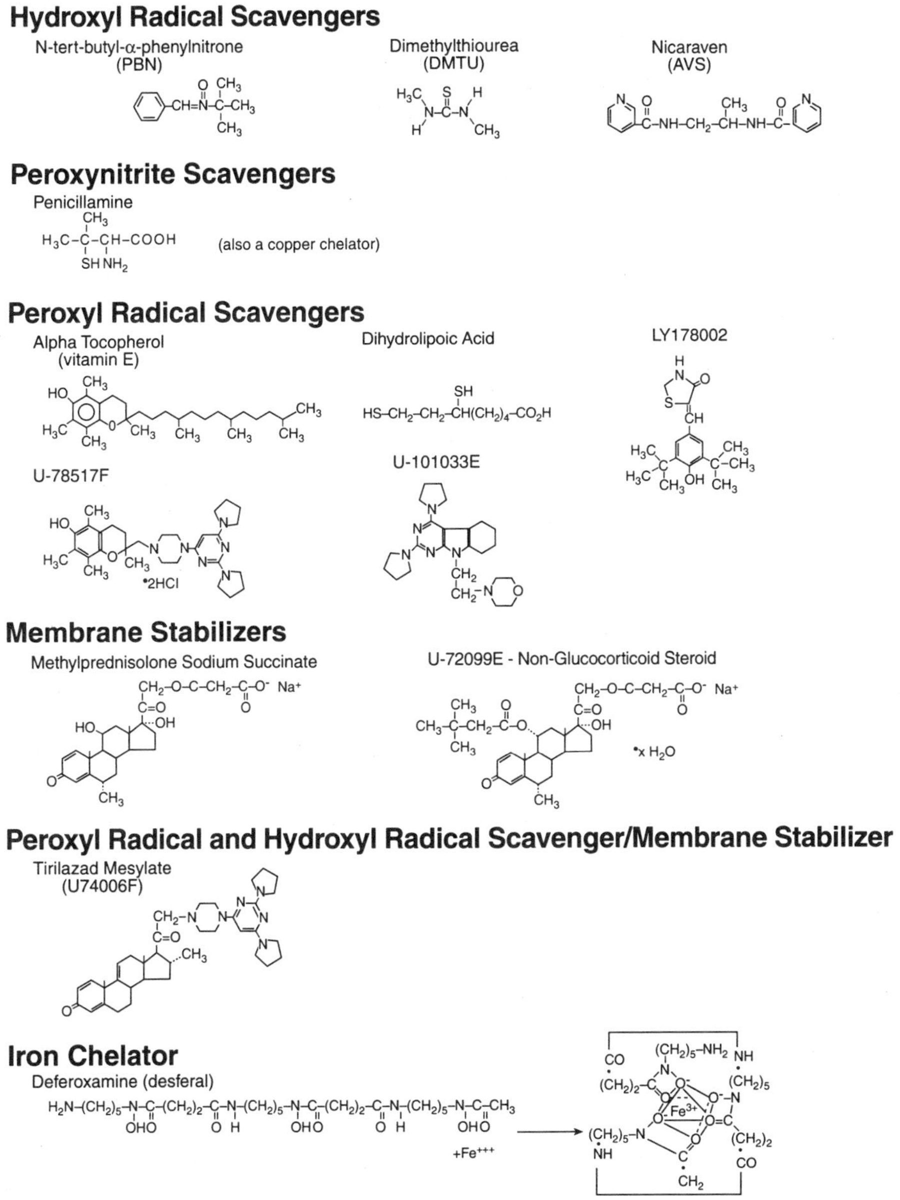

Hydroxyl Radical Scavengers

N-tert-butyl-α-phenylnitrone
(PBN)

Dimethylthiourea
(DMTU)

Nicaraven
(AVS)

Peroxynitrite Scavengers

Penicillamine

(also a copper chelator)

Peroxyl Radical Scavengers

Alpha Tocopherol
(vitamin E)

Dihydrolipoic Acid

LY178002

U-78517F

U-101033E

Membrane Stabilizers

Methylprednisolone Sodium Succinate

U-72099E - Non-Glucocorticoid Steroid

Peroxyl Radical and Hydroxyl Radical Scavenger/Membrane Stabilizer

Tirilazad Mesylate
(U74006F)

Iron Chelator

Deferoxamine (desferal)

FIGURE 17.1. Chemical structures of various antioxidant compounds examined in models of acute and/or chronic neurodegenerative disorders.

models of central nervous system injury, ischemia, and SAH (Hall *et al.*, 1994). One of these compounds, tirilazad mesylate (original name: U-74006F) (Figure 17.1) is in phase III clinical development for brain and spinal cord injury, ischemic stroke, and SAH. The rationale for the design of tirilazad was based on earlier work demonstrating the antioxidant neuroprotective properties of certain glucocorticoid steroids, such as methylprednisolone, which were found to be independent of steroid receptor pharmacology (Hall, 1992). Thus, it was reasoned that LP-inhibiting steroids that were chemically designed to be devoid of steroid receptor potential, while perhaps at the same time possessing greater antioxidant efficacy, might offer a safer and more effective way to aggressively treat acute cerebral insults.

17.4.1. Tirilazad Lipid Peroxidation-Inhibiting Mechanisms

Tirilazad is a non-glucocorticoid 21-aminosteroid that is a highly lipophilic, potent inhibitor of oxygen radical-induced, iron-catalyzed LP (Hall *et al.*, 1994). For example, in a model that used synaptic membranes prepared from rat brain as a lipid target and 200 μM ferrous chloride to initiate and catalyze the LP reactions, tirilazad inhibited LP with and IC_{50} as low as 10 μM (Braughler *et al.*, 1987). Tirilazad has also been shown to inhibit iron-catalyzed LP-dependent degeneration in murine neocortical (Monyer *et al.*, 1990) and spinal (Hall *et al.*, 1991a) neuronal cultures. The compound exerts its anti-LP action through cooperative mechanisms: a radical scavenging action (i.e., chemical antioxidant effect) and a physicochemical interaction with the cell membrane that serves to decrease membrane phospholipid fluidity (i.e., membrane stabilization).

Tirilazad has been reported to scavenge linoleic acid peroxyl radicals generated upon exposure to the free radical generator 2,2′-azobis(2,4-dimethylvaleronitrile), although it possesses a slower rate constant in this non-membranous environment than the prototypical peroxyl radical scavenger vitamin E. However, when the two antioxidants are studied simultaneously, tirilazad acts to slow the oxidation of vitamin E during linoleic acid peroxidation and potentiate vitamin E's antioxidant efficacy (Braughler and Pregenzer, 1989). Consistent with this, tirilazad has been found to attenuate the loss of tissue vitamin E that occurs in models of severe traumatic (Hall *et al.*, 1989) and ischemic (Hall *et al.*, 1991b) CNS injury.

In addition to scavenging of lipid peroxyl radicals, tirilazad can also react with hydroxyl radicals generated during *in vitro* Fenton reactions (i.e., $Fe^{+2} + H_2O_2 \rightarrow Fe^{+3}\ OH^- +\ \dot{O}H$) (Althaus *et al.*, 1993). *In vivo* studies employing the salicylate hydroxyl radical trapping method have demonstrated that tirilazad administration decreases brain hydroxyl radical levels in models of concussive head injury in mice (Hall *et al.*, 1992) and global cerebral ischemia/reperfusion injury in gerbils (Althaus *et al.*, 1993). Tirilazad has also been reported to lessen the increase in hydroxyl radical concentration in rat brain produced by infusion of glutamate (Boisvert and Schreiber, 1992). This may be due to either direct scavenging of hydroxyl radical, as discussed in section 17.3.3, or perhaps a decrease in its formation.

Reflecting its lipophilicity and consequent membrane affinity, tirilazad has been shown to exert significant effects on cell membrane phospholipid fluidity. For instance, in bovine brain microvessel endothelial cells that were labeled with diphenylhexatriene

(DPH) fluorophores, tirilazad interactions with cell membranes were characterized with fluorescence anisotropy and fluorescence lifetimes. Tirilazad preferentially altered the fluorescence anisotropy and lifetime parameters of the fluorescent DPH probe that distributed deep within the hydrophobic core of the membrane. Little or no effect was observed on the fluorescence parameters of the probe (TMA-DPH) that localized on the surface of endothelial plasma membrane. These experiments suggest that tirilazad selectively induces changes in the phospholipid packing order of the membrane hydrophobic domains (Audus et al., 1991).

Based on these and other membrane physicochemical studies, it has been hypothesized (Hall et al., 1994) that tirilazad resides in the cell membrane, and that the piperazine nitrogen, which is largely protonated (i.e., positively charged) at physiologic pH, orients among the head groups of the membrane bilayer by ionic interaction to the negatively-charged, phosphate-containing head groups. The steroid moiety, on the other hand, localizes within the hydrophobic core of the membrane. The positioning of the bulky 21-amino moiety toward the surface of the membrane then causes compression of the membrane phospholipid head groups. In addition to the chemical antioxidant properties (i.e., radical scavenging) of tirilazad described above, the membrane-stabilizing action of the compound may help to inhibit the propagation of LP by restricting the movement of lipid peroxyl radicals within the membrane so that they are less able to interact with other peroxidizable fatty acids in their immediate vicinity.

17.4.2. Protection of Endothelial Function

Tirilazad has been repeatedly demonstrated to exert a neuroprotective action in a wide variety of models of acute CNS injury, ischemia, and SAH (Hall et al., 1994). In many of these models, the actions of tirilazad have been demonstrated in terms of a preservation of normal cerebral microvascular perfusion, an attenuation of blood–brain barrier compromise, and in the case of SAH, a decrease in the incidence and severity of delayed spasm of the major cerebral arteries. Consistent with these actions, tirilazad is largely localized in the microvascular endothelium after intravenous administration (Raub et al., 1993). An important mechanism of some of the effects of endothelially-localized tirilazad may be the preservation of endothelial-dependent relaxing factor production that has, in fact, been reported (Mathews et al., 1992). In those studies, the multimechanistic tirilazad effectively protected the acetylcholine-induced (endothelium-dependent) relaxation of rabbit aortic strips from xanthine/xanthine oxidase-induced inhibition, while the monomechanistic peroxyl radical scavenger vitamin E was ineffective.

17.4.3. Protection of Calcium Homeostatic Mechanisms

Lipid peroxidation is a process that affects the functional and structural integrity of cell membranes. A critical aspect of this from the standpoint of neuronal survival relates to the sensitivity of membrane Ca^{++} ATPase (i.e., Ca^{++} pump) to LP-induced damage which results in intracellular Ca^{++} accumulation (Rohn et al., 1993). Similarly, oxidative inactivation of the membrane Na^+/K^+-ATPase (Rohn et al., 1993) can lead to intracellular Na^+ accumulation, which will then reverse the direction of the Na^+/Ca^{++} exchanger (antiporter) and further exacerbate intracellular Ca^{++} overload. Consequently, ischemia-damaged cerebral neurons suffer from abnormal membrane-dependent Ca^{++} and Na^+ homeostasis. As an example of this, gerbils subjected to 3 hrs of unilateral carotid occlusion manifest a pronounced deficit in the post-reperfusion recovery of cortical extracellular Ca^{++} levels (i.e., persistence of intracellular accumulation), which occurs simultaneously with post-ischemic LP (Hall *et al.*, 1991b). Administration of tirilazad has been shown to attenuate post-reperfusion LP (i.e., decreases LP-associated vitamin E depletion) and to facilitate recovery of extracellular Ca^{++} (Hall *et al.*, 1991b), together with a reduction in subsequent cortical neuronal damage (Hall *et al.*, 1988). The likelihood that the improved Ca^{++} recovery is due to an inhibition of peroxidative inactivation of the membrane Ca^{++} pumping mechanisms is based on the finding that tirilazad can protect red blood cell membrane Na^+/K^+- and Ca^{++}-ATPases from iron-induced inhibition simultaneous with inhibition of MDA formation (Rohn *et al.*, 1993). Similarly, superoxide dismutase administration has also been shown to improve the post-reperfusion recovery of cortical extracellular Ca^{++} after an episode of temporary middle cerebral artery occlusion (Araki *et al.*, 1992).

17.4.4. Tirilazad Clinical Neuroprotection Trials

Currently, tirilazad is being actively investigated in phase III clinical trials in head and spinal cord injury, ischemic stroke, and subarachnoid hemorrhage (SAH). The first completed phase III trial of the efficacy of tirilazad has been a multinational European/Australian/New Zealand trial in aneurysmal SAH that included 1023 patients (Kassell *et al.*, 1996). The patients were randomly assigned to either 0.6, 2, or 6 mg/kg/day, i.v., in four divided doses, or to placebo. All patients received the calcium channel blocker nimodipine based on the prior regulatory approval of this drug for aneurysmal SAH. Treatment was begun within 48 hrs after SAH and continued for 8–10 days.

In the 6 mg/kg dose group (compared to vehicle), there was a 43% reduction in 3-month mortality ($p < 0.01$), a 21% improvement in the incidence of "good" recovery (Glasgow Outcome Scale) ($p < 0.012$), a 28% decrease in the incidence of clinical vasospasm ($p < 0.047$), a 46% lesser use of therapeutic "triple H" hypertensive/hypervolemic/hemodilution therapy ($p < 0.006$) and, most impressively, a 48% increase in the number of patients who could return to full-time work (Kassell *et al.*, 1996). These results show that this antioxidant neuroprotective non-glucocorticoid 21-aminosteroid possesses significant neuroprotective efficacy in man.

17.5. 2-METHYLAMINOCHROMANS

Following the 21-aminosteroids and tirilazad, further discovery efforts were aimed at the possibility of further enhancing cerebral antioxidant activity by replacing the steroid functionality, which possesses only weak antioxidant activity in the absence of the 21-amino substitution, with a known antioxidant. A series of compounds was synthesized in which the steroid of tirilazad was replaced by the antioxidant ring structure (i.e., chromanol) of alpha-tocopherol (vitamin E). One of these, U-78517F (Figure 17.1), has been demonstrated to have predictably more potent *in vitro* lipid antioxidant and *in vivo* cerebroprotective activity in models of injury and global cerebral ischemia (Hall *et al.*, 1990, 1991). However, U-78517F does not possess significantly better blood–brain barrier penetration than tirilazad, and its development for other than acute indications was precluded by the observation of bone marrow suppression upon chronic treatment of animals.

17.6. PYRROLOPYRIMIDINES

As detailed above, the 21-aminosteroid lazaroid tirilazad mesylate has been demonstrated to be a potent inhibitor of lipid peroxidation and to reduce traumatic and ischemic damage in a number of experimental models (Hall *et al.*, 1994). Tirilazad acts, in large part, to protect the microvascular endothelium and consequently to maintain normal blood–brain barrier (BBB) permeability and cerebral blood flow autoregulatory mechanisms. However, due to its limited penetration into brain parenchyma, tirilazad has generally failed to affect delayed neuronal damage to the selectively vulnerable hippocampal CA1 and striatal regions (Hall *et al.*, 1994). Accordingly, the compound has limited application in chronic neurodegenerative disorders where penetration into brain parenchyma is an intuitive necessity.

Recently, a new group of antioxidant neuroprotective compounds, the pyrrolopyrimidines (e.g., U-101033E), have been discovered that possess significantly improved ability to penetrate the blood–brain barrier and gain direct access to neural tissue (Hall *et al.*, 1995; Hall *et al.*, in press). U-101033E has demonstrated greater ability to protect the CA1 region in the gerbil transient forebrain ischemia model, with a post-ischemic therapeutic window of at least 4 hrs (Andrus *et al.*, 1995). The compound has also been found to reduce infarct size in the mouse permanent middle cerebral artery occlusion model, in contrast to tirilazad, which is minimally effective (Hall *et al.*, 1995).

In addition to these experiments that suggest the utility of the pyrrolopyrimidines in acute ischemic insults, other data support their possible utility in the context of chronic neurodegenerative disorders. For instance, U-101033E has been demonstrated to limit the post-ischemic induction of amyloid precursor protein, β-amyloid, apolipoprotein E, and glial fibrillary acidic protein expression in the hippocampal CA1 region in the gerbil 5-min forebrain ischemia model, together with a preservation of the highly vulnerable CA1 neurons (Oostveen *et al.*, 1995b). This may indicate the compound's use in Alzheimer's disease. U-101033E also retards delayed ischemic degeneration of the nigrostriatal dopamine neurons in the gerbil 10-min forebrain ischemia paradigm (Oostveen *et al.*, 1995a), which implies possible neuroprotective efficacy in Parkinson's disease. Furthermore, U-101033E has been observed to retard axotomy-induced apoptotic retrograde degeneration of facial motor neurons in the neonatal rat, perhaps indicative of pyrrolopyrimidine application in motor neuron diseases (Hall *et al.*, unpublished observations). Thus, this novel class of brain-penetrating and orally-active antioxidants may prove to be effective neuroprotective agents in a wide spectrum of neurodegenerative disorders where oxygen radical mechanisms are operative. On the other hand, microvascularly-localized antioxidant agents like tirilazad appear to have better ability to limit blood–brain barrier damage (Hall *et al.*, 1995). Thus, it is unlikely that a single type of antioxidant agent will address all pathophysiological features of acute or chronic neurodegenerative conditions. In some instances, combinations of agents with differing mechanisms and target sites may prove to be better than the use of single antioxidant compounds.

17.7. REFERENCES

Althaus, J. S., Andrus, P. K., Williams, C. M., VonVoigtlander, P. F., Cazers, A. R., and Hall, E. D., 1993, The use of salicylate hydroxylation to detect hydroxyl radical generation in ischemic and traumatic brain injury: Reversal by tirilazad mesylate (U-74006F), *Molec. Chem. Neuropath.* **20:**147–162.

Althaus, J. S., Oien, T. T., Fici, G. J., Scherch, H. M., Sethy, V. H., and VonVoigtlander, P. F., 1994, Structure-activity relationships of peroxynitrite scavengers: An approach to nitric oxide neurotoxicity, *Res. Comm. Chem. Path. Pharmacol.* **83:**243–254.

Andrus, P. K., Fleck, T. J., and Hall, E. D., 1995, Neuroprotective efficacy of the novel brain-penetrating antioxidant U-101033E in the gerbil forebrain ischemia model, *J. Neurotrauma* **12:**967.

Araki, N., Greenberg, J. H., Uematsu, D., Sladky, J. T., and Reivich, M., 1992, Effect of superoxide dismutase on intracellular calcium in stroke, *J. Cereb. Blood Flow Metab.* **12:**43–52.

Asano, T., Johshita, H., Koide, T., and Takakura, K., 1984, Amelioration of ischaemic cerebral oedema by a free radical scavenger, AVS:1,2-bis(nicotinamido)-propane. An experimental study using a regional ischaemia model in cats, *Neurol. Res.* **6:**163–168.

Asano, T., Matsui, T., and Takuwa, Y., 1991, Lipid peroxidation, protein kinase C and cerebral vasospasm, *Crit. Rev. Neurosurg.* **1:**361–379.

Audus, K. L., Guillot, F. L., and Braughler, J. M., 1991, Evidence for 21-aminosteroid association with the hydrophobic domains of brain microvessel endothelial cells, *Free Rad. Biol. Med.* **11:**361–371.

Beckman, J. S., 1991, The double-edged role of nitric oxide in brain function and superoxide-mediated injury, *J. Devel. Physiol.* **15:**53–59.

Boisvert, D. P. C., and Schreiber, C., 1992, Interrelationship of excitotoxic and free radical mechanisms, in: *Pharmacology of Cerebral Ischemia* (J. Krieglstein and H. Oberpichler, eds.), Wassenschaftliche Verlaggesellschaft, Stuttgart, pp. 1–10.

Braughler, J. M., and Hall, E. D., 1989, Central nervous system trauma and stroke: I. Biochemical considerations for oxygen radical formation and lipid peroxidation, *Free Rad. Biol. Med.* **6:**289–301.

Braughler, J. M., and Pregenzer, J. F., 1989, The 21-aminosteroid inhibitors of lipid peroxidation: Reactions with lipid peroxyl and phenoxyl radicals, *Free Rad. Biol. Med.* **7:**125–130.

Braughler, J. M., Pregenzer, J. F., Chase, R. L., Duncan, L. A., Jacobsen, E. J., and McCall, J. M., 1987, Novel 21-aminosteroids as potent inhibitors of iron-dependent lipid peroxidation, *J. Biol. Chem.* **262:**10438–10440.

Clemens, J. A., Ho, P. P. K., and Panetta, J. A., 1991, LY178002 reduces rat brain damage after transient global ischemia, *Stroke* **22:**1048–1052.

Cohen, G., 1986, Monoamine oxidase, hydrogen peroxide and Parkinson's disease, *Adv. Neurol.* **45:**119–125.

Grammas, P., Liu, G.-J., Wood, K., and Floyd, R. A., 1993, Anoxia/reoxygenation induces hydroxyl free radical formation in brain microvessels, *Free Rad. Biol. Med.* **14:**553–557.

Gurney, M. E., Cutting, F. B., Zhai, P., Doble, A., Taylor, C. P., Andrus, P. K., and Hall, E. D., 1996, Antioxidants and inhibitors of glutamatergic transmission have therapeutic benefit in a transgenic model of familial amyotrophic lateral sclerosis, *Ann. Neurol.* **39:**147–157.

Hall, E. D., 1992, Neuroprotective pharmacology of methylprednisolone: A review, *J. Neurosurg.* **76:**13–22.

Hall, E. D., and Braughler, J. M., 1989, Central nervous system trauma and stroke: II. Physiological and pharmacological evidence for the involvement of oxygen radicals and lipid peroxidation, *Free Rad. Biol. Med.* **6**:303–313.

Hall, E. D., and Braughler, J. M., 1993, Free radicals in CNS injury, in: *Molecular and Cellular Approaches to the Treatment of Neurological Diseases* (S. G. Waxman, ed.), Raven Press, New York, pp. 81–105.

Hall, E. D., Pazara, K. E., and Braughler, J. M., 1988, The 21-aminosteroid lipid peroxidation inhibitor U-74006F protects against cerebral ischemia in gerbils, *Stroke* **19**:997–1002.

Hall, E. D., Yonkers, P. A., Horan, K. L., and Braughler, J. M., 1989, Correlation between attenuation of post-traumatic spinal cord ischemia and preservation of vitamin E by the 21-aminosteroid U-74006F: Evidence for an in vivo antioxidant action, *J. Neurotrauma* **6**:169–176.

Hall, E. D., Pazara, K. E., Braughler, J. M., Linseman, K. L., and Jacobsen, E. J., 1990, Non-steroidal lazaroid U-78517F in models of focal and global cerebral ischemia, *Stroke* **21**(Suppl III):83–87.

Hall, E. D., Braughler, J. M., Yonkers, P. A., Smith, S. L., Linseman, K. L., Means, E. D., Scherch, H. M., Jacobsen, E. J., and Lahti, R. A., 1991a, U-78517F: A potent inhibitor of lipid peroxidation with activity in experimental brain injury and ischemia, *J. Pharmacol. Exp. Ther.* **258**:688–694.

Hall, E. D., Pazara, K. E., and Braughler, J. M., 1991b, Effects of tirilazad mesylate on post-ischemic brain lipid peroxidation and recovery of extracellular calcium in gerbils, *Stroke* **22**:361–366.

Hall, E. D., Yonkers, P. A., Andrus, P. K., Cox, J. W., and Anderson, D. K., 1992, Biochemistry and pharmacology of lipid antioxidants in acute brain and spinal cord injury, *J. Neurotrauma* **9**(Suppl 2):425–442.

Hall, E. D., Andrus, P. K., and Yonkers, P. A., 1993, Brain hydroxyl radical generation in acute experimental head injury, *J. Neurochem.* **60**:588–594.

Hall, E. D., McCall, J. M., and Means, E. D., 1994, Therapeutic potential of the lazaroids (21-aminosteroids) in CNS trauma, ischemia and subarachnoid hemorrhage, *Adv. Pharmacol.* **28**:221–268.

Hall, E. D., Andrus, P. K., Smith, S. L., Oostveen, J. A., Scherch, H. M., Lutzke, B. S., Raub, T. J., Sawada, G. A., Palmer, J. R., Banitt, L. S., Tustin, J. M., Belonga, K. L., Ayer, D. E., and Bundy, G. L., 1995, Neuroprotective efficacy of microvascularly-localized versus brain-penetrating antioxidants, *Acta Neurochir.* **66**(Suppl.):107–113.

Halliwell, B., and Gutteridge, J. M. C., 1991, *Free Radicals in Biology and Medicine*, Oxford University Press, New York, pp. 1–543.

Imaizumi, S., Woolworth, V., Fishman, R. A., and Chan, P. H., 1990, Liposome-entrapped superoxide dismutase reduces cerebral infarction in cerebral ischemia in rats, *Stroke* **21**:1312–1317.

Jenner, P., Schapira, A. H. V., and Marsden, C. D., 1992, New insights into the cause of Parkinson's disease, *Neurology* **42**:2241–2250.

Kassell, N., Haley, E. C., Apperson-Hansen, C., and Alves, W. M., 1996, A randomized double-blind, vehicle-controlled trial of tirilazad mesylate in patients with aneurysmal subarachnoid hemorrhage: A cooperative study in Europe/Australia/New Zealand, *J. Neurosurg.* **84**:221–228.

Kukreja, R. C., Kontos, H. A., Hess, M. L., and Ellis, E. F., 1986, PGH synthase and lipoxygenase generate superoxide in the presence of NADH and NADPH, *Circ. Res.* **59**:612–619.

Kumar, K., White, B. C., Krause, G. S., Indrieri, R. J., Evans, A. T., Hoehner, T. J., Garritano, A. M., and Koestner, A., 1988, A quantitative morphological assessment of lidoflazine and deferoxamine therapy in global brain ischemia, *Neurol. Res.* **10**:136–140.

Lin, Y., and Phillis, J. W., 1992, Deoxycoformycin and oxypurinol: Protection against focal ischemic brain injury in the rat, *Brain Res.* **571**:272–280.

MacDonald, R. L., Weir, B. K. A., Runzer, T. D., Grace, M. G. A., and Poznansky, M. J., 1992, Effect of intrathecal superoxide dismutase and catalase on oxyhemoglobin-induced vasospasm in monkeys, *Neurosurgery* **30**:529–539.

Mathews, W. R., Marschke, C. K., Jr., and McKenna, R., 1992, Tirilazad mesylate protects endothelium from damage by reactive oxygen, *J. Mol. Cell Cardiology* **24**(suppl. III):517.

McCay, P. B., 1985, Vitamin E: Interactions with free radicals and ascorbate, *Ann. Rev. Nutr.* **5**:323–340.

McCord, J. M., 1985, Oxygen-derived radicals in postischemic tissue injury, *New Engl. J. Med.* **312**:159–163.

Monyer, H., Hartley, D. M., and Choi, D. W., 1990, 21-Aminosteroids attenuate excitotoxic neuronal injury in cortical cell cultures, *Neuron* **5**:121–126.

Oostveen, J. A., Andrus, P. K., and Hall, E. D., 1995a, Attenuation of retrograde degeneration of nigrostriatal dopamine neurons in the gerbil forebrain ischemia model, *J. Neurotrauma* **12**:967.

Oostveen, J. A., Carter, D. B., Dunn, E. J., and Hall, E. D., 1995b, Effects of U-101033E on the expression of amyloid protein precursor, apolipoprotein E, glial fibrillary acidic protein and β-amyloid expression following a bilateral carotid occlusion in the gerbil, *Neurosci. Abs.* **21**:1979.

Pahlmark, K., Folbergrova, J., Smith, M.-L., and Siesjo, B. K., 1993, Effects of dimethylthiourea on selective neuronal vulnerability in forebrain ischemia in rats, *Stroke* **24**:731–737.

Panter, S. S., Braughler, J. M., and Hall, E. D., 1992, Dextran-coupled deferoxamine improves outcome in a murine model of head injury, *J. Neurotrauma* **9**:47–53.

Phillis, J. W., 1989, Oxypurinol attenuates ischemia-induced hippocampal damage in the gerbil, *Brain Res. Bull.* **23**:267–470.

Phillis, J. W., and Cao, X., 1994, N-t-butyl-α-phenylnitrone (PBN) attenuates focal cortical injury in the rat, *Stroke* **25**:262.

Phillis, J. W., and Clough-Helfman, C., 1990, Protection from cerebral ischemic injury in gerbils with the spin trap agent N-tert-butyl-alpha nitrone, *Neurosci. Letts.* **116**:315–319.

Prehn, J. H. M., Karkoutly, C., Nuglisch, J., Peruche, B., and Krieglstein, J., 1992, Dihydrolipoate reduces neuronal injury after cerebral ischemia, *J. Cereb. Blood Flow Metab.* **12**:78–87.

Raub, T. J., Barsuhn, C. L., Williams, L. R., Decker, D. E., Sawada, G. A., and Ho, N. F. H., 1993, Use of a biophysical-kinetic model to understand the roles of protein binding and membrane partitioning on passive diffusion of highly lipophilic molecules across cellular barriers, *J. Drug Targeting* **1**:269–286.

Rohn, T. T., Hinds, T. R., and Vincenzi, F. F., 1993, Ion transport ATPases as targets for free radical damage: Protection by an aminosteroid of the Ca^{2+} pump ATPase and Na^+/K^+ pump ATPase of human red blood cell membranes, *Biochem. Pharmacol.* **46**:525–534.

Rosen, D. R., Siddique, T., Patterson, D., Figlewicz, D. A., Sapp, P., Hentati, A., Donaldson, D., Goto, J., O'Regan, J. P., Deng, H.-X., Rahmani, Z., Krizus, A., McKenna-Yasek, D., Cayabyab, A., Gaston, S. M., Berger, R., Tanzi, R. E., Halperin, J. J., Herzfeldt, B., Van den Bergh, R., Hung, W.-Y., Bird, R., Deng, G., Mulder, D. W., Smyth, C., Laing, N. G., Soriano, E., Pericak-Vance, M. A., Haines, J., Rouleau, G. A., Gusella, J. S., Horvitz, H. R., and Brown, R. H., 1993, Mutations in Cu/Zn superoxide dismutase gene are associated with familial amyotrophic lateral sclerosis, *Nature* **362**:59–62.

Rosenthal, R. E., Chanderbhan, R., Marshall, G., and Fiskum, G., 1992, Prevention of post-ischemic brain lipid conjugated diene production and neurological injury by hydroxyl-ethyl starch-conjugated deferoxamine, *Free Rad. Biol. Med.* **12**:29–33.

Sano, K., Asano, T., Tanishima, T., and Sasaki, T., 1980, Lipid peroxidation as a cause of cerebral vasospasm, *Neurol. Res.* **2**:253–272.

Siesjo, B. K., Agardh, C.-D., and Bengtsson, F., 1989, Free radicals and brain damage, *Cerebrovasc. Brain Metab. Rev.* **1**:165–211.

Smith, C. D., Carney, J. M., Starke-Reed, P. E., Oliver, C. N., Stadtman, E. R., and Floyd, R. A., 1991, Excess brain protein oxidation and enzyme dysfunction in normal aging and in Alzheimer's disease, *Proc. Natl. Acad. Sci. USA* **88**:10540–10543.

Smith, M. A., Richey, P. L., Taneda, S., Kutty, R. K., Sayre, L. M., Monnier, V. M., and Perry, G., 1994, Advanced Maillard reaction end products, free radicals, and protein oxidation in Alzheimer's disease, *Ann. N.Y. Acad. Sci.* **738**:447–454.

Spina, M. B., and Cohen, G., 1989, Dopamine turnover and glutathione oxidation: Implications for Parkinson's disease, *Proc. Natl. Acad. Sci. USA* **86**:1398–1400.

Subbarao, K. V., Richardson, J. S., and Ang, L. C., 1990, Autopsy samples of Alzheimer's cortex show increased peroxidation in vitro, *J. Neurochem.* **55**:342–345.

Traystman, R. J., Kirsch, J. R., and Koehler, R. C., 1991, Oxygen radical mechanisms of brain injury following ischemia and reperfusion, *J. Appl. Physiol.* **71**:1185–1195.

Uyama, O., Shiratsuki, N., Matsuyama, T., Nakanishi, T., Matsumoto, Y., Yamada, T., Narita, M., and Sugita, M., 1990, Protective effects of superoxide dismutase on acute cerebral injury of gerbil brain, *Free Rad. Biol. Med.* **8**:265–268.

Uyama, O., Matsuyama, T., Michishita, H., Nakamura, H., and Sugita, M., 1992, Protective effects of human recombinant superoxide dismutase on transient ischemic injury of CA1 neurons in gerbils, *Stroke* **23:**75–81.

Vollmer, D. G., Hongo, K., Ogawa, H., Tsukahara, T., and Kassell, N. F., 1991, A study of the effectiveness of the iron-chelating agent deferoxamine as vasospasm propylaxis in a rabbit model of subarachnoid hemorrhage, *Neurosurgery* **28:**27–32.

White, D. A., 1973, The phospholipid composition of mammalian tissues, in: *Function of Phospholipids* (G. B. Ansell, J. N. Hawthorne, and R. M. Dawson, eds.), Elsevier, Amsterdam, pp. 441–482.

Youdim, M. B. H., Ben-Schachar, D., and Riederer, P., 1993, The possible role of iron in the etiopathology of Parkinson's disease, *Movement Disorders* **8:**1–12.

Zaleska, M. M., and Floyd, R. A., 1985, Regional lipid peroxidation in rat brain in vitro: Possible role of endogenous iron, *Neurochem. Res.* **10:**397–410.

Zhang, J., and Piantodosi, C. A., 1994, Prolonged production of hydroxyl radical in rat hippocampus after brain ischemia-reperfusion is decreased by 21-aminosteroids, *Neurosci. Letts.* **177:**127–130.

Chapter 18

Oxidative Stress-Induced Cell Damage in the CNS

A Proposal for a Final Common Pathway

Sara J. Robb-Gaspers and James R. Connor

18.1. INTRODUCTION

The previous chapters in this book have reviewed the involvement of transition metals in diseases of the central nervous system. While each of the diseases described was unique in its symptoms, one common feature is present: Transition metals contribute to the pathological process by facilitating generation of reactive oxygen species (ROS). This concluding chapter will examine the cells and organelles at risk for oxidative damage and propose a final common pathway in neural pathogenesis of oxidative stress.

18.2. CHEMISTRY OF FREE RADICALS

The chemistry of radicals involves three steps: initiation, propagation and termination. Initiation is the step in which a species with an unpaired electron is generated, often by way of a reductive electron transfer. Propagation is the reaction of a radical with a substrate to produce a second radical by donating or abstracting an electron. Uniting of two radicals to form a nonradical species describes termination. Thus, once a radical is formed *in vivo,* propagation can lead to increasing levels of damage until

Sara J. Robb-Gaspers and James R. Connor George M. Leader Family Laboratory for Alzheimer's Disease Research, Department of Neuroscience and Anatomy, Pennsylvania State University College of Medicine, M. S. Hershey Medical Center, Hershey, Pennsylvania 17033.

Metals and Oxidative Damage in Neurological Disorders, edited by Connor. Plenum Press, New York, 1997.

the final step of termination. In situations of oxidative stress, ROS are formed, including both radical and nonradical derivatives of oxygen. Hydrogen peroxide (H_2O_2) is an example of a ROS that is not a radical, whereas, superoxide ($\cdot O_2^-$) and hydroxyl radical ($\cdot OH$) are more volatile, potentially damaging, radicals. However, when H_2O_2 is generated *in vivo,* it can cross membranes easily (Halliwell and Gutteridge, 1989) and through a reduction reaction involving one electron can lead to generation of $\cdot OH$. The highly reactive $\cdot OH$ in turn can cause lipid peroxidation and DNA damage. The key components in the generation of free radicals are availability of oxygen and presence of a transition metal. The most abundant transition metal in the brain is iron. Loss of iron balance in brain is present in a number of neurodegenerative diseases, reviewed in Chapter 2. Other metals such as copper (Chapter 4), manganese (Chapter 5) and zinc (Chapter 6) can also promote free radical formation and have also been investigated as components of pathogenesis in neurological disorders.

18.3. UNIQUE VULNERABILITY OF THE BRAIN

The brain has a number of unique qualities that make it more vulnerable than other organs to oxidative stress. It has the highest rate of oxygen consumption, is rich in lipids, and is relatively low in antioxidants. The brain does not have immediate access to plasma borne nutrients (which includes metals, see Chapter 7) because of the blood–brain barrier. Consequently, the cells in the brain must store metals in a bioavailable form. Furthermore, the brain differs from other organs in its consumption of oxygen, getting much of its energy from the mitochondrial respiratory chain. This combination of high oxidative metabolism and the low concentrations of endogenous antioxidants further increases the brain's susceptibility of oxidative cell death.

The brain contains a number of types of cells that vary in their oxygen consumption, iron requirements, and functional responsibilities, and as such may range in vulnerability to oxidative stress. Although the neuron represents the response mechanism in the brain, numerous glial cells are also required for normal neurological function. Neurons and glia contain protective mechanisms to combat oxidative stress, including mechanisms to scavenge ROS or convert them to less reactive products. However, there is not an even distribution of the protective antioxidants across all cells in the brain, suggesting that different cells may range in their vulnerability to oxidative damage. For example, catalase is found predominantly in oligodendrocytes (McKenna *et al.,* 1976), whereas glutathione is found predominantly in astrocytes (Raps *et al.,* 1989). Also, transition metals are not distributed evenly, neither by region, nor by cell (see Chapter 2 for discussion). Arguably the best examples of oxidative damage in CNS diseases are Parkinson's disease (PD) and amyotrophic lateral sclerosis (ALS). The regional specificity of neurodegeneration in PD is well established. ALS has been shown to possibly involve a mutation in the superoxide dismutase (SOD) gene. Although this mutation would affect all cells expressing this enzyme, only the central nervous system, and select regions and population of cells within the nervous system are affected. Clearly, selective vulnerability to oxidative injury occurs among different organs and among specific regions within the brain. Understanding the factors that increase the vulnerability of specific areas and/or cells will increase the likelihood of designing successful preventative and palliative treatments.

The neurons in the brain, which clearly undergo oxidative injury, have been discussed in numerous chapters throughout this book. However, neurons have a number of protective mechanisms, including secretions from other cells, available to them. In this chapter, we will consider the vulnerability of glial cells to oxidative injury and propose that damage to these cell types may, over time be of greater importance to pathogenic processes. There are three general types of glial cells: microglia, oligodendrocytes, and astrocytes.

Microglia, a non-neuronal cell type in the brain, are similar to macrophages and play the role of primary scavenger cells in the brain. Rather than being susceptible to oxidative damage, these cells generate free radicals as part of their defense mechanism (Takemura and Werb, 1984). Indeed it has been proposed that microglia (and macrophages) may exacerbate degenerative processes in brain by promoting the generation of free radicals. Microglia require iron for the generation of free radicals (Takemura and Werb, 1984). As these iron-rich cells are present in abundance around lesions in multiple sclerosis (MS) and neuritic plaques in AD, their role in pathogenesis or, perhaps, a more direct role in exacerbating the disease processes is intriguing. For example, deposition of AB4 increases in response to oxidative stress in AD (Mantyh *et al.*, 1993) and free radical generating microglia could increase the myelin degeneration in MS lesions (Mitrovic *et al.*, 1994). Hemosiderotic microglia are also found in white matter of AIDS patients, again raising the question of how much iron contributes, even secondarily, to pathogenic processes (Gelman *et al.*, 1992).

Oligodendrocytes are glial cells in the brain whose primary known function is to make myelin, the lipid-rich substance that insulates neuronal axons. However, other roles for oligodendrocytes, such as a sustentacular role for perineuronal oligodendrocytes, have not been vigorously explored. Oligodendrocytes are particularly vulnerable to oxidative damage (Oka *et al.*, 1993; Kim and Kim, 1991). *In vitro* studies (Mitrovic *et al.*, 1994, 1995) demonstrate oligodendrocytes are more susceptible to nitric oxide induced DNA damage and cell death than are astrocytes or microglia. Presumably consistent with this observation is the high concentration of iron in the cytoplasm of these cells (Benkovic and Connor, 1993). Examples of treatment regimes for MS, the prototypic demyelinating disease, involving the use of antioxidants are difficult to find. Experimental allergic encephalomyelitis (EAE), an autoimmune demyelinating disease in animals, responds favorably to antioxidant therapy and iron chelation (Bowern *et al.*, 1984; Hartung *et al.*, 1988). Loss of myelin results in inappropriate firing of neurons, which can lead to increased oxidative stress in neurons (Coyle and Puttfarcken, 1993). Thus, damage to oligodendrocytes must be considered as contributory to neuronal death seen in oxidative stress.

A third type of glial cell, astrocytes, monitor the environment within the brain and serve to restrict the exposure of other cells to various compounds, including ions and neurotransmitters. *In vitro* studies have demonstrated the protective roles of astrocytes for both neurons and other glial cell types. Norepinephrine and epinephrine are cytotoxic to oligodendrocytes in culture unless oligodendrocytes were cocultured on a layer of astrocytes (Noble *et al.*, 1994). Furthermore, astrocytes exposed to low levels of

H_2O_2 increase message transcription of neuroprotective proteins such as nerve growth factor (NGF) (Pechan *et al.*, 1992; Naveilhan *et al.*, 1994) and basic fibroblast growth factor (bFGF) (Pechan *et al.*, 1992). However, when exposed to higher levels of peroxide, protein synthesis of NGF and bFGF was inhibited (Naveilhan *et al.*, 1994). It is likely that when astrocytes are exposed to increased levels of peroxide, they become compromised in their ability to perform protective functions. Astrocytes treated with H_2O_2 were compromised in their ability to transport glutamate via the high-affinity glutamate transporter (Volterra *et al.*, 1994). Glutamate uptake is a critical function of astrocytes that when compromised, leads to neuronal excitotoxicity.

Astrocytes, like oligodendrocytes, also exist in a proximal arrangement with neurons. Astrocytes appear most potentially protective of other cell types and, thus, if damaged by enduring oxidative stress, will promote cell death in other cells, particularly neurons and oligodendrocytes. Oxidative stress can lead to cell death, with three events reported to occur prior to loss of cell viability: loss of ATP, decreased levels of NADH, and changes in Ca^{2+} homeostasis (Kehrer and Lund, 1994). Each of these changes involve the mitochondria and are relevant to CNS pathologies involving metals.

18.4. MITOCHONDRIA AS MEDIATORS OF OXIDATIVE DAMAGE IN ASTROCYTES

As electrons are shuffled during normal mitochondrial respiration, ROS are produced including $\cdot O_2^-$ and H_2O_2 (Halliwell and Gutteridge, 1989). Reducing equivalents are used to pass electrons from lower to higher potential by the proteins in four complexes in mitochondrial respiration chain. The chain can be "leaky," leading to the one electron reduction of diatomic oxygen to $\cdot O_2^-$ (Halliwell and Gutteridge, 1989). Two $\cdot O_2^-$ are united by mitochondrial SOD to produce H_2O_2. The peroxide produced is then detoxified by the actions of catalase and glutathione peroxidase, leading to the production of water and diatomic oxygen. This detoxification system is efficient as long as the system remains balanced; loss of balance results in oxidative stress. As previously mentioned, astrocytes are low in catalase and must rely primarily on glutathione peroxidase for detoxification of H_2O_2. Additionally, astrocytic monoamine oxidase (MAO) increases the intracellular concentration of H_2O_2 during monoamine metabolism, and the electron transport chain leaks more with age. Finally, astrocytes also have little intracellular ferritin for sequestering iron. These situations potentiate oxidative stress by increasing peroxide concentrations, possibly in the presence of bioavailable iron, and suggest mitochondria are a major source of intracellular ROS.

Mitochondria have all three components necessary for the Fenton reaction to occur: They actively take up iron (Flatmark and Romslo, 1975), they have a source of reactive electrons, and a source of H_2O_2. Leakage of $\cdot O_2^-$ can further increase the pool of Fe^{2+} by donating an electron to Fe^{3+}. This pool of Fe^{2+} is then free to react with H_2O_2 to produce $\cdot OH$ (Halliwell *et al.*, 1992). When $\cdot OH$ is generated within the mitochondria, it may be able to react with lipids in the inner mitochondrial membrane, causing lipid peroxidation and loss of the proton gradient. In situations where inner membrane integrity is lost, the mitochondria could become uncoupled such that oxidative phosphorylation would no longer lead to the production of ATP, yet still continue to reduce oxygen to water and generate $\cdot O_2^-$. As the rate of respiration tried to compensate for low levels of ATP, further production of $\cdot O_2^-$, H_2O_2, and $\cdot OH$ would occur.

The hydroxyl radical is highly reactive and attacks DNA. Support for the argument that mitochondria may be particularly vulnerable to oxidative inquiry is in the demonstration that mitochondrial DNA (mDNA) has a 10-fold greater mutation rate than that seen in nuclear DNA (nDNA) (Mecocci et al., 1993). Mutation of mDNA can be problematic, as many subunits of the respiratory chain complexes are coded for by mDNA. A high mutation rate could lead to increased inefficiency in respiration and perpetuate the cycle of ROS production and DNA mutations. Additionally, it has been noted that with aging there is a 15-fold increase in oxidized nucleotides in brain mitochondria (Mecocci et al., 1993).

Mitochondria have their own antioxidant system that includes glutathione (GSH), their own pools of NADPH and NADH, SOD, and glutathione peroxidase (Kehrer and Lund, 1994). Mitochondrial superoxide dismutase protects the mitochondria by dismutation of two $\cdot O_2^-$ to H_2O_2. Glutathione (GSH) has the ability to protect against free radical, peroxides, and other toxins due to its ability to be oxidized and reduced (O'Connor et al., 1995), with the reduction of oxidized glutathione (GSSG) concomitantly leading to oxidation of NADPH. GSH has also been implicated in the chelation, detoxification, and delivery of certain transition metals, and in maintaining oxidative states (Ballatori, 1994). Evidence suggests that GSH maintains copper in Cu I by providing reducing equivalents for Cu II and that Cu I-GSH was a source of Cu I for apo-metallothionien (Freedman et al., 1989). GSH metal complexes have also been suggested as a source for apo-superoxide dismutase (Steinkuhler et al., 1991). As a metal transporter, GSH contains six potential coordination sites (Ballatori, 1994) and has been shown to facilitate the transport of methyl mercury across the blood–brain barrier (Kerper et al., 1992). Formation of GSH-metal mercaptides is favorable under physiological conditions with metals including: Hg, Cd, Cu, Zn, Ag, As, Pb (Ballatori, 1994). In addition to metal buffering, GSH provides a redox buffering system that can maintain protein thiols, and levels of ascorbate and α-tocophorel (Meister, 1991).

As mentioned, GSH is found in the brain specifically within astrocytes. Astrocytic GSH is used by the enzyme glutathione peroxidase to remove intracellular H_2O_2. In situations where GSH is depleted, the peroxide concentration will remain high and is more available to react with metals. Astrocytic GSH concentration also has been reported to vary among subpopulations of astrocytes, possibly contributing to differential susceptibility to oxidative stress among the same cells within different regions (Devesa et al., 1993). There is evidence of at least organ specific susceptibility for GSH: Brain mitochondrial GSH has been found to be more oxidized than liver mitochondrial GSH (Ravindranath and Reed, 1990).

Finally, it has been demonstrated that astrocytes express inducible nitric oxide synthase (iNOS) and are capable of generating nitric oxide ($\cdot NO$). When astrocytes are exposed to cytokines they begin to express the iNOS protein, yet $\cdot NO$ production is delayed until the protein is in the proper conformation and L-arginine and tetrahydrobioptern (BH_4) are available (Albakri and Stuehr, 1996). Cytokines further contribute to the induction of iNOS by increasing BH_4, providing a necessary cofactor for the synthesis of ˙NO (Werner et al., 1993). Once the iNOS protein is expressed and in a functional conformation, in the presence of BH_4, the synthesis of NO from L-arginine begins. One physiological affect of astrocytes synthesizing ˙NO is the endogenously produced ˙NO reversibly inhibits the astrocytes own mitochondrial respiration through ˙NO binding to cytochrome oxidase (Brown, 1995). Such an inhibition can lead to increased electron leakage from the electron transport chain and subsequent increases in ˙O_2^-, H_2O_2, and ˙OH concentrations. Short-term inhibition of mitochondrial respiration at cytochrome oxidase is reversible, and therefore not necessarily fatal because astrocytes can switch to glycolysis to maintain energy supplies (Bolanos et al., 1994). However, increasing glycolysis can lead to an increase in lactate production and acidification of the astrocyte (Bolanos et al., 1994), increasing the protonation of $ONOO^-$ and subsequent formation of ˙OH. In situations where astrocytes are subjected to long-term exposure to ˙NO, cytotoxicity results. Death is believed to be a result of damage to mitochondrial enzymes as well as DNA (Brown et al., 1995). Astrocytes subjected to long-term exposure are irreversibly inhibited at cytochrome oxidase and complex I by ˙NO (Bolanos et al., 1994). Furthermore, ˙NO and O_2^- can react to produce peroxynitrite ($ONOO^-$), which is also capable of mitochondrial respiration inhibition (Radi et al., 1994) and formation of a mitochondrial permeability transition (MPT) pore, while neither ˙NO and O_2^- induce MPT alone (Packer and Murphy, 1995). Long-term exposure to ˙NO may also lead to enduring exposure to $ONOO^-$, potentiating inhibition of the mitochondrial respiratory chain.

Cellular energy loss by an additional mechanism involving H_2O_2 and iron uptake has been described that is believed to result from calcium cycling (Lehninger et al., 1978). The brain expends much of its energy to maintain ion gradients across membranes and sequester calcium. Calcium extrusion is achieved by sodium calcium exchangers and calcium ATPases (Richter and Kass, 1991). The direction of the calcium exchanger is dependent upon transmembrane potential such that depolarization or collapse of potential leads to increased intracellular calcium (Richter and Kass, 1991). Situations in which there is an increase in hydrogen peroxide production may further deplete cellular ATP. Mitochondria are capable of calcium influx and efflux across their inner membrane. A state of oxidized pyridine nucleotide hydrolysis can induce the mitochondrial release of calcium, with hydrolysis occurring between the nicotinamide and ADP ribose bond (Richter and Kass, 1991). ADP ribosylation of a protein in the inner mitochondrial membrane may be responsible for the calcium release as inhibition of ADP-ribosylation also inhibits calcium release (Richter et al., 1990). Hydroperoxides can induce calcium release from brain mitochondria, which is also paralleled by oxidation and hydrolysis of pyridine nucleotides (Gotz et al., 1994). Calcium cycling leads to the consumption of energy, loss of ATP and reducing equivalents, and, eventually, the collapse of the mitochondrial membrane potential ($\Delta\Psi$) (Richter and Kass, 1991). These calcium changes affects energy and metabolism of the whole cell by depletion of ATP.

Calcium ATPases hydrolyze ATP to transport calcium. In situations of ATP depletion, calcium ATPases are unable to function. A specialized calcium ATPase is located in the endoplasmic reticulum, the site of intracellular calcium stores. In situations of oxidative stress and depleted cellular ATP, calcium homeostasis can become seriously disrupted. Furthermore, the generation of ˙OH reaches high levels in the cytosol; lipid peroxidation of both endoplasmic reticulum and plasma membrane can cause increased cytosolic calcium through leaky membranes. Mitochondria have been shown to take up calcium in a manner that may serve to provide calcium buffering to the cell (Richter and Kass, 1991). Hydroperoxides oxidize pyridine nucleotides (Oshino and Chance, 1977) and have been shown to cause release of calcium from brain mitochondria (Satrustegui and Richter, 1984). As mentioned above, this release has been correlated with hydrolysis of NAD and ADP ribosylation of the inner mitochondria membrane (Richter and Kass, 1991). Iron gluconate also induces calcium efflux from mitochondria by a similar mechanism involving NAD hydrolysis (Masini *et al.*, 1987). In situations of increased metal availability, the cell may also experience these same effects, including loss of ATP. This decrease in ATP further affects calcium homeostasis, changing levels of activity of proteases, nucleases, kinases, and phosphatases and leading to disruptions in the cytoskeleton, DNA fragmentation, and changes in signal transduction cascades.

The increase in cellular calcium and subsequent loss of cellular energy also have consequences outside the mitochondria. The increase in cellular calcium can activate numerous calcium dependent proteins, some of which may cause further oxidative stress. Calpain I is a calcium dependent protease that is catalytically able to convert xanthine dehydrogenase to xanthine oxidase (Coyle and Puttfarcken, 1993), changing the function of this enzyme and enabling it to catabolize purine bases and generate superoxide. ROS with unpaired electrons are able to liberate iron from ferritin, including ˙NO (Reif and Simmons, 1990) and $˙O_2^-$ (Samokyszyn *et al.*, 1991) and 6 hydroxydopamine (Monteiro and Winterbourn, 1989), providing a pool of iron necessary for the Fenton reaction to proceed. Interestingly, other mechanisms of generating reactive oxygen species, such as glutamate activation of NMDA receptors via activation of phospholipase A_2 and cleavage of arachadonic acid, produces radicals (Dumuis *et al.*, 1988) and also involve calcium. Dysregulation and toxicity could be hypothesized to be a result of a similar mechanism.

18.5. SUMMARY

We have made a case for selective vulnerability of cells and organelles in the brain to oxidative injury. From this discussion astrocytes emerge as the most vulnerable cell type to enduring oxidative stress in the brain because they lack intracellular iron management proteins and generate high levels of hydrogen peroxide. Astrocytes contain relatively little catalase for reducing hydrogen peroxide, and although they contain elevated levels of glutathione, this protein can transport metals and may in some cases promote oxidative injury. The mitochondria are the probable source of much of the ROS that are formed during normal metabolic processing and are also the organelles most likely to dysfunction because of oxidative injury. Thus, we conclude by proposing that astrocytes become metabolically incapable over time as a result of oxidative

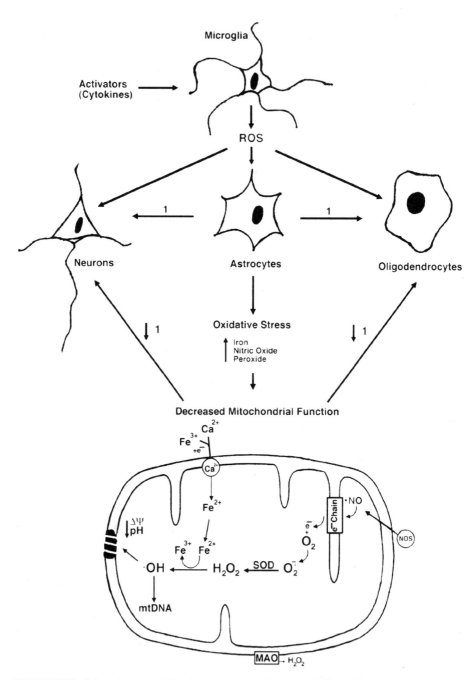

FIGURE 18.1. Schematic summarizing the concept suggesting that enduring oxidative stress to astrocytes, specifically to astrocytic mitochondria, will lead to increased vulnerability of neurons and oligodendrocytes to oxidative injury. This schematic recognizes that activation of microglia will increase reactive oxygen species (ROS) and oxidative stress in the brain, but considers this an acute process. Enduring oxidative stress

damage to their mitochondria. The decreased function of astrocytes results in loss of neurotransmitter uptake, ionic regulation, and general loss of maintenance of the extracellular milieu in the brain (Figure 18.1). Consequently, neuronal dysfunction results and neurological impairment is manifested.

18.6. REFERENCES

Albakri, Q., and Stuehr, D., 1996, Intracellular assembly of inducible NO synthase is limited by NO mediated changes in heme insertion and availability, *J. Biol. Chem.* **271**:5414–5421.

Ballatori, N., 1994, Glutathione mercaptides as transport forms of metals, *Adv. Pharmacol.* **27**:271–298.

Benkovic, S. A., and Connor, J. R., 1993, Ferritin, transferrin, and iron in selected regions of the adult and aged rat brain, *J. Comp. Neurol.* **338**:97–113.

Bolanos, J. P., Peuchen, S., Heales, S. J., Land, J. M., and Clark, J. B., 1994, Nitric oxide-mediated inhibition of the mitochondrial respiratory chain in cultured astrocytes, *J. Neurochem.* **63**:910–916.

Bowern, N., Ramshaw, I. A., Clark, I. A., and Doherty, P. C., 1984, Inhibition of autoimmune neuropathological process by treatment with an iron-chelating agent, *J. Exper. Med.* **160**:1532–1543.

Brown, G. C., 1995, Nitric oxide regulates mitochondrial respiration and cell functions by inhibiting cytochrome oxidase, *FEBS Lett.* **369**:136–139.

Brown, G. C., Bolanos, J. P., Heales, S. J., and Clark, J. B., 1995, Nitric oxide produced by activated astrocytes rapidly and reversibly inhibits cellular respiration, *Neurosci. Lett.* **193**:201–204.

Coyle, J. T., and Puttfarcken, P., 1993, Oxidative stress, glutamate, and neurodegenerative disorders, *Science* **262**:689–695.

Devesa, A., O'Connor, J. E., Garcia, C., Puertes, I. R., and Vina, J. R., 1993, Glutathione metabolism in primary astrocyte cultures: Flow cytometric evidence of heterogeneous distribution of GSH content, *Brain Res.* **618**:181–189.

Dumuis, A., Sebben, M., Haynes, L., Pin, J. P., and Bockaert, J., 1988, NMDA receptors activate the arachidonic acid cascade system in striatal neurons, *Nature* **336**:68–70.

Flatmark, T., and Romslo, I., 1975, Energy-dependent accumulation of iron by isolated rat liver mitochondria. Requirement of reducing equivalents and evidence for a unidirectional flux of Fe(II) across the inner membrane, *J. Biol. Chem.* **250**:6433–6438.

Freedman, J. H., Ciriolo, M. R., and Peisach, J., 1989, The role of glutathione in copper metabolism and toxicity, *J. Biol. Chem.* **264**:5598–5605.

Gelman, B. B., Rodriguez-Wolf, M. G., Wen, J., Kumar, S., Campbell, G. R., and Herzog, N., 1992, SideBritic cerebral macrophages in the acquired immunodeficiency syndrome, *Arch. Pathol. Lab. Med.* **116**:509–516.

Gotz, M. E., Kunig, G., Riederer, P., and Youdim, M. B., 1994, Oxidative stress: Free radical production in neural degeneration, *Pharmacol. Therap.* **63**:37–122.

---

resulting from iron accumulation, nitric oxide production, and peroxide production in astrocytes will eventually decrease the normal protective functions of astrocytes (denoted as the numeral 1) such as ionic balance and neurotransmitter uptake. Cytoplasmic iron accumulation can increase iron flow into the mitochondria possibly using the calcium uniporter. Nitric oxide will inhibit the electron transport chain increasing the leakage of electrons, which can combine with oxygen to form oxygen superoxide radicals. Monoamine oxidase activity in the mitochondrial membrane can increase H_2O_2. The elevated levels of iron and H_2O_2 can drive the reaction toward production of hydroxyl radicals ($^{\cdot}OH$), which can induce lipid peroxidation in the mitochondrial membrane resulting in a decrease in pH and membrane potential. The $^{\cdot}OH$ can also generate oxidative damage to the mitochondrial DNA (mtDNA), which will increase the potential for long-term dysfunction in the mitochondria.

Halliwell, B., and Gutteridge, J., 1989, *Free Radicals in Biology and Medicine,* Clarendon Press, Oxford.

Halliwell, B., Gutteridge, J. M., and Cross, C. E., 1992, Free radicals, antioxidants, and human disease: where are we now?, *J. Lab. Clin. Med.* **119:**598–620.

Hartung, H. P., Schafer, B., Heininger, K., and Toyka, K. V., 1988, Suppression of experimental autoimmune neuritis by the oxygen radical scavengers superoxide dismutase and catalase, *Ann. Neurol.* **23:**453–460.

Kehrer, J. P., and Lund, L. G., 1994, Cellular reducing equivalents and oxidative stress, *Free Rad. Biol. Med.* **17:**65–75.

Kerper, L. E., Ballatori, N., and Clarkson, T. W., 1992, Methylmercury transport across the blood-brain barrier by an amino acid carrier, *Am. J. Physiol.* **262:**R761–765.

Kim, Y. S., and Kim, S. U., 1991, Oligodendroglial cell death induced by oxygen radicals and its protection by catalase, *J. Neurosci. Res.* **29:**100–106.

Lehninger, A. L., Vercesi, A., and Bababunmi, E. A., 1978, Regulation of Ca^{2+} release from mitochondria by the oxidation-reduction state of pyridine nucleotides, *Proc. Natl. Acad. Sci. USA* **75:**1690–1694.

Mantyh, P. W., Ghilardi, J. R., Rogers, S., DeMaster, E., Allen, C. J., Stimson, E. R., and Maggio, J. E., 1993, Aluminum, iron and zinc ions promote aggregation of physiological concentrations of beta-amyloid peptide, *J. Neurochem.* **61:**1171–1174.

Masini, A., Trenti, T., Ceccarelli, D., and Muscatello, U. 1987, The effect of a ferric iron complex on isolated rat-liver mitochondria. III. Mechanistic aspects of iron-induced calcium efflux, *Biochim. Biophys. Acta* **891:**150–156.

McKenna, O., Arnold, G., and Holtzman, E., 1976, Microperoxisome distribution in the central nervous system of the rat, *Brain Res.* **117:**181–94.

Mecocci, P., MacGarvey, U., Kaufman, A. E., Koontz, D., Shoffner, J. M., Wallace, D. C., and Beal, M. F., 1993, Oxidative damage to mitochondrial DNA shows marked age-dependent increases in human brain, *Ann. Neurol.* **34**(4):609–16.

Meister, A., 1991, Glutathione deficiency produced by inhibition of its synthesis, and its reversal: Applications in research and therapy, *Pharmacol. Therap.* **51:**155–194.

Mitrovic, B., Ignarro, L. J., Montestruque, S., Smoll, A., and Merrill, J. E., 1994, Nitric oxide as a potential pathological mechanism in demyelination: its differential effects on primary glial cells in vitro, *Neuroscience* **61:**575–585.

Mitrovic, B., Ignarro, L. J., Vinters, H. V., Akers, M. A., Schmid, I., Uittenbogaart, C., and Merrill, J. E., 1995, Nitric oxide induces necrotic but not apoptotic cell death in oligodendrocytes, *Neuroscience* **65:**531–539.

Monteiro, H. P., and Winterbourn, C. C., 1989, 6-Hydroxydopamine releases iron from ferritin and promotes ferritin-dependent lipid peroxidation, *Biochem. Pharmacol.* **38:**4177–4182.

Naveilhan, P., Neveu, I., Jehan, F., Baudet, C., Wion, D., and Brachet, P., 1994, Reactive oxygen species influence nerve growth factor synthesis in primary rat astrocytes, *J. Neurochem.* **62:**2178–2186.

Noble, P. G., Antel, J. P., and Yong, V. W., 1994, Astrocytes and catalase prevent the toxicity of catecholamines to oligodendrocytes, *Brain Res.* **633:**83–90.

O'Connor, E., Devesa, A., Garcia, C., Puertes, I. R., Pellin, A., and Vina, J. R., 1995, Biosynthesis and maintenance of GSH in primary astrocyte cultures: role of L-cystine and ascorbate, *Brain Res.* **680:**157–163.

Oka, A., Beliveau, M. J., Rosenberg, P. A., and Volpe, J. J., 1993, Vulnerability of oligodendroglia in glutamate: Pharmacology, mechanisms and prevention, *J. Neurosci.* **13:**1441–1453.

Oshino, N., and Chance, B., 1977, Properties of glutathione release observed during reduction of organic hydroperoxide, demethylation of aminopyrine and oxidation of some substances in perfused rat liver, and their implications for the physiological function of catalase, *Biochem. J.* **162:**509–25.

Packer, M. A., and Murphy, M. P., 1995, Peroxynitrite formed by simultaneous nitric oxide and superoxide generation causes cyclosporin-A-sensitive mitochondrial calcium efflux and depolarisation, *Eur. J. Biochem.* **234:**231–239.

Pechan, P. A., Chowdhury, K., and Seifert, W., 1992, Free radicals induce gene expression of NGF and bFGF in rat astrocyte culture, *Neuroreport* **3:**469–472.

Radi, R., Rodriguez, M., Castro, L., and Telleri, R., 1994, Inhibition of mitochondrial electron transport by peroxynitrite, *Arch. Biochem. Biophys.* **308:**89–95.

Raps, S. P., Lai, J. C., Hertz, L., and Cooper, A. J., 1989, Glutathione is present in high concentrations in cultured astrocytes but not in cultured neurons, *Brain Res.* **493**:398–401.

Ravindranath, V., and Reed, D. J., 1990, Glutathione depletion and formation of glutathione-protein mixed disulfide following exposure of brain mitochondria to oxidative stress, *Biochem. Biophys. Res. Commun.* **169**:1075–1079.

Reif, D. W., and Simmons, R. D., 1990, Nitric oxide mediates iron release from ferritin, *Arch. Biochem. Biophys.* **283**:537–541.

Richter, C., and Kass, G. E., 1991, Oxidative stress in mitochondria: its relationship to cellular Ca^{2+} homeostasis, cell death, proliferation, and differentiation, *Chem. Biol. Interact.* **77**:1–23.

Richter, C., Theus, M., and Schlegel, J., 1990, Cyclosporine A inhibits mitochondrial pyridine nucleotide hydrolysis and calcium release, *Biochem. Pharmacol.* **40**:779–782.

Samokyszyn, V. M., Reif, D. W., Miller, D. M., and Aust, S. D., 1991, Effects of ceruloplasmin on superoxide-dependent iron release from ferritin and lipid peroxidation, *Free Rad. Res. Commun.* **12–13**:153–159.

Satrustegui, J., and Richter, C., 1984, The role of hydroperoxides as calcium release agents in rat brain mitochondria, *Arch. Biochem. Biophys.* **233**:736–740.

Steinkuhler, C., Sapora, O., Carri, M. T., Nagel, W., Marcocci, L., Ciriolo, M. R., Weser, V., and Rotilio, G., 1991, Increase of Cu,Zn-superoxide dismutase activity during differentiation of human K562 cells involves activation by copper of a constantly expressed copper-deficient protein, *J. Biol. Chem.* **266**:24580–24587.

Takemura, R., and Werb, Z., 1984, Secretory products of macrophages and their physiological functions, *Am. J. Physiol.* **246**:C1–9.

Volterra, A., Trotti, D., Tromba, C., Floridi, S., and Racagni, G., 1994, Glutamate uptake inhibition by oxygen free radicals in rat cortical astrocytes, *J. Neurosci.* **14**:2924–2932.

Werner, E. R., Werner-Felmayer, G., and Wachter, H., 1993, Tetrahydrobiopterin and cytokines, *Proc. Soc. Exper. Biol. Med.* **203**:1–12.

Appendix

Accepted Biomarkers of Oxidative Damage in Tissues

Leslie A. Shinobu and M. Flint Beal

This table is not intended as a comprehensive list. Markers have been chosen for their relative ease of isolation and identification and the fact that there is already established precedent for their measurement in body fluids or brain homogenates.

All compounds have been quantitated using one or more of the following techniques: spectrophotometry, gas chromatography–mass spectrophotometry, or high pressure liquid chromatographic techniques coupled with a suitable detector (e.g., fluorimeter, spectrophotometer, luminescence detector, electrochemical detector).

Markers of Oxidative Damage to DNA		
8-Hydroxy-2′-deoxyguanosine (8OH-dG)	Urine	Lunec *et al.*, 1994
		Loft *et al.*, 1992
		Park *et al.*, 1992
		Shigenaga and Ames, 1991
	Plasma	Park *et al.*, 1992
	Leukocytes	Takeuchi *et al.*, 1994
	Brain	Mecocci *et al.*, 1993
Malondialdehyde-deoxyguanosine adduct	Liver	Chaudhary *et al.*, 1994
8-Hydroxyguanine	Urine	Simic, 1992
Thymine glycol	Urine	Simic, 1992
Markers of Oxidative Damage to Proteins		
4-Hydroxy-2-nonenal-modified proteins	Renal cell	Okamoto *et al.*, 1994
	cancer	Lovell *et al.*, 1995
	CSF	Esterbauer *et al.*, 1991
		(continued)

Leslie A. Shinobu and M. Flint Beal   Neurology Service, Massachusetts General Hospital, Boston, Massachusetts 02114.

Metals and Oxidative Damage in Neurological Disorders, edited by Connor. Plenum Press, New York, 1997.

3-Nitrotyrosine	Plasma	Ohshima *et al.*, 1990
	Brain	Mathews and Beal, 1996
DiTyrosine	Leukocytes	Heinecke *et al.*, 1993
Protein carbonyls	Plasma	Levine *et al.*, 1994
	Brain	Smith *et al.*, 1991

Markers of Oxidative Damage to Lipids

Malondialdehyde- TBARS	Plasma	Ihara *et al.*, 1995
	Urine	Nacitarhan *et al.*, 1995
	Brain	Subbarao *et al.*, 1990
	CSF	Hunter *et al.*, 1985
4-Hydroxy-nonenal	CSF	Lovell *et al.*, 1995
F2-Isoprostanes	Plasma	Morrow *et al.*, 1995
Lipid hydroperoxides		
cholesterol ester	Plasma	Holley and Slater, 1991
		Frei *et al.*, 1989
	Brain	Dexter *et al.*, 1994
phosphatidylcholine	Brain	Miyazawa *et al.*, 1993
C3-10 straight chain aldehydes	Plasma	Esterbauer *et al.*, 1991

Hydroxyl Radical Trapping

Salicylate (2-hydroxy-benzoate)	Plasma	Ghiselli *et al.*, 1992
2,3-dihydroxybenzoate	Plasma	Grootveld and Halliwell, 1986
2,5-dihydroxybenzoate		
Electron spin resonance-spin	Brain	Dexter *et al.*, 1994
t-phenyl-a-butyl-nitrone		

REFERENCES

Chaudhary, A. K., Nokubo, M., Reddy, G. R., Yeola, S. N., Morrow, J. D., Blair, I. A., and Marnett, L. J., 1994, Detection of endogenous malondialdehyde-deoxyguanosine adducts in human liver, *Science* **265:**180–1481.

Chirico, S., 1994, High-performance liquid chromatography-based thiobarbituric acid tests, *Methods in Enzymology* **233:**314–318.

Dexter, D. R., Holley, A. E., Flitter, W. D., Slater, T. F., Wells, F. R., Daniel, S. E., Lees, A. J., Jenner, P., and Marsden, C. D., 1994, Increased levels of lipid hydroperoxides in the Parkinsonian substantia nigra: an HPLC and ESR study, *Movement Disorders* **9:**92–97.

Esterbauer, H., Schaur, R. J., and Zollner, H., 1991, Chemistry and biochemistry of 4-hydroxynonenal, malonaldehyde and related aldehydes, *Free Rad. Biol. Med.* **11:**81–128.

Frei, B., England, L., and Ames, B. N., 1989, Ascorbate is an outstanding antioxidant in human blood plasma, *Proc. Natl. Acad. Sci. USA* **86:**6377–6381.

Ghiselli, A., Laurenti, O., De Mattia, G., Maiani, G., and Ferro-Luzzi, A., 1992, Salicylate hydroxylation as an early marker of in vivo oxidative stress in diabetic patients, *Free Rad. Biol. Med.* **13:**621–626.

Grootveld, M., and Halliwell, B., 1986, Aromatic hydroxylation as a potential measure of hydroxyl radical formation in vivo. Identification of hydroxylated derivatives of salicylate in human body fluids, *Biochem. J.* **237:**499–504.

Heinecke, J. W., Li, W., Daehnke, H. L. III, and Goldstein, J. A., 1993, Dityrosine, a specific marker of oxidation, is synthesized by the myeloperoxidase-hydrogen peroxide system of human neutrophils and macrophages, *J. Biol. Chem.* **268:**4069–4077.

Holley, A. E., and Slater, T. F., 1991, Measurement of lipid hydroperoxides in normal human blood plasma using HPLC-chemiluminescence linked to a diode array detector for measuring conjugated dienes, *Free Rad. Res. Comms.* **15:**51–63.

Hunter, M. I. S., Nlemadim, B. C., and Davidson, D. L. W., 1985, Lipid peroxidation products and antioxidant proteins in plasma and cerebrospinal fluid from multiple sclerosis patients, *Neurochem. Res.* **10:**1645–1652.

Ihara, Y., Mori, A., Hayabara, T., Namba, R., Nobukuni, K., Sato, K., Miyata, S., Edamatsu, R., Liu, J., and Kawai, M., 1995, Free radicals, lipid peroxides and antioxidants in blood of patients with myotonic dystrophy, *J. Neurol.* **242:**119–122.

Kaur, H., and Halliwell, B., 1994, Evidence for nitric oxide-mediated oxidative damage in chronic inflammation, *FEBS Lett.* **350:**9–12.

Levine, R. L., Williams, J. A., Stadtman, E. R., and Shacter, E., 1994, Carbonyl assays for determination of oxidatively modified proteins, *Methods in Enzymology* **233:**346–357.

Loft, S., Vistisen, K., Ewertz, M., Tjonneland, A., Overvad, K., and Poulsen, H. E., 1992, Oxidative DNA damage estimated by 8-hydroxydeoxyguanosine excretion in humans: influence of smoking, gender, and body mass index, *Carcinogenesis* **13:**2241–2247.

Lunec, J., Herbert, K., Blount, S., Griffiths, H. R., and Emery, P., 1994, 8-hydroxydeoxyguanosine. A marker of oxidative DNA damage in systemic lupus erythematosus, *FEBS Lett.* **348:**131–138.

Lovell, M. A., Ehmann, W. D., Butler, S. M., and Markesbery, W. R., 1995, Elevated thiobarbituric acid-reactive substances and antioxidant enzyme activity in the brain in Alzheimer's disease, *Neurology* **45:**1594–1601.

Mathews, R. T., and Beal, M. F., 1996, Increased 3-nitrotyrosine in brains of Apo E-deficient mice, *Brain Res.* **718:**181–184.

Mecocci, P., MacGarvey, U., Kaufman, A. E., Koontz, D., Shoffner, J. M., Wallace, D. C., and Beal, M. F., 1993, Oxidative damage to mitochondrial DNA shows marked age-dependent increases in human brain, *Ann. Neurol.* **34:**609–616.

Morrow, J. D., Frei, B., Longmire, A. W., Gaziano, J. M., Lynch, S. M., Shyr, Y., Strauss, W. E., Oates, J. A., and Roberts, L. J. II., 1995, Increase in circulating products of lipid peroxidation (F2-isoprotanes) in smokers, *New Eng. J. Med.* **332:**1198–1203.

Miyazawa, T., Suzuki, T., and Fujimoto, K., 1993, Age-dependent accumulation of phosphaticylcholine hydroperoxide in the brain and liver of the rat, *Lipids* **28:**789–793.

Nacitarhan, S., Ozben, R., and Tuncer, N., 1995, Serum and urine malondialdehyde levels in NIDDM patients with and without hyperlipidemia, *Free Rad. Biol. Med.* **19:**893–896.

Ohshima, H., Friesen, M., Brouet, I., and Bartsch, H., 1990, Nitrotyrosine as a new marker for endogenous nitrosation and nitration of proteins, *Fd. Chem. Toxic.* **28:**647–652.

Okamoto, K., Toyokuni, S., Ushida, K., Ogawa, O., Takenewa, J., Kakehi, Y., Kinoshita, H., Hattori-Nakakuki, Y., Hiai, H., and Yoshida, O., 1994, Formation of 8-hydroxy-2'-deoxyguanosine and 4-hydroxy-2-nonenal-modified proteins in human renal-cell carcinoma, *Int. J. Cancer* **58:**825–829.

Park, E. M., Shigenaga, M. K., Degan, P., Korn, T. S., Kitzler, J. W., Wehr, C. M., Kolachana, P., and Ames, B. N., 1992, Assay of excised oxidative DNA lesions: Isolation of 8-oxoguanine and its nucleoside derivatives from biological fluids with a monoclonal anitbody column, *Proc. Natl. Acad. Sci. USA* **89:**3375–3379.

Shigenaga, M. K., and Ames, B. N., 1991, Assays for 8-hydroxy-2'-deoxyguanosine: A biomarker of in vivo oxidative DNA damage, *Free Rad. Biol. Med.* **10:**211–216.

Simic, M. G., 1992, Urinary biomarkers and the rate of DNA damage in carcinogenesis and anticarcinogenesis, *Mutat. Res.* **267:**277–290.

Smith, C. D., Carney, J. M., Starke-Reed, P. E., Oliver, C. N., Stadtman, E. R., Floyd, R. A., and Markesbery, W. R., 1991, Excess brain protein oxidation and enzyme dysfunction in normal aging and in Alzheimer disease, *Proc. Natl. Sci. USA* **88:**10540–10543.

Subbarao, D. V., Richardson, J. S., and Ang, L. C., 1990, Autopsy samples of Alzheimer's cortex show increased peroxidation in vitro, *J. Neurochem.* **55:**342–345.

Takeuchi, T., Nakajima, M., Ohta, Y., Mure, K., Takeshita, T., and Morimoto, K., 1994, Evaluation of 8-hydroxydeoxyguanosine, a typical oxidative DNA damage, in human leukocytes, *Carcinogenesis* **15:**1519–1523.

Uchida, K., and Kawakishi, S., 1993, 2-Oxo-histidine as a novel biological marker for oxidatively modified proteins, *FEBS* **332:**208–210.

Index

Abetolipoproteinemia, vitamin E deficiency associated with, 176–179

n-Acetylaspartate, in Alzheimer's disease, 245

Acidosis, ischemic, 326

Aconitase, 132, 134, 142, 193; *see also* Iron regulatory/response element-binding proteins

Acquired immunodeficiency syndrome (AIDS), microglia in, 343

Acrodermatitis enteropathica, 97

Actin, 5

Adenosine triphosphate (ATP)
   calcium cycling-related inhibition, 346–347
   mitochondrial production, 238–240
   nitric oxide-related inhibition, 193

Adenylate cyclase
   aluminum-related modulation, 256–257
   in iron deficiency, 10–11

Adriamycin, 134

Aging
   brain iron content during, 25
   free radical involvement in, 243
   genetic factors in, 237, 238
   mitochondrial changes during
      DNA structural changes, 240–242
      functional changes, 242–243
   neurofibrillary tangle formation during, 41
   oxidative processes in, 238–243
   rapid, Down's syndrome-related, 105
   somatic theories of, 237, 238
   superoxide dismutase activity during, 88
   vitamin E deficiency associated with, 178

Albumin, zinc binding by, 106, 114

Alcohol abstinence syndrome, 101

Alcoholism, zinc deficiency in, 101

Alkaline phosphatase
   in axonal myelination, 102
   in zinc deficiency, 97

Alkoxyl radicals, in lipid peroxidation, 327

Aluminum
   in Alzheimer's disease, 105–108, 127, 140–141, 256–257
      effect on β-amyloid peptide aggregation, 141
      distribution patterns, 139
      ferritin-bound, 139–141
      as neuritic plaque component, 141
      as neurofibrillary tangle component, 41–42
   blood–brain barrier transport, 123, 126, 127
   in manganese intoxication, 288
   neurotoxicity, 113, 140–141
   in Parkinson's disease, 285–286
   serum and blood concentrations, 115
   transport system, 139

Alzheimer's disease
   age-related frequency, 41
   aluminum in, 105, 106–108, 127, 140–141, 256–257
      effect on β-amyloid peptide aggregation, 141
      distribution patterns, 139
      ferritin-bound, 139–141
      as neuritic plaque component, 141
      as neurofibrillary tangle component, 41–42
   apoptosis in, 253
   autosomal dominant, 103
   axonal transport impairment in, 5
   blood–brain barrier in, 105, 127
   bromide in, 255
   cerebral blood flow in, 245
   choline acetyltransferase decrease in, 244

Alzheimer's disease (*cont.*)
  chromosome 19 in, 104
  clinical features, 244
  copper in, 68–69, 257
  cytochrome oxidase in, 247
  deferoxamine treatment, 141, 261
  dementia of, 141, 245, 259
  deprenyl treatment, 261
  endogenous antioxidants in, 248–250
  energy metabolism impairment in, 245
  familial, 103, 104, 244–246
  glutamate in, 194
  iron in, 24–26, 31–32, 41–42, 105, 108, 127,
      140, 141, 257–258
    colocalization with aluminum, 140
    as lipid peroxidation catalyst, 258
    magnetic resonance imaging of, 43, 49–52
    neuritic plaque content, 258
    relationship to ferritin content, 31–32
  iron regulatory protein activity in, 32–33
  lipid peroxidation in, 179, 248, 326
  oxidative markers of, 250–251
  mercury in, 255
  mitochondrial DNA mutations/deletions in,
      245–247
    COX IV defects, 253–255
  neuritic plaques, 141
    β-amyloid sheet structure, 103–104
    β-amyloid toxicity and, 251–254
    components, 244, 251
    ferritin content, 31–32, 140, 258
    iron content, 258
    microglia associated with, 343
    as oxidative stress site, 248, 252–254
  neurofibrillary tangles
    aluminum content, 41–42, 256
    composition, 244, 250–251
    ferritin content, 258
    iron content, 41–42
    nitrostyrosine staining, 282
    as oxidative stress site, 248, 254–255
    relationship to dementia severity, 41
    tau protein content, 248, 254–255, 261
    zinc deficiency and, 104–105
  nitric oxide toxicity in, 196–197, 199
  nuclear magnetic resonance studies of, 245
  oxidative damage markers for, 250–251,
      254–255
    implication for antioxidant treatment, 260–
      261
  oxidative stress in, 247–250
    as neuritic plaque cause, 252–254
  risk factors, 244

Alzheimer's disease (*cont.*)
  selenium in, 255–256
  silicon in, 105–108, 255–256
  SPECT studies of, 245
  superoxide dismutase in, 70
  transferrin receptor expression in, 28, 127
  trimethyl tin in, 255–256
  vitamin E deficiency associated with, 184, 248,
      249
  vitamin E treatment, 178–180
  zinc deficiency associated with, 103–104, 259
  zinc in, 259–260
  zinc treatment, 105
Amine oxidases, 60–61
Amino acids, copper complexes, 64
γ-Aminobutyric acid (GABA), synthesis, iron
      requirements of, 23
γ-Aminobutyric acidergic (GABAergic) neuronal
      system, 3–4
  in tardive dyskinesia, 318
  zinc-related inhibition, 97
γ-Aminobutyric acid (GABA) mimetics, as tardive
      dyskinesia treatment, 317
γ-Aminobutyric acid receptor-A (GABA-A)
      activation, 9
  by benzodiazepines, 12
  effect of zinc on, 12, 98
γ-Aminobutyric acid receptor-B (GABA-B), 98
γ-Aminobutyric acid (GABA) transaminase, in
      iron deficiency, 3
Aminoguanidine, 191
Aminolevulonic acid, 85
21-Aminosteroids, as hydroxyl radical scavengers,
      329
Amphetamine analogs, 158
Amphetamines
  effect on ascorbate release, 152–154, 162,
      163
  dopamine-related toxicity, 157–158
  neuroleptics-induced sensitivity to, 317
Amygdala
  in Alzheimer's disease, 25–26, 49, 50, 244
  ascorbate content, 150
  iron content, in Alzheimer's disease, 25–26, 49,
      50
β-Amyloid, 141, 179
  aluminum-induced precipitation, 256, 257
  in Alzheimer's disease, 25, 103–104
  β-sheet structure, 103–104
  microglia-associated deposition, 343
  as neuritic plaque component, 251
  nitric oxide-associated toxicity, 197
  oxidative processes associated with, 251–254

β-Amyloid (*cont.*)
  post-ischemic inhibition, 335
  zinc deficiency and, 104–106
  zinc-related solubility decrease, 260
Amyloid precursor protein, 64, 103
  interaction with copper, 257
  iron-modulated cleavage, 258
  effect of oxidative processes on, 254–255
  zinc binding sites, 260
Amyloid precursor protein gene
  isoforms, 252
  mutations, 251
Amyotrophic lateral sclerosis, familial, superoxide
      dismutase 1-linked, 279, 295–296
  copper bioavailability in, 305–307
  disease severity, 307
  etiological factors/mechanisms of, 298–300
  hydroxyl radical formation in, 305–306
  lipid peroxidation in, 326
  managanese exposure-related, 81
  neuropathological phenotypes of, 296–298
  superoxide dismutase gene mutations associated
      with, 63, 282, 342
  toxic property, 305–307
  transgenic models of, 303–305, 307, 308
Anorexia, zinc deficiency-related, 97–98, 101
Anoxia, ascorbate levels in, 159
α-Antichymotrypsin, 141
Antioxidants; *see also specific antioxidants*
  action mechanisms of, 156
  in Alzheimer's disease, 248–250
  brain content, 342
  in cerebral ischemia, 211–212
  interactions among, 184–185
  effect on lifespan, 243
  lipid peroxidation-inhibiting mechanisms of,
      327–330
    therapeutic applications, 330–336
  neonatal deficiency of, 207–208
  synthetic, 184, 185
  in tardive dyskinesia, 318
Apoferritin, 132–135
Apolipoprotein E, 104, 335
Apomorphine, neuroleptics-related sensitivity to, 317
Apoptosis
  in aging, 243
  in Alzheimer's disease, 253
  definition, 243
  neonatal hypoxic–ischemic brain injury-related,
      206, 217
  neuroleptics-related, 321
  in Parkinson's disease, 289–290
  superoxide dismutase mutations-related, 282

Apo-superoxide dismutase, 345
Arachidonic acid
  central nervous system content, 325
  in oxygen radical production, 82
  phospholipase-related release, 82
L-Arginine, as nitric oxide precursor, 220, 345–
      346
L-Arginine analogs, 190–191
Arsenate, 138–139
Ascorbate, 149–173
  brain uptake, 150–151
  in copper deficiency, 66
  glutathione-induced maintenance, 345
  heteroexchange with glutamate transporters,
      154–155
  neonatal levels, 208
  neostriatal effects, 163–164
  neostriatal release, 152–155, 161–163
  neuroprotective antioxidant activity, 149–150,
      155–165, 279
    in amphetamine toxicity, 157–158
    dopaminergic modulation in, 161–164
    glutaminergic modulation in, 159, 163, 164
    in Parkinson's disease, 156–157
    in schizophrenia, 160
    in tardive dyskinesia, 160–161
  neurotoxicity, 161
  prooxidant activity, 161, 162
  redox potential, 149
  release into extracellular fluid, 151–155
Ascorbate deficiency, 151, 160
Astrocytes
  glutathione content, 342, 345, 347
  iron regulatory protein content, 32
  manganese uptake by, 80–81
  in neural transmission, 14, 15
  neuroprotective activity, 343–344
  nitric oxide synthase expression by, 345–346
  oxidative damage susceptibility, 343–349
    hydrogen peroxide-related, 343–344
    mitochondrial mediation of, 344–349
Astrocytic growth-inhibitory factor, 259
Ataxia
  familial, vitamin E deficiency associated with,
      176–177, 184
  olivopontocerebellar, 181
Attentional disorders, iron deficiency-related, 23–
      24, 25
Axonal growth, 2, 4–5
Axonal myelinated fibers, in amyotrophic lateral
      sclerosis, 297–298
Axonal spheroids, in amyotrophic lateral sclerosis,
      296–297

Axonal transport
  iron in, 5–6
  in vitamin E deficiency, 177

Barbiturates, 3
Basal ganglia
  γ-Aminobutyric acid content, 3
  ascorbate content, 150
  copper content, 68
  iron content, 24, 25, 27
    in Alzheimer's disease, 50
  manganese content, 81, 84
Basic fibroblast growth factor, 343–344
Bcl-2, 243, 253–254
Bedlington terriers, hepatic copper accumulation
      in, 69
Benzodiazepines, 3, 98
Beryllium, 135–137
Blood–brain barrier, 113–114
  in Alzheimer's disease, 105, 127
  anatomy, 113–116
  astrocytes in, 14
  in Down's syndrome, 105
  in encephalopathies, 105
  metal transport systems, 113–130
    in Alzheimer's disease, 127
    copper, 60
    iron, 24
    kinetic analysis of, 116–119
    manganese, 77, 79–80
    transport mechanisms, 121–126
    transport methods, 116–119
    transport rates, 119–121
  in nutrient access, 342
  perinatal closure, 29
  permeability, 116
Brain; see also specific areas of brain
  cell membrane phospholipids, 325
  oxidative stress vulnerability, 342–343
  oxygen consumption, 83, 342
Brain-derived neurotrophic factor, 14
Brain injury; see also Head injury; Neonatal
      hypoxic–ischemic brain injury
  antioxidant treatment, 330, 332
Brain malformations, zinc deficiency-related, 96,
      101, 107
Brain surgery, ischemia prevention during, 181
Bromide, in Alzheimer's disease, 255
Bunina bodies, in amyotrophic lateral sclerosis,
      296–297
Butylated hydroxytoluene, 327
  effect on lifespan, 243
BX-9, 141

Cadherins, 5
Cadmium, toxicity, 136, 138
Calcitonin, 62
Calcium
  in axonal sprouting, 5
  blood–brain barrier transport, 125, 126
  in brain development and function, 113
  in cerebral ischemia, 214–215
  homeostatic mechanisms, tirilazad protection of,
      333–334
  interaction with nitric oxide synthases, 189–
      190
  in neurotransmitter translocation, 7–8
  in Parkinson's disease, 281
  serum and blood concentrations, 115
Calcium adenosine triphosphatases, 346–347
Calcium cycling, as cellular energy loss cause,
      346–347
Calmodulin
  interaction with nitric oxide synthases,
      189
  neuroleptics-related decrease, 320–321
Calpain I, 86, 347
CaM-dependent enzymes, 10
Carbamazepine, effect on plasma zinc concentra-
      tion, 102
Cardiolipin, 242
Catalase, 134
  in Alzheimer's disease, 248, 249
  in cerebral ischemia, 211
  effect on lifespan, 243
  manganese-related decrease, 81
  oligodendrocyte content, 342
  peroxide detoxifying activity, 344
  in tardive dyskinesia, 320
Catecholamines
  in actin modification, 5–6
  biosynthesis, 3
  uptake into vesicles, 6–7
Cations, toxicity, 131
Caudate, iron content
  in Alzheimer's disease, 49
  magnetic resonance imaging of, 46–50
  in Parkinson's disease, 29, 50
Caudate-putamen, iron content, in Alzheimer's dis-
      ease, 26
Cell cultures, effect of vitamin E on, 182
Cell death, programmed: see Apoptosis
Cell division, zinc in, 96, 107
Cell membranes, tardive dyskinesia-related destabi-
      lization, 318
Central nervous system; see also Brain
  intercellular interactions in, 14–15

Cerebellum
  iron content, 25
  nitric oxide synthase content, 190
  vitamin E content, 182–183
Cerebral arteries
  nitric oxide synthase localization in, 190
  occlusion
    calcium homeostasis in, 334
    focal ischemia associated with, 194–195
Cerebral cortex
  ascorbate content, 150
  nitric oxide synthase content, 190
Cerebrospinal fluid
  calcium uptake into, 125
  iron transport in, 27
  manganese transport in, 79
  metals uptake into, 117, 119–121
  transferrin content, 29
  transferrin receptors, 27
  zinc content, in Alzheimer's disease, 106, 107, 259
  zinc elimination in, 97
Ceruloplasmin, 24
  in Alzheimer's disease, 68–69, 250
  as copper-dependent antioxidant protein, 61, 65
  ferrous iron oxidation by, 65
  ferroxidase activity, 24, 61
  neonatal levels, 208
  in Wilson's disease, 288
  zinc binding to, 106
Ceruloplasmin gene, mutations of, 69
Chemical radical scavenging, 328–329
Chlorpromazine, neurotoxicity, 181, 316
Cholecystokinin, 62
Cholestasis, vitamin E deficiency associated with, 176
Cholesterol, oxygen radical-induced damage to, 325
Choline acetyltransferase, in Alzheimer's disease, 244
Chorea
  drug-induced, 315
  Huntington's: see Huntington's disease
Choroid plexus
  manganese transport across, 79
  zinc transport across, 97
Chromatolysis, neuroleptics-related, 316, 317
Chromosome 19, in late-onset Alzheimer's disease, 104
Ciliary neurotrophic factor, 14
Circumventricular ograns, 116
Cis-aconitase, 193; see also Aconitase

L-Citrulline, 189–190
Cobalt, in brain development and function, 113
Coenzyme Q10, as Alzheimer's disease treatment, 261
Communication, intercellular/intracellular, neuro-transmitters in, 1
Copper, 57–75
  in Alzheimer's disease, 68–69, 257
  antioxidant activity, 58, 65
  biological reactivity, 57
  in brain development and function, 113
  as ceruloplasmin component, 61
  distribution in brain, 57–61
  in familial amyotrophic lateral sclerosis, 295–296
  glutathione-induced maintenance, 345
  neurochemical functions, 60–65
  neurotoxicity, 58, 67–70
    in genetic diseases, 68–70
  as oxidative reaction catalyst, 278
  prooxidant activity, 58, 64–65
  reactive oxygen species of, 58
  serum and blood concentrations, 115
  types 1–3, 58
  in Wilson's disease, 288
Copper deficiency, 57–58, 65–67
  ethane production in, 65–66
  genetic, 66–67
  neurodevelopmental effects, 59, 62, 66
Corpus striatum, in copper deficiency, 62, 66
Corticosteroids, nitric oxide synthase-inhibiting activity, 190–191
Creatine kinase
  in Alzheimer's disease, 247–248
  β-amyloid protein-related decrease, 252–253
Cuproenzymes, 60–65, 70–71
2,3-Cyclic nucleotide phosphohydratase, 97
2,3-Cyclic nucleotide 3'-phosphohydrogenase, 102
Cyclooxgenases, in arachidonic acid bioconver-sion, 82
Cystic fibrosis, vitamin E deficiency associated with, 176
Cytochrome a, 61
Cytochrome b, age-related changes in, 242
Cytochrome c oxidase, 60, 61, 67
Cytochrome III, age-related changes in, 242
Cytochrome oxidase, in Alzheimer's disease, 247
Cytochrome P-450, interaction with manganese, 84–85
Cytokines; see also specific cytokines
  in iron metabolism, 15
  in neural transmission, 14, 15
  nitric oxide synthase-inhibiting activity, 190–191

DATATOP study, 289
Deferoxamine
    as Alzheimer's disease treatment, 141, 261
    as cerebral ischemia preventive, 217, 218
    as cerebral ischemia treatment, 220
    as hypoxic–ischemic cerebral injury preventive,
        219
    as lipid peroxidation blocker, 217
    structure, 331
Dementia
    Alzheimer's disease-related, 245
        deferoxamine treatment, 141
        zinc deficiency-related, 259
    Boxer's, 105
    Down's sydnrome-related, 251
Demyelinating diseases, 5
    oxidative stress associated with, 33
Dentate gyrus
    manganese content, 79
    nitric oxide synthase content, 190
Deoxyribonucleic acid: see DNA
Deprenyl
    as Alzheimer's disease treatment, 261
    as Parkinson's disease treatment, 289–290
Desferal: see Deferoxamine
Desferrioxamine: see Deferoxamine
Diethyldithiocarbamate, copper toxicity and, 60,
        67–68
Dihydrolipoic acid, 329–330
Dihydrophenylacetic acid, 3
2,3-Dimethoxy-1,4-naphthoquinone, 243
Dimethylnitrosamine, 137
Dimethylthiourea, 328
D_1 receptors, in iron deficiency, 9–10
Disulfiram, effect on copper uptake, 60
DNA
    free radical-related cross-linkage, 99
    mitochondrial
        age-related changes, 238, 240–243
        Alzheimer's disease-related defects, 245–247,
            253–255
        mutations, 345
        oxidative damage, 349
        oxidative damage markers, 353
        random mutations, 238
DNA repair, nitric oxide-related inhibition, 193–194
DNA synthesis, nitric oxide-related inhibition,
        193–194
Dopamine
    ascorbate-regulatory activity, 152–153
    autooxidation
        manganese-induced, 85–86
        in Parkinson's disease, 277–278, 289

Dopamine (cont.)
    autooxidation products, 277–278
        enzyme-related inhibition, 279
    in copper deficiency, 61–62, 66
    in manganese-induced neurotoxicity, 86
    oxidative activity, 149–150
    in schizophrenia, 319
    synthesis, 3
        iron requirements for, 23
    in zinc deficiency, 97
Dopamine agonists, ascorbate as, 161–164
Dopamine antagonists, ascorbate as, 161–164
Dopamine decarboxylase, 3
Dopamine-β-hydroxylase, 102
Dopamine-β-monooxygenase, 60–62, 67
Dopamine receptor hypothesis, of tardive dyskine-
        sia, 317
Dopaminergic neurons
    effect of amphetamines on, 157–160, 162, 163
    effect of manganese on, 81
    in manganese deficiency, 78
    in Parkinson's disease, 29
Dopaminergic receptors, neuroleptics-related
        blockage, 159–161
Dopaminergic transmission, effect of metals on,
        10
Dorsal tegmentum, in Alzheimer's disease, 244
Down's syndrome, 105
    neuritic plaques/neurofibrillary tangles of, 251
    superoxide dismutase expression in, 63, 70
Dyskinesia
    levodopa-induced, 315
    spontaneous, in schizophrenia, 319
Dyskinesie facio-bucco-lingui-masticatrice, 315
Dystonia, manganese and, 78, 81

Encephalomalacia, vitamin E deficiency-related,
        175–176
Encephalomyelitis, experimental allergic/autoim-
        mune, 198, 199, 343
Encephalopathia saturnica, 105
Encephalopathy, Guam's, 105
Endocytosis, as manganese transport mechanism,
        138
Endothelial-dependent relaxing factor, 333
Endothelium
    nitric oxide synthase localization in, 190
    tirilazad localization in, 333, 335
Entorhinal cortex, in Alzheimer's disease, 244, 247
Enzymatic radical scavenging, 328
Enzymes
    copper-dependent, 60–65, 70–71
    zinc-dependent, 95, 96

Epilepsy
  iron-related, 4
  manganese deficiency-related, 80–81
  zinc deficiency and, 101–102
Ethane, 65–66
Excitotoxic injury, cerebral ischemia-related, 215
Exocytosis, copper release during, 63–64
Extracellular fluid, ascorbate release into, 151–155
Extrapyramidal motor system disorder, 78

$FADH_2$, as Krebs cycle product, 238
Fenton reaction, 24, 278
  copper in, 305
  iron in, 65, 84, 347
  manganese in, 84, 86
  tirilazad in, 332
Ferritin, 131–147
  in Alzheimer's disease, 31–32, 41, 248
    neuritic plaque content, 31, 140, 258
    neurofibrillary tangle content, 258
  astrocyte content, 344
  in cerebral ischemia, 220, 218
  functions, 133
  H-chain, 27, 30–31, 132
    in Alzheimer's disease, 258
    ferroxidase activity, 30
    mRNA, 139, 142
    iron-regulatory function, 30–34, 132–133, 279
    zinc-related inhibition, 137
  iron release from, nitric oxide-induced, 192
  as ischemic heart disease risk factor, 42
  L-chain, 30, 31, 132
  in lipid peroxidation, 326, 327
  magnetic resonance imaging of, 44–45
  in manganese neurotoxicity, 84, 138
  metal-binding activity, 134–141
    aluminum, 136, 139–141
    arsenate, 138–139
    beryllium, 135–137
    cadmium, 136, 138
    manganese, 84, 138
    zinc, 136–138
  neonatal levels, 217, 218
  nitric oxide-related inhibition, 221–222
  in Parkinson's disease, 27, 31, 32
  mRNA, 221
  structure, 132
  transport across blood–brain barrier, 123, 126
Ferritin/microglia system, 258

Ferritin receptors, 27
Ferroxidase
  expression by ceruloplasmin, 24, 61
  expression by ferritin, 30
  neonatal levels, 208
Fetus, manganese blood–brain passage in, 77
Fibroblast growth factor, 14
Fibronectin, 5
Flavin adenine dinucleotide, as Krebs cycle component, 238
Forebrain, ascorbate release in, 150
Free-radical hypothesis, of tardive dyskinesia, 318–321
Free radicals; *see also specific free radicals*
  chain reactions of, 156
  definition, 82, 206
  detoxification, 279
  initiation, 341
  manganese-induced, 81–87
  oxidative reactions of, 278
  propagation, 341–342
  redox reactions, 206
  termination, 341–342
Free radical scavengers, 134, 279, 328–330; *see also specific free radical scavengers*
Frontal cortex, iron content, in Alzheimer's disease, 25–26
Frontal white matter, iron content, in Alzheimer's disease, 49

Glial cells; *see also* Astrocytes; Microglia; Oligodendrocytes
  manganese content, 80–81
  oxidative injury susceptibility, 343–349
  oxidative stress resistance, 342
Glial fibrillary acid, 335
Gliosis, neuroleptics-related, 316, 317
Globus pallidus
  iron content
    in Alzheimer's disease, 25–26, 49
    magnetic resonance imaging of, 46–50
    in Parkinson's disease, 50
  manganese content/toxicity, 80–81, 83, 287–288
Glucose-6-phosphate dehydrogenase, in Alzheimer's disease, 250
Glutamate, neurotoxicity, 159
  in Alzheimer's disease, 194
  ascorbate in, 164
  in cerebral ischemia, 215
  manganese in, 80–81
  nitric oxide in, 194–195
  in oligodendrocytes, 216, 217

Glutamate decarboxylase
  in γ-Aminobutyric acid synthesis, 102
  zinc-related inhibition, 102
Glutamate dehydrogenase
  in iron deficiency, 3
  in NADH formation, 240
Glutamate-gated cation channels, in manganese-
  induced neurotoxicity, 81–82, 86–87
Glutamate receptors, oxidative activity, 149–150
Glutamate synthetase, interaction with manganese,
  77, 80–81
Glutamate transporters, 159
  in amyotrophic lateral sclerosis, 306
  heteroexchange with ascorbate, 154–155
Glutamic acid decarboxylase
  interaction with zinc, 98–99
  in tardive dyskinesia, 318
  in zinc deficiency, 102
Glutamic acid dehydrogenase, interaction with
  zinc, 97, 98, 107
Glutaminergic synaptic transmission
  ascorbate in, 164
  nitric oxide in, 194
Glutamine synthetase
  in Alzheimer's disease, 247–248
  β-amyloid protein-related decrease, 252–253
  astrocytic, 78
  manganese-related, 78
γ-Glutamyl transpeptidase, in Parkinson's disease,
  281
Glutathione
  astrocytic, 342, 345, 347
  in copper deficiency, 66
  functions, 345
  Lewy body content, 286
  mitochondrial content, 345
  neonatal deficiency of, 207–208
  in Parkinson's disease, 280, 281, 286–287
  in tardive dyskinesia, 320
Glutathione-metal mercaptides, 345
Glutathione peroxidase
  in Alzheimer's disease, 248, 249
  in hydrogen peroxide reduction, 279
  mitochondrial content, 345
  neonatal deficiency of, 207–208
  in Parkinson's disease, 157
Glutathione reductase, in Alzheimer's disease, 248,
  249
Glyceraldehyde-3-phosphate dehydrogenase, 193,
  221
Glycoproteins, in axonal growth, 5
Golgi apparatus, in amyotrophic lateral sclerosis,
  296–298

G proteins, 10–11, 14
  aluminum-induced activation, 256–257
  receptor coupling by, 12
Guanosine triphosphatases, 6

Haber Weiss reaction, 84
Hallervorden–Spatz disease, 24, 69, 288
Haloperidol, 162, 181, 316
Haptoglobulin, zinc binding to, 106
Head injury, as Alzheimer's disease risk factor,
  244, 247, 251
Heavy metals, as memory deficit cause, 106–107
Heme-oxygenase-1, in Alzheimer's disease, 248
Hemeproteins, 84–85
Hemochromatosis, 31
Hemoglobin, in lipid peroxidation, 327
Hemorrhage, cerebral
  perinatal, 181
    ischemic brain injury-related, 218
  subarachnoid, 325, 326
    antioxidant treatment, 330, 332–334
Hippocampus
  in Alzheimer's disease, 25–26, 104, 105, 244
  ascorbate content, 150
  iron content, in Alzheimer's disease, 25–26
  manganese content, 79
  metallothionein content, 100
  mossy fibers
    in prenatal alcohol exposure, 101
    zinc content, 97, 98, 107
  zinc content, 100, 259
    in Alzheimer's disease, 104, 105
    heavy metals-related displacement, 106–107
    of mossy fibers, 97, 98, 107
Holoferritin, 133, 135
Homovanillic acid, 3
Huntington's disease, 315
  brain iron content, 24, 42
    magnetic resonance imaging of, 43–44
  nitric oxide toxicity in, 196
Hydrogen peroxide, 206, 342
  effect on astrocytes, 343–344
  as cellular energy loss cause, 346
  in cerebral ischemia, 215
  detoxification, 344
  as dopamine autooxidation product, 278
  as hydroxyl radical source, 87
  in Parkinson's disease, 280
  production of, 87
    manganese-related, 81
    mitochondrial, 344, 345
    NADPH oxidation in, 100
  reaction with iron, 131–132

Hydroperoxides
  in mitochondrial calcium release, 346, 347
  vitamin E and, 183
3-Hydroxyacyl-CoA-dehydrogenase, 240
β-Hydroxybutyrate-dehydrogenase, 240
8-Hydroxy-2-deoxyguanosine
  in Alzheimer's disease, 247–248
  in mitochondrial DNA deletions, 241
6-Hydroxydopamine
  as amine oxidase component, 61
  amphetamine-induced production, 158
  in dopaminergic neurons, 156, 157
  manganese-induced production, 81–82
  as parkinsonism cause, 286, 287
  vitamin E-related modulation, 180
4-Hydroxyl-2-nonenal, 250–251
Hydroxyl radical
  brain content, 342
  in cerebral ischemia, 209, 215–216
  cuprous ion precursor, 64, 65
  as dopamine autooxidation product, 278
  iron-related production, 24
  in lipid peroxidation, 192, 326
  manganese-related production, 81
  markers of, 354
  measurement of, 212
  mitochondrial production, 345
  neurotoxicity, 342
  in Parkinson's disease, 197
  peroxynitrous acid precursor, 279
  reactivity, 24, 345
Hydroxyl radical scavengers, 328–329
  chemical structure, 331
  tirilazad as, 332
Hypoperfusion, post-ischemic, 210
Hypothalamus, copper content, 59
Hypoxic–ischemic brain injury, neonatal: see Neonatal hypoxic–ischemic brain injury

Immune functions, zinc in, 95
Immunoglobulin G, zinc binding to, 106
1,4,5-Inositol triphosphate, 11
Insulin-like growth factor, 14
Integrines, 5
Interleukins, 14, 88
Iron, brain content, 342
  acquisition, 27–29
  age-related increase, 25, 41, 44–50
  in Alzheimer's disease, 24, 25–26, 31–32, 41–42, 105, 108, 127, 140, 141, 257–258
    colocalization with aluminum, 140
    as lipid peroxidation catalyst, 258

Iron, brain content (*cont.*)
  in Alzheimer's disease (*cont.*)
    magnetic resonance imaging of, 43, 49–52
    neuritic plaque content, 258
    relationship to ferritin content, 31–32
  effect on γ-Aminobutyric acid, 3–4
  effect on β-amyloid peptide aggregation, 141
  in axonal transport, 5–6
  in brain development and function, 113
  in catecholamine biosynthesis, 3
  in cerebral ischemia, 208
  excess, storage of, 132
  ferritin regulation of, 30–34, 132–133, 137, 279
  as free radical source, 131–132
  homeostasis, effect of nitric oxide on, 192–193
  in Huntington's disease, 24, 42–44
  intracellular management, 30–32, 84
  as ischemic heart disease risk factor, 42
  in lipid peroxidation, 7, 217–218, 258, 325–327, 330
  in manganese intoxication, 288
  mobilization, 29–30
    post-ischemic, 218–219
  in multiple system atrophy, 288–289
  neonatal levels, 27, 208–209
    in neonatal hypoxic–ischemic brain injury, 216–220
  neurofibrillary tangle content, 41–42
  neuronal cellular functions, 1
  neurotoxicity, 1
  effect on neurotransmitters
    neural signaling, 13
    protein phosphorylation, 11–12
    receptor expression regulation, 9–10
    receptor sensitivity and desensitization, 12–13
    second messenger systems, 10–11
    translocation, 6–8
  oxidation states, 131
  as oxidative reaction catalyst, 278
  in Parkinson's disease, 24, 26–27, 50, 280–281
    microprobe studies of, 282–287
  post-ischemic, 218–220
  protein-bound intracellular release, 84
  reductive release, 133–134
  relationship to transferrin levels, 33–34
  serum and blood concentrations, 115
  in tardive dyskinesia, 318
  tissue concentration, 131
  transport at blood–brain barrier, 121–124
  uptake mechanisms, 27–29
Iron chelation, antioxidant activity, 330

Iron citrate, as Alzheimer's disease treatment, 261
Iron deficiency
  γ-Aminobutyric acid in, 3–4
  catecholamine turnover in, 3
  cerebral metabolism in, 23
  cognitive performance in, 23–24
  D_1 receptors in, 9–10
Iron regulatory factor B, 142
Iron regulatory proteins, 24, 25, 32–33
Iron regulatory/response element, 25, 132, 138
Iron regulatory/response element-binding proteins, 132, 134, 142
  interaction with nitric oxide, 192
  phosphorylation, 13
  of *Xenopus laevis* ferritin, 138
Ischemia, cerebral; *see also* Neonatal hypoxic–ischemic brain injury
  antioxidant treatment, 328–330, 335
  focal, 194–195
  superoxide dismutase in, 328
  during surgery, 181
Ischemic heart disease, 42
Isocitrate, 240
Isoniazid, convulsant activity, 101
Itai-itai disease, 138

Janus kinase, 15

Karyorrhexis, 304
Kearns–Sayre syndrome, 241
α-Ketoglutarate, 240
2-α-Ketoglutarate dehydrogenase, in Alzheimer's disease, 245–246
Kinesin, 6
Krebs cycle, 3, 4, 238, 239

Lactoferrin, in Alzheimer's disease, 258
Lactoferrin receptors, 27, 29
Laminin, 5
Lazaroids, 329, 330
L-DOPA: *see* Levodopa
Lead
  blood–brain barrier transport, 125, 126
  ferritin binding by, 137
  heme enzyme-inhibiting activity, 131
  neurotoxicity, 113
  in protein phosphorylation, 11
  as seizure cause, 101
  serum and blood concentrations, 115
Learning deficits, zinc deficiency-related, 100
Leukotrienes, production of, 82

Levodopa, 157
  decarboxylation, 3
  as dyskinesia cause, 315
  quinone autooxidation products, 85
Lewy bodies
  glutathione deficiency in, 287
  in multiple system atrophy, 288
  in Parkinson's disease, 277
  transition metals associated with, 66
Lifespan
  free radical inhibition and, 243
  metabolic rate and, 238
Linoleic acid, 325
Linoleic acid hydroperoxide, 183
Lipid hydroperoxides, in Parkinson's disease, 282
Lipid peroxidation, 156
  in Alzheimer's disease, 179, 248, 326
    oxidative damage markers of, 250–251
  ascorbate-induced, 157
  β-amyloid-induced, 253
  in cerebral ischemia, 212
  chemistry of, 326–327
  in copper deficiency, 65–67
  copper toxicity-induced, 67, 68
  definition of, 325
  free radical scavenger-related prevention, 279
  inhibition
    inhibitors, 330–336
    mechanisms, 327–330
  iron-induced, 7, 258, 325–327, 330
    in Alzheimer's disease, 258
    neonatal, 217–218
    in neurotransmitter translocation, 7
  manganese in, 88
  mitochondrial, 344
  neuroleptics-induced, 160
  in Parkinson's disease, 282, 326
  peroxynitrite-induced, 192
  superoxide-induced, 82
  in zinc deficiency, 100
Lipids, oxidative damage markers of, 354
Lipofuscin
  aging-related deposition, 178
  in Alzheimer's disease, 105, 247–248, 256
Lipooxygenase, in arachidonic acid bioconversion, 82
5-Lipoxygenase, 327–328
  inhibitors, 328
Lisuride, 86
Liver
  copper content, 69
  ferritin content, 137
  glutathione content, 345

Liver disease
  alcoholic, 11
  manganese-related, 287
Locura manganica, 78
Locus ceruleus
  in Alzheimer's disease, 244
  copper content, 59, 61
LY178002, 329–331
Lysyl oxidase, 60, 62

α2-Macroglobulin, metal-binding property, 106, 114
Magnesium
  in brain development and function, 113
  serum and blood concentrations, 115
Magnesium-guanosine triphosphatases, 10
Magnetic resonance imaging (MRI), of brain iron
      content, 25, 41–56
  in Alzheimer's disease, 49–52
  field dependent R_2 increase method, 45–50
  therapeutic implications, 51–52
  transverse relaxation rates and times, 43–45
Malondialdehyde
  in Alzheimer's disease, 247–248
  in diethyldithiocarbamate intoxication, 68
  as lipofuscin component, 105
  in Parkinson's disease, 157, 282
  in zinc deficiency, 100
Manganese
  antioxidant activity, 87–88
  in brain development and function, 113
  excess, 77, 78, 80, 81
  ferritin-associated transport and storage, 138
  neurotoxicity, 77, 78, 80–88, 138, 287–288
  oxidation states, 78–79
  as oxidative reaction catalyst, 278
  in Parkinson's disease, 287–288
  prooxidant activity, 81–87
  serum and blood concentrations, 115
  transport into brain, 78–80, 118–119, 123, 124
Manganese dioxide, 77
Manganese-guanosine triphosphatases, 10
Manganese-pyrophosphate, 88
Manganese-tartarate, 88
Mass spectrometry, inductively coupled plasma, 106
Membrane stabilizers, antioxidant, 330, 331
Memory deficits
  heavy metals-related, 106–107
  zinc deficiency-related, 100, 107
Menkes' disease/syndrome, 66–68, 257
Mental lethargy, zinc deficiency-related, 95, 101,
      103, 107
Mental retardation, as tardive dyskinesia risk fac-
      tor, 319

Mercaptoethylamine, effect on lifespan, 243
Mercaptopropionic acid, convulsant activity, 101
Mercury
  in Alzheimer's disease, 108
  blood–brain barrier transport, 126
  neurotoxicity, 113
Metabolic rate, relationship to lifespan, 238
Metal ions, multivalent, 82, 84
Metallothionein
  in Alzheimer's disease, 106, 259
  copper binding by, 62
  as copper-dependent antioxidant protein, 65
  as free radical scavenger, 100
  metal ion-induced synthesis, 137–138
  zinc binding by, in Alzheimer's disease, 259
  zinc-induced synthesis, 100, 107
Metallothionein gene family, 104
Metallothionein I, 60, 62, 67
Metallothionein II, 60, 62, 67, 250
Metallothionein III, 60, 62
Metals; see also specific metals
  blood–brain barrier transport systems, 113–130
    in Alzheimer's disease, 127
    transport mechanisms, 121–126
    transport rates, 119–121
  brain functions of, 113
  organic acids-binding property, 115
  protein-binding property, 114–115
Methamphetamine, 158
2-Methylaminochromans, 334–335
Methylcyclopentadienyl manganese tricarbonyl, 88
N-Methyl-D-asparate
  blockade, 259
  nitric oxide-mediated neurotoxicity, 221
N-Methyl-D-aspartate receptors, 158, 159
  blockade, 220, 288
  calcium influx, 9
  in cerebral ischemia, 215
  desensitization, 12–13
  glycine-related activation, 12
  in manganese-induced neurotoxicity, 86–87
  effect of zinc on, 98
  in zinc deficiency, 104, 259
Methyl mercury
  blood–brain barrier transport, 126
  serum and blood concentrations, 115
1-Methyl-4-phenyl-1,2,3,6-tetrahydropyridine, 14,
      29, 280, 286
1-Methyl-4-phenyl-pyridinium, 280, 287
Methylprednisolone
  chemical structure, 331
  as lipid peroxidation inhibitor, 330, 332
α-Methyltyrosine, 86

Microglia
    amyloid fibers of, 258
    in cerebral ischemia, 216, 218
    ferritin content, in amyloid formation, 258
    in free radical formation, 343, 348
    oxidative damage susceptibility, 343
    transferrin receptor expression on, 28
Microtubule-associated proteins, 6
Microtubules, 6
    zinc-dependent assembly, 96
Middle East, zinc deficiency in, 95, 96, 101
Mitochondria
    age-related changes, 242–243
    antioxidant system, 345
    of astrocytes, oxidative damage to, 344–349
    in cerebral ischemia, 215–216
    copper-related damage to, 68
    DNA
        age-related changes, 238, 240–242
        in Alzheimer's disease, 245–247
    Krebs cycle in, 238, 239
    manganese effects on, 83–85
    manganese uptake by, 83
    oxidative phosphorylation in, 238–240
    oxygen consumption by, 83
    in Parkinson's disease, 281
    as reactive oxygen species source, 344–345,
        347, 349
    vitamin E oxidation in, 183–184
Mitochondrial myopathies, 241–242
Mitochondrial permeability respiration pore, 346
Mitochondrial respiration chain: see Respiratory
        chain complex
Mixed function monooxidases, 84–85
Molybdenum, in brain development and function, 113
Monoamine oxidase, astrocytic, 344
Monoamine oxidase-A, 278
Monoamine oxidase-B, 278
Monoamine oxidase-B inhibitors
    as Alzheimer's disease treatment, 261
    as Parkinson's disease treatment, 289–290
N^GMonomethyl-L-arginine, 190–191
Motor cortex, iron content, in Alzheimer's disease,
        25–26
Motor neuron disease
    antioxidant treatment, 335–336
    neuropathology, 296
Motor neurons, dysfunction and death, 298–299
Multiple sclerosis, 5
    antioxidant treatment, 33, 343
    iron in, 24, 33
    microglia in, 343
    nitric oxide-related cytokine release in, 198

Multiple system atrophy, 288–289
Myelin
    iron-induced oxidation, 5
    synthesis, 23, 343
Myelination
    in copper deficiency, 66, 67
    zinc in, 97, 107
Myo-inositol, 245

NADH
    as Krebs cycle product, 238–240
    mitochondrial content, 345
NADH dehydrogenase, age-related changes, 242
NADPH
    mitochondrial content, 345
    zinc-related inhibition, 100
NADPH diaphorase, 196
Na^+K^+ATPase, inhibition, 213, 217
Neonatal hypoxic–ischemic brain injury, 205–236
    apoptosis associated with, 206, 217
    iron in, 216–220
    microvascular injury of, 209–212, 217–218
        effect of antioxidants on, 211–212
    nitric oxide in, 220–222
    parenchymal injury of, 212–217
    in premature infants, 207–208, 218–219
    reactive oxygen species in, 206–209
    rescue therapies, 223–224
Neonates
    antioxidant deficiency in, 207–208
    iron levels in, 27, 208
        in hypoxic–ischemic brain injury, 216–220
Neostriatum, ascorbate content and release in,
        152–155, 161–163
    during anoxia, 159
Nerve growth factor, 14, 343–344
Network, neural, 1
Neural transmission, 1–2
Neural tube, zinc-dependent development, 96
Neurocuprein, 60, 62
Neurodegenerative disorders; see also Alzheimer's
        disease; Huntington's disease; Parkin-
        son's disease
    nitric oxide-related toxicity in, 196–199
Neurofibrillary tangles, 104, 105
    age-related increase, 41
    in Alzheimer's disease
        aluminum content, 41–42, 256
        composition, 244, 250–251
        ferritin content, 258
        iron content, 41–42
        nitrotyrosine staining in, 282
        as oxidative stress site, 248, 254–255

Neurofibrillary tangles (*cont.*)
  in Alzheimer's disease (*cont.*)
    relationship to dementia severity, 41
    tau protein content, 248, 254–255, 261
    zinc deficiency and, 104–105, 108
Neurofilaments, in amyotrophic lateral sclerosis,
    296–297
Neuroleptics, as tardive dyskinesia cause,
    180–181, 315–324
  dopamine receptor up-regulation in, 160–161
  free radical hypothesis of, 318–321
  models of, 317–318
  movement abnormalities associated with, 315,
    316
  neuropathology of, 316–317
  oxidative stress mechanisms of, 318–321
Neuromelanin, in Parkinson's disease, 28–29, 280
  aluminum accumulation with, 285–286
  iron accumulation with, 283, 284–285, 286
Neuromelanin-containing cells, of substantia nigra,
    27
Neuromuscular disorders, zinc deficiency-related,
    102, 107
Neuronal cell/tissue cultures, effect of vitamin E
    on, 182
Neurons
  network organization, 1
  oxidative stress resistance, 342, 343
Neuropeptide Y, 62
Neurotransmitters
  biosynthesis, 3–4
  in inter-/intra-cellular communication, 1
  effect of iron on
    neural signaling, 13
    protein phosphorylation, 11–12
    receptor expression regulation, 9–10
    receptor sensitivity and desensitization, 12–13
    second messenger systems, 10–11
    translocation, 6–8
  metabolism, 3–4
  zinc-mediated release, 102–103
Neutrophils, in cerebral ischemia, 210–211, 218
Nicotinamide adenine dinucleotide
  as Krebs cycle product, 238–240
  mitochondrial content, 345
Nicotinamide adenine dinucleotide dehydrogenase,
    age-related changes, 242
Nicotinamide adenine dinucleotide phosphate
  mitochondrial content, 345
  zinc-related inhibition, 100
Nicotinamide adenine dinucleotide phosphate dia-
    phorase, 196
Nigrostriatum, effect of vitamin E deficiency on, 179

Nigro-thalamo-cortical system, ascorbate in,
    153–155
Nitric oxide, 189–203, 206
  in cerebral ischemia, 208, 214–216, 218,
    220–222
  ferritin-inhibiting activity, 221–222
  interaction with iron, 13, 14
  neonatal levels, 209
  neuroprotective effects, 220
  neurotoxicity, 189, 190
    in Alzheimer's disease, 196–197, 199
    in astrocytes, 346
    in cerebral ischemia, 220–222
    mechanisms of, 191–194
    in multiple sclerosis, 33, 197–199
    in Parkinson's disease, 197, 199
    in stroke, 194–195, 199
  as N-methyl-D-asparate toxicity mediator, 221
  as peroxynitrite anion component, 329
  in receptor desensitization, 12–13
  as superoxide scavenger, 192
  synthesis, 189, 220
Nitric oxide synthase, 189–191, 220, 329
  astrocytic expression, 345–346
  in cerebral ischemia, 220
  constitutive, 189, 190
    in oxidative injury, 191–192
  in free radical production, 282
  inducible, 189–190
    in oxidative injury, 191–192
  inhibition, 190–191
    neuroleptics-induced, 320–321
  neuroprotective activity, 189, 222
  overlap with iron regulatory proteins, 32
Nitric oxide synthase inhibitors, 190–191, 198
  as Alzheimer's disease treatment, 199
L-Nitroarginine-methyl-ester, 222
Nitrogen monoxide, 220
7-Nitro-indazole, 191, 195
N^{G}-Nitro-L-arginine, 190–191
Nitrotyrosine
  antibody, 282
  immunoreactivity, in amyotrophic lateral sclero-
    sis, 306
  in superoxide detoxification, 279
Norepinephrine
  in copper deficiency, 61–62
  β-receptor, 9
  synthesis, 3
    iron requirements of, 23
  in zinc deficiency, 97
Nuclear factor-k, 190
Nuclear transcription factor NF-kB, 217

Nucleoproteins, zinc-containing, 95
Nucleus basalis, iron content, in Alzheimer's disease, 25–26
Nucleus basalis of Meynert, in Alzheimer's disease, 244

Occipital cortex, in Alzheimer's disease, 26, 31–32
Oligodendrocytes, 14
    catalase content, 342, 342
    ferritin expression by, 31
    glutamate toxicity in, 216, 217
    iron content, 25
        in multiple sclerosis, 33
    manganese uptake by, 80
    oxidative damage susceptibility, 343
    transferrin production by, 29–30
    transferrin receptor expression on, 28–29
    in Parkinson's disease, 28–29
Oligodendroglial precursors, 218
Opioids, interaction with zinc, 97–98, 107
Oxidative damage markers, 353–354
Oxidative phosphorylation
    manganese-related inhibition, 83–84
    in mitochondria, 238–240
Oxidative reactions, catalysts, 278
Oxidative stress, brain's selective vulnerability to, 342–344
Oxygen
    dismutation, 301–302
    mitochondrial production, 344, 345
Oxygen consumption, by brain, 83, 342
Oxygen consumption rate, relationship to protein turnover, 238
Oxygen free radicals
    antioxidant-induced inhibition, 327–328
    causes, 82–83
    as central nervous system injury cause, 325–327
    inhibition, 327–330
    as lipid damage cause: see Lipid peroxidation
    as manganese-superoxide dismutase activator, 88
    neurodegenerative effects, 149–150
    production, 82
    scavengers, 82, 328–329
Oxytocin, 62

Pallidum, manganese content, 80–81
Paraquat, 134, 303
Parkinson's disease
    antioxidant treatment, 156–157, 178–179, 289
    comparison with manganism, 81
    copper toxicity in, 68–69
    definition, 277
    deprenyl treatment, 289–290

Parkinson's disease (cont.)
    idiopathic, relationship to tardive dyskinesia, 318–319
    iron in, 24, 26–27, 50
        microprobe studies of, 282–287
    iron regulatory protein activity in, 32–33
    lipid peroxidation in, 282, 326
    manganese deficiency associated with, 78
    manganese in, 77, 81, 287–288
    metals-related oxidative damage in, 113
    mitochondrial dysfunction in, 246
    neurodegeneration in, regional specificity of, 342
    neuromelanin in, 28–29, 280
        aluminum accumulation with, 285–286
        iron accumulation with, 283, 284–286
    nitric oxide-induced dopaminergic neuronal degeneration in, 197
    oxidative damage associated with, 178
    oxidative stress in, 277–290
        iron in, 26–27, 50, 280–287
        LAMMA studies of, 280, 282–288
        neurotoxic mechanisms, 280–282
        therapeutic implications, 289–290
    signs and symptoms, 277
    superoxide dismutase activity in, 88
    trace elements in, 282–290
        microprobe studies of, 282–287
        therapeutic implications, 289–290
    transferrin receptor expression in, 28–29
    tyrosine hydroxylase activity in, 3
    vitamin E deficiency associated with, 184
    vitamin E treatment, 178–179, 289
Pcr/Pi ratio, in Alzheimer's disease, 245
Peptidylglycine-α-amidating monooxygenase, 60, 62–63
Peripheral nerve injury
    transferrin receptor expression in, 28
    vitamin E deficiency-related, 177
Peroxidases, 134
    ALS-mutant superoxide dismutase-1 as, 305–306
    manganese-related decrease, 81
Peroxyl radicals, in lipid peroxidation, 327
Peroxyl radical scavengers, 329–332
Peroxynitrite
    as aconitase inactivator, 134
    as antiprotease inactivator, 221
    formation, 279
    as hydroxyl radical source, 216
    as lipid peroxidative cause, 192
    manganese-induced degeneration, 84
    in Parkinson's disease, 197, 282

Peroxynitrite (*cont.*)
  superoxide dismutase-related inhibition, 279
  in tardive dyskinesia, 321
  tyrosine residue nitrating activity, 221
Peroxynitrite scavengers, 329, 331
Peroxynitrous acid, in Parkinson's disease, 282
Phencyclidine psychosis, 160
Phenobarbital, effect on plasma zinc concentration, 102
Phenothiazines, 181
Phenylethanolamine-N-methyl transferase, 102
Phenytoin, effect on plasma zinc concentration, 102
Phosphocreatine/phosphatidyl inositol (Pcr/Pi)
  ratio, in Alzheimer's disease, 245
6-Phosphogluconate dehydrogenase, in
  Alzheimer's disease, 250
Phospholipases
  in arachidonic acid release, 82
  in cerebral ischemia, 214–215
Phospholipids, of cell membranes, effect of
  tirilazad on, 332–333
Phospholipase A$_2$, 210
Phosphoplipase C, 11
Pick's disease, 24
Picrotoxin, convulsant activity, 101
Platelet-activating factor, 210
Poly(ADP-ribose)polymerase, 221
Poly(ADP-ribose)synthase, 191, 193–194
Polyethylene glycol, 211
Polypeptides, mutant, 302–303
Polyunsaturated fatty acids, as target of oxygen
  radicals, 325
Porphyrins, manganic, 88
Premature infants, hypoxic–ischemic brain injury
  in, 207–208, 218–219
Prostacyclin, 210
Prostaglandins, production of, 82
Prostaglandin synthase, 327–328
Prostaglandin synthase inhibitors, 328
Proteases, in cerebral ischemia, 214–215
Protein
  metal binding by, 120–121
  oxidative damage markers for, 353–354
  synthesis, 99
Protein carbonyl, in Alzheimer's disease, 247–248
  β-amyloid protein-related increase, 252–253
Protein kinase C, 11–13
  in cerebral ischemia, 214–215
Protein phosphorylation, 9, 11–12
Psychosis, phencyclidine, 160
Putamen, iron content
  in Alzheimer's disease, 49
  magnetic resonance imaging of, 46–50

Pyknosis, 304
Pyridine nucleotides, in calcium release, 346, 347
Pyridoxal phosphate, 98–99
Pyrrolopyrimidines, antioxidant activity, 329–330,
  335–336
Pyruvate carboxylase, manganese-dependent, 77
Pyruvate dehydrogenase, in Alzheimer's disease,
  245–246

Quinones
  as apoptosis cause, 243
  manganese-induced production, 81–82, 85–86

Reactive oxygen species; *see also specific reactive*
  oxygen species
  effect on γ-Aminobutyric acid, 3
  effect on second messenger systems, 11
  formation, 342
Receptor binding, zinc in, 102–103
Reserpine, as tardive dyskinesia treatment, 317
Respiration, cellular, nitric oxide-related inhibition, 193
Respiratory chain complex, 239
  age-related functional changes, 242
  deficiency, 242
  electron leakage, 344, 346
  inhibition, 193, 346
  polypeptide structure, 240

Satellitosis, neuroleptics-related, 316, 317
Schizophrenia, 159–160, 319
Schut–Swier family, 181
Seizures
  alcohol abstinence syndrome-related, 101
  lead-related, 101
  zinc deficiency-related, 101–102
Selenium
  in Alzheimer's disease, 255–256
  as tardive dyskinesia treatment, 320
Semiquinone radicals, 278
Sendai cocktail, 181
Serotonin
  in actin modification, 5–6
  synthesis, 4
Short bowel syndrome, vitamin E deficiency asso-
  ciated with, 176
Sickle cell disease, 107
Silicon, in Alzheimer's disease, 105–108, 255–
  256
SOD: *see* Superoxide dismutase
Sodium, potassium-ATPase, inhibition, 213, 217
S 182 proteins, 103
Spinal cord, cell membrane phospholipids of, 325

Spinal cord injury, antioxidant treatment, 330, 332
Striatum
γ-Aminobutyric acid content, 3
manganese content, 83–84
Stroke
glutamate in, 194
ischemic, antioxidant treatment, 330, 332
nitric oxide synthase inhibitor treatment, 195, 199
nitric oxide toxicity in, 194–195
Strychnine, convulsant activity, 101
Substance P, 62
Substantia nigra
aminobutyric acid content, 3
γ-ascorbate toxicity in, 161
copper content, 59
hydrogen peroxide content, 85
iron content, 23, 25
in Alzheimer's disease, 26
in Parkinson's disease, 28, 29, 32, 284–286
manganese content, 80–83, 88, 287–288
neuromelanin content, 27, 277–278
in Parkinson's disease, 26–27, 156–157, 161, 178, 277–278, 280
iron content, 28, 29, 32, 284–286
oxidative damage to, 280–282
transferrin receptor expression in, 28–29
effect of vitamin E on, 178–179
Succinic semialdehyde dehydrogenase, 3–4
Superoxide, 206, 206
brain content, 342
catalyzing proteins for, 63
in cerebral ischemia, 208, 209, 217, 218, 221
as dopamine autooxidation product, 278
in hydroxyl radical production, 65
neurotoxicity, 82
reaction with nitric oxide, 192
Superoxide dismutase, 300, 301
apo-, 345
effect on calcium homeostasis, 334
in cerebral ischemia, 211, 328
copper/zinc, 60
in aging, 99
in Alzheimer's disease, 70, 248
antioxidant activity, 63, 70
expression in transgenic mice, 70
gene, Down's syndrome-associated, 105
localization, 87
in Menkes' disease, 67
neuroprotective effects, 99
in postnatal development, 99
in zinc deficiency, 99

Superoxide dismutase (cont.)
as copper-dependent antioxidant protein, 65
dismutation reaction, 87
extracellular, 60, 63
effect on lifespan, 243
interaction with nitric oxide, 192
manganese, 77, 78, 82–83, 87–88, 99
in Alzheimer's disease, 248
mitochondrial, 344, 345
neonatal deficiency of, 207–208
in subarachnoid hemorrhage, 328
as superoxide anion detoxification catalyst, 279
in tardive dyskinesia, 320
Superoxide dismutase-1
familial amyotrophic lateral sclerosis-related mutant, 279, 295–296, 299–300
copper bioavailability and, 305–307
disease severity and, 307
etiological factors/mechanisms of, 298–300
hydroxyl radical formation and, 305–306
lipid peroxidation and, 326
manganese exposure and, 81
mutations, 63, 282, 342
neuropathological phenotypes of, 296–298
toxic property, 305–307
transgenic models of, 303–305, 307, 308
normal functions and properties, 302–303
Superoxide radicals, manganese-related production, 81

Tardive dyskinesia, 315–324
dopamine receptor hypothesis of, 317
iron in, 24
neuroleptics-related, 180–181, 315–324
dopamine receptor up-regulation in, 160–161
free radical hypothesis of, 318–321
models of, 317–318
movement abnormalities associated with, 315, 316
neuropathology of, 316–317
oxidative stress mechanisms, 318–321
vitamin E treatment, 181
Tardive dysmentia, 319
Tau protein, 248, 254–255, 261
N-Tert-butyl-α-phenol nitrone, 328, 331
Tetrabenazine, as tardive dyskinesia treatment, 317
Tetrahydrobioptern, 345–346
Thalamus
in Alzheimer's disease, 244
manganese content, 80–81
Thiothixine, 181
Thromboxanes, production of, 82
Thymidine kinase, 96, 107

Tirilazad mesylate
   antioxidant activity, 329–335
   as peroxyl radical scavenger, 332
   as hydroxyl radical scavenger, 329
   structure, 331
Tissue cultures, effect of vitamin E on, 182
α-Tocopherol: *see* Vitamin E
Tocopherol-binding protein, 176, 177
α-Tocopherol transfer protein, 176, 177
Toxic-milk mice, 69–70
Transferrin
   age-related decrease, 33
   in aluminum transport, 126
   in Alzheimer's disease, 127, 248, 258
   function, 30
   in iron binding and transport, 29–30, 122–124,
        127, 132, 279
   in lipid peroxidation, 326, 327
   in manganese binding and transport, 79–80, 84,
        124, 287
   metal-binding property, 114
   mRNA, perinatal, 29
   neonatal levels, 208, 217
   synthesis, 122
   zinc binding to, 106
Transferrin receptors, 24, 27–29
   in Alzheimer's disease, 127, 258
   of cerebrospinal fluid, 27
   density, 27–28
   iron regulation by, 9
   in iron uptake, 27
   in manganese uptake, 287
   mRNA, 32
   in Parkinson's disease, 27
Transgenic mice models
   of Alzheimer's disease, 254
   of cerebral hypoxia–ischemia, 211
   copper/zinc-superoxide dismutase expression in,
        70
   of superoxide dismutase 1-linked familial amyot-
        ropic lateral sclerosis, 303–305, 307, 308
Tricarboxylic acid cycle: *see* Krebs cycle
Trimethyl tin, in Alzheimer's disease, 255–256
Trophic factors, transferrin-coupled, 28, 30
Tryptophane hydroxylase, 4
Tubulin
   aggregation, 6
   zinc-related polymerization of, 96
Tumor necrosis factor, 88
Tyrosinase, 60, 63
Tyrosine, 3, 279
Tyrosine-dependent nitration, 221
Tyrosine hydroxylase, 3

U-101033E, 329–331, 335–336
U-72099E, 331
U-74389G, 212, 329–330
U-74500A, 329–330
U-785517F, 331, 335
Ubiquinone, 239, 240
Ubiquitin-positive inclusions, in amyotrophic lat-
     eral sclerosis, 296–297

Vasoactive intestinal peptide, 62
Vasopressin, 62
Vitamin A, in Alzheimer's disease, 248, 249
Vitamin B$_6$, as Alzheimer's disease treatment,
     261
Vitamin C
   in Alzheimer's disease, 248, 249
   interaction with antioxidants, 184
   effect on lifespan, 243
Vitamin E, 175–188
   in Alzheimer's disease, 248, 249
   antioxidant activity, 175, 329–330
   binding protein, 176, 177
   as cell/tissue culture component, 182
   chemical structure, 331
   in copper excess, 67
   deficiency, 175–178
   distribution in central nervous system, 182–184
   as free radical scavenger, 279
   glutathione-induced maintenance, 345
   effect on lifespan, 243
   interaction with ascorbate, 156
   in manganese-induced neurotoxicity, 86
   oxidizing agents for, 183
   as Parkinson's disease treatment, 157, 289
   as peroxyl radical scavenger, 332
   as tardive dyskinesia treatment, 181, 317, 320
   transfer protein, 176, 177
Vitamin E deficiency, Alzheimer's disease-associ-
     ated, 184, 248, 249

Wallerian degeneration, in amyotrophic lateral scle-
     rosis, 297–298
Wedler, Frederick, 78
White matter
   iron content
      in Alzheimer's disease, 49
      magnetic resonance imaging of, 46–49
   myelogenic foci, 218
Wilson's disease, 68, 69, 71, 257, 288
   zinc treatment, 107

Xanthine, 82, 133–134
Xanthine dehydrogenase, 86

Xanthine oxidase, 3, 86, 328
in cerebral ischemia, 209, 210, 218
in reductive release of iron, 133–134
synthesis, 347
Xenopus laevis, cadmium toxicity in, 138

Zinc, 95–111
in Alzheimer's disease, 259–260
as Alzheimer's disease treatment, 105, 108
effect on β-amyloid, 141
in brain development and function, 102–103,
113
in cell division, 96
central nervous system transport mechanisms,
96–99
ferritin binding by, 137–138
metallothionein binding by, 137
opioid interactions, 97–98, 107
plasma content, 106
in protein phosphorylation, 11–12

Zinc (*cont.*)
serum and blood concentrations, 115
transport across blood–brain barrier, 124–126
as Wilson's disease treatment, 107
Zinc-binding proteins, 102
Zinc deficiency, 95, 97–99
alcoholism-related, 101
as Alzheimer's disease dementia cause, 259
amyloid-induced, 104–106, 108
behavioral effects, 100, 107
cognitive effects, 100–101, 107
Down's syndrome-related, 259
epilepsy associated with, 101–102
fetal, 101
in Middle East, 95, 96, 101
neurodevelopmental effects, 96, 98, 99, 101–103
neuromuscular disorders associated with, 102, 107
teratogenicity, 96, 107
Zinc supplements, contraindication in Alzheimer's
disease, 106